Calculations for A-level Physics

T L LOWE BSc, PhD

Department of Applied Sciences
Paddington College, London

J F ROUNCE BSc

School of Science
Gloucestershire College of Arts and Technology, Gloucester

Stanley Thornes (Publishers) Ltd

First published in 1987 by

Stanley Thornes (Publishers) Ltd
Old Station Drive
Leckhampton
CHELTENHAM GL53 0DN

Reprinted 1987
Reprinted 1989

British Library Cataloguing in Publication Data

Lowe, T. L.
Calculations for A-level physics
1. Mathematical physics
I. Title II. Rounce, J. F.
510.2453 QC20

ISBN 0-85950-144-2

Typeset in Times and Helvetica by KEYTEC, Bridport, Dorset.
Printed and bound in Great Britain at The Bath Press, Avon

CONTENTS

SECTION H Atomic and nuclear physics 265

SECTION I Calculations involving graphs 299

Appendices 309

LIST OF EXERCISES

PREFACE

Worked examples, numerous exercises and questions of A-level standard for practice, together with a thorough treatment of the relevant theory, are all to be found here under one cover. This book will be valuable to all A-level physics students. Those who have had little experience or success with physics calculations will particularly appreciate guidance given on how to approach calculations and will obtain greatest help from the hints given for the A-level standard questions. All necessary mathematics is explained in the text.

A large number of questions from past GCE examinations are used both as exercises and as worked examples. The authors are most grateful to the examining boards concerned. The answers given and working out of answers are the responsibility of the authors, and the examination boards are in no way responsible for these.

The authors extend their thanks to the following examination boards for allowing them to print questions from recent A-level papers.

The Associated Examining Board [AEB]
The Joint Matriculation Board [JMB]
The Oxford and Cambridge Schools Examination Board [O & C Nuff]
The Oxford Delegacy of Local Examinations [O]
The Southern Universities Joint Board [SUJB]
The University of Cambridge Schools Local Examinations Syndicate [C]
The University of London School Examinations Council [L]
The Welsh Joint Education Committee [WJEC]

T L Lowe
J F Rounce

Section A
Basic Ideas

1

HOW TO APPROACH CALCULATIONS

Getting started

When difficulty is experienced with physics calculations it is often not so much carrying out the calculation, i.e. the mathematical part of the problem, that causes the trouble but rather deciding what calculations are needed. How do we get started?

Easy calculations — one formula only is needed

For most students the easiest problem to get started on is the one that requires only the use of one simple, well-known formula or equation, into which numbers provided have to be substituted for the symbols in the formula. The formula tells us exactly how addition, multiplication or division of these numbers must be carried out to obtain the answer, and there is very little thinking to do. Example 1 is of this kind.

Example 1

Calculate (a) the voltage across a 5.0 Ω resistance that has a current of 2.0 ampere flowing through it, (b) the heat produced in this resistance per second and (c) the current that would flow if the resistance were changed to 6.0 Ω while the voltage is unchanged.

Method

(a) The relation between current I, potential difference (voltage) V and resistance R is $V = I \times R$, i.e. $V = IR$. Putting the given values of I and R into this formula for V, we get $V = 2.0 \times 5.0$. Therefore

$V = 10$, but we must not forget the question of units, without which the answer is incomplete.

The unit for potential difference is the volt (V), so that the answer is 10 volt (10 V).

(b) The heat produced per second $= V \times I = 10 \times 2.0 = 20$, and the unit is joule per second (or watt). See Table I in the Appendix (p.333). Thus our answer is 20 watt (20 W).

(c) Here we obtain a formula for I by rewriting the equation $V = IR$ as $I = V/R$ (i.e. V divided by R).

This gives $I = 10/6.0 = 1\frac{4}{6} = 1\frac{2}{3} = 1.67$, i.e. close to 1.7.*

*See Chapter 4.

The unit for current is the ampere (A). Thus $I = 1.7$ ampere (1.7 A).

The above example might be found in a GCSE examination. The following GCE A-level question is also of this simple type.

Answer
10 V, 20 W, 1.7 A.

Example 2
What is the force per unit length on each of two long parallel wires 20 cm apart in which currents of 2.0 A and 1.5 A flow in opposite directions?
($\mu_0 = 4\pi \times 10^{-7}$ H per m). [L 81]

Method
The formula for the force per unit length, which should be learnt by the student, is

$$\frac{\text{Force}}{\text{Length}} = \frac{F}{L} = \frac{\mu_0 I_1 I_2}{2\pi d}$$

Where I_1 and I_2 are the currents, 2.0 A and 1.5 A here, d is the separation of the wires, given as 20 cm, and the value of μ_0 is given.

It is important to note that formulae give correct answers only if the correct units are employed for all the quantities in the formula. If we use the internationally agreed SI units,* then all the formulae we meet will apply. The SI unit for a distance is the metre. Consequently $d = 20$ cm must first be written as 0.20 metre. Now we write

$$\frac{F}{L} = \frac{4 \times \pi \times 10^{-7} \times 2.0 \times 1.5}{2 \times \pi \times 0.20}$$

The meaning of 10^{-7} is explained in Chapter 2.

(Writing 2.0×1.5 in the numerator is better than 2.0 1.5 or 2 1.5 because of the possibility of reading this as 2.01 or 21.5. Writing 4π would be as good as $4 \times \pi$.)

Rewriting our equation and cancelling† we get

$$\frac{F}{L} = \frac{4\not{\pi} \times 10^{-7} \times \not{2.0} \times 1.5}{\not{2}\not{\pi} \times 0.20}$$

$$= \frac{6.0 \times 10^{-7}}{0.20}$$

$$= \frac{60 \times 10^{-7}}{2}$$

$$= 30 \times 10^{-7}$$

$$= 3.0 \times 10^{-6}\ \text{N per m}$$

or 3.0 μN per m or 3.0 μN m^{-1} (see Tables I and II p.333).

*See Chapter 3.
†See Chapter 2.

Answer
3.0 μN m^{-1}.

Note the following points in the above example:

(1) The cancelling was delayed until the second writing of the figures in the formula because cancelling can make it difficult for the reader (e.g. the examiner) to see exactly what values have been used.
(2) The calculation is performed and written down in simple steps gradually leading to the final answer.
(3) Don't forget the units.

Multiples and submultiples like kilo (k) and μ (μ meaning 1 millionth and not to be confused with the μ or μ_0 in our formula above, which is a symbol for a quantity, not a unit) are listed in Table II, p.333.

We wrote m^{-1}, meaning 'per metre', and this is explained in Chapter 2.

Translating the question

Some physics calculation questions can be read and read again, but still it is not obvious what calculations must be done. The situation is like reading a passage in a foreign language. It needs to be translated so that we can see what it means. Only then can we really get started. Quite often the difficulty lies in the fact that by the time one has read the last line of the question the first line has been forgotten. The method here is to translate the question into *either* a labelled diagram *or* equations accompanied by listing of the relevant measurements (quantities) given in the question, *or both*.

A diagram allows a lot of information to be displayed quickly, and the information put into it can be seen at a glance. Diagrams are sometimes of no help, but a diagram usually takes very little time to sketch and is always worth trying. As an illustration of this consider Example 3 below.

Example 3
A uniform horizontal rod AB, which is 1.4 m long and weighs 30 kg, is supported by two vertical wires attached at 0.20 m from A and 0.20 m from B. If loads of 10 kg

and 40 kg are hanging from the rod at distances of 0.40 m and 1.0 m from A, what are the forces provided by the supporting wires?

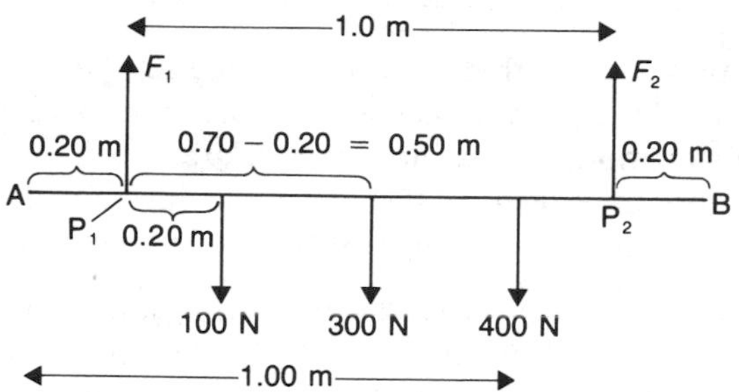

Fig 1.1 Diagram for Example 3 (1 kilogram force has been taken as equal to 10 N)

Method

Because the rod is uniform (the same all along its length), its weight of 300 N acts downwards at its middle (centre of gravity), as shown in Fig 1.1.

Since no movement is occurring anywhere, there is no rotation about the point P_1, so that the clockwise turning effect (or 'moment') about this point must equal the anticlockwise moment.

The clockwise moment is 100×0.20 plus 300×0.50 plus 400×0.80, which equals $20 + 150 + 320$ or 490.

The anticlockwise moment is $F_2 \times 1.00$.

Equating clockwise and anticlockwise moments we get

$$F_2 = 490\ \mathrm{N} = 0.49\ \mathrm{kN}, \quad \text{to two significant figures.}$$

Note that moments about P_1 were considered, a point where one of the unknown forces, namely F_1, was acting. This has the advantage that it keeps F_1 out of the calculations while we evaluate F_2. This keeps the method simple.

To find the value of F_1 we could repeat the above procedure for the point P_2. However, since all the other forces are now known, there is a simpler way to find F_1, as follows. We use the fact that F_1 plus F_2 upwards must equal the total downward forces.

$$\therefore \quad F_1 + F_2 = 100 + 300 + 400 = 800\ \mathrm{N}$$

$$\therefore \quad F_1 + 490 = 800$$

$$\therefore \quad F_1 = 800 - 490 = 310\ \mathrm{N} = 0.31\ \mathrm{kN}$$

Answer

0.31 kN and 0.49 kN.

Is the diagram necessary?

Well, in the last example the diagram made the problem far easier than without it. In Example 4 below, the diagram does not help much and the question must be translated into equations. However, since it is desirable to have a picture in one's mind of the practical situation to which the problem applies, the diagram may have some use.

Example 4

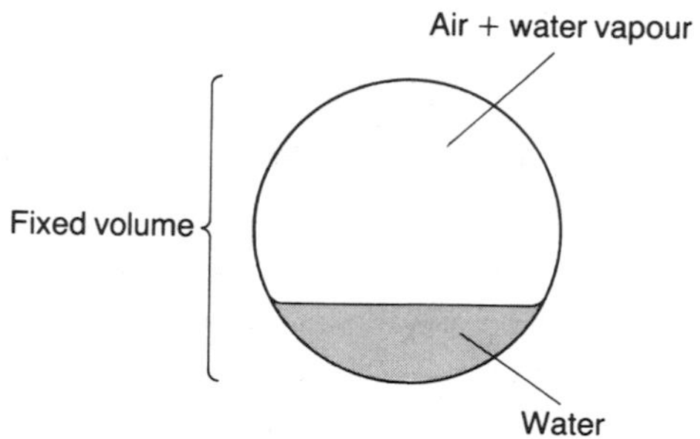

Fig 1.2 Diagram for Example 4

A sealed vessel contains a mixture of air and water vapour in contact with water. The total pressures in the vessel at 27 °C and 60 °C are respectively 1.0×10^5 Pa and 1.3×10^5 Pa. If the saturated vapour pressure at 60 °C is 2.0×10^4 Pa, what is its value at 27 °C? (1 Pa = 1 N m^{-2}.) [L 77]

Method

For temperature T_1 = 27 °C (= 273 + 27 = 300 K), we have total pressure = 1.0×10^5, SVP = ? (Call it x.)

For temperature T_2 = 60 °C (= 273 + 60 = 333 K), we have total pressure = 1.3×10^5, SVP = 2.0×10^4.

The important fact here is that the total pressure equals the pressure of the air plus the pressure (SVP) due to the saturated vapour.

At 27 °C, $1.0 \times 10^5 = P_1 + x$

At 60 °C, $1.3 \times 10^5 = P_2 + 2.0 \times 10^4$

From the second equation

$$P_2 = 13 \times 10^4 - 2.0 \times 10^4 = 11 \times 10^4$$

But, for a fixed mass of air whose volume is fixed, the pressure is proportional to the absolute temperature (the 'constant volume law' or 'pressure law', Chapter 25).

$$\frac{P_1}{P_2} = \frac{T_1}{T_2} = \frac{300}{333}$$

$$P_1 = \frac{300}{333} \times 11 \times 10^4 = 9.9 \times 10^4$$

$$1.0 \times 10^5 = (0.99 \times 10^5) + x$$

$$x = 1.0 \times 10^5 - 0.99 \times 10^5$$

$$= 0.01 \times 10^5 \text{ Pa}$$

Answer

0.01×10^5 Pa.

Reading the question

The first reading of a question may give only an impression of the kind of situation involved. It may tell the reader that it is a question on the expansion of metal, or image formation by a lens.

If a reasonably clear picture of the situation is immediately created in your mind, then reading once more at least as a check is recommended.

Whether this is so or not, we have said that our aim must be to 'translate' the written question into labelled diagrams and equations, and during the next reading this should be attempted. Even further reading may be necessary to achieve this and, finally, the question must be carefully read through, giving attention to every word, to check that the diagram and equation are precisely correct.

Often we find that the questions themselves include diagrams. There may then be no need to draw a diagram, or it may be found useful to copy the diagrams or draw a similar one and add extra labelling such as values of currents, lengths, temperature, etc. Example 5 is an illustration of this.

Progress with the calculation — which route to take

Suppose now that you have read the question carefully, drawn a diagram labelled with data and written down equations that relate the items of information given in the question, but there is no immediately obvious method of getting the answer.

Experience of similar calculations is important at this stage. In this book a large number of common types of A-level calculations are shown, and familiarity with the techniques used in these examples will prove most valuable. No route is easier than one you have trodden before.

Nevertheless there are still those problems where new ideas are needed, where the question is of a novel type. Consider Example 5 below.

Example 5

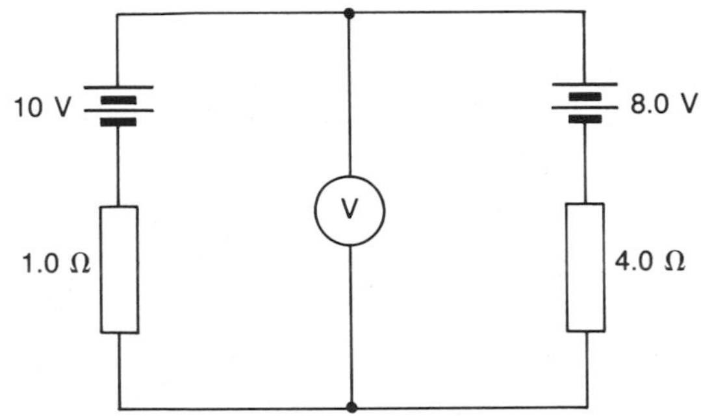

Fig 1.3 Diagram for Example 5

In the circuit in Fig 1.3 the batteries have negligible internal resistance and the voltmeter has a very high resistance. What would be the reading of the voltmeter? [L 77]

Method

Here the diagram is given in the question and all the data are included in the diagram.

First we must note the crucial fact that the voltmeter has a very high resistance. This means that any current flowing through the voltmeter will be so small as to be negligible. Thus we can neglect the presence of the voltmeter and its connecting wires and merely calculate the potential difference (call this V) between the top and bottom lines of the circuit. So we want the value of V.

Ask yourself, perhaps, what factors are likely to decide V. Well, there are two batteries with 2.0 volts difference between their EMF values. Could this be the answer? As a guess, yes, but this is not good enough. Doesn't it seem that V should be near to 8.0 or 10 volts or somewhere between? What about the resistances? In A-level questions one expects that figures given are necessary figures. But really there must be a current flowing because the 10 volt EMF exceeds the 8.0 volt EMF which is opposing it. So this current (call it I and mark it on your diagram if you have drawn it on your paper) will flow (clockwise in the diagram since the 10 volt EMF is the dominant battery) and passes through the 4.0 ohm resistor, showing that there must be, across this resistance, a PD of $I \times 4.0$. Because of the current direction, the upper end of the 4.0 ohm resistance must be at the greater (+) potential. The conclusion is (and this is what we need to write down)

$$V = 8.0 + 4.0I$$

and

$$I = \frac{\text{Total voltage supply}}{\text{Total resistance}}$$

$$= \frac{(10 - 8.0)}{(1.0 + 4.0)} = \frac{2}{5}$$

$$= 0.4\ \text{A}$$

$$\therefore \quad V = 8.0 + (4.0 \times 0.4)$$

$$= 9.6\ \text{volt}$$

Answer

9.6 V.

This seems a reasonable value for the answer. But haven't we given more attention to the right-hand side of the circuit than to the left? Is this justified? Surely we could have given greater prominence to the left-hand side? If we try this, we equate V to the total of the 10 volt of the left-hand battery and the PD $I \times 1.0\,\Omega$ (i.e. 0.4 volt) across the $1.0\,\Omega$ resistance. Here we must realise that the polarity of the 0.4 volt is negative at the upper end so that $V = 10 - 0.4 = 9.6$ V. This has in fact checked the correctness of our first answer and has shown two ways of getting the answer. If time allows, a check like this is always worthwhile.

So what ideas can be used to help us answer a question like this? We ask what factors will affect the answer. What is happening in the example? Have we met a similar question before? Are there crucial words in the question? Remember too that all figures given in the question are usually needed.

Problems in two parts

Looking back at Example 4 we observe that the question describes two experimental results, one at 27 °C and the other at 60 °C. Two similar equations were written, one for each of the temperatures, and we could even have drawn two separate diagrams. (The information given for 60 °C simply enabled us to calculate the gas pressure P_1 which was not given, but was needed, at 27 °C.) Questions like this become much simpler once they are regarded as made up of two parts.

As explained later, in Chapter 2, these two-part situations are a common feature of A-level questions, and examples will be shown in Chapter 2 and in other chapters.

Some calculations can be divided into two or more parts for convenience. The following calculation, Example 6, involves a number of energy conversions, so that each of these can be considered as a step in the calculations.

Example 6

A bullet fired from a gun strikes a block of wood of mass 1.0 kg and specific heat capacity $8 \times 10^2\ \text{J kg}^{-1}\,\text{K}^{-1}$ and produces a temperature rise in the wood of 0.4 K. There were 2.0 g of explosive before firing. If the energy per unit mass released by the explosive is $8 \times 10^6\ \text{J kg}^{-1}$, estimate the percentage of energy released in the firing which is converted to kinetic energy of the bullet. State the assumptions you have made. [L 77]

Method

The situation is illustrated in Fig 1.4 (overleaf).

Step 1. Energy (call it E) is released by the explosive.

$$E = \frac{2.0}{1000}\ \text{kg} \times 8 \times 10^6\ \text{J kg}^{-1}$$

$$= \frac{2.0 \times 8 \times 10^6}{10^3} = 16 \times 10^3\ \text{J}$$

Step 2. A fraction f of this explosion energy becomes kinetic energy of the bullet.

$$KE = f \times 16 \times 10^3 \text{ J}$$

Step 3. As the bullet comes to rest inside the wood all of this kinetic energy becomes heat (internal energy) within the wood. Here we are making a bold assumption (see below).

So $16 \times 10^3 f$ = mass of wood × specific heat capacity × temperature rise.

$$16 \times 10^3 f = 1.0 \times 8 \times 10^2 \times 0.4$$
$$= 3.2 \times 10^2$$
$$f = \frac{320}{16000}$$
$$= \frac{2}{100} \quad \text{i.e. } 2\%$$

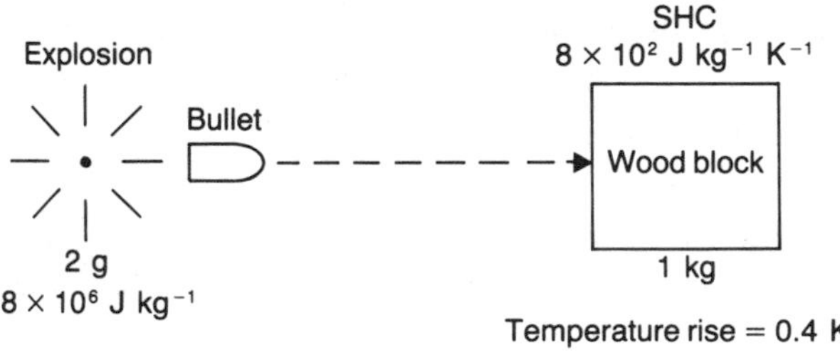

Fig 1.4 Diagram for Example 6

The assumptions made in the above calculation include:

1. That none of the kinetic energy ends up as KE of the block, i.e. the block is held still.
2. That none of the energy given to the wood has become potential energy of strain.
3. That none of the energy left the block as sound energy.
4. That all of the heat produced enters the wood and spreads evenly throughout, no significant amount remaining in the bullet and none escaping into the surrounding air or the block's support.

Answer
2%, and see above.

The language of physics problems

A word like 'calculate' is unambiguous, 'determine' is occasionally written instead and 'evaluate' is an alternative. 'Find' is another possibility (inviting the facetious reply 'When was it lost?'). It is usually obvious that 'What is the . . . ? really means 'What is the size of . . . ?' if sufficient information is given for the size to be calculated.

The physical situation to which the calculation applies may require the reader to understand the meanings of the expressions 'in the plane of', 'coaxial with' or 'perpendicular to' describing positions; while 'simultaneously', 'suddenly', 'quickly' or 'slowly' can be most important as regards the timing of events. 'Quickly' may imply that there is insufficient time for heat to escape from an object. Something lifted 'slowly' means that it moves so slowly that all the energy used in the lifting becomes gravitational potential energy as the object rises and none becomes kinetic energy because the object acquires no significant speed.

Words like these are not for picturesque description but convey essential information and they must be accorded full attention.

Already we have seen (in Example 5) the meaning of a meter with 'very high resistance' and batteries with 'negligible internal resistance'. Similarly lenses may be 'thin', current-carrying wires may be 'long' (having infinite length) and sources of radiation (e.g. lamps giving out light) may be 'point sources' because their sizes are very small compared with other lengths being considered. The true meanings of these terms are obtained if we add 'sufficiently . . . to simplify the calculation': e.g. a lens 'sufficiently thin to simplify the problem'.

These peculiarities of physics questions will become familiar as the many exercises in this book are studied.

Assumptions

2 apples + 3 apples = 5 apples

This is true—exactly true—under all circumstances. However, most rules that relate quantities in physics are exactly true only under particular circumstances. Quite often a rule (or law) of physics will be used in a calculation when there is some uncertainty as to whether the required conditions are satisfied.

In Example 4 use was made of the constant volume law and it was necessary to assume that the volume of the vessel was fixed, i.e. of constant size. We tacitly assumed that the change of temperature did not cause the vessel to expand appreciably. The expansion of a solid vessel (perhaps a glass one) would be small but (what is more important) we could not obtain an answer if an unknown volume change occurred.

In Example 6 we had to assume that all of the energy of the bullet became heat in the wood and remained there. In electric circuit problems we repeatedly assume that the connecting wires have negligible resistance. This is usually justified, because a copper connecting wire 1 mm in diameter, while being thin enough to be flexible, is thick enough for a 1 metre length to have a resistance of only about one-fiftieth of an ohm.

2 ESSENTIAL MATHEMATICS

Simple rules for handling equations

If the whole of one side of an equation is multiplied, divided, added to or reduced by any number, then the equation will remain true if the same is done to the other side.

Examples are

- $x + 2 = 5$ gives $x = 5 - 2$ by subtracting 2 from each side, i.e. $x = 3$
- $4x = 8$ gives $x = 2$ when each side is divided by 4
- $6x = 4$ or $x = 4/6$ can be written as $3x = 2$ or $x = 2/3$

The popular rules which express these facts are

- A number multiplying one side can be moved to the other side if it is made to divide that side, and vice versa
- A number adding on one side can be moved to the other side if its sign is changed
- A fraction is unchanged if its top and bottom are both multiplied or divided (but not added to or reduced) by the same number

When we use these rules numbers can be removed from the equation or fraction and the process is then called 'cancelling'.

Examples are

- $x + \cancel{2} = 3 + \cancel{2}$ gives $x = 3$
- $x = \dfrac{7 \times \cancel{2}}{\cancel{2}}$ gives $x = 7$
- $x = 8/4$ gives $x = {}^{2}\cancel{8}/\cancel{4}$, i.e. $x = 2$

Example 1

At a pressure of $10^5\ \text{N m}^{-2}$, and temperature of 0 °C, the density of oxygen is $1.4\ \text{kg m}^{-3}$. Calculate the root mean square velocity of oxygen molecules under these conditions.

Method

The equation relating pressure P, density ρ and mean square velocity $\overline{c^2}$ for a perfect gas is

$$P = \tfrac{1}{3}\rho\overline{c^2}$$

We have $P = 10^5\ \text{N m}^{-2}$ and $\rho = 1.4\ \text{kg m}^{-3}$.

$$\therefore \quad 10^5 = \tfrac{1}{3} \times 1.4 \times \overline{c^2}$$

Multiplying both sides by 3 (or taking the 3 to the left-hand side) we get

$$3 \times 10^5 = 1.4\,\overline{c^2}$$

and taking the 1.4 to the left-hand side gives us

$$\frac{3 \times 10^5}{1.4} = \overline{c^2}$$

$$\therefore \overline{c^2} = 2.14 \times 10^5$$

The aim has been to get the unknown quantity on one side of the equation and all the figures on the other. Hence

$$\sqrt{\overline{c^2}} = \sqrt{(2.14 \times 10^5)} = 463\ \mathrm{m\,s^{-1}}$$

Note too that

$$\sqrt{\overline{c^2}} = \sqrt{(21.4 \times 10^4)} = \sqrt{21.4} \times \sqrt{10^4}$$
$$= 4.63 \times 10^2 = 463\ \mathrm{m\,s^{-1}}$$

Rules for working with numbers like 10^4 and 10^5 are discussed below.

Other examples of using the above rules have already been given in Chapter 1.

Answer
0.46 km s^{-1}

Exponents

a^2 means $a \times a$, a^3 means $a \times a \times a$, etc., so that $10^2 = 100$, $10^3 = 1000$, etc. The small superscript numbers are called exponents or powers.

When two numbers multiply, their exponents add, e.g. $10^2 \times 10^3 = 10^5$ (or 100 000).

When two numbers divide, their exponents subtract, e.g. $10^3/10^2 = 10^{3-2} = 10^1$, or 10.

The reciprocal of a number (meaning 1/the number) is obtained by making the exponent negative, e.g. $1/100 = 10^{-2}$ ('ten to the power of minus two'). Another useful fact is $(a^x)^y = a^{xy}$, e.g. $(10^3)^2 = 10^6$.

Logarithms

If $10^L = x$, then L is called the logarithm of x or, more exactly, the logarithm to the base 10 of x.

For example $10^3 = 1000$ so that log (to the base 10) of 1000 is 3. We write this as $\log_{10} 1000 = 3$.

If the base is not specified, then we assume it to be 10, so that log 2 is taken to mean $\log_{10} 2$.

The number 2.718 has a special significance in many physical processes and is denoted by the symbol e. It is called the exponential function and is frequently used as the base for logarithms, which are then called natural logarithms.

For example $e^{2.3} = 10$ and the natural logarithm (denoted by ln) of 10 is therefore 2.3. i.e. $\ln 10 = 2.3$.

To convert between $\log_{10}$ and ln values we can use the fact that $\ln x = 2.3 \log_{10} x$.

Exercise 1

1 Evaluate x:
(a) $2.7x = 8.1$ (b) $x/1.5 = 22$
(c) $x/3 + 5 = 7$ (d) $2(x - 4) = 5$
(e) $2(x + 4) = 5$

2 Simplify:
(a) $1000/10^2$ (b) $10^2/10^3$ (c) $10^{18}/10^2$
(d) $10^2/10^{17}$ (e) $0.01/10^{18}$ (f) $10^8 \times 10^2$
(g) $10^8 \times 10^{-2}$ (h) $(100^2)^3$ (i) $\sqrt{10^2}$
(j) $\sqrt{10^4}$ (k) $(10^4)^{1/2}$ (l) $\sqrt{(9 \times 10^4)}$

3 Write down the value of $\log_{10} x$ when
(a) $x = 100$ (b) $x = 10^9$ (c) $x = 1/100$

4 (a) Fill in the missing terms:
(i) $\log_{10} 1000 + \log_{10} 100 = \ldots$
(ii) $\log_{10} 1000 = \log_{10} 100 + \ldots$
(iii) $\log_{10} 20 = \log_{10} 2 + \ldots$
(b) Given that $\log_{10} 2 = 0.301$, evaluate
(i) $\log_{10} 20$
(ii) $\log_{10} 2000$

5 Given that $e^{2.3} = 10$, evaluate ln 100.

6 Given $e^{2.3} = 10$ and $\log_{10} 3 = 0.477$, calculate
(a) ln 3 (b) ln 300

Some rules of geometry

Angles may be measured either in degrees (one revolution is 360 degrees (360°)) or in radians (rad), whose size is such that 2π rad equals one revolution. Some useful facts about angles are shown in Fig 2.1 (where degrees are used).

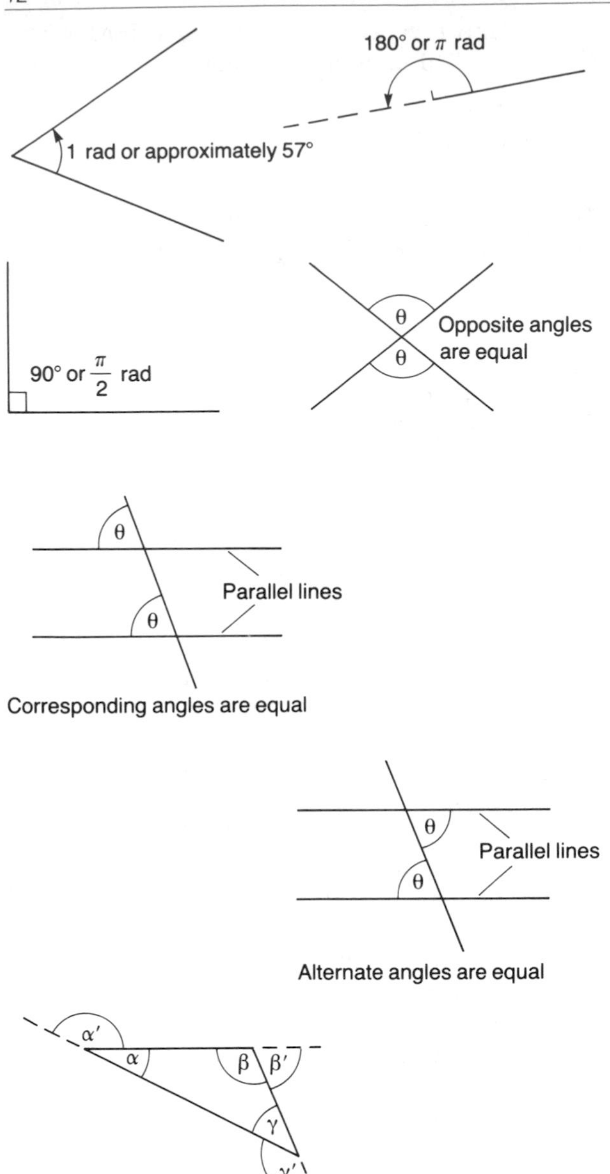

Fig 2.1 Useful information concerning angles

Pythagoras' theorem

In a right-angled triangle (Fig 2.2) the longest side (the hypotenuse) has a length c related to the lengths a and b by

$$c^2 = a^2 + b^2 \tag{2.1}$$

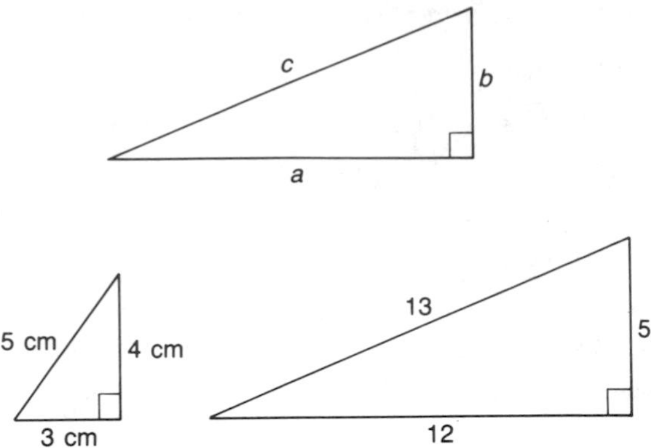

Fig 2.2 Right-angled triangles

Well-known examples of right-angled triangles are the 3, 4, 5 and 5, 12, 13 triangles shown in Fig 2.2.

Isosceles and equilateral triangles

The isosceles triangle has two sides of equal length and so two of the angles are equal (Fig 2.3a).

An equilateral triangle has three sides of equal length and each angle equals 60° (Fig 2.3b).

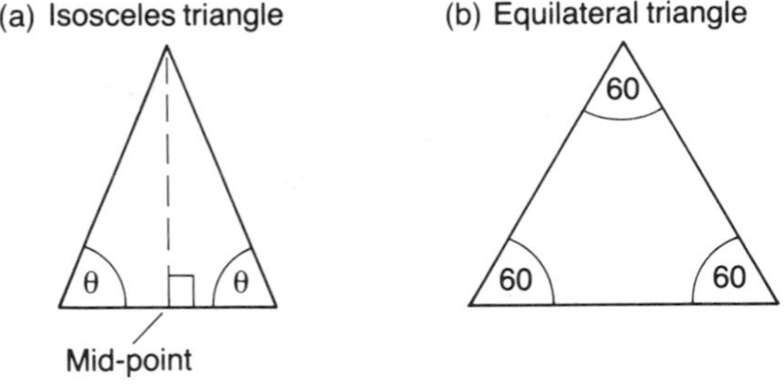

Fig 2.3 Isosceles and equilateral triangles

Some properties of circles, discs and spheres

The circumference of a circle is $2\pi R$ (where R is its radius) or π times the diameter.

As shown in Fig 2.4a, an angle of 1 radian subtends, at any radius R, an arc equal to a fraction $1/2\pi$ of the circumference, i.e. it subtends an arc of length R.

(a) Arc = radius for 1 radian. (b) Angle θ = arc/radius

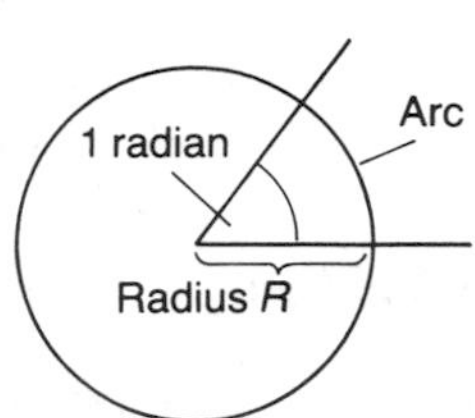

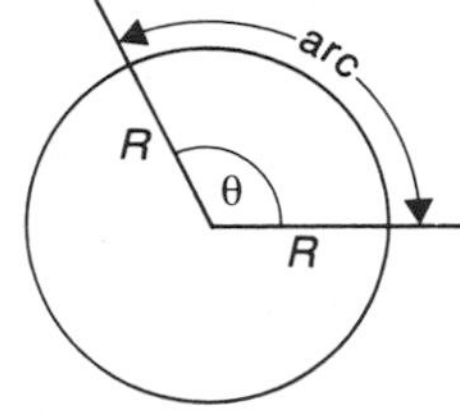

Fig 2.4 Using radians

The size of any angle in radians equals the arc it subtends divided by the radius (Fig 2.4b).

$$\theta = \frac{\text{arc}}{R} \qquad (2.2)$$

Exercise 2

1 In Fig 2.5:
(a) How is the sum of r_1 plus r_2 related to (i) α, (ii) γ?

(b) How is the angle BCD related to (i) γ, (ii) α?

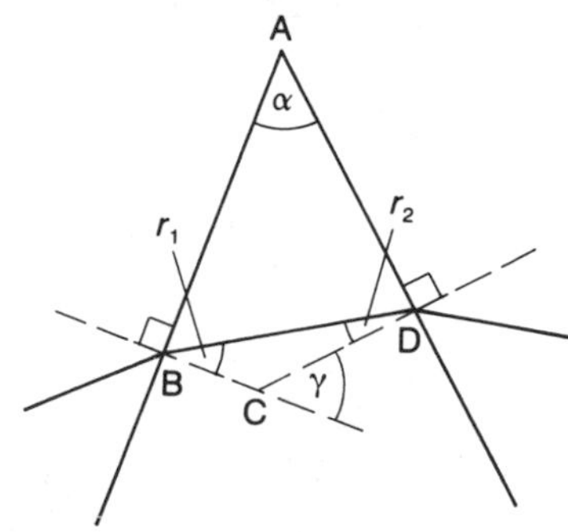

Fig 2.5 Diagram for Question 1

2 In Fig 2.6:
How is (a) γ related to angle i, and (b) α related to angle i?

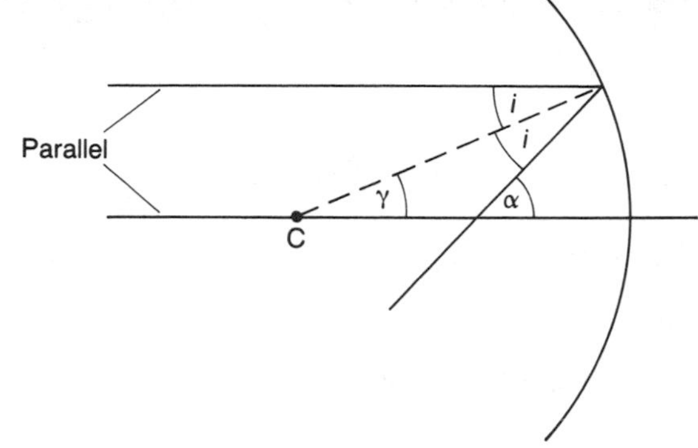

Fig 2.6 Diagram for Question 2

3 In Fig 2.7:
Express β and γ in terms of α and i.

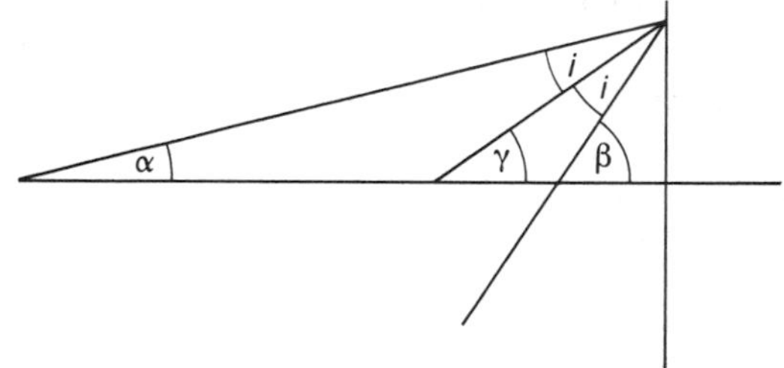

Fig 2.7 Diagram for Question 3

4 If an object moves steadily in a circular path of 20 cm diameter, what length of arc does it cover in a time of 4 s if it takes 12 s for every revolution? ($\pi = 3.142$.)

5 Convert the following angles in degrees into radians ($\pi = 3.142$):
(a) 60° (b) 45° (c) 4 revolutions
(d) 90° (e) 180°

Trigonometrical ratios

The size of any angle θ can be specified by imagining it to be part of a right-angled triangle and then describing the resulting shape of the triangle as shown in Fig 2.8. For example when $\theta = 60°$, the ratio of the *adjacent side* to the *hypotenuse* is $\frac{1}{2}$. So b/c, which we call cosine θ, is 0.5.

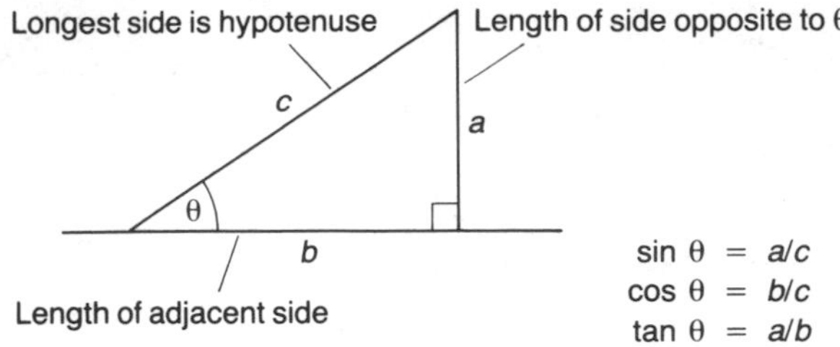

Fig 2.8 Trigonometrical ratios

The most useful ratios are

$$\text{sine } \theta \text{ (or } \sin\theta) = a/c$$

$$\text{cosine } \theta \text{ (or } \cos\theta) = b/c$$

$$\text{tangent } \theta \text{ (or } \tan\theta) = a/b$$

For a given θ (in degrees or radians) we can get $\sin\theta$, $\cos\theta$, etc. using suitable electronic calculators or tables and similarly can deduce θ from any given trigonometrical ratio.

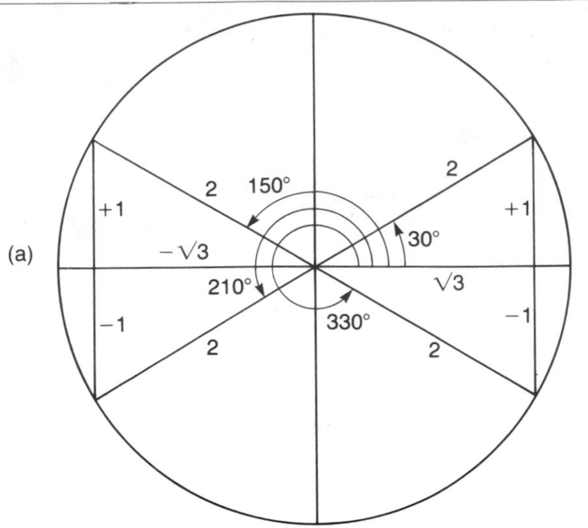

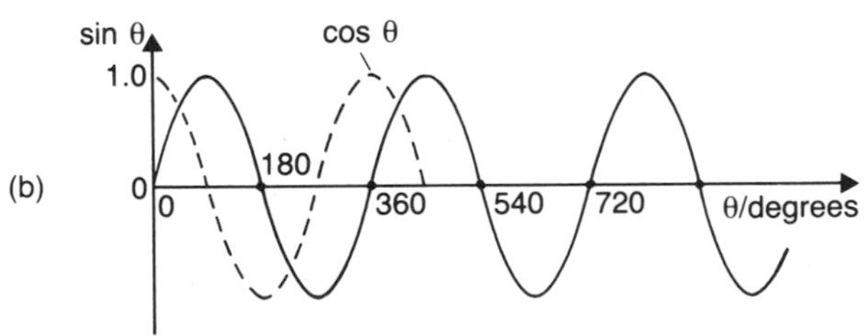

Fig 2.9 Trigonometrical ratios for large angles

Small angles

For a small angle (θ about 5° or less), $\tan\theta \simeq \sin\theta \simeq \theta$ in radians and $\cos\theta \simeq 1$, to better than 1%.

Large angles

For $\theta = 90°$, $\sin\theta = 1$, $\cos\theta = 0$ and $\tan\theta = \infty$. For $\theta > 90$, we can still use $\sin\theta$, $\cos\theta$, etc. if we apply suitable rules as illustrated in Fig 2.9a. This shows that a negative sign must be given to an opposite side that is below the horizontal axis and to an adjacent side if it is to the left. The relationship between $\sin\theta$ and θ is shown as a graph in Fig 2.9b.

The cosine rule

This rule is an extension of the rule (or theorem) of Pythagoras and applies to a triangle of any shape. It relates the lengths a, b and c of the triangle's sides (see Fig 2.10).

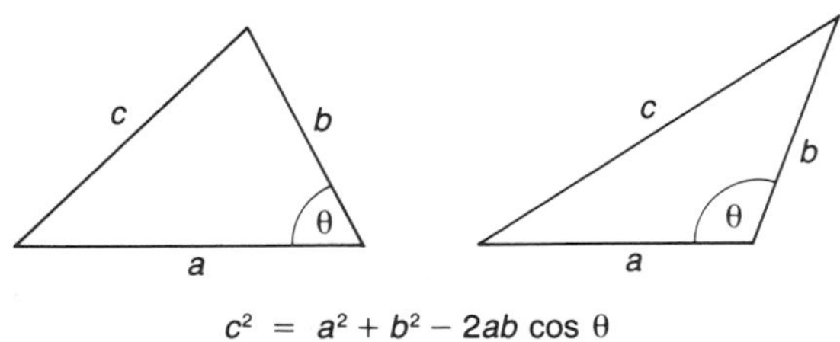

Fig 2.10 The cosine rule

The sine rule

This rule states that $\dfrac{a}{\sin A} = \dfrac{b}{\sin B} = \dfrac{c}{\sin C}$ (see Fig 2.11).

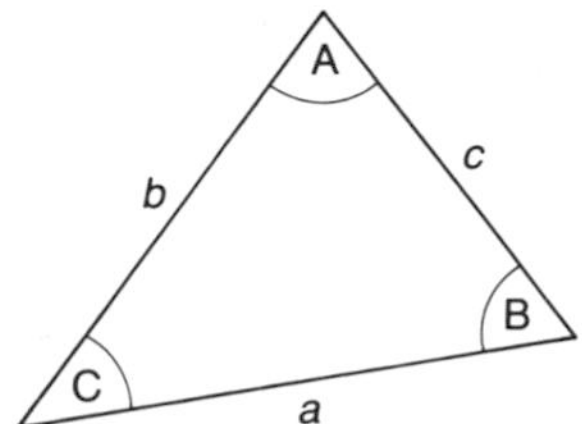

Fig 2.11 The sine rule

Expressions and their factors

An expression consists of numbers adding, multiplying, etc. together. Numbers (or other expressions) which multiply to produce a given expression are called its factors.

For example $(x - a)$ multiplied by $(x + a)$ gives $x^2 - a^2$, so that they are factors of this expression. This particular example can be very useful.

Factorising can, again, be most useful in solving some equations.

Consider $x^2 - 3x = -2$ for example. This can be written as $(x - 2)(x - 1) = 0$, and it is obvious that either $x = 2$ or $x = 1$.

These are the two possible answers ('solutions') for the value of x.

Simultaneous equations

If an equation contains two unknown quantities (say x and y), then it cannot by itself tell us either x or y. However, both x and y can be evaluated if they are the unknowns in *two* different equations. (Three equations would be needed for three unknowns.)

Example 2

A liquid is flowing through a pipe with a heater inside. The flow rate is 0.90 kg per second, the heat supply is 2000 J per second (i.e. the power is 2.0 kilowatt) and the liquid's temperature rises by 1.0 °C (1.0 K) as the result of passing the heater. If the power supplied is increased to 3.0 kilowatt, the same temperature rise is obtained, with a flow of 1.4 kg per s. Calculate the specific heat capacity of the liquid. (It should be assumed that the heat loss per second from the outside of the pipe into the surrounding air does not change.)

Method

$$\begin{pmatrix}\text{Heat}\\ \text{supplied}\\ \text{per s}\end{pmatrix} = \begin{pmatrix}\text{Heat per s}\\ \text{entering and}\\ \text{warming liquid}\end{pmatrix} + \begin{pmatrix}\text{Heat loss}\\ \text{per s}\\ \text{into air}\end{pmatrix}$$

$$2000 = (\text{Mass per s} \times \text{SHC} \times \text{Temp rise}) + h$$

$$2000 = (0.9 \times \text{SHC} \times 1.0) + h \qquad \text{(i)}$$

and similarly

$$3000 = (1.4 \times \text{SHC} \times 1.0) + h \qquad \text{(ii)}$$

Subtracting (i) from (ii) gives

$$1000 = (1.4 - 0.9) \times \text{SHC}$$

$$\text{SHC} = \frac{1000}{0.5} = 2000 \text{ J per kg per degree}$$

Now suppose that, instead, we were asked to evaluate h. We have to make the SHC terms equal so that subtracting eliminates these terms. So we can multiply (i) by 1.4 and (ii) by 0.9. Then we subtract.

(i) becomes $2800 = (0.9 \times 1.4 \times \text{SHC}) + 1.4h$

and (ii) becomes $2700 = (1.4 \times 0.9 \times \text{SHC}) + 0.9h$

Subtracting gives

$$100 = 0.5h, \quad \text{so that } h = 200 \text{ J per second.}$$

If both SHC and h are required, we can evaluate one of them, say SHC, as above and then quickly use this value in one of the equations to find the other unknown.

Answer

2.0×10^3 J kg^{-1} K^{-1}.

Exercise 3

1 Multiply out each of the following expressions, i.e. put each into the form $ax^2 + bx + c$:
(a) $(x + 4)(x + 2)$ (b) $(x - 4)(x + 2)$
(c) $(x - 4)(x - 2)$ (d) $(x - 4)(x - 4)$
(e) $(x + 2)(3x - 5)$

2 Find two possible values of x given by each of the following equations:
(a) $x^2 - 2x - 8 = 0$ (b) $6x^2 - 11x = 10$
(c) $x^2 - 9 = 0$

3 If $2y + x = 17$ and $y + 2x = 13$, what is the value of y and what is the value of x?

3 UNITS AND DIMENSIONS

Measuring a quantity

When a length is measured as 7 feet it means 7 times the length of a foot. What is measured (i.e. the quantity) consists of a number (7) multiplied by the chosen unit (foot, metre, etc.).

Fundamental and derived quantities

Several quantities, like mass, length, time, temperature, are called fundamental or base quantities while others are derived from these. For example a velocity equals a length divided by a time.

SI units

The International System of Units suggests a set of convenient units (and suitable abbreviations for them) that have been widely accepted. These SI units are used in A-level physics.

All equations given in this book will work with SI units.

The system uses seven base units including the kilogram (kg), metre (m) and second (s), and all other SI units are derived from these: e.g. metre per second for velocity.

Unit abbreviations remain unaltered in the plural.

A unit is not spelt with a capital letter even when the unit is named after a person, e.g. the SI unit for force is the newton (N). The recommended abbreviations, however, are often capital letters.

Recommended multiples and submultiples for units are given in Table II (p.333.)

Dimensions

Regardless of the units employed, a velocity is always a length divided by a time, and a force (= mass × acceleration) is always a mass times a length divided by time squared, i.e.

$$F = [M][L][T]^{-2}$$

The powers of the multiplying base quantities are called the dimensions of the derived quantity concerned. Square brackets are often used to indicate that we are discussing dimensions only.

The '≡' sign indicates an identity. It tells us more than an equation. Each dimension is the same on both sides of the identity. Capital letters M, L, T are used as abbreviations for mass, length and time, θ for temperature change, Q for electric charge, when we are dealing with dimensions.

Some quantities are dimensionless (i.e. the dimensions are zero). Angles are an example.

Important properties of dimensions

'Three plus two equals five' is always true but 'three pints plus 2 kilograms' is meaningless in an equation since all of the terms must have the same dimensions, i.e. must be the same kind of quantity.

Units must have the same dimensions as the quantity to which they apply. Thus the newton is $\text{kg} \times \text{m} \times \text{s}^{-2}$ (written kg m s^{-2}) because the dimensions of force are MLT^{-2}.

A ratio of two like quantities is a number, without dimensions, e.g. 7.0 m divided by 2.0 m is 3.5. The metres have cancelled.

Example 1

What are the dimensions of (a) force, (b) moment, (c) work, (d) pressure?

Method

We need to relate each of these quantities to quantities whose dimensions are known.

(a) Force = Mass × Acceleration

$$F = M \times \frac{L}{T^2} \quad \text{or} \quad MLT^{-2}$$

(b) Moment = Force × Perpendicular distance

$$\therefore \text{ Moment} = MLT^{-2} \times L \quad \text{or} \quad ML^2T^{-2}$$

(c) $$\text{Work} = \text{Force} \times \text{Distance} = MLT^{-2} \times L \quad \text{or} \quad ML^2T^{-2}$$

(d) $$\text{Pressure} = \frac{\text{Force}}{\text{Area}}$$

$$P = \frac{MLT^{-2}}{L^2} \quad \text{or} \quad ML^{-1}T^{-2}$$

Answer

(a) MLT^{-2} (b) ML^2T^{-2}
(c) ML^2T^{-2} (d) $ML^{-1}T^{-2}$.

Example 2

The dimensions of capacitance in terms of mass M, length L, time T and charge Q are

A $ML^2T^{-1}Q$
B $ML^2T^{-2}Q^2$
C $M^{-1}L^{-1}Q^2$
D $M^{-1}L^{-2}T^2Q^2$
E $M^{-1}L^{-2}T^3Q$ [AEB 82]

Method

Some relationships that might be useful are $C = Q/V$ and $\frac{1}{2}CV^2$ = work done (or energy stored); see Chapter 31.

Neither of these gives an immediate answer because we don't know the dimensions of V. However, this is no problem because V = work/charge; see Chapter 27.

$$V = \frac{\text{Force} \times \text{Distance}}{Q} \quad \text{or} \quad MLT^{-2}LQ^{-1}$$

and $$C = \frac{Q}{V} \quad \text{or} \quad \frac{Q}{ML^2T^{-2}Q^{-1}}$$

$$\text{or} \quad Q^2M^{-1}L^{-2}T^2$$

Answer

D.

Checking equations and units

All terms in an equation must have the same dimensions, and this can be useful for checking the correctness of an equation. For example the lens equation

$$\frac{1}{u} + \frac{1}{v} = \frac{1}{f}$$

(see Chapter 20) might, by mistake, be written as

$$\frac{v}{u} + 1 = \frac{1}{f}$$

instead of

$$\frac{v}{u} + 1 = \frac{v}{f}$$

The mistake is obvious if dimensions are considered because v/u and 1 are dimensionless but $1/f$ has the dimension L^{-1}.

As regards checking units, an example of a unit which is difficult to remember is that for thermal conductivity, k; see Chapter 24. We need an equation containing k. Now k is given by $Q = kA(\theta_2 - \theta_2)/l$ whence $k = Ql/A(\theta_2 - \theta_1)$ and the units are $\text{W} \times \text{m}/(\text{m}^2 \times \text{K})$ or $\text{W m}^{-1}\,\text{K}^{-1}$.

Exercise 4

1 What are the dimensions of:
(a) density, (b) area, (c) cubic feet per minute, (d) power?

2 What are the dimensions of:
(a) distance/velocity, (b) force × time, (c) angle moved through per second?

3 What are the dimensions of magnetic flux density? (Chapter 33 gives $\Phi = BA$, PD = Blv, PD = $d\Phi/dt$ and Chapter 27 gives PD = W/Q.)

4 The equation relating current I through a semiconductor diode (see Chapter 36) to the applied potential difference V at temperature T is

$$I = I_0 e^{-eV/kT}$$

where e is the electron charge and k is the Boltzmann constant. What are the dimensions of k?

5 The surface tension of a liquid is measured in N m^{-1}. What are the dimensions of surface tension?

Deriving an equation by use of dimensions

Suppose it is known or thought likely that a quantity w is related to quantities x, y and z by an equation of the form

$$w = Cx^p y^q z^r$$

where C is a dimensionless constant, i.e. just a number.

We can use our knowledge of dimensions to determine p, q and r (although it will not determine C).

Example 3

The mass M of gas escaping in a short time t through a small hole of area A in the wall of a gas cylinder is given by

$$\frac{M}{t} = Cp^\alpha \rho^\beta A^\gamma$$

where C is a constant, and p and ρ are the pressure and density of the gas. Give the dimensions of p and ρ, and determine the values of the constants α, β, γ.

[SUJB 79]

Method

The dimensions are

For p, MLT^{-2}/L^2 or $ML^{-1}T^{-2}$

For ρ, $\dfrac{M}{L^3}$ or ML^{-3}

For A, L^2

For $\dfrac{M}{t}$, MT^{-1}

$$\therefore \quad [MT^{-1}] = [ML^{-1}T^{-2}]^\alpha [ML^{-3}]^\beta [L^2]^\gamma$$

Equating M dimensions gives

$$1 = \alpha + \beta + 0$$

Equating L dimensions gives

$$0 = -\alpha - 3\beta + 2\gamma$$

Equating T dimensions gives

$$-1 = -2\alpha + 0 + 0$$

$$\therefore \quad \alpha = \tfrac{1}{2} \qquad 1 = \tfrac{1}{2} + \beta \qquad \text{and so} \quad \beta = \tfrac{1}{2}$$

$$0 = -\tfrac{1}{2} - 3 \times \tfrac{1}{2} + 2\gamma$$

$$2\gamma = 2 \qquad \text{or} \qquad \gamma = 1$$

Answer

$\alpha = \frac{1}{2}$, $\beta = \frac{1}{2}$, $\gamma = 1$.

Example 4

It is suggested that pressure P at depth h in a liquid of density ρ is $P = Ch\rho g$, where g is the acceleration due to gravity. Show that this equation is dimensionally correct.

Method

We need to show that, writing $P = Ch^p\rho^q g^r$, the values of p, q and r are each 1.

The dimensions are

For P, $\dfrac{MLT^{-2}}{L^2}$ or $ML^{-1}T^{-2}$

For h^p, L^p

For ρ^q, $(ML^{-3})^q$

For g^r, $(LT^{-2})^r$

$$\therefore \quad ML^{-1}T^{-2} = L^p(ML^{-3})^q(LT^{-2})^r$$

Equating the M dimensions gives

$$1 = q$$

Equating the L dimensions gives

$$-1 = p - 3q + r$$

Equating the T dimensions gives

$$-2 = -2r$$

$$q = 1, \quad r = 1 \quad \text{and} \quad -1 = p - 3 + 1$$

$$\text{or} \quad p = 1$$

Answer

$p = 1, q = 1, r = 1$ so that $p = Ch\rho g$. (As we know, $p = h\rho g$, i.e. $C = 1$ if SI units are used, but dimensional analysis cannot tell us this.)

Exercise 5

1 Evaluate α and β in the equation $E = Cm^{\alpha}v^{\beta}$, where E is kinetic energy, m is mass, v is velocity and C is a dimensionless constant.

2 The force of attraction F between two particles of masses m_1 and m_2 situated a distance d apart is given by $F = Gm_1m_2/d^2$. Show that the dimensions of G are $M^{-1}L^3T^{-2}$.

3 The minimum velocity needed for a body to escape from the earth is given by $v = \sqrt{(2GM/R)}$ where M is the mass of the earth and R is its radius. Show that the equation is dimensionally correct.

Displaying units

When a quantity is given verbally or in writing its units must be given.

The units should preferably be written in the singular, i.e. 7 metre rather than 7 metres, and although metre per second may be written as m per s or m/s, it is preferable to use $\mathrm{m\,s^{-1}}$ (verbally 'metre per second' is more popular). Only the proper SI abbreviations should be used (see Tables I and II, p.333).

Carelessness with units can lead to confusion. Consider the meaning of 9 N per m per s. This could be interpreted as $9\,\mathrm{N\,m^{-1}\,s^{-1}}$ (which is correct) or $9\,\mathrm{N/m\,s^{-1}}$ which is $9\,\mathrm{N\,m^{-1}\,s}$. Writing $9\,\mathrm{N\,m^{-1}\,s^{-1}}$ overcomes this problem.

A single space should be left between each unit: $6\,\mathrm{m\,s^{-1}}$ rather than $6\,\mathrm{ms^{-1}}$ because $6\,\mathrm{ms^{-1}}$ means 6 millisecond^{-1}. No space is left between a multiple or submultiple and the unit it applies to (e.g. centimetre is cm), and $\mathrm{cm^{-1}}$ means reciprocal cm, i.e. 1/cm (not one-hundredth of a reciprocal metre).

In a graph it is essential to label each axis with the quantity being plotted and its units. Note too that we like to mark along the axis with numbers. It is usual therefore to label the axis with the quantity divided by its units, the division being shown by a dividing line, e.g. volume/$\mathrm{cm^3}$ or $V/\mathrm{cm^3}$. The $V/\mathrm{cm^3}$ values are numbers and this agrees with having the axis marked off with numbers.

Whereas 'volume in $\mathrm{cm^3} \times 10^6$' might be read quite differently as volume $\times 10^6$ in $\mathrm{cm^3}$ or, correctly, as volume in $\mathrm{m^3}$, *there is no ambiguity when we write quantity/unit*. When we have, say, for an area A

$$A \times 10^4/\mathrm{m^2}$$

this is the same as

$$A/10^{-4}\,\mathrm{m^2} \qquad \text{or} \qquad A/\mathrm{cm^2}$$

The same system is recommended for tables of results. At the top of a column of results we might read 'velocity/$\mathrm{m\,s^{-1}}$'. This may be seen in questions 3, 4 and 5 in Exercise 147 (p.307).

Nevertheless few people will complain if 'velocity in $\mathrm{m\,s^{-1}}$' is written, or '$v \times 10^3$ in km per h'.

Conversion of units

Students usually remember conversion factors, e.g. 1000 for changing metres to millimetres; but it is not always obvious whether to divide or multiply by a factor. Common sense should be used. 'Am I changing to smaller units? Will I therefore get more of them?' 1 metre changed to the smaller millimetre units will become 1000 units. A density of $1\,\mathrm{g\,cm^{-3}}$ (1 g per $\mathrm{cm^3}$ of substance) will give many more (100^3 times more) when volume is $1\,\mathrm{m^3}$, i.e. $1\,\mathrm{g\,cm^{-3}}$ is equivalent to $10^6\,\mathrm{g\,m^{-3}}$. Changing to kg the answer will become smaller by 1000, i.e. $10^3\,\mathrm{kg\,m^{-3}}$.

Exercise 6

1 Convert
(a) 30 km h^{-1} to m s^{-1} (b) 0.01 m^2 to mm^2
(c) 400 nm to μm (d) 120 000 min^{-1} to s^{-1}

2 The conductance σ of a conductor is 0.01 Ω^{-1}. Convert this to mΩ^{-1}.

Equations where conversion factors cancel

Consider the Boyle's law equation

$$P_1V_1 = P_2V_2$$

where P_1 and V_1 are initial pressure and volume of a gas and P_2 and V_2 are new values. Perhaps $V_1 = 3.0\,m^3$, with $P_1 = 1.0$ atmosphere (1 bar) and then the pressure is changed to 2.0 atmosphere (i.e. 2 bar). We are asked to calculate V_2.

All our equations work with SI units. Now 1 bar = 10^5 SI units of pressure (N m^{-2} or Pa).

$$1.0 \times 10^5 \times 3.0 = 2.0 \times 10^5 \times V_2$$

But the 10^5 on each side cancels, so that we get $V_2 = 1.5\,m^3$, whether P_1 and P_2 are in Pa or bar. All that is necessary here is that P_1 and P_2 have the same units.

An unusual unit — the mole

This is one of the base units of the SI system. It is an amount of substance that contains the Avogadro number* of particles. The particles concerned have to be named. For example 1 mole of carbon atoms has a mass of 12.000 g, but 1 mole of hydrogen atoms has a mass of 1.008 g, and 1 mole of hydrogen H_2 molecules has a mass of 2.016 g. The SI abbreviation for mole is mol.

*6.022×10^{23}.

Exercise 7: Questions of A-level standard

Questions 1 to 5
The following are sets of dimensions of physical quantities involving the fundamental dimensions mass M, length L and time T.

A ML^2T^{-2} **B** MLT^{-2} **C** $ML^{-1}T^{-2}$
D MT^{-2} **E** $ML^{-1}T^{-1}$

Which of the above is the dimension of
1 force?
2 work?
3 pressure?
4 couple (moment of force)?
5 longitudinal stress? [L 82]

6 The value of the gravitational constant G is 6.7×10^{-11} N m^2 kg^{-2}, so the dimensions of G are
A $ML^{-1}T^{-1}$ **B** MLT^{-2} **C** $M^{-1}L^3T^{-2}$
D ML^2T^{-3} **E** ML^2T^{-2} [O 80]

7 What is meant by the dimensions of a physical quantity? Give the dimensions of velocity, acceleration and force.
The acceleration due to gravity g_r at a point outside the earth, a distance r from the centre of the earth, is given by $g_r = g(R/r)^2$, where g is the acceleration due to gravity at the earth's surface and R is the earth's radius. A satellite of mass m is in a circular orbit of radius r. It is thought that the orbital time T should depend on m, r and g_r in the form $T = km^ar^bg_r^c$, where k, a, b and c are dimensionless constants. Use the method of dimensional analysis to find a, b and c. Hence state how the orbital time varies with the orbital radius. [WJEC 78]

4 ERRORS AND MISTAKES

Introduction

In physics we like to reserve the term 'errors' for inaccuracies which occur when a quantity is measured and when calculations are made using measurements of limited accuracy, i.e. experimental errors. For slips such as we might make in a calculation we can use the word 'mistakes'. Errors and mistakes can both lead to wrong answers.

Errors of measurement

When we count a number of items, the answer is a whole number and it can be faultlessly accurate. When a quantity is measured there is always some uncertainty left as to its value, depending upon the measuring instrument and the skill with which it is used. Thus any answer written down is subject to some error.

Systematic and random errors

If the measuring technique gives answers which are constantly greater or constantly smaller than they should be then we have a *systematic error*. Repeating the measurements and averaging the answers does not remove the error.

Random errors have equal probability of giving higher or lower answers (i.e. being + or −). Averaging a sufficient number of answers will make these errors negligible.

Number of significant figures

When we write 2.43 cm it is assumed that the measurement shows the answer to be closer to 2.43 than 2.42 or 2.44. We can write 2.43 ± 0.005 cm. The 2.43 consists of 'three significant figures'. Similarly 2.00 cm implies greater accuracy than 2 cm. If one centimetre was measured with a ruler and then divided by three using a calculator we might get 0.333 333 3 cm, but all the threes would not be significant. A ruler cannot give an answer of such accuracy. Not more than two figures would be justified (0.33 or .33). A value like 1200 m would be better written as 1.2 km or 12×10^2 m if the accuracy justifies two, not four, significant figures.

Maximum possible error

For many measurements the maximum possible error can be easily estimated. A ruler marked off in millimetres should give an answer correct to within 1 mm unless the ruler is badly made or is not seen clearly. The percentage (maximum) possible error for say 25 mm ± 1 mm is $\pm (1/25) \times 100$ or ±4%.

Possible error for a calculated answer

Suppose w is calculated from an equation of the form $w = xy$, and that the errors in x and y are δx and δy. The error in the calculated w is δw, given by

$$w + \delta w = (x + \delta x)(y + \delta y)$$
$$= xy + \delta x\delta y + y\delta x + x\delta y$$

If the errors δx and δy are both small, then $\delta x\delta y$ is negligible. Now we subtract w from the left-hand side and the equal xy from the right to get

$$\delta w = y\delta x + x\delta y$$

Dividing by w on the left and xy on the right and multiplying all by 100 gives

$$\frac{\delta w}{w} \times 100 = \frac{\delta x}{x} \times 100 + \frac{\delta y}{y} \times 100$$

This shows that the percentage error in w (let's call it $\%w$) equals the sum of the percentage errors in x and y. Thus

$$\%w = \%x + \%y \qquad (4.1a)$$

When $w = x/y$ we get

$$\%w = \%x - \%y \qquad (4.1b)$$

For $w = x + y$,

$$\delta w = \delta x + \delta y \qquad (4.1c)$$

For $w = x - y$,

$$\delta w = \delta x - \delta y \qquad (4.1d)$$

It should be noted that when quantities add or subtract, their actual (not percentage) errors add or subtract.

For systematic errors the signs of the errors may be known and the above equations may be used directly.

For random errors the signs of the errors are unknown and for these and systematic errors of unknown sign we have, for multiplying and dividing quantities

$$\%w = \pm\%x \pm \%y \qquad (4.2a)$$

and for quantities adding or subtracting

$$\delta w = \pm\delta x \pm \delta y \qquad (4.2b)$$

Mostly we work out the maximum possible errors so that, instead of Equations 4.2a and b, we use

$$\%w = \pm(\%x + \%y) \qquad (4.3a)$$

for quantities that multiply or divide, and

$$\delta w = \pm(\delta x + \delta y) \qquad (4.3b)$$

for quantities that add or subtract.

Example 1

The resistance R of a uniform conducting wire is calculated using measurements of its length l and radius r. The instrument for measuring l has a systematic error of +1% and that for measuring r has a systematic error of −1%. If there is no error in the value of the resistivity, the calculated value of R will have a systematic error of

A +3% **B** +2% **C** 0
D −1% **E** −2% [O 80]

Method

As shown in Chapter 27, the resistance is given by $R = \rho l/A = \rho l/\pi r^2$, where ρ is the resistivity of the wire. The % error in r^2 is given by Equation 4.1a as

$$\%r^2 = \%r + \%r = -1\% - 1\% = -2\%$$

The % error in l/r^2 is given by Equation 4.1b as

$$\%l/r^2 = \%l - \%r^2 = +1\% - -2\% = +3\%$$

(The error in ρ which adds to the 3% is zero.)

Answer

3% (answer **A**).

Example 2

The density ρ of a metal cube is given by ρ = mass/volume and is calculated from the following data:

mass $m = 12.2 \pm 0.1$ g
cube side $l = 2.1 \pm 0.05$ cm

What is the maximum possible error in ρ (a) as a percentage of ρ and (b) absolute?

Method

(a) The maximum % error in volume is given by Equation 4.3a as

$$\%V = \pm 3\%l = 3 \times \pm \frac{0.05}{2.1} \times 100\%$$

$$= 7.1\%$$

Using Equation 4.3a again gives

$$\%\rho = \pm(\%m + \%V)$$

$$= \pm\left(\frac{0.1}{12.2} \times 100 + 7.1\right)\%$$

$$= \pm(0.82 + 7.1)$$

$$= \pm 7.92$$

$$= \pm 8\%$$

(b) The absolute possible error (not compared with the value of ρ) is 8% times ρ, which is

$$\frac{8}{100} \times \frac{12.2}{2.1^3} = \frac{8}{100} \times 1.32 = 0.106 \text{ g cm}^{-3}$$

We write this as 0.1 g cm^{-3}, so that

$$\rho = 1.3 \pm 0.1 \text{ g cm}^{-3}.$$

Answer

(a) ±8%, (b) 0.1 g cm^{-3}.

A rough guide to possible error

The calculations in Example 2 could be simplified by using round figures, perhaps 12 for 12.2 and 2 for 2.1. Even then the calculations may be considered too lengthy. For a quick rough guide we can say that the percentage possible error of the answer cannot be less than that of the least accurate of the data used (assuming random errors). So if a quantity is given to two significant figures,* e.g. 2.3 or 0.0023, the answer calculated from it should not be given to a greater accuracy.

During a calculation, however, the number of significant figures should *not* be reduced at any stage to less than two more than the number required for the answer. (A greater number might be too cumbersome.)

**For two significant figures the percentage possible error is at least 1/99 × 100, i.e. at least 1%.*

Subtraction of similar values

When two similar values are subtracted, the answer is small. The possible random error in the answer, however, is not made small by the subtraction operation. In fact, as we have seen, random errors add when values are subtracted. For example subtraction of 9.83 ± 0.01 from 9.87 ± 0.01 gives 0.04 ± 0.02. The percentage possible error in the answer is 50%, compared with little more than 1% for each of the original quantities! This problem can produce answers with possible errors greater than the answer itself. Always be wary of this pitfall.

Mistakes

A slip in one's mental arithmetic or the pressing of the wrong key on a calculator can ruin a calculation. Consequently the final answer should be checked. Is the answer ridiculous (e.g. size of molecule calculated as so many metres)? If so, check each step of the working.

Try a rough check, like the following example. Suppose

$$F = \frac{3.1 \times 10^{-9} \times 4.2 \times 10^{-8}}{4\pi \times 10 \times 10^{-12} \times (1.1 \times 10^{-2})^2}$$

and the answer is calculated as 8.56×10^{-3} newton.

Roughly we have

$$F = \frac{3 \times 4}{4\pi} \times \frac{10^{-17}}{10^{-15}} \quad \text{or} \quad \frac{12}{12} \times 10^{-2}$$

$$\text{or} \quad 10^{-2} \text{ approx.}$$

This agrees with the 8.6×10^{-3} N.

Exercise 8

1 A resistance R is calculated from a potential difference V and current I using the formula $R = V/I$. If V is 12.6 V and I = 6.0 A, which is the best answer?

A 2 **B** 2.1 **C** 2.10
D 210×10^{-2} **E** 20×10^{-1}

2 The surface area A of a sphere of radius r is given by $A = 4\pi r^2$. If r measures 10.0 ± 0.2 cm, what is the percentage possible error in A?

3 A pressure P is calculated from force F and radius R using the formula $P = F/\pi R^2$. If the % possible errors are ±2% for F and ±1% for R, what is the % possible error for P?

4 A micrometer measures length l_1 as 0.80 mm and length l_2 as 0.50 mm. If it has a zero error that causes it to read low by 0.01 mm, what % error results in the calculation of the ratio l_2/l_1 if the zero error is overlooked?

Exercise 9: Questions from A-level papers

1 An experimenter wishing to determine the volume of glass in a length of glass tubing obtains the following readings:

length l	40 ± 1 mm
external diameter D	12.0 ± 0.2 mm
internal diameter d	10.0 ± 0.2 mm

The volume V of the glass is calculated using the formula

$$V = \tfrac{1}{4}\pi l(D^2 - d^2)$$

(a) What is the greatest possible percentage error in each reading?
(b) Calculate the volume V of the glass.
(c) Estimate the greatest possible percentage error in the value of V calculated from these readings.
[O 81, p]

2 A quantity P is calculated from measured values of x, y and z using the formula $P = kx^{1/2}yz^{-3/2}$, where k is a constant whose value is known exactly. The uncertainties ('errors') in the measured values are:

for x	±0.6%
for y	±0.3%
for z	±0.4%

The uncertainty ('error') in the calculated value of P is:

A ±0.50% **B** ±0.54% **C** ±0.70%
D ±1.2% **E** ±1.3% [O 82]

5 PRESENTATION AND EXAMINATION TECHNIQUE

Being prepared

When arriving for an A-level physics examination it is essential that definitions are well known or, at worst, temporarily memorised. Some formulae must be remembered too; but for most students the number of these should be kept to a minimum. Equations for particular situations, e.g. for particular experiments, can be constructed as needed from the more general formulae which are learnt. Rules (or laws) should be known largely by their repeated use, e.g. the laws of conservation of energy and of momentum. For calculation questions plenty of experience is needed of the style, terminology and methods of handling A-level calculations. Familiarity with descriptive work also helps.

Choice of questions

Most students have their favourite topics with which they are familiar, but a question on any topic can be made unusual so that the calculation is not straightforward. Give a little thought to how you will handle the calculation before embarking on a lengthy description preceding it. The calculation could be done first, provided that the order of your work is made clear to the examiner.

Translating the question

This has been discussed in Chapter 1. A simple, clear diagram with labelling may be used, equations may be stated and data written down with symbols to relate these to the equations or diagrams. Example 1 in this chapter is a further illustration of this approach.

Remember the importance of careful reading of the question with special attention to words like 'small', 'long', 'maximum', 'slow', or 'approximate'.

Presentation

This is of great importance both in examinations and course work.

Begin by making clear that you are starting the calculation, e.g. by giving the part number of the question or by writing 'calculation'.

Diagrams must be of sufficient size (you are not charged for using more paper in an examination). Writing must be easily readable. (Examiners do look for marks, sometimes like a detective looking for clues, but trying their patience with poor writing or untidiness can lose you their sympathy and the benefit of any doubts in the marks deserved.) Figures must be clear, not only to the examiner but also to you when you look back at earlier stages of your calculation.

'Rough work', perhaps written in pencil to distinguish it from the final work, can be done on a separate page (e.g. opposite page). It should not in fact be rough so much as brief. An example is given for the worked example below. The skeleton of the whole calculation is obtained easily in this example and it inspires the confidence needed in completing a rather long calculation like this one.

Expect the examiner to recognise the well-known equations but use common sense. If in doubt, name the symbols employed. He is unlikely to have memorised formulae for individual experiments. It is easy to see that

$$Q = \frac{kA(\theta_2 - \theta_1)}{l} = \frac{mc}{t}(\theta_4 - \theta_3)$$

applies to the Searle's bar experiment (Chapter 24) but

$$c = \frac{kAt(\theta_2 - \theta_1)}{ml(\theta_4 - \theta_3)}$$

is not so obvious.

When an equation has been written down using symbols, introduce figures as you wish but show where the figures have come from. It may be helpful, for example, to show that 5 mm has been converted to 5×10^{-3} m before using the 5×10^{-3} in the equation.

Do not cancel so that the figures used cannot be read:

not $$x = \frac{\cancel{9.0} \times 10^{\cancel{-9}} \times \cancel{8.0} \times \cancel{10^{-8}}}{{}_2\cancel{4}\pi \times \cancel{9.0} \times 10^{\cancel{-12}} \times \cancel{4.0} \times \cancel{10^{-4}}}$$

$$= \frac{0.1}{2\pi}$$

but $$x = \frac{9.0 \times 10^{-9} \times 8.0 \times 10^{-8}}{4\pi \times 9.0 \times 10^{-12} \times 4.0 \times 10^{-4}}$$

$$= \frac{\cancel{9.0} \times {}^2\cancel{8.0}}{4\pi \times \cancel{9.0} \times \cancel{4}} \times \frac{10^{-17}}{10^{-16}}$$

$$= \frac{2}{4\pi} \times 10^{-1}$$

$$= \frac{0.1}{2\pi}$$

Cancelling in steps is safer. Of course cancelling is not needed so much when calculators are used, but small, tidy steps help when a quick check is done after completion of the calculation. Equals signs must be used only between expressions that are equal:

not $\frac{1}{R} = \frac{1}{6} = R = 6$

$\left(\text{seeing } \frac{1}{6} = R \text{ could lead to a mistake}\right)$

but $\frac{1}{R} = \frac{1}{6} \qquad \therefore \quad R = 6$

A stage in a calculation often begins with a formula and ends with a number. The unit should be added to this number before another quantity is discussed. For example

$$\frac{1}{R} = \frac{1}{R_1} + \frac{1}{R_2}, \qquad \frac{1}{R} = \frac{1}{2} + \frac{1}{3} = \frac{3}{6} + \frac{2}{6} = \frac{5}{6}$$

$$\therefore R = \frac{6}{5} = 1.2\,\Omega$$

It is neat and helpful to align all the equals signs if this is convenient. (In Example 1 below, a student would probably keep to equals signs rather than write 'i.e.' and 'or' and this is quite acceptable.)

Comments such as 'using equation 1' or 'C_1 and C_2 are now in parallel' (Example 1) are sometimes essential and usually helpful. If space is left between stages, comments could be inserted afterwards. However, comments from students are not expected to be as detailed as some shown in textbooks.

Example 1

Copper plates and glass plates are placed alternately on top of each other to form a stack. There are three copper plates and two glass plates. The stack has copper plates at the two ends. All the plates have the same area of cross-section, $2.5 \times 10^{-2}\,\text{m}^2$, and the glass plates are 1.5×10^{-3} m thick. Consider, separately, the two arrangements explained below.

(a) Connecting leads are joined to the two end copper plates only. Calculate the capacitance between these two ends. The maximum electric field which glass can sustain without breakdown is $10^7\,V\,m^{-1}$. Calculate the maximum energy which can be stored in this system.

(b) The two outer copper plates are joined by a copper strip. One connecting lead is taken to the copper strip and the other to the central copper plate. Calculate the capacitance between the two leads.

(ε_0 is numerically equal to 8.85×10^{-12}. ε_r for glass = 5.) [WJEC 77]

Method

Rough work:

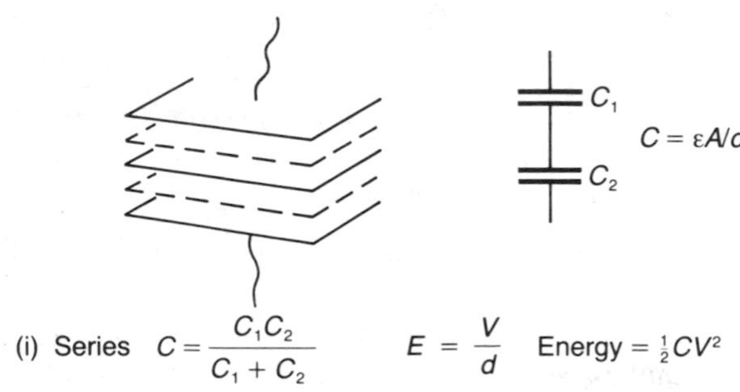

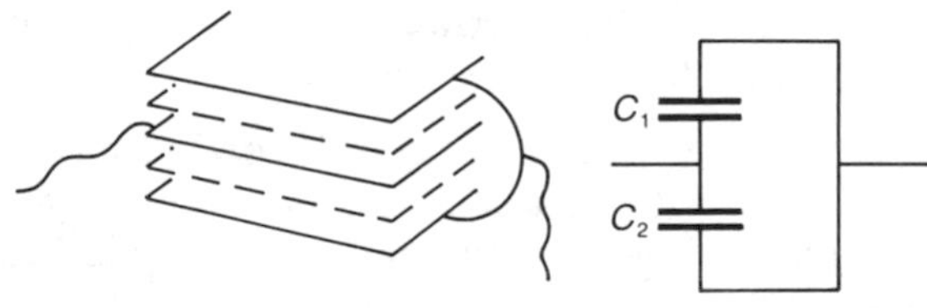

(ii) Parallel $C = C_1 + C_2$

Calculation proper:

The arrangement of the plates is drawn (Fig 5.1).

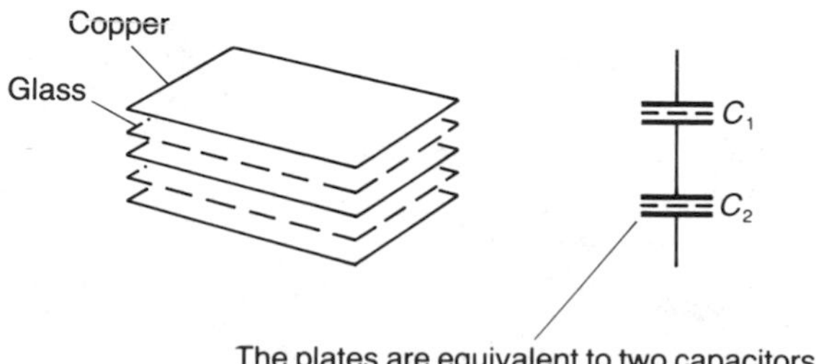

Fig 5.1 Diagram for Example 1

Using the formula $C = \varepsilon A/d$ (see Chapter 31) we have $\varepsilon = \varepsilon_r\varepsilon_0 = 5 \times 8.85 \times 10^{-12}$ numerically (the unit is in fact $F\,m^{-1}$); $A = 2.5 \times 10^{-2}$ and $d = 1.5 \times 10^{-3}$.

$$C_1 = C_2 = \frac{5 \times 8.85 \times 10^{-12} \times 2.5 \times 10^{-2}}{1.5 \times 10^{-3}}$$

$$= \frac{5 \times 8.85 \times 2.5}{1.5} \times 10^{-11}$$

$$= 73.75 \times 10^{-11}\,F$$

(a) C_1 and C_2 are in series and their combined capacitance is

$$C = \frac{C_1C_2}{C_1 + C_2} = \frac{73.75^2 \times 10^{-22}}{2 \times 73.75 \times 10^{-11}}$$

$$= 36.87 \times 10^{-11}\,F$$

i.e. $37 \times 10^{-11}\,F$ to two significant figures.

(We check that we have answered the question part (a), and note that more is required.)

$E\ (= V/d) = 10^7\,V\,m^{-1}$ maximum, and maximum V is given by

$$10^7 = \frac{V}{1.5 \times 10^{-3}} \qquad \therefore \quad V = 1.5 \times 10^4\,V$$

The equal PDs across C_1 and C_2 will both have this value.

For the whole capacitor ($37 \times 10^{-11}\,F$) the PD is $2 \times 1.5 \times 10^4\,V$ and the energy stored is given by

$$\tfrac{1}{2}CV^2 = \tfrac{1}{2} \times 37 \times 10^{-11} \times (3 \times 10^4)^2$$

or $$\frac{37 \times 9}{2} \times 10^{-3}$$

i.e. $166 \times 10^{-3}\,J$ or 17×10^{-2} or 0.17 J, to two significant figures.

(b) C_1 and C_2 are now in parallel and

$$C = C_1 + C_2 = 2 \times 73.75 \times 10^{-11} = 147 \times 10^{-11}\,F$$

or $15 \times 10^{-10}\,F$ to two significant figures.

Answer

(a) $37 \times 10^{-11}\,F$, 0.17 J, (b) $15 \times 10^{-10}\,F$.

Multiple-choice calculations

When selecting the answer to a multiple-choice calculation question, students too often avoid written work unless it is obviously essential. The calculation is achieved instead by mental arithmetic and/or guesswork. However, space is available in examination papers for rough work and one should not hesitate to use it.

Consider the following example.

Example 2

A wire X is suspended vertically and has a mass m attached to its lower end. A second wire Y of the same material but of twice the length and twice the radius of X is also suspended vertically with a mass m attached to its lower end.

Assuming that the elastic limit is not exceeded in either case, the ratio $\dfrac{\text{Extension of X}}{\text{Extension of Y}}$ will be

A $\frac{1}{4}$ **B** $\frac{1}{2}$ **C** 1 **D** 2 **E** 4

Method

As explained in Chapter 13, the extension per unit length is proportional to the extending force divided by the area of cross-section, i.e. $e/l \propto mg/\pi r^2$ so that e = Constant × l/r^2 or

$$e_1/e_2 = l_1 r_2^2/r_1^2 l_2 = l_1(2r_1)^2/r_1^2 2l_1 = 4/2 = 2$$

Answer

D.

Without writing anything down we can say that e doubles if l doubles, while increasing the thickness reduces e. This reduction is proportional to the cross-section area (πr^2), which quadruples when r is doubled. Doubling then dividing by 4 means halving, i.e. the first extension (wire X) is twice that of Y.

Writing down is not necessary if a convincing answer can be obtained with clear thinking based on sound physics principles. Writing may take a little longer but it can be more reliable.

Some 'golden rules' for examinations

Try not to spend too long on any one question. (When the marks possible are indicated, think of no more that 1 or 2 minutes per mark.)

Attempt the full number of questions required even if the last question is answered with a few main points, or the calculation is not completed to the final answer. ('The first few marks of a question are easier to gain than the last few'.)

Check that you have answered the question, no more and no less.

Check that the answer for a calculation is displayed clearly, first with all available figures shown, and then with the significant figures that are justified.

Don't forget the units, e.g. at the end of each step in a calculation.

Section B
Mechanics

6 VECTORS — ADDITION AND RESOLUTION

Vector addition

A vector quantity has magnitude and direction—examples are force and velocity. We represent a vector by a line in the appropriate direction and of length proportional to the magnitude of the vector (see Fig 6.1).

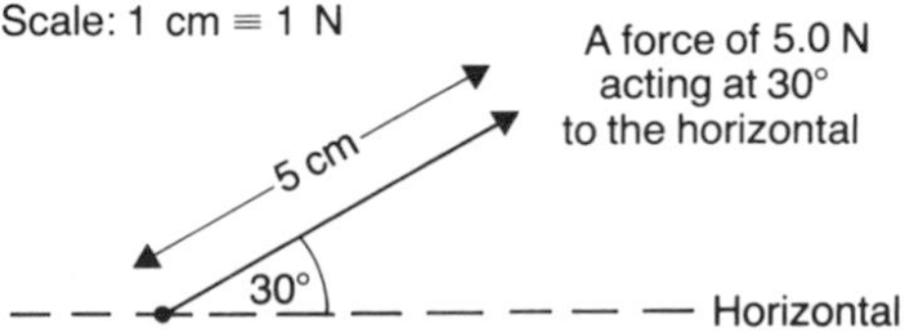

Fig 6.1 Representation of a vector

Vector quantities are added using the parallelogram rule (see Fig 6.2) — the resultant is the appropriate diagonal of the parallelogram.

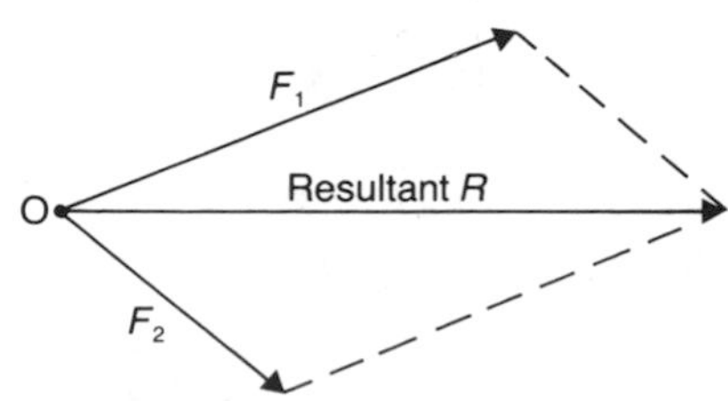

Fig 6.2 Addition of vectors

Example 1

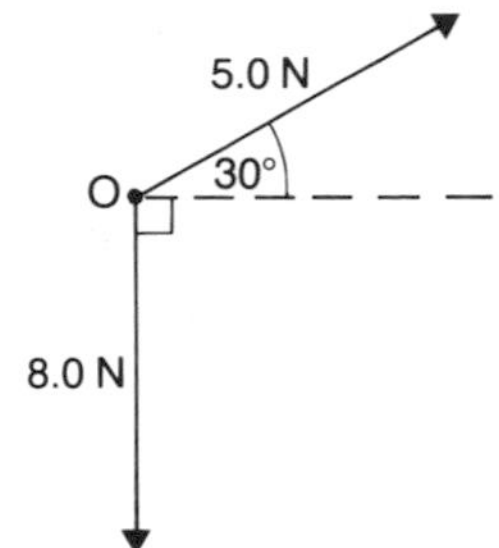

Fig 6.3 Information for Example 1

Fig 6.3 shows two forces acting at a point O. Find the magnitude and direction of the resultant force.

Method

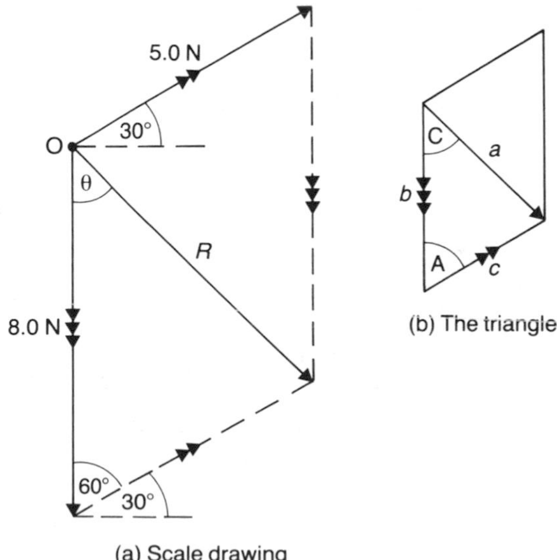

Fig 6.4 Solution to Example 1

Referring to Fig 6.4a, we could find the resultant R and angle θ by scale drawing, but for A-level calculations this is usually not accurate enough. Referring to Fig 6.4b (see Chapter 2) we see that

$$a^2 = b^2 + c^2 - 2bc \cos A$$

We have $a = R$, $b = 8.0$, $c = 5.0$ and $\mathrm{A} = 60°$. So

$$R^2 = 8^2 + 5^2 - 2 \times 8 \times 5 \times \cos 60° = 49$$

$$\therefore \quad R = 7.0\,\mathrm{N}$$

To find θ, we know (see Chapter 2)

$$\frac{a}{\sin \mathrm{A}} = \frac{c}{\sin \mathrm{C}}$$

We have $a = 7.0$, $\mathrm{A} = 60°$, $c = 5.0$ and $\mathrm{C} = \theta$. So

$$\frac{7}{\sin 60°} = \frac{5}{\sin \theta}$$

$$\therefore \quad \theta = 38.2°$$

Answer

The resultant is of magnitude 7.0 N at an angle of 38° to the 8.0 N force, as shown in Fig 6.4a.

Example 2

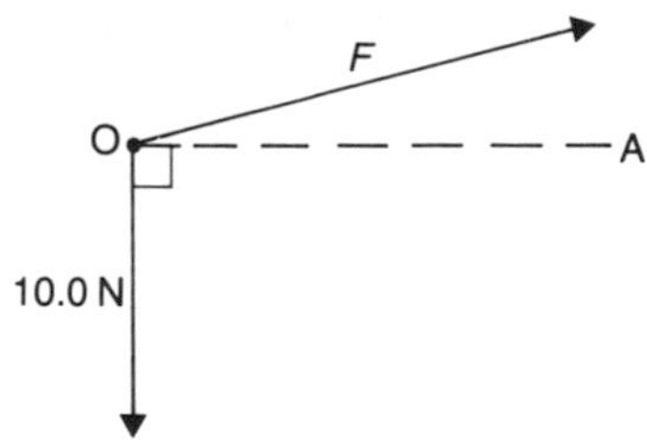

Fig 6.5 Information for Example 2

Refer to Fig 6.5. Two forces of magnitude 10.0 N and F newtons produce a resultant of magnitude 30.0 N in the direction OA. Find the magnitude and direction of F.

Method

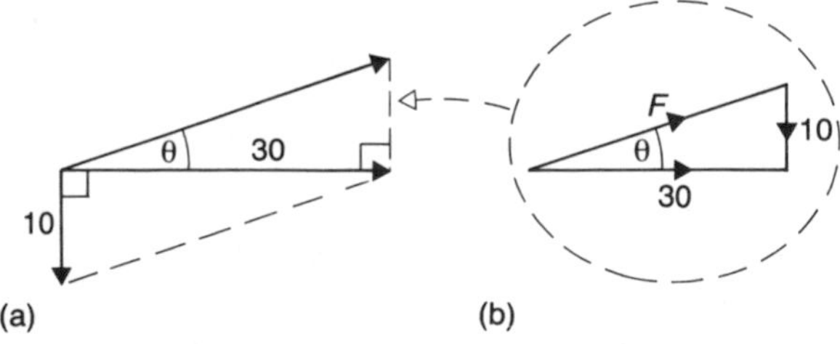

Fig 6.6 Solution to Example 2

Referring to Fig 6.6 then, in diagram (b):

$$F^2 = 10^2 + 30^2 = 1000$$

$$\therefore \quad F = 31.6\,\mathrm{N}$$

Also

$$\tan \theta = \frac{10}{30} = 0.3333$$

or $$\theta = 18.4°$$

Answer

F is of magnitude 31.6 N at an angle of 18.4° to the resultant force as shown.

Exercise 10

1

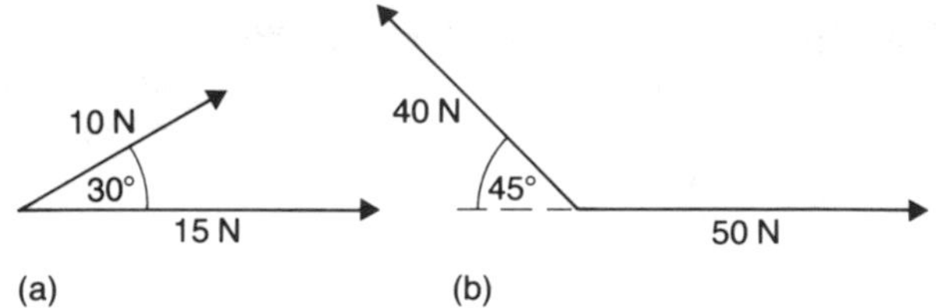

Fig 6.7 Information for Question 1

Find the resultant of the forces in (a) Fig 6.7a, (b) Fig 6.7b.
Note that for $\theta > 90°$, $\cos \theta = -\cos(180 - \theta)$.

2

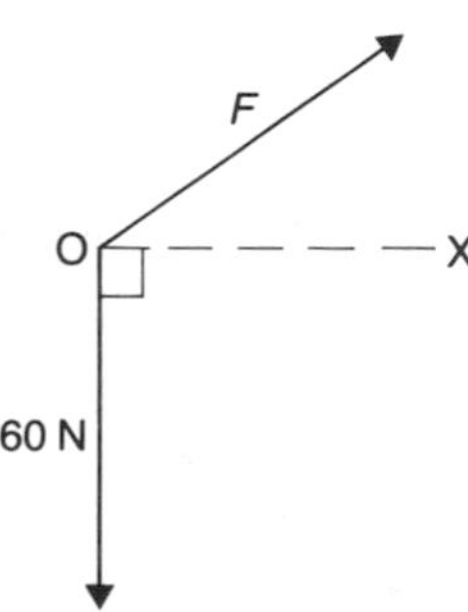

Fig 6.8 Information for Question 2

Refer to Fig 6.8. Forces of 60.0 N and F newtons act at a point O. Find the magnitude and direction of F if the resultant force is of magnitude 30.0 N along OX.

3

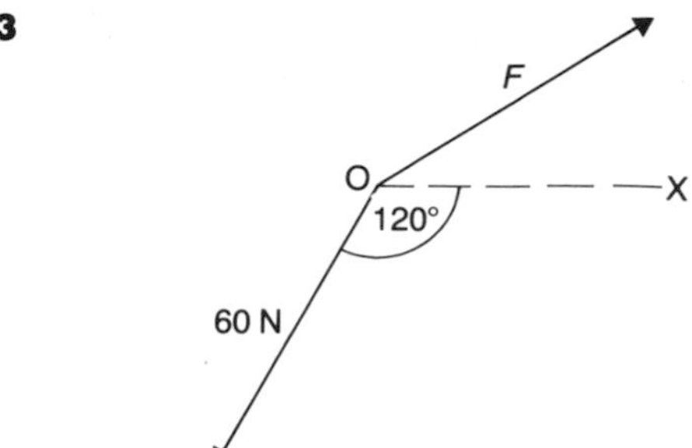

Fig 6.9 Information for Question 3

Refer to Fig 6.9 and repeat Question 2.

Resolution of vectors

A single vector can be formed by combining two (or more) vectors so it follows that a single vector can be replaced by, or *resolved* into, two components. This is usually done at right angles (see Fig 6.10) because the separate components V and H have no effect on each other — e.g. V has no effect in the direction of H.

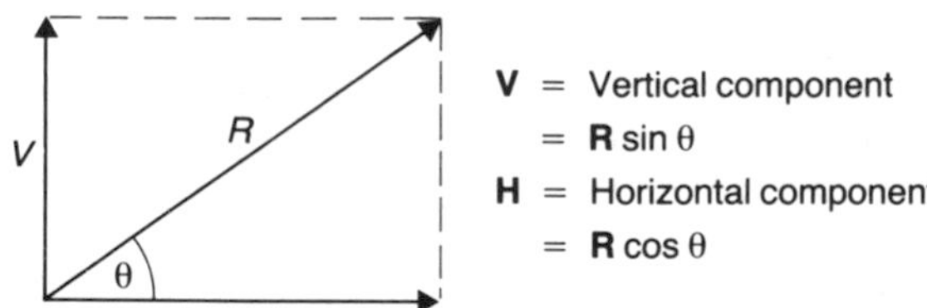

Fig 6.10 Components of a vector

Example 3

A shell is fired at 400 m s^{-1} at an angle of 30° to the horizontal. If the shell stays in the air for 40 s, calculate how far it lands from its original position. Assume that the ground is horizontal and that air resistance may be neglected.

Method

Refer to Fig 6.11. We require the range S. The horizontal component of the initial velocity is

$$H = 400 \cos 30° = 347 \text{ m s}^{-1}$$

The horizontal component H remains unchanged if air resistance is negligible. So range S is given by

$$S = H \times \text{Time of flight}$$
$$= 347 \times 40 = 13\,880 \text{ m}$$

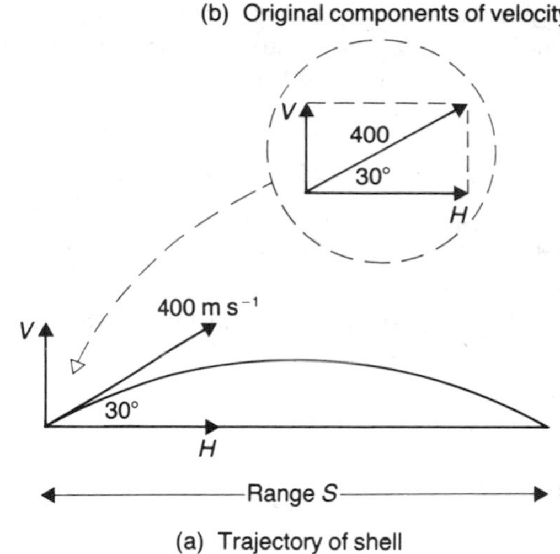

Fig 6.11 Solution to Example 3

Answer

The shell lands 14 km from its original position.

Example 4

A body of mass 1.5 kg is placed on a plane surface inclined at 30° to the horizontal. Calculate the friction and normal reaction forces which the plane must exert if the body is to remain at rest. Assume $g = 10 \text{ m s}^{-2}$.

Method

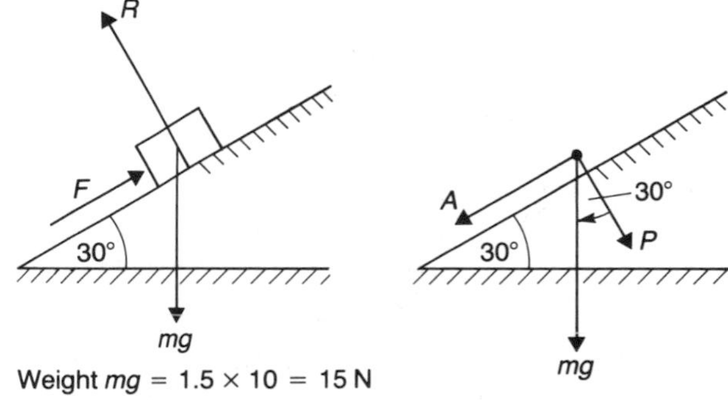

Fig 6.12 Solution to Example 4

The body exerts a downward force mg on the plane, as shown in Fig 6.12, so the plane must exert an equal and opposite (upwards) force if the body is to remain at rest.

It is convenient to resolve mg into a component P, perpendicular to the plane, and a component A, along the plane, as shown in Fig 6.12b. Now

$$P = mg\cos 30 = 15 \times 0.866 = 13\text{ N}$$

$$A = mg\sin 30 = 15 \times 0.500 = 7.5\text{ N}$$

So, as shown in Fig 6.12a, the plane must provide a normal reaction R equal to 13 N and a force F, due to friction, equal to 7.5 N. When R and F are added vectorially, they provide a vertically upwards force equal to mg.

Answer
7.5 N, 13 N.

Example 5

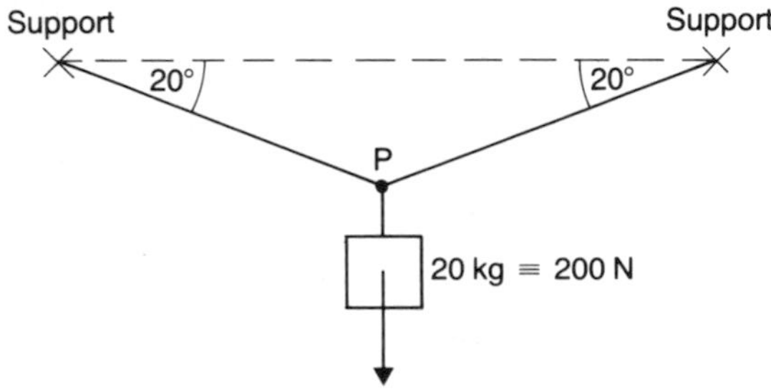

Fig 6.13 Information for Example 5

A mass of 20.0 kg is hung from the midpoint P of a wire, as shown in Fig 6.13. Calculate the tension in the wire. Assume $g = 10\text{ m s}^{-2}$.

Method

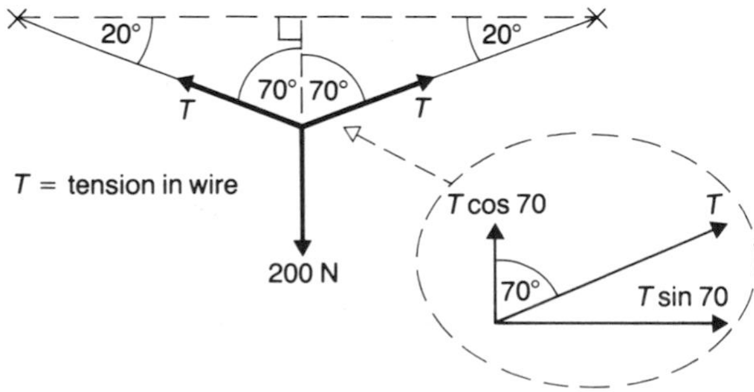

Fig 6.14 Solution to Example 5

Fig 6.14 shows the forces acting at the point P. The vertical component of tension T is $T\cos 70°$ in each case, so for equilibrium in a vertical direction

$$2T\cos 70° = 200$$

$$\therefore \quad T = 292\text{ N}$$

Note that the horizontal component of tension is $T\sin 70°$ in each case, but these forces are in opposite directions and so cancel each other. This ensures equilibrium in the horizontal direction.

Answer
292 N

Exercise 11

(Assume $g = 10\text{ m s}^{-2}$.)

1 A shell is fired at 500 m s^{-1} at an angle of θ degrees to the horizontal. The shell stays in the air for 80 s and has a range of 24 km. Assuming that the ground is horizontal and that air resistance may be neglected, calculate (a) the horizontal component of the velocity, (b) the value of θ.

2 A body of mass 3.0 kg is placed on a smooth (i.e. frictionless) plane inclined at 20° to the horizontal. A force of (a) 5.0 N, (b) 20 N is applied to the body parallel to the line of greatest slope of the plane and in a direction up the plane. Calculate the net force acting on the body in each case.

3 Refer to Fig 6.15 and calculate (a) the tension in the string, (b) the value of m.

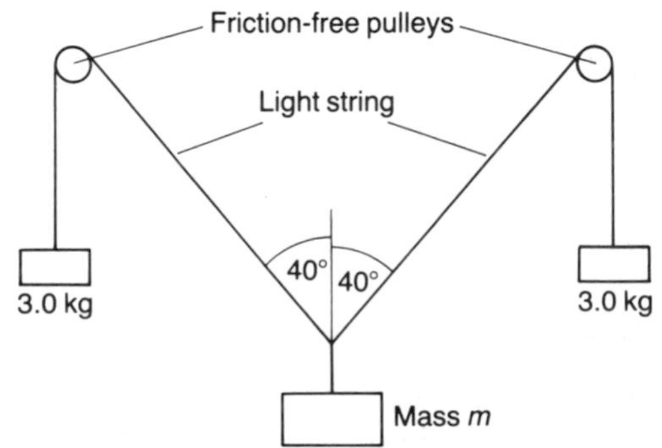

Fig 6.15 Information for Question 3

Exercise 12: Questions of A-level standard

(Assume $g = 10\,\text{m s}^{-2}$.)

1

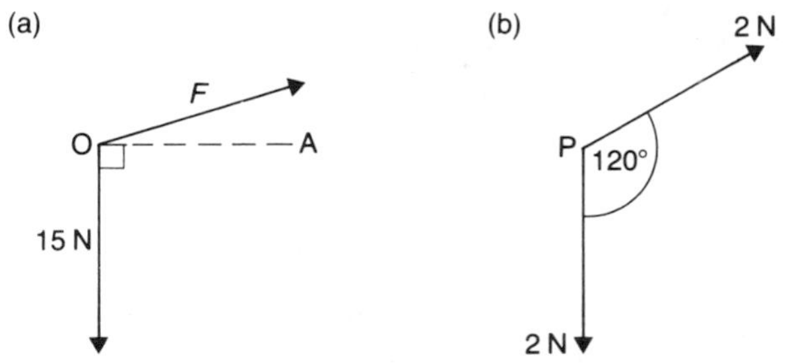

Fig 6.16 Diagram for Question 1

(a) Fig 6.16a shows two forces acting on a mass Q of 5 kg at O. If Q has an acceleration of $4\,\text{m s}^{-2}$ in the direction OA, determine the magnitude and direction of the force F.

(b) Fig 6.16b shows two forces acting on a mass M of 3 kg at P. Determine the magnitude and direction of the acceleration of M. [WJEC 78, p]

2 Raindrops of mass 5×10^{-7} kg fall vertically in still air with a uniform speed of $3\,\text{m s}^{-1}$. If such drops are falling when a wind is blowing with a speed of $2\,\text{m s}^{-1}$, what is the angle which the paths of the drops make with the vertical? What is the kinetic energy of the drop? [SUJB 79]

3 The components of a particle's velocity in the three directions at right angles are $3\,\text{km s}^{-1}$, $4\,\text{km s}^{-1}$ and $12\,\text{km s}^{-1}$ respectively. The actual velocity of the particle, in km s^{-1}, is

A 5 **B** 7 **C** 13 **D** 17 **E** 19

[AEB 82]

4 Distinguish between scalar and vector quantities, giving two examples of each. Explain with the aid of a diagram how a vector can be resolved into components in two given directions. Consider, with diagrams, whether a vector of magnitude 12 units can be resolved into the following pairs of components (a) 15 and 8 units, (b) 8 and 4 units, (c) 6 and 4 units. [WJEC 77, p]

5

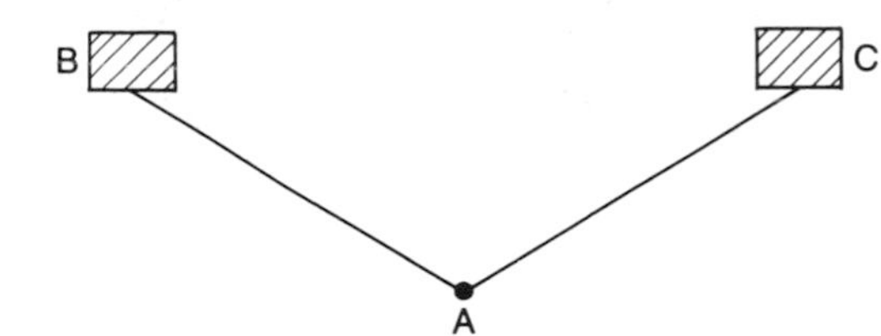

Fig 6.17 Diagram for Question 5

A small sphere A of mass 2 kg is suspended by two strings AB and AC (Fig 6.17), each 0.5 m long and each inclined at 30° to the horizontal. Calculate the tension in each string. [SUJB 82, p]
(Note: length of string not required for this question.)

6 State the conditions in which a rigid body may be in equilibrium under the action of three forces.

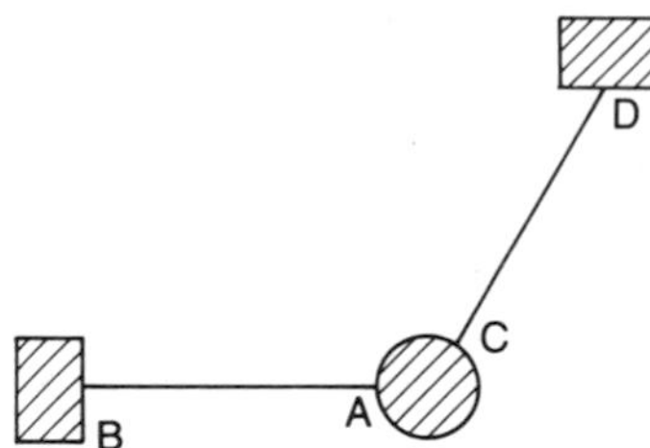

Fig 6.18 Diagram for Question 6

A uniform sphere of mass 2 kg is kept in position as shown in Fig 6.18 by two strings AB and CD. AB is horizontal and CD is inclined at 30° to the vertical. Calculate the tension in each string. [SUJB 80, p]

7 VELOCITY, ACCELERATION AND FORCE

Velocity and speed

Velocity is a vector and speed is a scalar. Sometimes this difference is not properly recognised, so we must remember it.

Example 1

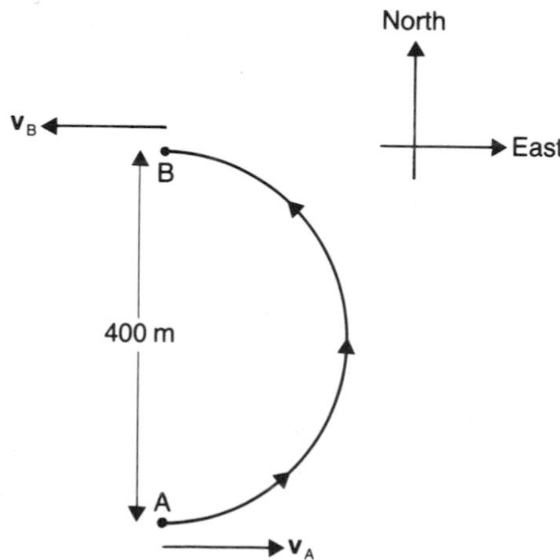

Fig 7.1 Information for Example 1

A car takes 80 s to travel at constant speed in a semicircle from A to B as shown in Fig 7.1. Calculate (a) its speed, (b) its average velocity, (c) the change in velocity from A to B.

Method

(a) $\text{Speed} = \dfrac{\text{Distance}}{\text{Time}}$

$= \dfrac{\text{Length of semicircular arc}}{\text{Time}}$

$= \dfrac{\pi \times 200}{80}$

$= 2.5\pi \text{ m s}^{-1}$

(b) $\text{Average velocity} = \dfrac{\text{Total displacement}}{\text{Time}}$

$= \dfrac{400}{80}$

$= 5.0 \text{ m s}^{-1}$ north

(c) The speed at A and B is $2.5\pi \text{ m s}^{-1}$, but velocities are different. Taking velocity to the 'right' (east) as positive, then

Velocity at A, $v_A = +2.5\pi$

Velocity at B, $v_B = -2.5\pi$

Change in velocity $= v_B - v_A = -5.0\pi \text{ m s}^{-1}$

Note: the negative sign indicates the change is to the left.

Answer

(a) $2.5\pi \text{ m s}^{-1}$, (b) 5.0 m s^{-1} north, (c) $5.0\pi \text{ m s}^{-1}$ to the left.

Example 2
A ship travels due east at $3.0\,\mathrm{m\,s^{-1}}$. If it now heads due north at the same speed, calculate the change in velocity.

Method*

Initial velocity $\vec{u} = 3\,\mathrm{m\,s^{-1}}$ east

Final velocity $\vec{v} = 3\,\mathrm{m\,s^{-1}}$ north

The change in velocity is

$$\vec{v} - \vec{u} = \vec{v} + (-\vec{u})$$

We have $\vec{v} = 3\,\mathrm{m\,s^{-1}}$ north and $-\vec{u} = 3\,\mathrm{m\,s^{-1}}$ *west*. Fig 7.2 shows that vector addition of $\vec{v}$ and $-\vec{u}$ is a vector of magnitude $\sqrt{18} = 4.2$ in direction north-west.

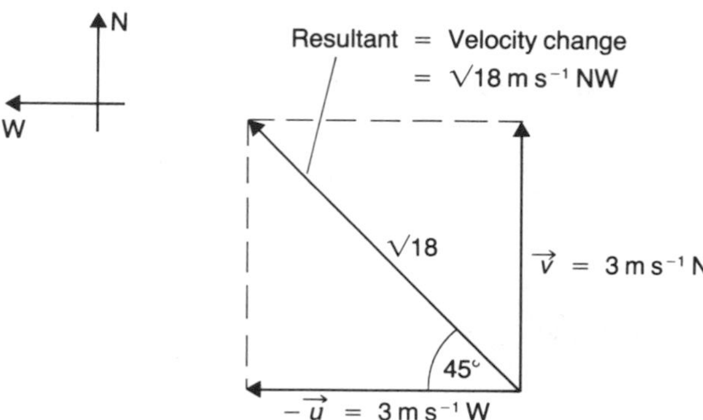

Fig 7.2 Solution to Example 2

Answer
Velocity change = $4.2\,\mathrm{m\,s^{-1}}$ north-west.

Exercise 13

1

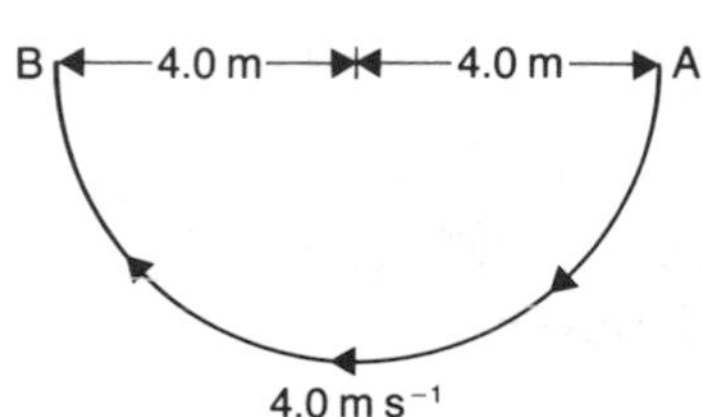

Fig 7.3 Diagram for Question 1

An object moves along a semicircular path AB of radius 4.0 m as shown in Fig 7.3, at a constant speed of $4.0\,\mathrm{m\,s^{-1}}$. Calculate (a) the time taken, (b) the average velocity, (c) the change in velocity.

**To denote the vector nature of velocity we sometimes put an arrow above it.*

2

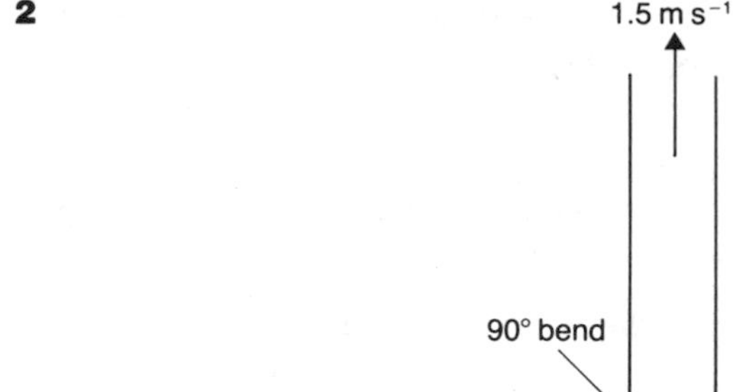

Fig 7.4 Diagram for Question 2

Water enters and leaves a pipe, as shown in Fig 7.4, at a steady speed of $1.5\,\mathrm{m\,s^{-1}}$. Find the change in velocity.

Acceleration

Uniform acceleration means a constant rate of change of velocity—for example $4\,\mathrm{m\,s^{-1}}$ per second ($4\,\mathrm{m\,s^{-2}}$).

Example 3
A car moving with velocity $5.0\,\mathrm{m\,s^{-1}}$, in some direction, accelerates uniformly at $2.0\,\mathrm{m\,s^{-2}}$ for 10 s. Calculate (a) the final velocity, (b) the distance travelled.

Method

(a) Increase in velocity $= 2 \times 10 = 20\,\mathrm{m\,s^{-1}}$

Original velocity $u = 5.0\,\mathrm{m\,s^{-1}}$

$\therefore$ Final velocity $v = 5 + 20 = 25\,\mathrm{m\,s^{-1}}$

Alternatively, to find v, use

$$v = u + at \tag{7.1}$$

where acceleration $a = +2.0$, time $t = 10\,\mathrm{s}$ and initial velocity $u = 5.0$. So

$$v = u + at = 5 + 2 \times 10$$
$$= 25\,\mathrm{m\,s^{-1}}$$

(b) Distance travelled s = Average velocity × Time, or

$$s = \tfrac{1}{2}(u + v) \times t \tag{7.2}$$
$$= \tfrac{1}{2}(5 + 25) \times 10 = 1.5 \times 10^2\,\mathrm{m}$$

Answer
(a) $25\,\mathrm{m\,s^{-1}}$, (b) $1.5 \times 10^2\,\mathrm{m}$.

Exercise 14

1 A body starts from rest ($u = 0$) and accelerates at $3.0\ \text{m s}^{-2}$ for 4.0 s. Calculate (a) its final velocity, (b) the distance travelled.

2 Calculate the quantities indicated:
(a) $u = 0$, $v = 20$, $t = 8.0$, $a =$ ______
(b) $u = 10$, $v = 22$, $a = 1.5$,
$t =$ ______, $s =$ ______
(c) $u = 15$, $v = 10$, $a = -0.5$,
$t =$ ______, $s =$ ______

Equations of motion

These are:

$$v = u + at \qquad (7.1)$$

$$v^2 = u^2 + 2as \qquad (7.3)$$

$$s = ut + \tfrac{1}{2}at^2 \qquad (7.4)$$

The meaning of the symbols was given earlier. These equations are obtained by combining Equations 7.1 and 7.2, and are recommended because they are more convenient to use.

Example 4

A car moving with velocity $10\ \text{m s}^{-1}$ accelerates uniformly at $2.0\ \text{m s}^{-2}$. Calculate its velocity after travelling 200 m.

Method

We have $u = 10$, $a = 2.0$ and $s = 200$. We require v, so Equation 7.3 is used

$$v^2 = u^2 + 2as = 10^2 + 2 \times 2 \times 200$$

$$= 900$$

$$\therefore \quad v = 30\ \text{m s}^{-1}$$

Note that since t is unknown it would be more difficult to use Equations 7.1 and 7.2.

Answer

Velocity acquired $= 30\ \text{m s}^{-1}$.

Example 5

How far does a body travel in the fourth second if it starts from rest with a uniform acceleration of $2.0\ \text{m s}^{-2}$?

Method

We have $u = 0$, $a = 2.0$ and require distance travelled between $t_1 = 3.0$ s and $t_2 = 4.0$ s. Let s_1 and s_2 be distances travelled in 3 s and 4 s respectively. From equation 7.4

$$s_1 = ut_1 + \tfrac{1}{2}at_1^2 = 0 \times 3 + \tfrac{1}{2} \times 2 \times 3^2$$

$$= 9.0\ \text{m}$$

$$s_2 = ut_2 + \tfrac{1}{2}at_2^2 = 0 \times 4 + \tfrac{1}{2} \times 2 \times 4^2$$

$$= 16.0\ \text{m}$$

$\therefore$ Distance travelled $s_2 - s_1 = 7.0$ m.

Answer

The body travels 7.0 m in the fourth second.

Exercise 15

1 It is required to uniformly accelerate a body from rest to a velocity of $12\ \text{m s}^{-1}$ in a distance of 0.20 m. Calculate the acceleration.

2 Calculate the quantities indicated (assume that all quantities are in SI units):
(a) $u = 0$, $a = 10$, $s = 45$, $t =$ ______
(b) $u = 15$, $a = -1.5$, $v = 6$, $s =$ ______
(c) $u = 20$, $a = -2.0$, $s = 84$, $t =$ ______

3 In an electron gun, an electron is accelerated uniformly from rest to a velocity of $4.0 \times 10^7\ \text{m s}^{-1}$ in a distance of 0.10 m. Calculate the acceleration.

Motion under gravity — vertical motion

Gravitational attraction produces a force which, on earth, causes a free-fall acceleration g of approximately $9.8\ \text{m s}^{-2}$. For simplicity we take $g = 10\ \text{m s}^{-2}$ here.

'Free' vertical motion is simply uniformly accelerated motion, assuming negligible opposing forces, in which $a = g = \pm 10\ \text{m s}^{-2}$ depending on the direction chosen as positive.

Example 6

An object is dropped from a height of 45 m. Calculate (a) the time taken to reach the ground, (b) its maximum velocity. Neglect air resistance. (Assume $g = 10\,\text{m s}^{-2}$.)

Method

We have $u = 0, s = +45$ and $a = g = +10\,\text{m s}^{-2}$ if we take downwards as the positive direction. We require t and v.

(a) To find t, rearrange Equation 7.4

$$t^2 = \frac{2s}{a} = \frac{2 \times 45}{10}$$

$$\therefore \quad t = 3.0\,\text{s}$$

(b) To find v, use Equation 7.1

$$v = u + at = 0 + 10 \times 3$$

$$\therefore \quad v = 30\,\text{m s}^{-1}$$

Answer

(a) 3.0 s, (b) $30\,\text{m s}^{-1}$.

Example 7

A cricket ball is thrown vertically upwards with a velocity of $20\,\text{m s}^{-1}$. Calculate (a) the maximum height reached, (b) the time taken to return to earth. Neglect air resistance.

Method

We now take upwards as positive. So

$$u = +20, \qquad a = g = -10.$$

(a) At the maximum height, distance s from ground level, the velocity v is zero.
From Equation 7.3

$$v^2 = u^2 + 2as$$

$$0^2 = 20^2 + 2 \times (-10) \times s$$

$$\therefore \quad s = 20\,\text{m}$$

(b) On its return to earth, after time t, we have $s = 0$.
So, using Equation 7.4

$$s = ut + \tfrac{1}{2}at^2$$

$$0 = 20 \times t + \tfrac{1}{2} \times (-10) \times t^2$$

$$\therefore \quad t = 4.0\,\text{s}$$

(Note that $t = 0$ is also, obviously, a solution when $s = 0$.)
Alternatively find the time to reach its maximum height, (when its velocity is zero) which is 2.0 s, and double it.

Answer

(a) 20 m, (b) 4.0 s.

Exercise 16

(Assume $g = 10\,\text{m s}^{-2}$.)

1 A ball is dropped from a cliff top and takes 3.0 s to reach the beach below. Calculate (a) the height of the cliff, (b) the velocity acquired by the ball.

2 With what velocity must a ball be thrown upwards to reach a height of 15 m?

3 A man stands on the edge of a cliff and throws a stone vertically upwards at $15\,\text{m s}^{-1}$. After what time will the stone hit the ground 20 m below?

Motion under gravity — projectile motion

This includes objects which have horizontal as well as vertical motion, e.g. shells and bullets. We resolve any initial velocity into its horizontal and vertical components, which are then *treated separately. The vertical component determines the time of flight* (and any vertical distances) *and the horizontal component determines the range.*

Example 8

A stone is projected horizontally with velocity $3.0\,\text{m s}^{-1}$ from the top of a vertical cliff 200 m high. Calculate (a) how long it takes to reach the ground, (b) its distance from the foot of the cliff, (c) its vertical and horizontal components of velocity when it hits the ground. Neglect air resistance.

Method
As in Chapter 6, resolve initial velocity into its components (Fig 7.5a):

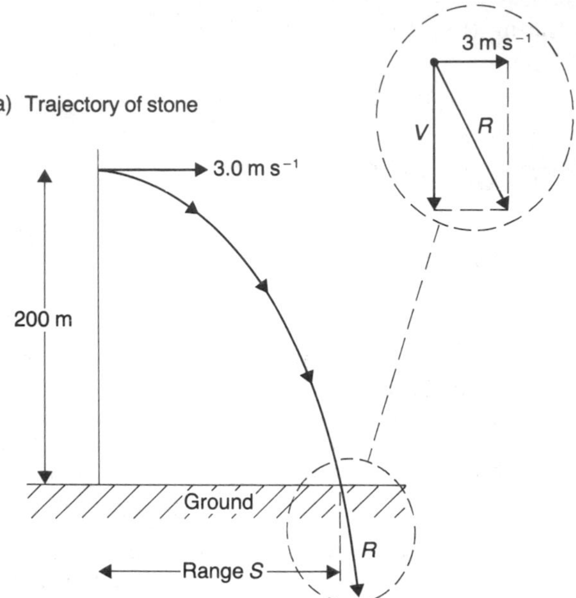

Fig 7.5 Solution to Example 8 (not to scale)

initial vertical component = 0

initial horizontal component = $3.0\,\mathrm{m\,s^{-1}}$

(a) The vertical motion decides the time of flight. Taking downwards as positive we have $u = 0$, $s = 200$, $a = g = +10\,\mathrm{m\,s^{-2}}$. To find t use

$$s = ut + \tfrac{1}{2}at^2$$

$$\therefore\ 200 = 0 \times t + \tfrac{1}{2} \times 10 \times t^2$$

$$\therefore\ t = \sqrt{40} = 6.3\,\mathrm{s}$$

(b) The horizontal component of velocity is unchanged (see Chapter 6). So

$$\text{Range } S = \text{Horizontal velocity} \times \text{Time}$$
$$= 3.0 \times \sqrt{40} = 19\,\mathrm{m}$$

(c) The vertical component of velocity when the stone hits the ground is required. From part (a)

$$v^2 = u^2 + 2as$$
$$= 0^2 + 2 \times 10 \times 200$$
$$\therefore\ v = \sqrt{4000} = 63\,\mathrm{m\,s^{-1}}$$

The horizontal component remains at $3.0\,\mathrm{m\,s^{-1}}$.

Note that to find the resultant velocity R of the stone on hitting the ground we must add the components vectorially, as shown in Fig 7.5b.

Answer
(a) 6.3 s, (b) 19 m, (c) $63\,\mathrm{m\,s^{-1}}$, $3.0\,\mathrm{m\,s^{-1}}$.

Exercise 17

(Neglect air resistance.)

1 Repeat Example 8 for a stone having a horizontal velocity of $4.0\,\mathrm{m\,s^{-1}}$ and a cliff which is 100 m high.

2

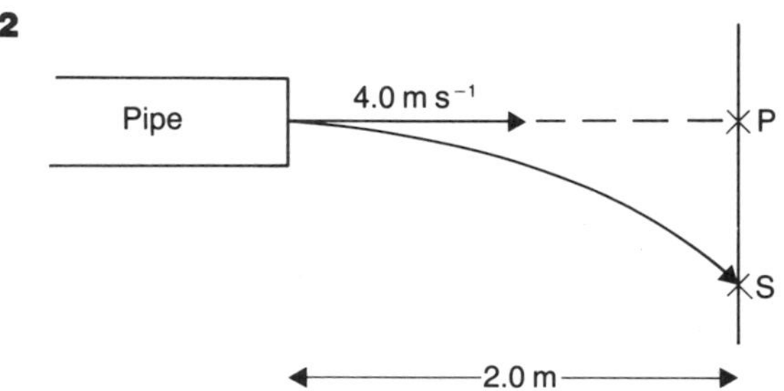

Fig 7.6 Diagram for Question 2

Water emerges horizontally from a hose pipe with velocity of $4.0\,\mathrm{m\,s^{-1}}$ as shown in Fig 7.6. The pipe is pointed at P on a vertical surface 2.0 m from the pipe. If the water strikes at S, calculate PS.

3 A shell is fired from a gun with a velocity of $400\,\mathrm{m\,s^{-1}}$ at an angle of 30° to the ground which is horizontal. Calculate (a) the time of flight, (b) the range, (c) the maximum height reached.

Force, mass and acceleration

A net force F (N) applied to a mass m (kg) produces an acceleration a ($\mathrm{m\,s^{-2}}$) given by

$$F = ma \qquad (7.5)$$

By net force we mean the resultant force arising from applied forces, friction, gravitational forces and so on.

Example 9
A car of mass 900 kg is on a horizontal and slippery road. The wheels slip when the total push of the wheels on the road exceeds 500 N. Calculate the maximum acceleration of the car.

Method
It is the push of the road on the car wheels which is responsible for acceleration. This is equal in magnitude, but opposite in direction, to the push of the wheels on the road. We have $m = 900$ and $F = 500$ so, from Equation 7.5

$$a = \frac{F}{m} = \frac{500}{900} = 0.556 \text{ m s}^{-2}$$

Answer
The maximum acceleration is 0.556 m s^{-2}.

Example 10
A car of mass 1000 kg tows a caravan of mass 800 kg and the two have an acceleration of 2.0 m s^{-2}. If the only external force acting is that between the driving wheels and the road, calculate (a) the value of this force and (b) the tension in the coupling between the car and the caravan.

Method
(a) For car and caravan combined we have

$$m = 1000 + 800 = 1800$$

and $a = 2.0$. From Equation 7.5 the force F required is

$$F = ma = 1800 \times 2 = 3600 \text{ N}$$

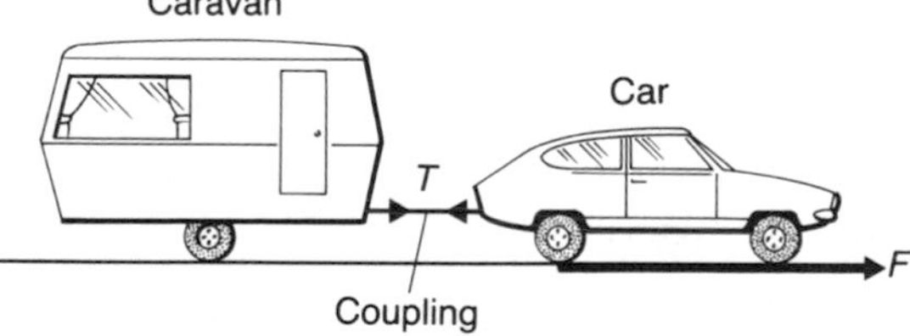

Fig 7.7 Solution to Example 10

(b) Refer to Fig 7.7. T is the tension in the coupling and is the net force accelerating the caravan. So for the caravan alone we have $m = 800$ kg and $a = 2.0$. So

$$T = ma = 800 \times 2 = 1600 \text{ N}$$

Note that the net force on the car alone is $F - T = 3600 - 1600 = 2000$ N. This gives the car an acceleration of 2.0 m s^{-2}.

Answer
(a) 3.6 kN, (b) 1.6 kN.

Example 11
An aircraft of mass 20×10^3 kg lands on an aircraft-carrier deck with a horizontal velocity of 90 m s^{-1}. If it is brought to rest in a distance of 100 m, calculate the (average) retarding force acting on the plane.

Method
We must first find the (negative) acceleration a of the plane, We have $u = 90$, $v = 0$, $s = 100$ and from Equation 7.3

$$v^2 = u^2 + 2as$$

$$\therefore \quad 0^2 = 90^2 + 2 \times a \times 100$$

$$\therefore \quad a = -40.5 \text{ m s}^{-2}$$

The negative sign indicates the plane is slowing down. Force F required, since $m = 20 \times 10^3$, is given by

$$F = ma = 20 \times 10^3 \times (-40.5)$$

$$= -81 \times 10^4 \text{ N}$$

The negative sign indicates that the force is in the opposite direction to the original velocity.

Answer
Retarding force is 81×10^4 N on average.

Exercise 18

1 Calculate the quantities indicated (assume that all quantities are in SI units):
(a) $a = 2.5$, $m = 3.0$, $F =$ ——
(b) $F = 15$, $m = 30$, $a =$ ——
(c) $a = 2.5$, $F = 7.5$, $m =$ ——

2 A force of 24 N acts on a mass of 6.0 kg initially at rest. Calculate (a) the acceleration, (b) the distance travelled prior to achieving a velocity of 20.0 m s^{-1}.

3 A lorry of mass 3.0×10^3 kg pulls two trailers each of mass 2.0×10^3 kg along a horizontal road. If the lorry is accelerating at 0.8 m s^{-2}, calculate (a) the net force acting on the whole combination, (b) the tension in the coupling between lorry and first trailer, (c) the tension in the coupling between first and second trailers.

4 A metal ball of mass 0.5 kg is dropped from the top of a vertical cliff of height 90 m. When it hits the beach below it penetrates to a depth of 6.0 cm. Calculate (a) the velocity acquired by the ball just as it hits the sand, (b) the (average) retarding force of the sand. Neglect air resistance; $g = 10 \text{ m s}^{-2}$.

5 What net force must be applied to an object of mass 5.0 kg, initially at rest, for it to acquire a velocity of 12 m s^{-1} over a distance of 0.10 m?

Exercise 19: Questions of A-level standard

(Assume $g = 10\,\mathrm{m\,s^{-2}}$ except where stated.)

1 A car is braked so that it slows down with uniform retardation from $30\,\mathrm{m\,s^{-1}}$ to $15\,\mathrm{m\,s^{-1}}$ over a distance of 75 m. If the car continues to slow down in this way, after what further distance will it be brought to rest?

A 25 m **B** 50 m **C** 75 m
D 100 m **E** 150 m [O 82]

2 A motorist travelling at $13\,\mathrm{m\,s^{-1}}$ approaches traffic lights which turn red when he is 25 m away from the stop line. His reaction time (i.e. the interval between seeing the red light and applying the brakes) is 0.7 s and the condition of the road and his tyres are such that the car cannot slow down at a rate of more that $4.5\,\mathrm{m\,s^{-2}}$. If he brakes fully, how far from the stop line will he stop, and on which side of it? [C 82]

3 A cricketer throws a ball vertically upwards and catches it 3.0 s later. Neglecting air resistance, find (a) the speed with which the ball leaves his hands, (b) the maximum height to which it rises. [C 81, p]

4 A stone of mass 80 g is released at the top of a vertical cliff. After falling for 3 s it reaches the foot of the cliff and penetrates 9 cm into the ground. What is (a) the height of the cliff, (b) the average force resisting penetration of the ground by the stone? [SUJB 80]

5 A lunar landing module is descending to the moon's surface at a steady velocity of $10\,\mathrm{m\,s^{-1}}$. At a height of 120 m a small object falls from its landing gear. Taking the moon's gravitational acceleration as $1.6\,\mathrm{m\,s^{-2}}$, at what speed, in $\mathrm{m\,s^{-1}}$, does the object strike the moon?

A 202 **B** 22 **C** 19.6
D 16.8 **E** 10 [AEB 80]

6 A lift is travelling with a downward acceleration of $5.8\,\mathrm{m\,s^{-2}}$. A ball, held 2 m above the floor of the lift and at rest with respect to the lift is released. The time, in seconds, taken by the ball to reach the floor of the lift is

A 1 **B** $\frac{1}{\sqrt{2}}$ **C** $\frac{2}{\sqrt{5.8}}$
D $\frac{2}{\sqrt{9.8}}$ **E** $\frac{2}{\sqrt{15.6}}$

($g = 9.8\,\mathrm{m\,s^{-2}}$.) [O & C 81]

7 A helicopter is flying in a straight line at a speed of $20\,\mathrm{m\,s^{-1}}$ and at a constant height of 0.18 km. If a small object falls from it, calculate (a) the time it takes to reach the ground, (b) the horizontal distance it travels in doing so.
Neglect air resistance.

8

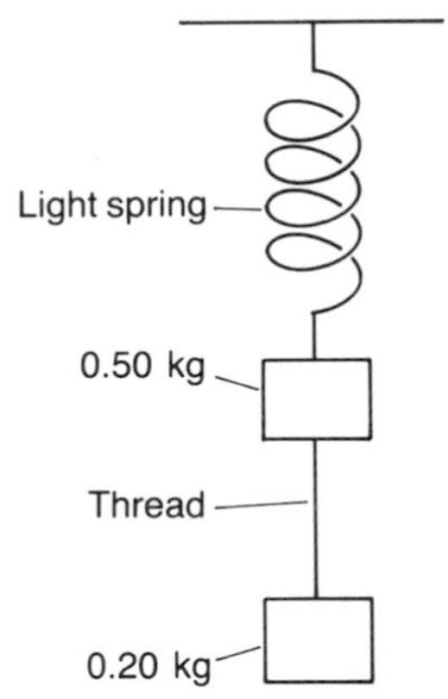

Fig 7.8 Diagram for Question 8

Fig 7.8 shows a system in static equilibrium, in which two masses, joined by a thread, are suspended from a vertical light spring. If the thread is burned through, calculate the magnitude and direction of the acceleration of the 0.50 kg mass.

9 A car of mass 1100 kg tows a caravan of mass 750 kg. The total resistance to their motion has a constant value of 2000 N and a half of this acts on the caravan. Calculate the force exerted on the caravan by the tow bar (a) when the acceleration of car and caravan is $1.0\,\mathrm{m\,s^{-2}}$; (b) when the car and caravan move at a constant speed of $10\,\mathrm{m\,s^{-1}}$.

10 A bullet of mass 0.020 kg is fired from a rifle. The barrel of the rifle is 0.50 m long and has an internal diameter of 8.0 mm, and the average excess pressure of the gas in the barrel is 5.0×10^3 atmospheres. Assuming that the recoil velocity of the rifle is negligible, and neglecting friction and any rotational energy acquired by the bullet,

(a) show that the average force on the bullet is 3.5×10^4 N;

(b) calculate the acceleration of the bullet;

(c) calculate the muzzle velocity of the bullet.

Assume 1 atmosphere = 1.4×10^5 Pa. [AEB 80]

8 ENERGY, WORK AND POWER

Energy

Mechanical energy exists in two basic forms:

(1) Kinetic energy (KE) is energy due to motion. KE $= \frac{1}{2}mv^2$ (m is the mass, v the velocity of the body.)

(2) Potential energy (PE) is energy stored — e.g. in a compressed spring or due to the position of a body in a force field. Gravitational PE $= mgh$ (m is mass, g is acceleration due to gravity, h is height above a datum), and is energy stored in a gravitational field (see Chapter 12) due to the elevated position of the body.

Work and energy

Work is done when energy is transferred from one system to another. It involves a force acting over a distance. We define

$$\text{Work done (J)} = \text{Force (N)} \times \text{Distance (m)} \tag{8.1}$$

Also Work done = Energy transferred (J)

Example 1

A body of mass 5.0 kg is initially at rest on a horizontal frictionless surface. A force of 15 N acts on it and accelerates it to a final velocity of 12 m s^{-1}. Calculate (a) the distance travelled, (b) the work done by the force, (c) the final KE of the body.
Compare (b) and (c) and comment.

Method

(a) We have $m = 5.0$, $u = 0$, $F = 15$ and $v = 12$. To find distance s we must find acceleration a. From Equation 7.5

$$a = \frac{F}{m} = \frac{15}{5} = 3.0\,\text{m s}^{-2}$$

Using Equation 7.3

$$v^2 = u^2 + 2as$$

$$\therefore \quad 12^2 = 0^2 + 2 \times 3 \times s$$

$$\therefore \quad s = 24\,\text{m}$$

(b) From Equation 8.1

$$\text{Work done} = F \times s = 15 \times 24 = 360\,\text{J}$$

(c) $\text{KE} = \frac{1}{2}mv^2 = \frac{1}{2} \times 5 \times 12^2$

$= 360\,\text{J}$

The answers to (b) and (c) are the same. This is because, in the absence of friction forces and on a horizontal surface, all the work done by the force becomes KE of the moving body.

Answer

(a) 24 m, (b) 3.6×10^2 J, (c) 3.6×10^2 J.

Example 2

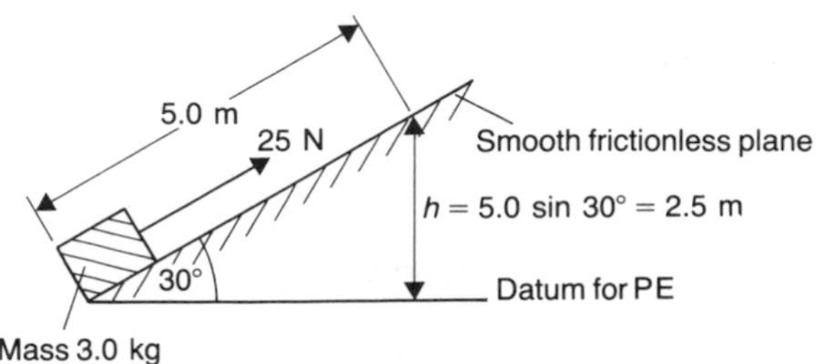

Fig 8.1 Information for Example 2

Refer to Fig 8.1. A block of mass 3.0 kg is pulled 5.0 m up a smooth plane, inclined at 30° to the horizontal, by a force of 25 N parallel to the plane. Find the velocity of the block when it reaches the top of the plane.

Method

Work done on the body becomes KE and PE. So, if final velocity of block is v, then

Work done = (KE + PE) gained by block, or

$$F \times s = \tfrac{1}{2}mv^2 + mgh$$

We have $F = 25$, $s = 5.0$, $m = 3.0$, $g = 10$ and $h = 5.0 \sin 30° = 2.5$. So

$$25 \times 5 = \tfrac{1}{2} \times 3 \times v^2 + 3 \times 10 \times 2.5$$

$$\therefore \quad v = \sqrt{\frac{100}{3}} = 5.8 \text{ m s}^{-1}$$

Note that in the above we do *not* subtract from the 25 N force the component of weight acting 'down' the plane ($mg \sin 30°$, from Chapter 6) since gravitational effects have been accounted for in the PE term.

Answer

5.8 m s^{-1}.

Exercise 20

(Assume $g = 10 \text{ m s}^{-2}$.)

1 Calculate the kinetic energy of the following:
(a) a car: mass 900 kg, velocity 20 m s^{-1};
(b) an aeroplane: mass 20×10^3 kg, velocity 200 m s^{-1};
(c) an electron: mass 9.0×10^{-31} kg, velocity $20 \times 10^6 \text{ m s}^{-1}$.

2 A force of 15 N is applied to a body of mass 3.0 kg, initially at rest on a smooth horizontal surface, for a time of 3.0 s. Calculate (a) the final velocity, (b) the distance travelled, (c) the work done, (d) the final KE of the body.

3 A block of mass 10 kg is pulled 20 m up a smooth plane inclined at 45° to the horizontal. The block is initially at rest and reaches a velocity of 2.0 m s^{-1} at the top of the plane. Calculate the magnitude of the force required, assuming it acts parallel to the plane.

4 A body of mass 5.0 kg is pulled 4.0 m up a *rough* plane, inclined at 30° to the horizontal, by a force of 50 N parallel to the plane. Find the velocity of the block when it reaches the top of the plane if the frictional force is of magnitude 12 N.

Energy interchange

The principle of conservation of energy states that the total amount of energy in an isolated system remains constant.

If dissipative effects, e.g. friction, are ignored then we have simply KE and PE interchange.

Example 3

A ball of mass 0.50 kg falls from a height of 45 m. Calculate (a) its initial PE, (b) its final KE, (c) its final velocity. Neglect air resistance. (Assume $g = 10 \text{ m s}^{-2}$.)

Method

(a) We have $m = 0.50$, $h = 45$, $g = 10$. So

$$\text{PE} = mgh = 0.5 \times 10 \times 45 = 225 \text{ J}$$

(b) All the PE has been converted to KE just prior to striking the ground. So

$$\text{Final KE} = 225 \text{ J}$$

(c) Let final velocity be v. Since $\text{KE} = \tfrac{1}{2}mv^2$ then

$$225 = \tfrac{1}{2} \times 0.5 \times v^2$$

$$\therefore \quad v = 30 \text{ m s}^{-1}$$

Note that the final velocity does not depend on the mass because it cancels out (since $\tfrac{1}{2}mv^2 = mgh$, $v^2 = 2gh$).

Note that, as in Example 6, Chapter 7, we could have solved this using the equations of motion with acceleration $a = g = 10 \text{ m s}^{-2}$. This is because air resistance is negligible.

Answer

(a) 2.3×10^2 J, (b) 2.3×10^2 J, (c) 30 m s^{-1}.

Example 4

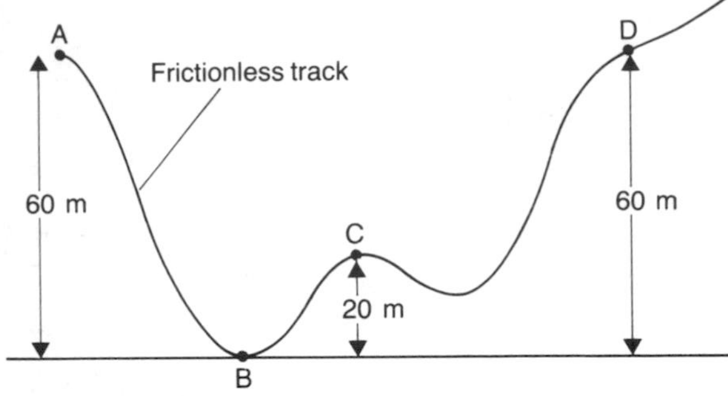

Fig 8.2 Information for Example 4

Refer to Fig 8.2. A truck of mass 150 kg is released from rest at A and moves along the frictionless track. Calculate (a) its maximum KE, (b) its maximum velocity, (c) its velocity at C. (Assume $g = 10\,\mathrm{m\,s^{-2}}$.)

Explain what happens when it reaches D.

Method

(a) Maximum KE is when PE is a minimum. This occurs at B. We have

$$\text{Gain in KE} = \text{Loss in PE } (mgh)$$

Since $m = 150$, $g = 10$, $h = 60$,

$$\text{Loss in PE} = 150 \times 10 \times 60 = 90\,000\,\mathrm{J}$$

$$\therefore \quad \text{KE gain} = 90\,\mathrm{kJ}$$

(b) Maximum velocity v occurs for maximum KE.

$$\therefore \quad \tfrac{1}{2}mv^2 = \tfrac{1}{2} \times 150 \times v^2 = 90\,000$$

$$\therefore \quad v = \sqrt{1200} = 34.6\,\mathrm{m\,s^{-1}}$$

(c) At C, the drop in height h_1 is 40 m below A. So if velocity at C is v_1 then

$$\text{Gain in KE } (\tfrac{1}{2}mv_1^2) = \text{Loss in PE } (mgh_1)$$

$$\tfrac{1}{2} \times 150 \times v_1^2 = 150 \times 10 \times 40$$

$$\therefore \quad v_1 = \sqrt{800} = 28.3\,\mathrm{m\,s^{-1}}$$

Note, once again, that v_1 does not depend on the mass of the truck.

The truck arrives at D with zero KE, hence zero velocity. since its PE at D equals its PE at A. It then starts to roll back to C, B and A.

Answer

(a) 90 kJ, (b) 35 $\mathrm{m\,s^{-1}}$, (c) 28 $\mathrm{m\,s^{-1}}$.

Exercise 21

(Assume $g = 10\,\mathrm{m\,s^{-2}}$.)

1 An object of mass 0.30 kg is thrown vertically upwards and reaches a height of 8.0 m. Calculate (a) its final PE, (b) the velocity with which it must be thrown, neglecting air resistance.

2 A cricket of mass 2.5 g has a vertical velocity of 2.0 $\mathrm{m\,s^{-1}}$ when it jumps. Calculate (a) its maximum KE and (b) the maximum vertical height it could reach.

3 A ball of mass 0.20 kg drops from a height of 10 m and rebounds to a height of 7.0 m. Calculate the energy lost on impact with the floor. Neglect air resistance.

4 A simple pendulum oscillates with an amplitude of 30°. If the length of the string is 1.0 m, calculate the velocity of the pendulum bob at its lowest point.

5

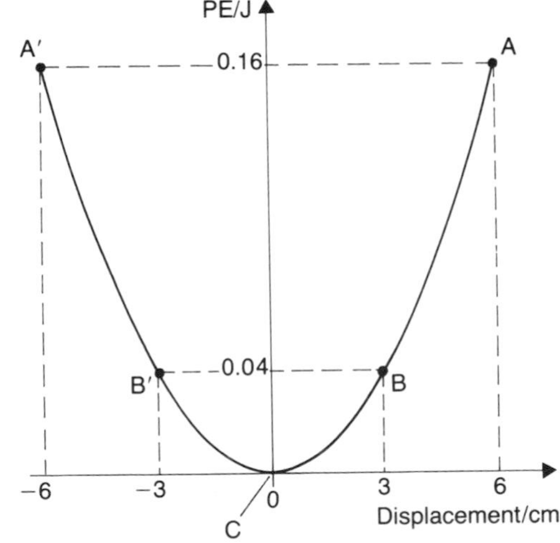

Fig 8.3 Information for Question 5

Fig 8.3 shows the PE versus displacement graph for a body, of mass 0.10 kg, oscillating about the point C. If the body has total energy 0.16 J, calculate its velocity at (a) A and A′, (b) B and B′, (c) C. Neglect friction, air resistance, etc.

'Lost' energy

If dissipative forces, e.g. friction and air resistance, cannot be neglected, then some energy will be 'lost' in the sense that it is converted to other forms (e.g. heat energy).

Example 5

A ball of mass 0.20 kg is thrown vertically upwards with a velocity of $15\,\mathrm{m\,s^{-1}}$. If it reaches a height of 10 m, calculate the percentage loss in energy caused by air resistance. (Assume $g = 10\,\mathrm{m\,s^{-2}}$.)

Method

The final PE is less than the initial KE due to transfer of energy to the surrounding air. We have $m = 0.20$, $u = 15$, $h = 10$ and $g = 10$, so

$$\text{Initial KE} = \tfrac{1}{2}mu^2 = \tfrac{1}{2} \times 0.2 \times 15^2 = 22.5\,\mathrm{J}$$

$$\text{Final PE} = mgh = 0.2 \times 10 \times 10 = 20.0\,\mathrm{J}$$

$$\therefore \quad \text{Energy transfer} = 22.5 - 20.0 = 2.5\,\mathrm{J}$$

$$\text{Percentage loss} = \frac{\text{Energy transfer}}{\text{Initial KE}} \times 100 = \frac{2.5}{22.5} \times 100 = 11.1$$

Answer

11% of the initial KE is transferred to the surrounding air.

Example 6

A block of mass 6.0 kg is projected with a velocity of $12\,\mathrm{m\,s^{-1}}$ up a rough plane inclined at 45° to the horizontal. If it travels 5.0 m up the plane, calculate (a) the energy dissipated via frictional forces, (b) the magnitude of the (average) friction force. (Assume $g = 10\,\mathrm{m\,s^{-2}}$.)

Method

(a)

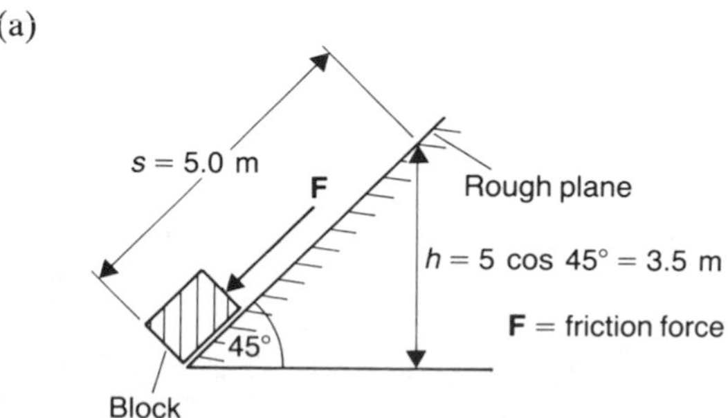

Fig 8.4 Solution to Example 6

Refer to Fig 8.4. Initial KE becomes PE and energy dissipated via work done ($F \times s$) against friction force F. So

$$\tfrac{1}{2}mu^2 = mgh + F \times s$$

We have $m = 6.0$, $u = 12$, $g = 10$ and $h = 3.5$, so

$$\tfrac{1}{2} \times 6 \times 12^2 = 6 \times 10 \times 3.5 + F \times s$$

$$\therefore \quad F \times s = 222\,\mathrm{J}$$

(b) $F \times s = 222\,\mathrm{J}$ and $s = 5.0\,\mathrm{m}$

$$\therefore \quad F = 44\,\mathrm{N}$$

Answer

(a) 0.22 kJ, (b) 44 N.

Exercise 22

(Assume $g = 10\,\mathrm{m\,s^{-2}}$.)

1 An object of mass 1.5 kg is thrown vertically upwards with a velocity of $25\,\mathrm{m\,s^{-1}}$. If 10% of its initial energy is dissipated against air resistance on its upward flight, calculate (a) its maximum PE, (b) the height to which it will rise.

2 A cricket ball of mass 0.20 kg is thrown vertically upwards with a velocity of $20\,\mathrm{m\,s^{-1}}$ and returns to earth at $15\,\mathrm{m\,s^{-1}}$. Find the work done against air resistance during the flight.

3

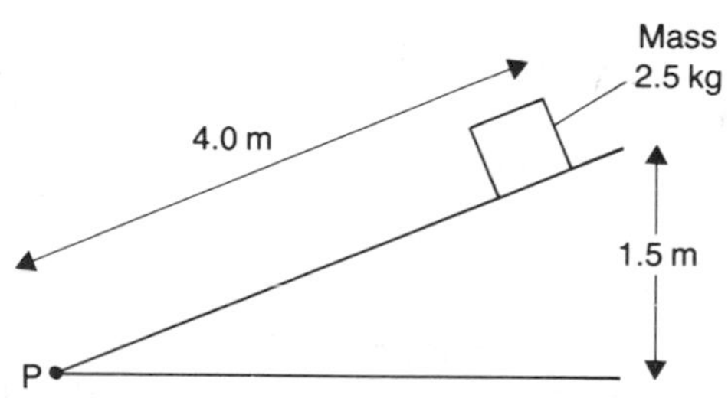

Fig 8.5 Diagram for Question 3

Fig 8.5 shows a mass of 2.5 kg initially at rest on a rough inclined plane. The mass is now released and acquires a velocity of $4.0\,\mathrm{m\,s^{-1}}$ at P, the base of the incline. Find (a) the work done against friction, (b) the (average) friction force.

Machines — efficiency and power

A machine is a device that serves to transfer energy from one system to another. The useful energy output will be less than the energy input due to energy 'lost', e.g. in work done against friction. We define

$$\text{Efficiency (\%)} = \frac{\text{Useful energy (power) output}}{\text{Energy (power input)}} (\times 100)$$

Power is the rate of transfer of energy—i.e. the work done in unit time.

Example 7

A 1.0 kW motor drives a pump which raises water through a height of 15 m. Calculate the mass of water lifted per second, assuming the system is (a) 100% efficient, (b) 75% efficient. (Assume $g = 10\,\text{m s}^{-2}$.)

Method

(a) Each second the motor supplies 1000 J of energy, which we assume is all converted to gravitational PE of the water. We have $h = 15$ and require mass m. So

$$1000 = mgh = m \times 10 \times 15$$

$$\therefore \quad m = 6.7\,\text{kg}$$

(b) Only 75% of 1000 J, that is 750 J, becomes available to lift water. So, if the new mass is m_1,

$$750 = m_1 gh = m_1 \times 10 \times 15$$

$$\therefore \quad m_1 = 5.0\,\text{kg}$$

Answer

(a) $6.7\,\text{kg s}^{-1}$, (b) $5.0\,\text{kg s}^{-1}$.

Example 8

20×10^3 kg of water moving at $2.2\,\text{m s}^{-1}$ is incident on a water wheel each second. Calculate the maximum power output from the mill, assuming 40% efficiency.

Method

Energy input is KE of the water. In *one second* we have $m = 20 \times 10^3$ and $v = 2.2\,\text{m s}^{-1}$. So

$$\text{KE input} = \tfrac{1}{2}mv^2 = \tfrac{1}{2} \times 20 \times 10^3 \times 2.2^2$$

$$= 48.4 \times 10^3\,\text{J}$$

Of this energy 40% becomes useful energy output. So

$$\text{Useful energy output} = \frac{40}{100} \times 48.4 \times 10^3$$

$$= 19.4 \times 10^3\,\text{J}$$

Answer

Maximum power output = 19 kW.

Example 9

A boat travels at a constant velocity of $8.0\,\text{m s}^{-1}$. If the engine develops a useful power output of 20 kW, calculate the push exerted by the propellor on the water. Why is the boat not accelerating?

Method

$$\text{Power } P = \text{Rate of doing work}$$

$$= \frac{\text{Force} \times \text{Distance}}{\text{Time}}$$

$$= \text{Force } F \times \text{Velocity } v \qquad (8.2)$$

We have $P = 20 \times 10^3$, $v = 8.0$, and require F. So

$$20 \times 10^3 = F \times 8$$

$$\therefore \quad F = 2.5 \times 10^3\,\text{N}$$

The boat is not accelerating because the resistance to motion of the boat as it passes through water is equal to 2.5 kN. The net force on the boat is thus zero, so it does not accelerate.

Answer

2.5 kN.

Example 10

A train of mass 10×10^3 kg, initially at rest, accelerates uniformly at $0.50\,\text{m s}^{-2}$. Calculate the power required at time 5.0 s and 8.0 s, assuming (a) no resistive forces, (b) resistive forces of 1.0 kN act.

Method

Equation 8.2 tells us we require force F and velocity v at a given time in order to calculate the instantaneous power P.

(a) To find F use Equation 7.5, with $a = 0.50$ and $m = 10 \times 10^3$. So

$$F = ma = 0.5 \times 10 \times 10^3 = 5 \times 10^3\,\text{N}$$

To find v after $t = 5.0$ s and $t = 8.0$ s use Equation 7.1 with $u = 0$ and $a = 0.50$. Thus

after 5 s, $v = 0 + 0.5 \times 5 = 2.5\,\text{m s}^{-1}$

after 8 s, $v = 0 + 0.5 \times 8 = 4.0\,\text{m s}^{-1}$

From Equation 8.2 we have

after 5 s, $P = F \times v = 5 \times 10^3 \times 2.5$

$= 12.5 \times 10^3$

after 8 s, $P = F \times v = 5 \times 10^3 \times 4 = 20 \times 10^3$

(b) The engine must apply an additional force of 1.0×10^3 N in excess of that in part (a). So the engine must apply a constant force of 6.0×10^3 N. From equation 8.2

after 5 s, $P = F \times v = 6 \times 10^3 \times 2.5 = 15 \times 10^3$

after 8 s, $P = F \times v = 6 \times 10^3 \times 4.0 = 24 \times 10^3$

Answer
(a) 12.5 kW, 20 kW, (b) 15 kW, 24 kW.

Example 11

A car of mass 1.2×10^3 kg moves up an incline at a steady velocity of $15\,\text{m s}^{-1}$ against a frictional force of 0.6 kN. The incline is such that it rises 1.0 m for every 10 m along the incline. Calculate the output power of the car engine. (Assume $g = 10\,\text{m s}^{-2}$.)

Method
The car engine does work against friction forces and in raising the PE of the car as it moves up the incline. So, referring to energy transfer per second gives

$$\text{Power } P = \left(\begin{matrix}\text{Rate of doing}\\ \text{work against friction}\end{matrix}\right) + \left(\begin{matrix}\text{Rate of}\\ \text{gain of PE}\end{matrix}\right)$$

$$= F \times v + mgh$$

where F (frictional force) $= 0.60 \times 10^3$
v (velocity) $= 15$
m (mass of car) $= 1.2 \times 10^3$
$g = 10$
h (gain in height per second) $= 15 \times \dfrac{1}{10} = 1.5$

$$\therefore \quad P = (0.6 \times 10^3 \times 15) + (1.2 \times 10^3 \times 10 \times 1.5)$$
$$= 27 \times 10^3\,\text{W}$$

Answer
Output power = 27 kW.

Exercise 23

(Assume $g = 10\,\text{m s}^{-2}$.)

1 Calculate the power rating of a pump if it is to lift 180 kg of water per minute through a height of 5.0 m, assuming
(a) 100%, (b) 50%, (c) 70% efficiency.

2 200 kg of air moving at $15\,\text{m s}^{-1}$ is incident each second on the vanes of a windmill. Estimate the maximum output power of the mill. Why is this not achieved in practice?

3 A hydroelectric power station is driven by water falling on to a system of wheels from a height of 100 m. If the output power of the station is 10 MW (10×10^6 W), calculate the rate at which water must impinge on the wheels, assuing (a) 100%, (b) 50% efficiency.

4 A car of mass 900 kg, initially at rest, accelerates uniformly and reaches $20\,\text{m s}^{-1}$ after 10 seconds. Calculate the power developed by the engine after (a) 5.0 s, (b) 10 s, (c) a distance of 50 m from the start position. Assume that resistive forces are negligible.

5 Repeat Question 4 with a constant resistive force of 0.50 kN acting.

6 A car pulls a caravan of mass 800 kg up an incline of 8% (8 up for 100 along the incline) at a steady velocity of $10\,\text{m s}^{-1}$. Calculate the tension in the tow bar, assuming (a) resistive forces are negligible, (b) a resistive force of 0.40 kN acts on the caravan.

7 A car has a maximum output power of 20 kW and a mass of 1500 kg. At what maximum velocity can it ascend an incline of 10%, assuming (a) no dissipative forces, (b) a constant resistance of 1.0 kN opposing its motion.

Exercise 24: Questions of A-level standard

(Assume $g = 10\,\mathrm{m\,s^{-2}}$.)

1 When a gun is fired from a stationary aeroplane, the bullet has a speed v and a kinetic energy K. The gun is again fired, straight ahead, when the plane is flying at a speed equal to v. The kinetic energy of the second bullet is

A K **B** $\sqrt{(2)}K$ **C** $2K$
D $3K$ **E** $4K$ [L 79]

2 A mass of 5 kg is moving in a straight line with a constant speed of $4\,\mathrm{m\,s^{-1}}$. A force of 10 N is then applied to the mass along the direction of motion and removed after the mass has travelled a distance of 2.5 m. The increase in kinetic energy of the mass, in joules, due to the action of the force is

A 10 **B** 12.5 **C** 20
D 25 **E** 32.5 [O & C 82]

3 A shell of mass 0.2 kg is fired vertically upwards with a speed of $100\,\mathrm{m\,s^{-1}}$ and 15 s later arrives back at ground level with a speed of $80\,\mathrm{m\,s^{-1}}$. Draw sketches showing the directions of the forces acting on the shell (a) during the upward flight and (b) during the downward flight.
Find the total work done against air resistance during the flight. [WJEC 81 p]

4 Explain briefly why a simple pendulum eventually comes to rest after being set in motion.

A simple pendulum of length 1 m and mass 0.2 kg is set in motion with amplitude 10 cm. After a time the amplitude has reduced to 5 cm. Find the energy lost. [WJEC 81]

5

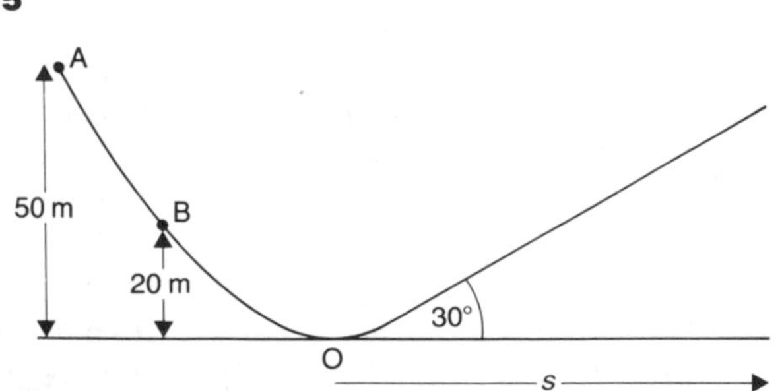

Fig 8.6 Diagram for Question 5

Fig 8.6 shows the vertical section through a ski track. A skier of mass 80 kg starts from rest at A and there is no friction. The down-slope is curved but the up-slope is plane. What is the greatest horizontal distance s from O that the skier reaches? What is his velocity at B? [WJEC 82 p]

6 Giant windmills have recently been proposed as power sources. One such windmill sweeps out a circle of 60 m radius. If 60% of the available energy is converted into useful energy, calculate the power produced in a wind of $8\,\mathrm{m\,s^{-1}}$. Density of air $= 1\,\mathrm{kg\,m^{-3}}$. [WJEC 82]

7 The outboard motor of a small boat has a propellor of diameter 0.2 m. If the boat is at rest, the propellor sends back a stream of water at a speed of $10\,\mathrm{m\,s^{-1}}$. One half of the work that is being done by the motor is transferred to this water as kinetic energy. What is the power of the motor? Take the density of water to be $1000\,\mathrm{kg\,m^{-3}}$.

A 1.25 kW **B** 6.50 kW **C** 15.7 kW
D 31.4 kW **E** 125 kW [O 81]

8 Here are some facts about the Inter City 125 train. At its top speed of $50\,\mathrm{m\,s^{-1}}$ the engine exerts a power of about 2 MW. The mass of train and engine is about 400 tonnes (400 000 kg).

A 20 **B** 40 **C** 100 **D** 500 **E** 4000

Which of **A** to **E** is

(a) the best estimate of the kinetic energy of train and engine at top speed, measured in MJ?
(b) the best estimate of the tractive force of the engine needed to overcome drag at top speed, measured in kN? [O & C Nuff 82]

9 (a) A particle of mass m, initially at rest, is acted upon by a constant force until its velocity is v. Show that the kinetic energy of the particle is $\frac{1}{2}mv^2$.
(b) A train of mass 2.0×10^5 kg moves at a constant speed of $72\,\mathrm{km\,h^{-1}}$ up a straight incline against a frictional force of 1.28×10^4 N. The incline is such that the train rises vertically 1.0 m for every 100 m travelled along the incline. Calculate
(i) the rate of increase per second of the potential energy of the train;
(ii) the necessary power developed by the train. [JMB 80]

9 LINEAR MOMENTUM

Momentum

The (linear) momentum of a body is defined by

$$\text{Momentum} = \text{Mass} \times \text{Velocity} \qquad (9.1)$$

Example 1

A body A of mass 5 kg moves to the right with a velocity of 4 m s^{-1}. A body of mass 3 kg moves to the left with a velocity of 8 m s^{-1}. Calculate (a) the momentum of A, (b) the momentum of B, (c) the total momentum of A and B.

Method

We use Equation 9.1

(a) Momentum of

$$\begin{aligned} A &= \text{Mass} \times \text{Velocity} \\ &= 5 \times 4 = +20\ \text{kg m s}^{-1} \end{aligned}$$

(b) Velocity, and hence momentum. are *vector* quantities. We assumed in part (a) that motion to the right is positive in sign. So motion to the left is negative.

$$\begin{aligned} \text{Momentum of B} &= \text{Mass} \times \text{Velocity} = 3 \times -8 \\ &= -24\ \text{kg m s}^{-1} \end{aligned}$$

(c) $$\begin{pmatrix}\text{Total momentum}\\ \text{of A and B}\end{pmatrix} = \begin{pmatrix}\text{Momentum}\\ \text{of A}\end{pmatrix} + \begin{pmatrix}\text{Momentum}\\ \text{of B}\end{pmatrix}$$
$$= 20 - 24 = -4\ \text{kg m s}^{-1}$$

Answer

(a) 20 kg m s^{-1}, (b) -24 kg m s^{-1}, (c) -4 kg m s^{-1}.

Exercise 25

1 A body has a mass of 2.5 kg. Calculate (a) its momentum when it has a velocity of 3.0 m s^{-1}, (b) its velocity when it has a momentum of 10.0 kg m s^{-1}.

2 An object A has mass 2 kg and moves to the left at 5 m s^{-1}. An object B has mass 4 kg and moves to the right at 2.5 m s^{-1}. Calculate (a) the momentum of A, (b) the momentum of B, (c) the total momentum of A and B.

Conservation of momentum

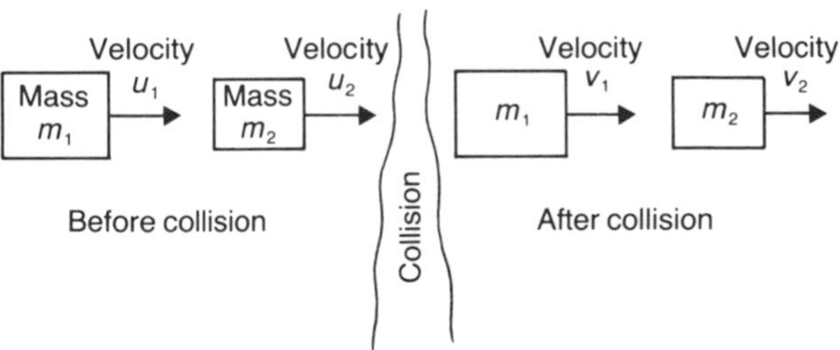

Fig 9.1 Conservation of linear momentum

Provided that no *external* forces (such as friction) are acting, then, when bodies collide, the total momentum before collision is the same as that after collision. With reference to Fig 9.1, this means

$$m_1u_1 + m_2u_2 = m_1v_1 + m_2v_2 \qquad (9.2)$$

Example 2

A 2.0 kg object moving with a velocity of 8.0 m s^{-1} collides with a 4.0 kg object moving with a velocity of 5.0 m s^{-1} along the same line. If the two objects join together on impact, calculate their common velocity when they are initially moving (a) in the same direction, (b) in opposite directions.

Method

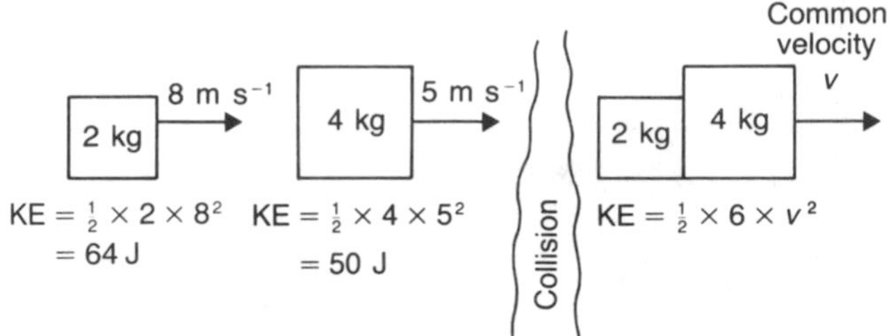

(a) Objects moving in same direction

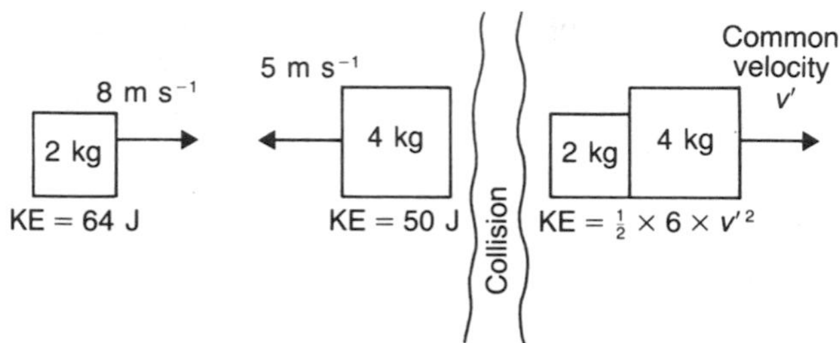

(b) Objects moving in opposite directions

Fig 9.2 Diagram for Examples 2 and 3

(a) Fig 9.2a shows the situation before and after impact. Since $v_1 = v_2 = v$, Equation 9.2 gives

$$m_1u_1 + m_2u_2 = m_1v + m_2v$$
$$= (m_1 + m_2)v$$

So

$$2 \times 8 + 4 \times 5 = 6v$$

$$\therefore \quad v = \frac{36}{6} = 6.0\,\mathrm{m\,s^{-1}}$$

(b) Fig 9.2b illustrates the situation. As in Example 1, the 4 kg mass now has a negative velocity, so $u_2 = -5$. Hence, if v' is the common velocity,

$$2 \times 8 - 4 \times 5 = 6v'$$

$$\therefore \quad v' = -\frac{4}{6} = -0.67\,\mathrm{m\,s^{-1}}$$

Answer
(a) $6.0\,\mathrm{m\,s^{-1}}$, (b) $-0.67\,\mathrm{m\,s^{-1}}$.

Note: The negative value of v' means that the combined masses move to the left after collision. This is because the momentum of the 4 kg mass is larger than that of the 2 kg mass. Since the objects *join together* and their *two momenta are about the same*, there is a *small common velocity* after impact. During this collision a large fraction of the initial kinetic energy is converted to other forms of energy.

Exercise 26

1 A truck of mass 1.0 tonne moving at $4.0\,\mathrm{m\,s^{-1}}$ catches up and collides with a truck of mass 2.0 tonne moving at $3.0\,\mathrm{m\,s^{-1}}$ in the same direction. The trucks become coupled together. Calculate their common velocity. (1 tonne = 1000 kg.)

2 Repeat Question 1 but assume the trucks are moving in the same line and in opposite directions.

3 A pile-driver of mass 380 kg moving at $20\,\mathrm{m\,s^{-1}}$ hits a stationary stake of mass 20 kg. If the two move off together, calculate their common velocity.

Collisions and energy

Momentum is conserved in a collision. Total energy is also conserved but kinetic energy might not be. In general some kinetic energy will be converted to other forms (e.g. sound, work done during plastic deformation).

> An *inelastic collision* is one in which kinetic energy is not conserved.
> An *elastic collision* is one in which kinetic energy is conserved.
> A *completely inelastic collision* is one in which the objects stick together on impact.

Example 3

Calculate the KE converted to other forms during the collisions in (a) and (b) of Example 2.

Method
Refer to Fig 9.2a and b which show the kinetic energy of the various objects before and after collision.

(a) Before collision, total KE = 64 + 50 = 114 J

After collision, since $v = 6$,

$$\text{Total KE} = \tfrac{1}{2} \times 6 \times 6^2 = 108\,\mathrm{J}$$

$$\therefore \quad \text{KE converted} = 114 - 108 = 6\,\mathrm{J}$$

(b) Before collision, total KE = 114 J

After collision, since $v' = -0.67$,

$$\text{Total KE} = \tfrac{1}{2} \times 6 \times (-0.67)^2 = 1.3\,\mathrm{J}$$

$$\therefore \quad \text{KE converted} = 114 - 1.3 = 112.7\,\mathrm{J}$$

Answer
(a) 6 J, (b) 113 J.

Example 4

A 2.0 kg object moving with velocity 6.0 m s^{-1} collides with a stationary object of mass 1.0 kg. Assuming that the collision is perfectly elastic, calculate the velocity of each object after the collision.

Method

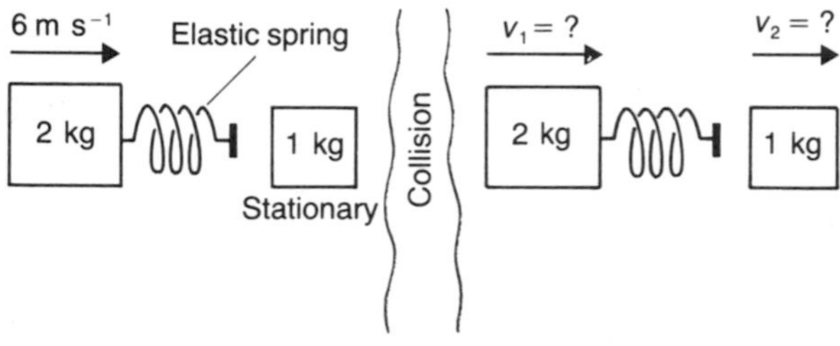

Fig 9.3 An elastic collision (for Example 4)

Fig 9.3 shows the situation before and after collision. We must find v_1 and v_2, the final velocities of the 2 kg and 1 kg objects respectively. This means we need two equations.

Since momentum is conserved, Equation 9.2 gives

$$m_1u_1 + m_2u_2 = m_1v_1 + m_2v_2$$

i.e. $2 \times 6 + 1 \times 0 = 2v_1 + v_2$

or $$v_2 = 12 - 2v_1 \qquad (9.3)$$

Kinetic energy is also conserved. So

$$\tfrac{1}{2}m_1u_1^2 + \tfrac{1}{2}m_2u_2^2 = \tfrac{1}{2}m_1v_1^2 + \tfrac{1}{2}m_2v_2^2$$

or $\tfrac{1}{2} \times 2 \times 6^2 + \tfrac{1}{2} \times 1 \times 0^2 = \tfrac{1}{2} \times 2 \times v_1^2 + \tfrac{1}{2} \times 1 \times v_2^2$

$$\therefore \quad 6^2 + 0 = v_1^2 + \tfrac{1}{2}v_2^2 \qquad (9.4)$$

Note that we have two simultaneous equations (see Chapter 2), so we can substitute for v_2 from Equation 9.3 into Equation 9.4. This gives an expression which can be factorised (see Chapter 2) to give $v_1 = 2.0$ m s^{-1} and hence $v_2 = 8.0$ m s^{-1}.

Answer
The velocities are 2.0 m s^{-1} and 8.0 m s^{-1} in the original direction. Note the following values of KE:

Before collision: m_1 has 36 J, m_2 has 0 J

After collision: m_1 has 4 J, m_2 has 32 J

So the total KE remains unchanged at 36 J before and after collision.

Energy interchange is via the elastic spring which stores energy on compression during impact. This potential energy is converted to KE when the objects separate.

Exercise 27

1 Calculate the KE converted to other forms during the collision in Question 1 Exercise 26.

2 Calculate the KE converted to other forms during the collision in Question 2 Exercise 26.

3 A 2.0 kg object moving with a velocity of 8.0 m s^{-1} collides with a 3.0 kg object moving with a velocity of 6.0 m s^{-1} along the same direction. If the collision is completely inelastic, calculate the decrease in KE during collision.

Explosions

When an object explodes it does so as a result of some *internal* force. Thus the total momentum of the separate parts will be the same as that of the original body. This is often zero.

Example 5

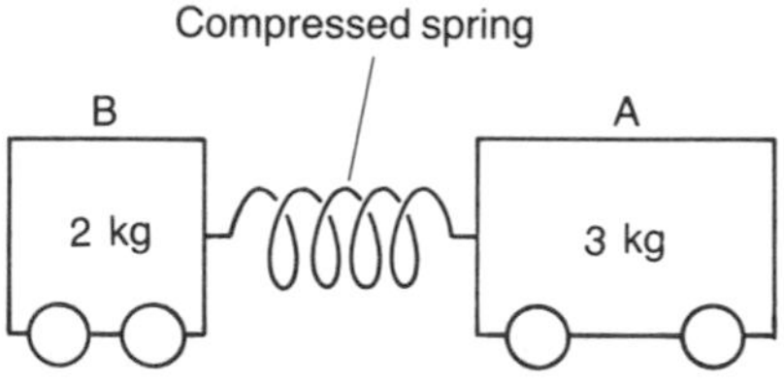

Fig 9.4 Information for Example 5

Fig 9.4 shows two trolleys A and B initially at rest, separated by a compressed spring. The spring is now released and the 3.0 kg trolley moves with a velocity of 1.0 m s^{-1} to the right. Calculate (a) the velocity of the 2.0 kg trolley, (b) the total KE of the trolleys.

Neglect the mass of the spring and any friction forces.

Method

(a) Both trolleys are initially at rest so their momentum is zero. So

$$0 = m_A v_A + m_B v_B$$

where $m_A = 3.0$, $v_A = 1.0$, $m_B = 2.0$ and v_B is required. So

$$0 = 3v_A + 2v_B$$

or $$v_B = -1.5\,\text{m s}^{-1}$$

The negative sign indicates that trolley B moves to the left.

(b) Total KE is the sum of the separate KE of each trolley. So

$$\text{Total KE} = \tfrac{1}{2}m_A v_A{}^2 + \tfrac{1}{2}m_B v_B{}^2$$
$$= \tfrac{1}{2} \times 3 \times 1^2 + \tfrac{1}{2} \times 2 \times (1.5)^2$$
$$= 3.75\,\text{J}$$

Note that the energy is of course positive in each case. The initial energy is zero, and the final energy comes from the potential energy stored in the compressed spring.

Answer

(a) $1.5\,\text{m s}^{-1}$ to the left, (b) 3.8 J.

Exercise 28

1 A shell of mass 1.6 kg is fired from a gun with a velocity of $250\,\text{m s}^{-1}$. Assuming that the gun is free to move, calculate its recoil velocity if it has a mass of 1000 kg.

2 A space satellite has total mass 500 kg. A portion of mass 20 kg is ejected at a velocity of $10\,\text{m s}^{-1}$. Calculate the recoil velocity of the remaining portion. Neglect the initial velocity of the satellite.

3 A radioactive nucleus of mass 235 units travelling at $400\,\text{km s}^{-1}$ disintegrates into a nucleus of mass 95 units and a nucleus of mass 140 units. If the nucleus of mass 95 units travels backwards at $200\,\text{km s}^{-1}$, what is the velocity of the nucleus of mass 140 units?

Impulse

If a force F (N) acts on a body of mass m (kg) for a time t (s) so that the velocity of the body changes from u (m s^{-1}) to v (m s^{-1}), then

$$F \times t = mv - mu \qquad (9.5)$$

The product $F \times t$ is called the *impulse* of the force. It equals the change of momentum of the body. Also

$$F = \frac{(mv - mu)}{t} = \left(\begin{matrix}\text{Rate of change}\\ \text{of momentum}\end{matrix}\right) \qquad (9.6)$$

Example 6

A stationary golf ball is hit with a club which exerts an average force of 80 N over a time of 0.025 s. Calculate (a) the change in momentum, (b) the velocity acquired by the ball if it has a mass of 0.020 kg.

Method

(a) Change of momentum = Impulse

$$= F \times t$$
$$= 80 \times 0.025$$
$$= 2.00\,\text{N s}$$

Note that the unit N s is the same as kg m s^{-1}.

(b) Change of momentum $= mv - mu$

We have $m = 0.020$, $u = 0$ and require v.

$$\therefore \quad 2.00 = 0.020 \times v - 0$$
$$\therefore \quad v = 100\,\text{m s}^{-1}$$

Answer

(a) $2.00\,\text{kg m s}^{-1}$, (b) $100\,\text{m s}^{-1}$.

Exercise 29

1 A squash ball of mass 0.024 kg is hit with a racket and acquires a velocity of $10\,\text{m s}^{-1}$. Its initial velocity is zero. If the time of contact with the racket head is 0.040 s, calculate the average force exerted on the ball.

2 A machine gun fires bullets at a rate of 360 per minute. The bullets have a mass of 20 g and a speed of $500\,\text{m s}^{-1}$. Calculate the average force exerted by the gun on the person holding it.

Exercise 30: Questions of A-level standard

1 A body of mass m travelling with speed $5u$ collides with and adheres to a body of mass $5m$ travelling in the same direction with speed u. The speed with which the two travel together in the original direction is

A zero **B** $\frac{8}{10}u$ **C** u
D $\frac{6}{5}u$ **E** $\frac{10}{6}u$ [O 80]

2 Four identical railway trucks, each of mass m, are coupled together and are at rest on a smooth horizontal track. A fifth truck of mass $2m$ and moving at $5\,\text{m s}^{-1}$ collides and couples with the stationary trucks. After impact the speed of the trucks is v, where v equals

A $\frac{5}{6}\,\text{m s}^{-1}$ **B** $1\,\text{m s}^{-1}$ **C** $\frac{5}{4}\,\text{m s}^{-1}$
D $\frac{5}{3}\,\text{m s}^{-1}$ **E** $\frac{5}{2}\,\text{m s}^{-1}$ [L 79]

3 A sphere of mass 3 kg moving with velocity $4\,\text{m s}^{-1}$ collides head-on with a stationary sphere of mass 2 kg and imparts to it a velocity of $4.5\,\text{m s}^{-1}$. Calculate the velocity of the 3 kg sphere after the collision and the amount of energy lost by the moving bodies in the collision. [SUJB 79]

4 In an elastic head-on collision, a ball of mass 1.0 kg moving at $4.0\,\text{m s}^{-1}$ collides with a stationary ball of mass 2.0 kg. Calculate the velocities of the balls after the collision, indicating the direction in which they are travelling. [AEB 80]

5 A puck collides perfectly inelastically with a second puck originally at rest and of three times the mass of the first puck. What proportion of the original kinetic energy is lost, and where does it go? [WJEC 79]

6

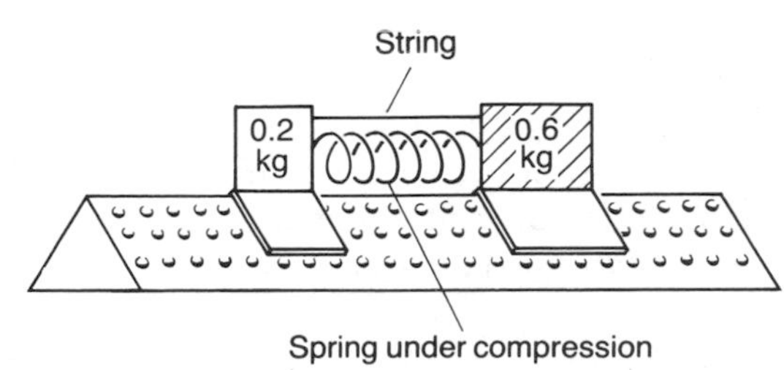

Fig 9.5 Diagram for Question 6

Fig 9.5 shows two riders on a linear track with a light spring under compression placed between them. Originally the arrangement is stationary, with a light string holding the riders in place; the spring is not attached to either rider. The string is then cut, and it is found that the heavier rider acquires a velocity of $0.03\,\text{m s}^{-1}$. Find the velocity acquired by the second rider and deduce the potential energy originally stored in the spring. [WJEC 80, p]

7 A stationary atomic nucleus disintegrates into an α-particle of mass 4 units and a daughter nucleus of mass 234 units. Calculate the ratio

$$\frac{\text{KE of } \alpha\text{-particle}}{\text{KE of daughter nucleus}}$$

8 An astronaut of mass 75 kg becomes detached from his space vehicle and is floating at a distance of 15 m from it. With what velocity (relative to the vehicle) must he throw away his pack of mass 25 kg in order to regain the vehicle in 30 s? How much energy does he expend in the process? [WJEC 81]

9

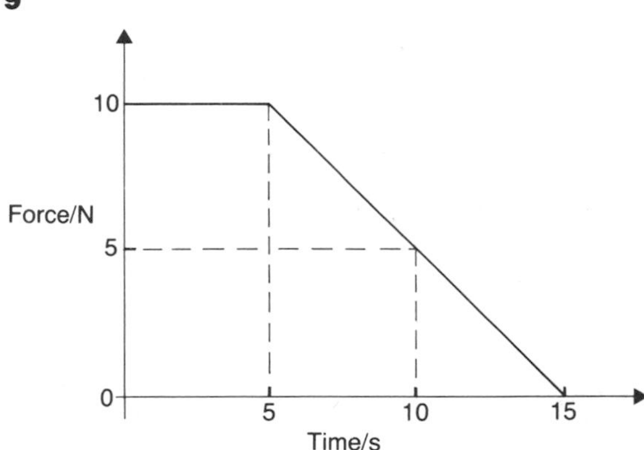

Fig 9.6 Diagram for Question 9

Fig 9.6 shows the variation with time of the resultant force acting on a body. Calculate the change in momentum of the body after (a) 5 s, (b) 10 s, (c) 15 s.

10 A helicopter of mass 810 kg supports itself in a stationary position by imparting a downward velocity v to all the air in a circle of area $30\,\text{m}^2$. Given that the density of air is $1.20\,\text{kg m}^{-3}$, calculate the value of v. What is the power needed to support the helicopter in this way, assuming no energy is lost? Assume $g = 10\,\text{m s}^{-2}$. [SUJB 80, p]

10 CIRCULAR MOTION

Uniform circular motion

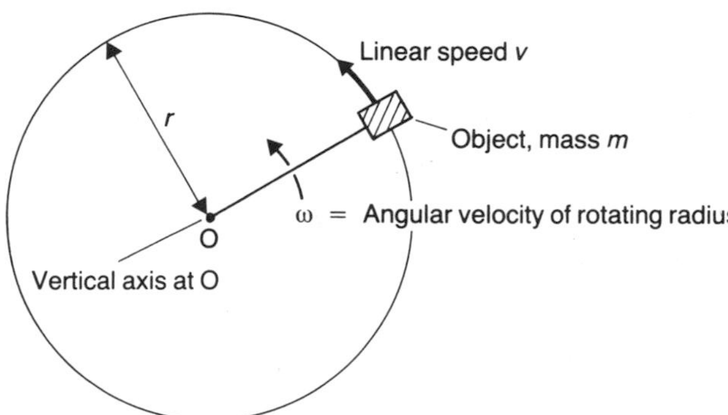

Fig 10.1 Object moving in uniform circular motion

Fig 10.1 shows a body moving with *uniform speed* at a *fixed distance* from a *fixed axis*. It is in uniform circular motion.

Linear and angular motion

In Fig 10.1 an object moves with uniform speed v ($\mathrm{m\,s^{-1}}$) around the circumference of a circle, centre O. The rotating radius, of length r (m), has angular velocity ω ($\mathrm{rad\,s^{-1}}$) such that

$$v = r\omega \qquad (10.1)$$

Example 1

A pulley wheel rotates at 300 rev min^{-1}. Calculate (a) its angular velocity in rad s^{-1}, (b) the linear speed of a point on the rim if the pulley has a radius of 150 mm.

Method

(a) $300\ \mathrm{rev\,min^{-1}} = \dfrac{300}{60}\ \mathrm{rev\,s^{-1}} = 5.00\ \mathrm{rev\,s^{-1}}$ (or Hz)

The frequency f of rotation is thus 5.00 Hz. Now in one revolution the radius rotates through 2π rad. Thus the angular velocity ω of the rotating radius is given by

$$\omega = 2\pi f = 2\pi \times 5 = 10\pi\ \mathrm{rad\,s^{-1}}$$

(b) Use Equation 10.1, in which $\omega = 10\pi$ and $r = 150\ \mathrm{mm} = 0.150\ \mathrm{m}$. Thus

$$v = r\omega = 0.15 \times 10\pi$$
$$= 1.50\pi\ \mathrm{m\,s^{-1}}$$

Answer

(a) 31.4 rad s^{-1}, (b) 4.71 m s^{-1}.

Exercise 31

1 The turntable on a record player rotates at 45 rev min^{-1}. Calculate (a) its angular velocity in rad s^{-1}, (b) the linear speed of a point 14 cm from the centre.

2 A car moves round a circular track of radius 1.0 km at a constant speed of 120 km h^{-1}. Calculate its angular velocity in rad s^{-1}.

Centripetal acceleration and force

The object in Fig 10.1 has uniform speed, but its velocity is constantly changing, since its direction is changing. It is constantly accelerating *towards the centre* O, with magnitude a ($\mathrm{m\,s^{-2}}$) given by

$$a = r\omega^2 = \frac{v^2}{r} \qquad (10.2)$$

A net inward force is needed to provide this

acceleration. For a body of mass m the magnitude of the 'centripetal' force F is given by

$$F = mr\omega^2 = m\frac{v^2}{r} \qquad (10.3)$$

This force can be provided, for example, by the tension in a string, by gravitational or electrostatic attraction, or by friction.

Example 2

An object of mass 0.30 kg is attached to the end of a string and is supported on a smooth horizontal surface. The object moves in a horizontal circle of radius 0.50 m with a constant speed of 2.0 m s^{-1}. Calculate (a) the centripetal acceleration, (b) the tension in the string.

Method

(a) Use Equation 10.2 with $v = 2.0$ and $r = 0.50$. The centripetal acceleration a is given by

$$a = \frac{v^2}{r} = \frac{2^2}{0.5} = 8.0\ \text{m s}^{-2}$$

(b) Use Equation 10.3 with $m = 0.30$, $v = 2.0$ and $r = 0.50$. So

$$F = m\frac{v^2}{r} = \frac{0.3 \times 2^2}{0.5} = 2.4\ \text{N}$$

This force is provided by the tension in the string.

Answer

(a) 8.0 m s^{-2}, (b) 2.4 N.

Example 3

An object of mass 4.0 kg is whirled round in a vertical circle of radius 2.0 m with a speed of 5.0 m s^{-1}. Calculate the maximum and minimum tension in the string connecting the object to the centre of the circle. Assume acceleration due to gravity $g = 10$ m s^{-2}.

Method

Use Equation 10.3 with $m = 4.0$, $v = 5.0$ and $r = 2.0$. Thus the centripetal force F is given by

$$F = m\frac{v^2}{r} = \frac{4 \times 5^2}{2} = 50\ \text{N}$$

Thus a net inward force of 50 N must act on the body during its rotation. In Fig 10.2a the body is at the bottom of the vertical circle. So

$$T_1 - mg = 50$$

$$\therefore \quad T_1 = 50 + mg = 50 + 40 = 90\ \text{N}$$

This is the maximum tension in the string.

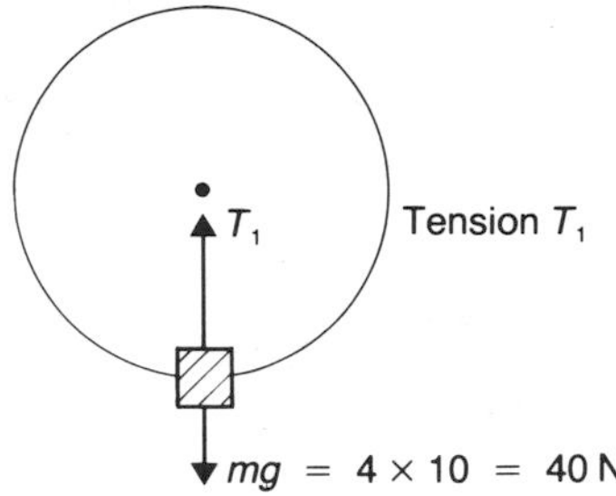

(a) Body at bottom of circle

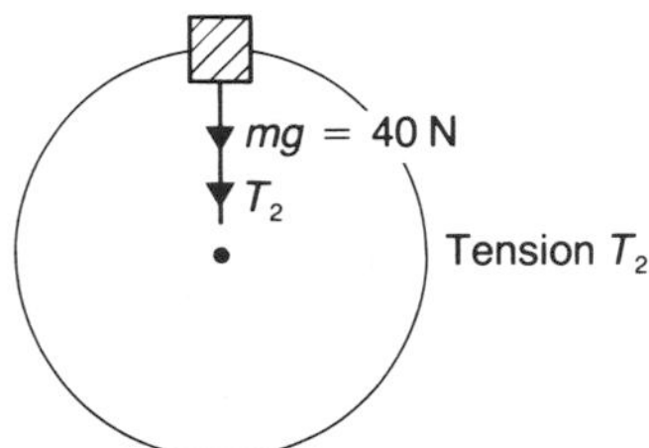

(b) Body at top of circle

Fig 10.2 Forces acting on a body moving in a vertical circle

At the top of the vertical circle, in Fig 10.2b,

$$T_2 + mg = 50$$

$$\therefore \quad T_2 = 50 - mg = 50 - 40$$

$$= 10\ \text{N}$$

This is the minimum tension in the string.

Answer

Maximum tension = 90 N,
Minimum tension = 10 N.

Example 4

A car travels over a humpback bridge of radius of curvature 45 m. Calculate the maximum speed of the car if its road wheels are to stay in contact with the bridge. Assume $g = 10$ m s^{-2}.

Method

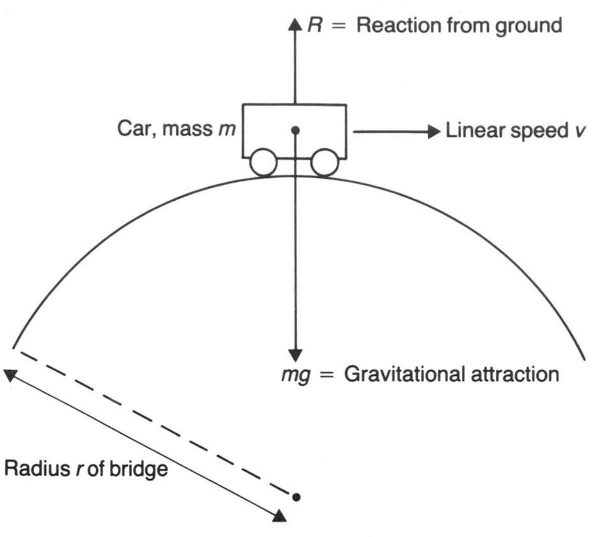

Fig 10.3 Forces acting on a car

Fig 10.3 shows the forces acting on the car when its wheels are in contact with the bridge. *A net inward force equal to mv^2/r must always exist.* So

$$mg - R = m\frac{v^2}{r}$$

As v increases, so R must decrease, since mg is constant. In the limiting case, when the wheels are just about to leave the ground, $R = 0$, so

$$mg = m\frac{v^2}{r}$$

The mass m cancels out and is not required. So maximum speed v is given by

$$v^2 = rg$$

We have $r = 45$ and $g = 10$, so

$$v = \sqrt{rg} = \sqrt{450} = 21.2$$

Answer
The maximum speed is $21\ \mathrm{m\,s^{-1}}$.

Exercise 32

1 A car of mass 1.0×10^3 kg is moving at $30\ \mathrm{m\,s^{-1}}$ around a bend of radius 0.60 km on a horizontal track. What centripetal force is required to keep the car moving around the bend, and where does this force come from?

2 An object of mass 6.0 kg is whirled round in a vertical circle of radius 2.0 m with a speed of $8.0\ \mathrm{m\,s^{-1}}$. Calculate the maximum and minimum tension in the string connecting the object to the centre of the circle.

If the string breaks when the tension in it exceeds 360 N, calculate the maximum speed of rotation, in $\mathrm{m\,s^{-1}}$, and state where the object will be when the string breaks.
Assume $g = 10\ \mathrm{m\,s^{-2}}$.

3 A car travels over a humpback bridge at a speed of $30\ \mathrm{m\,s^{-1}}$. Calculate the minimum radius of the bridge if the car road wheels are to remain in contact with the bridge. What happens if the radius is less than the limiting value? Assume $g = 10\ \mathrm{m\,s^{-2}}$.

The conical pendulum

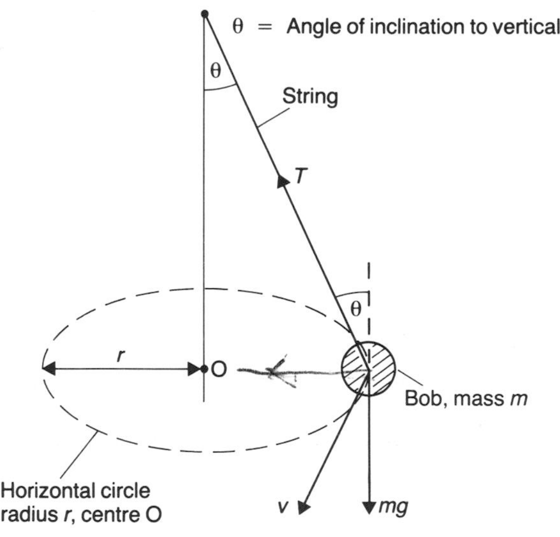

Fig 10.4 The conical pendulum

Fig 10.4 shows the forces acting on a conical pendulum in which the bob sweeps out a horizontal circle, centre O and radius r, with linear speed v. Resolving forces on the mass m gives

$$\text{(vertically)} \quad T\cos\theta = mg \qquad (10.4)$$

$$\text{(horizontally)} \quad T\sin\theta = m\frac{v^2}{r} \qquad (10.5)$$

Example 5

A conical pendulum consists of a small bob of mass 0.20 kg attached to an inextensible string of length 0.80 m. The bob rotates in a horizontal circle of radius 0.40 m, of which the centre is vertically below the point of suspension. Calculate (a) the linear speed of the bob in $\mathrm{m\,s^{-1}}$, (b) the period of rotation of the bob, (c) the tension in the string. Assume $g = 10\,\mathrm{m\,s^{-2}}$.

Method

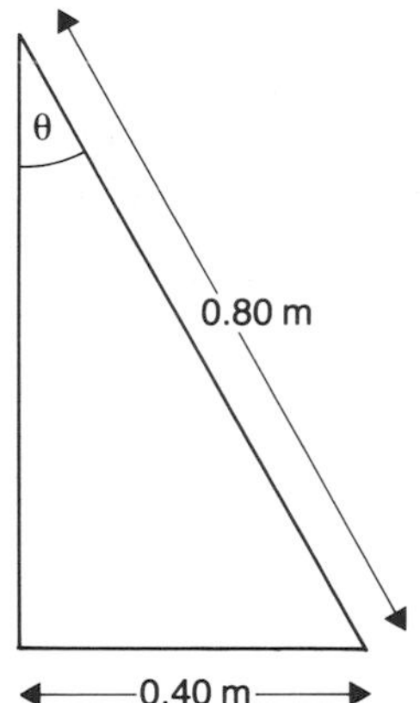

We are given $m = 0.20$, $r = 0.40$, $g = 10$. Also we are given θ since, from Fig 10.5,

$$\sin\theta = \frac{0.40}{0.80} = 0.50$$

&& $\therefore \quad \theta = 30°$

(a) To find v divide Equation 10.5 by 10.4 to give

$$\tan\theta = \frac{v^2}{rg}$$

$$\therefore \quad v^2 = rg\tan 30° = 0.4 \times 10 \times 0.577$$

$$\therefore \quad v = 1.52\,\mathrm{m\,s^{-1}}$$

(b) $$\text{Periodic time } T = \frac{\text{Circumference of circle}}{\text{Linear speed}}$$

$$\therefore \quad T = \frac{2\pi r}{v} = \frac{2 \times 3.14 \times 0.4}{1.52}$$

$$= 1.65\,\mathrm{s}$$

(c) Rearranging Equation 10.4 gives

$$T = \frac{mg}{\cos\theta} = \frac{0.2 \times 10}{\cos 30°}$$

$$= 2.31\,\mathrm{N}$$

Answer

(a) $1.5\,\mathrm{m\,s^{-1}}$, (b) 1.7 s, (c) 2.3 N.

Exercise 33

1 A conical pendulum consists of a bob of mass 0.50 kg attached to a string of length 1.0 m. The bob rotates in a horizontal circle such that the angle the string makes with the vertical is 30°. Calculate (a) the period of the motion, (b) the tension in the string. Assume $g = 10\,\mathrm{m\,s^{-2}}$.

Exercise 34: Questions of A-level standard

1 A man stands at the earth's equator. Find (a) his angular velocity, (b) his linear speed, (c) his acceleration due to the rotation of the earth's axis. (1 day = 8.6×10^4 s; radius of the earth = 6.4×10^6 m.) [C 81]

2 A particle moves with constant speed in a circular path of radius r and completes one revolution in time t. Its acceleration toward the centre of the circle is

A $\dfrac{2\pi r}{t}$ **B** $\dfrac{2\pi r}{t^2}$ **C** $\dfrac{4\pi^2 r^2}{t}$

D $\dfrac{4\pi^2 r}{t^2}$ **E** $\dfrac{4\pi^2}{rt^2}$ [O & C 81]

3 An aircraft flies horizontally at $70\,\mathrm{m\,s^{-1}}$ and its propellor makes 20 revolutions each second. Find the velocity of the upper tip of the propellor, situated 0.5 m from its (horizontal) axis, when the blade is vertical.

Find at this instant the force experienced by a mass of 0.2 g attached to the propellor tip. [WJEC 82]

4 A particle of mass 0.1 kg moves in a circular path on a horizontal frictionless plane, attached freely to a point O by an inelastic string of negligible mass which can support a maximum tension of 1 N without breaking.

Find the maximum angular velocity with which the particle can circulate about O in a path of radius 1 m without breaking the string. [O & C 81, p]

5 The natural length of a spring is 20 cm, and when allowed to hang vertically with a steel ball attached to it, it is extended by 2.0 cm. The spring is now laid on a smooth horizontal surface and the free end is attached to a fixed point in the surface. The ball is then made to rotate in a horizontal circular path on the surface with a uniform angular velocity of 2π rad s^{-1}. What is the radius of the circular path? Assume $g = 10\,\text{m s}^{-2}$; $\pi^2 = 10$. [AEB 80]

6 A car of mass 1000 kg travels over a humpback bridge of radius of curvature 50 m at a constant speed of 15 m s^{-1}. Calculate the magnitude and direction of the force exerted by the car on the road when it is at the top of the bridge. Assume $g = 10\,\text{m s}^{-2}$.

7

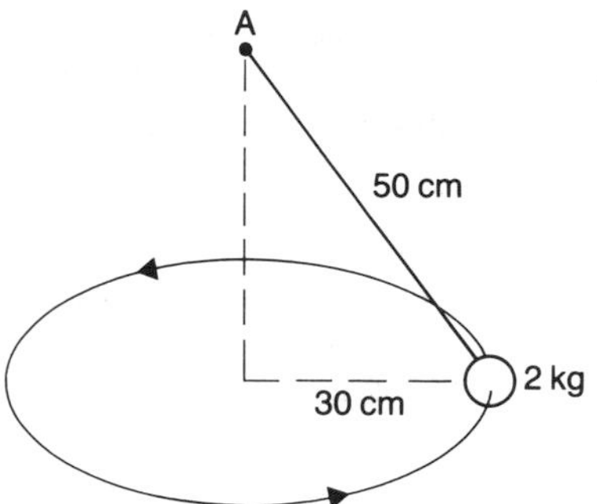

Fig 10.6 Diagram for Question 7

A mass of 2 kg, suspended from a fixed point A by a string 50 cm long, is moving in a horizontal circle of radius 30 cm at uniform speed (Fig 10.6). Calculate (a) the tension in the string, (b) the centripetal force acting on the mass. (c) Hence determine the period of the mass in its circular orbit. Assume $g = 10\,\text{m s}^{-2}$. [SUJB 81]

8 A special prototype model aeroplane of mass 400 g has a control wire 8 m long attached to its body. The other end of the control line is attached to a fixed point. When the aeroplane flies with its wings horizontal in a horizontal circle, making one revolution every 4 s, the control wire is elevated 30° above the horizontal. Draw a diagram showing the forces exerted on the plane, and determine (a) the tension in the control wire, (b) the lift on the plane. Assume that the acceleration of free fall, g, is $10\,\text{m s}^{-2}$ and π^2 is 10. [AEB 79]

11 ROTATIONAL DYNAMICS

Angular motion

A net force produces a linear acceleration such that (see Equation 7.5)

$$\text{Force } F \text{ (N)} = \text{Mass } m \text{ (kg)} \times \begin{pmatrix} \text{Linear} \\ \text{acceleration } a \\ (\text{m s}^{-2}) \end{pmatrix}$$

Similarly a net torque, or moment, produces an angular acceleration such that

$$\begin{pmatrix} \text{Torque} \\ \Gamma \text{ (N m)} \end{pmatrix} = \begin{pmatrix} \text{Moment of} \\ \text{inertia } I \\ (\text{kg m}^2) \end{pmatrix} \times \begin{pmatrix} \text{Angular} \\ \text{acceleration } \alpha \\ (\text{rad s}^{-1}) \end{pmatrix} \tag{11.1}$$

Comparing the two equations we see that Γ replaces *F*, *I* replaces *m* and α replaces *a*. Table 11.1 (p.62) lists a range of 'linear' quantities and gives their angular equivalents alongside.

Example 1

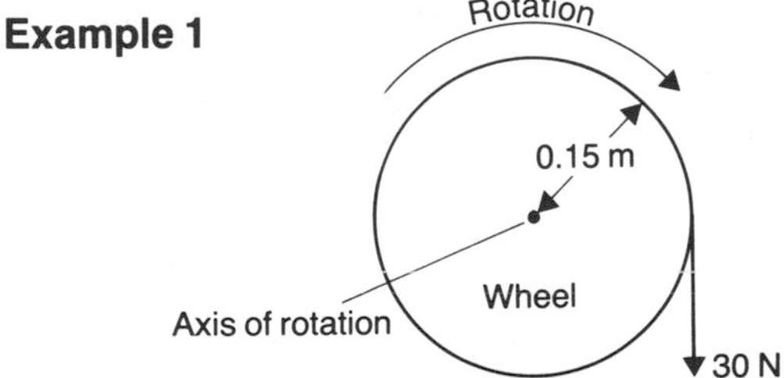

Fig 11.1 Information for Example 1

Refer to Fig 11.1. A constant tangential force of 30 N acts on a wheel of radius 0.15 m which rotates about its centre. Calculate (a) the torque acting on the wheel, (b) its angular acceleration if the moment of inertia of the wheel is 5.0 kg m². Neglect friction.

Method

(a)
$$\begin{pmatrix} \text{Torque} \\ \Gamma \text{ (N m)} \end{pmatrix} = \begin{pmatrix} \text{Force} \\ F \text{ (N)} \end{pmatrix} \times \begin{pmatrix} \text{Perpendicalar distance} \\ d \text{ (m) from axis of} \\ \text{rotation} \end{pmatrix} \tag{11.2}$$

We have $F = 30$ and $d = 0.15$, so

$$\Gamma = 30 \times 0.15 = 4.5 \text{ N m}$$

This torque causes the angular velocity of the wheel to increase in the clockwise direction, i.e. it has an angular acceleration in the clockwise direction.

(b) Equation 11.1 gives $\Gamma = I\alpha$. We have $\Gamma = 4.5$ and $I = 5.0$, so

$$\alpha = \frac{\Gamma}{I} = \frac{4.5}{5.0} = 0.90 \text{ rad s}^{-2}$$

Every second the angular velocity of the wheel increases by 0.90 rad s^{-1} in the clockwise direction.

Answer

(a) 4.5 N m, (b) 0.90 rad s^{-2}

Example 2

A flywheel on a motor increases its rate of rotation uniformly from 120 rev min^{-1} to 300 rev min^{-1} in 10 s. Calculate (a) its angular acceleration, (b) its angular displacement in this time.

Method

(a) We require initial angular velocity ω_0 and final angular velocity ω. In one revolution the angle swept out is 2π rad. So

$$\omega_0 = 120 \times 2\pi \text{ rad min}^{-1} = 4\pi \text{ rad s}^{-1}$$

$$\omega = 300 \times 2\pi \text{ rad min}^{-1} = 10\pi \text{ rad s}^{-1}$$

We have $t = 10$, so angular acceleration is

$$\alpha = \frac{\text{Change of angular velocity}}{\text{Time taken}} = \frac{\omega - \omega_0}{t}$$

$$= \frac{10\pi - 4\pi}{10} = 0.6\pi \text{ rad s}^{-2} \tag{11.3}$$

(b) Angular displacement θ is given by

$$\theta = \text{Average angular velocity} \times \text{Time}$$

$$= \tfrac{1}{2}(\omega + \omega_0) \times t \qquad (11.4)$$

$$= \tfrac{1}{2}(4\pi + 10\pi) \times 10 = 70\pi$$

Note that we can readily solve this problem using the equations of uniform angular acceleration (see below).

Answer

(a) 0.6π rad s^{-2}, (b) 70π rad.

Exercise 35

1 Calculate the required quantities:

	Γ/N m	I/kg m^2	α/rad s^{-2}
(a)	3.0	?	0.6
(b)	?	3.5	4.0
(c)	3.6	0.6	?

2 A torque of 15 N m acts on a wheel of moment of inertia 6.0 kg m^2, initially at rest. Calculate (a) its angular acceleration, (b) its angular velocity after 20 s, (c) its angular displacement in this time.

3 A flywheel of moment of inertia 0.40 kg m^2 is initally rotating at 90 rev min^{-1}. It is brought to rest in 45 s by a constant torque. Calculate (a) its initial angular velocity in rad s^{-1}, (b) its angular acceleration, (c) the magnitude of the torque, (d) its angular displacement in the first 15 s.

Equations of uniform angular acceleration

Table 11.1 lists 'linear' quantities and their 'angular' equivalents. The equations of uniform angular acceleration are

$$\omega = \omega_0 + \alpha t \qquad (11.5)$$

$$\omega^2 = \omega_0^2 + 2\alpha\theta \qquad (11.6)$$

$$\theta = \omega_0 t + \tfrac{1}{2}\alpha t^2 \qquad (11.7)$$

They can be obtained by analogy with Equations 7.1 to 7.3 or by combining Equations 11.3 and 11.4.

Example 3

A wheel is rotating initially at 90 rev min^{-1}. What torque is required to bring it to rest in 5.0 revolutions if the wheel has moment of inertia 0.80 kg m^2?

Method

We must first find α. We have

$$\omega_0 = 2\pi \times (90/60) = 3.0\pi$$

$\omega = 0$ and angular displacement $\theta = 5.0 \times 2\pi = 10\pi$.

So, from Equation 11.6

$$\omega^2 = \omega_0^2 + 2\alpha\theta$$

$$0^2 = (3\pi)^2 + 2 \times \alpha \times 10\pi$$

or $$\alpha = -0.45\pi \text{ rad s}^{-2}$$

Note the negative sign which indicates that the flywheel is slowing down (negative acceleration).

From Equation 11.1, with $I = 0.80$,

$$\Gamma = I\alpha = 0.80 \times (-0.45\pi)$$

$$= -0.36\pi \text{ N m}$$

The negative sign indicates that the torque is applied in the opposite direction to that of the original rotation.

Answer

-0.36π N m.

Table 11.1 Rotational and translational quantities

Translational quantity	*Rotational equivalent*
Force F (N)	Torque Γ (N m)
Mass m (kg)	Moment of Inertia I (kg m^2)
Acceleration a (m s^{-2})	Angular acceleration α (rad s^{-2})
Time t (s)	Time t (s)
Initial velocity u (m s^{-1})	Initial angular velocity ω_0 (rad s^{-1})
Final velocity v (m s^{-1})	Final angular velocity ω (rad s^{-1})
Distance s (m)	Angular displacement θ (rad)
Momentum mv (kg m s^{-1})	Angular momentum $I\omega$ (kg m^2 rad s^{-1})
Kinetic energy $\frac{1}{2}mv^2$ (J)	Angular kinetic energy $\frac{1}{2}I\omega^2$ (J)

Example 4

A torque of 40 N m is applied to a wheel of moment of inertia 25 kg m^2, initially at rest. Calculate (a) the time it takes to make 10 revolutions and (b) its angular velocity at that time.

Method

(a) We have $\Gamma = 40$ and $I = 25$. So

$$\alpha = \frac{\Gamma}{I} = \frac{40}{25} = 1.6 \text{ rad s}^{-2}$$

We use Equation 11.7, with $\omega_0 = 0$, $\alpha = 1.6$ and $\theta = 10$ revolutions $= 10 \times 2\pi$ rad, to find time t:

$$\theta = \omega_0 t + \tfrac{1}{2}\alpha t^2$$

$$20\pi = 0 \times t + \tfrac{1}{2} \times 1.6 \times t^2$$

$$\therefore \quad t = \sqrt{(25\pi)} = 8.9\,\text{s}$$

(b) We have $\omega_0 = 0$, $\alpha = 1.6$, $t = 8.9$ and require ω. So

$$\omega = \omega_0 + \alpha t = 0 + 1.6 \times 8.9$$
$$= 14$$

Answer

(a) 8.9 s, (b) 14 rad s^{-1}.

Example 5

A flywheel rotates on a bearing which exerts a constant frictional torque of 12 N m. An external torque of 36 N m acts on the flywheel for a time of 15 s, after which time it is removed. If the angular velocity of the flywheel increases from zero to 60 rad s^{-1} in the 15 s period, (a) calculate the moment of inertia of the flywheel, (b) find at what time the flywheel will come to rest.

Method

(a) To find I first find Γ and α. Net torque Γ is

$$\Gamma = \text{External torque} - \text{Frictional torque}$$
$$= 36 - 12 = 24\,\text{N m}$$

We have $\omega_0 = 0$, $\omega = 60$ and $t = 15$. Rearranging Equation 11.5 gives

$$\alpha = \frac{\omega - \omega_0}{t} = \frac{60 - 0}{15} = 4.0\,\text{rad s}^{-2}$$

So, since $I = \Gamma/\alpha = 24/4$, $I = 6.0\,\text{kg m}^2$.

(b) When the external torque is removed, the flywheel slows down since a net torque Γ' of -12 N m now acts on it due to friction. Since $I = 6.0$, the angular acceleration α' is given by

$$\alpha' = \frac{\Gamma'}{I} = \frac{-12}{6} = -2.0\,\text{rad s}^{-2}$$

Initial angular velocity $\omega_0' = 60$, final angular velocity (at rest) $\omega' = 0$ and $\alpha' = -2.0$. To find time t' to come to rest, we rearrange Equation 11.5:

$$t' = \frac{\omega' - \omega_0'}{\alpha} = \frac{0 - 60}{-2}$$
$$= 30\,\text{s}$$

Answer

(a) 6.0 kg m^2, (b) 30 s after the external torque is removed.

Exercise 36

1 Find α in each case, given the following:

(a) $\omega_0 = 10$, $\omega = 25$, $t = 5.0$
(b) $\omega_0 = 30$, $\omega = 5.0$, $t = 10$
(c) $\omega_0 = 30$, $\omega = 10$, $\theta = 200$
(d) $\omega_0 = 0$, $t = 5.0$, $\theta = 25$

2 Calculate the torque which must be applied to a flywheel of moment of inertia 3.0 kg m^2 if it is to be accelerated uniformly from rest to 300 rev min^{-1} in (a) 10 s, (b) 10 revolutions.

3 A flywheel of moment of inertia 20 kg m^2 is slowed down by a frictional torque of 8.0 N m. If it is initially rotating at 12 rad s^{-1}, calculate (a) the time it takes to stop, (b) the angular displacement in this time, (c) its angular velocity 15 s after it starts to slow down.

4 A torque of 15 N m is applied to a flywheel initially at rest, which then completes 5.0 revolutions in 2.0 s. Calculate (a) its angular acceleration, (b) its moment of inertia.

5 A wheel of moment of inertia 5.0 kg m^2 rotates on an axle which provides a constant frictional torque of 15 N m. (a) Calculate the external torque which must be supplied to increase its angular velocity from zero to 100 rad s^{-1} in 20 s. (b) If the external torque is removed, calculate (i) the time before it comes to rest, (ii) the angular displacement in this time.

Angular kinetic energy

The angular kinetic energy of a wheel of moment of inertia I rotating with angular velocity ω is $\frac{1}{2}I\omega^2$.

Example 6

Calculate the angular kinetic energy of a flywheel of moment of inertia 5.0 kg m^2 rotating at 120 rev min^{-1}.

Method

The rotation rate is 120 rev min^{-1} or 2 rev s^{-1}. So $\omega = 2 \times 2\pi = 4\pi$ rad s^{-1}, and since $I = 5.0$

$$\text{Angular KE} = \tfrac{1}{2}I\omega^2 = \tfrac{1}{2} \times 5 \times (4\pi)^2 = 395\,\text{J}$$

Answer

0.40 kJ.

Example 7

Calculate the kinetic energy of a cylinder of mass 12 kg and radius 0.20 m if it is rolling along a plane with a translational velocity of 0.30 m s^{-1}. The moment of inertia of the cylinder is 0.24 kg m^2.

Method

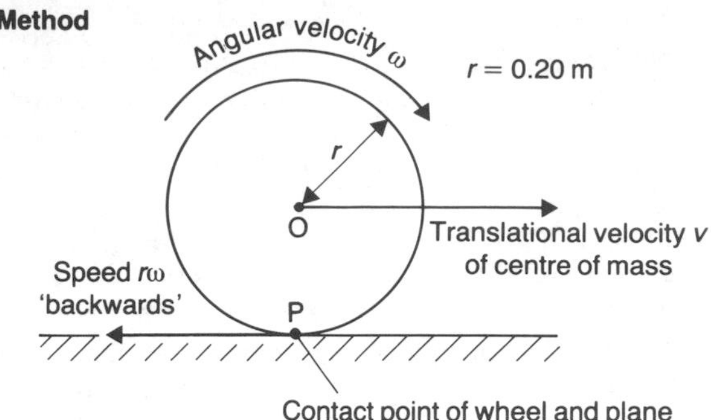

Fig 11.2 Solution to Example 7

Refer to Fig 11.2. For there to be no sliding between point P on the wheel and the plane it touches, P must be stationary at the instant of contact. This means the translational velocity v (forwards) must be cancelled by speed $r\omega$ in the opposite direction due to rotation of the wheel, i.e. $v = r\omega$, so $\omega = v/r$. Now

$$\begin{pmatrix}\text{Total}\\ \text{KE}\end{pmatrix} = \begin{pmatrix}\text{Translational}\\ \text{KE } (\frac{1}{2}mv^2)\end{pmatrix} + \begin{pmatrix}\text{Rotational}\\ \text{KE } (\frac{1}{2}I\omega^2)\end{pmatrix}$$

We have $m = 12$, $v = 0.30$, $I = 0.24$. Since $r = 0.20$ then $\omega = v/r = 0.3/0.2 = 1.5$. Hence

$$\text{Total KE} = \frac{1}{2} \times 12 \times (0.3)^2 + \frac{1}{2} \times 0.24 \times (1.5)^2$$

$$= 0.81 \text{ J}$$

Answer

0.81 J.

Exercise 37

1 A wheel possesses 200 J of angular kinetic energy and has a moment of inertia of 0.80 kg m^2. Calculate its rate of rotation (a) in rad s^{-1}, (b) in rev min^{-1}.

2 The wheels of a car rotate 8.0 times each second. Each wheel has mass 15 kg, radius 0.30 m and moment of inertia 0.27 kg m^2. Calculate (a) the translational speed of the car, (b) the total KE of the four wheels (combined).

Work and rotational energy

When work is done to make a body rotate, all of the work done will become rotational KE if the body is constrained to rotate about a fixed point and if friction can be neglected (see Example 8). Alternatively some of the work done can become translational KE and/or gravitational PE (see Example 9).

Example 8

A torque of 8.0 N m is applied to a flywheel, initially at rest, for 15 s. If the flywheel has moment of inertia 2.0 kg m^2, calculate (a) the angular velocity acquired, (b) the kinetic energy acquired, (c) the work done by the torque. Comment on the magnitude of (b) and (c). Neglect friction.

Method

(a) We have $\Gamma = 8.0$ and $I = 2.0$.
Thus $\alpha = \Gamma/I = 4.0$ rad s^{-2}. Now

$$\omega = \omega_0 + \alpha t$$

Since $\omega_0 = 0$, $\alpha = 4.0$ and $t = 15$, $\omega = 60$ rad s^{-1}.

(b) Since $I = 2.0$ and $\omega = 60$,

$$\text{KE} = \tfrac{1}{2}I\omega^2 = 3600 \text{ J}$$

(c)
$$\begin{pmatrix}\text{Work done}\\ \text{by torque (J)}\end{pmatrix} = \begin{pmatrix}\text{Torque}\\ \Gamma \text{ (N m)}\end{pmatrix} \times \begin{pmatrix}\text{Angular}\\ \text{displacement}\\ \theta \text{ (rad)}\end{pmatrix} \quad (11.8)$$

Using Equation 11.7 with $\omega_0 = 0$, $t = 15$ and $\alpha = 4.0$

$$\theta = \omega_0 t + \tfrac{1}{2}\alpha t^2 = 450 \text{ rad}$$

$$\therefore \quad \text{Work done} = \Gamma \times \theta = 8 \times 450$$

$$= 3600 \text{ J}$$

Note the answers to (b) and (c) are the same since all the work done by the couple becomes rotational KE.

Answer

(a) 60 rad s^{-1}, (b) 3600 J, (c) 3600 J

Example 9

A wheel of mass 4.0 kg is pulled up a plane, inclined at 30° to the horizontal, by a force of 45 N applied to the axle and parallel to the plane. If the wheel has radius 0.50 m and moment of inertia 0.50 kg m^2, calculate the translational velocity acquired after travelling 12 m up the plane, assuming the wheel is initially at rest. Take $g = 10$ m s^{-2}.

Method

Refer to Fig 11.3. Now

$$\begin{pmatrix}\text{Work}\\ \text{done by}\\ \text{force}\end{pmatrix} \xrightarrow{\text{becomes}} \begin{pmatrix}\text{Gravita-}\\ \text{tional PE}\end{pmatrix} + \begin{pmatrix}\text{Transla-}\\ \text{tional}\\ \text{KE}\end{pmatrix} + \begin{pmatrix}\text{An-}\\ \text{gular}\\ \text{KE}\end{pmatrix}$$

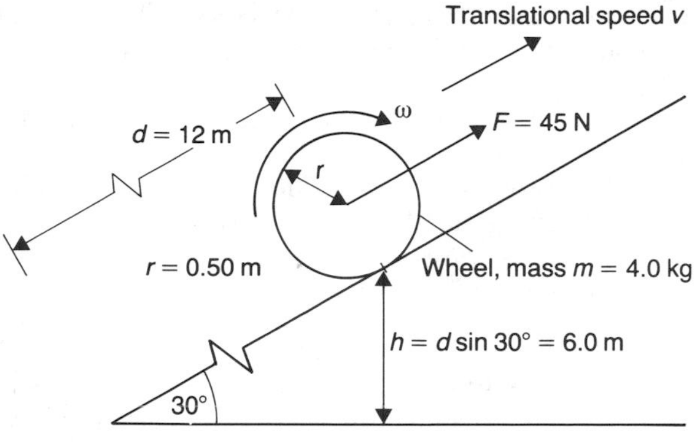

Fig 11.3 **Solution to Example 9**

We have $F = 45$, $d = 12$, $m = 4.0$, $g = 10$, $h = 6$ amd $I = 0.50$.

As $v = r\omega$ (see Example 7), $\omega = v/r = v/0.50$. Hence Equation 11.9 gives

$$45 \times 12 = (4 \times 10 \times 6) + (\tfrac{1}{2} \times 4 \times v^2) + \left(\tfrac{1}{2} \times 0.5 \times \frac{v^2}{0.25}\right)$$

$$\therefore \quad v = 10\,\text{m s}^{-1}$$

Answer

Translational speed $= 10\,\text{m s}^{-1}$.

Note that the translational speed would be greater if the wheel was dragged up the (smooth) incline on its side since there is no energy used to produce rotation.

Exercise 38

1

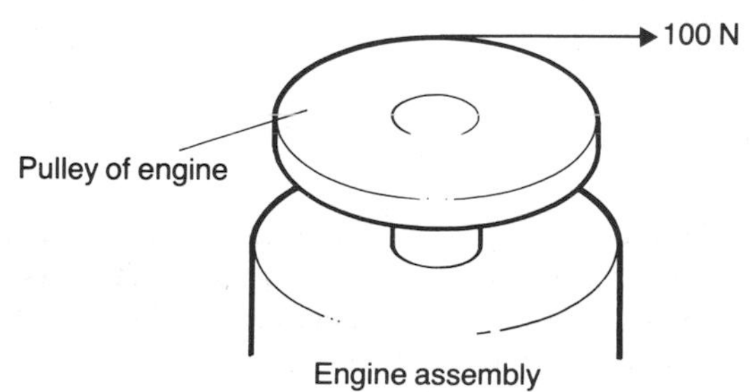

Fig 11.4 Information for Question 1

Refer to Fig 11.4. The engine of a lawn mower is turned over by applying a constant force of 100 N to a string wrapped round the pulley. If the moment of inertia of the lawn mower engine is $1.6\,\text{kg m}^2$, and the string is pulled out by 0.60 m, calculate (a) the work done by the force, (b) the angular velocity acquired by the pulley. Neglect friction.

2 A wheel of mass 12 kg, radius 0.20 m and moment of inertia $0.30\,\text{kg m}^2$ is pulled up a plane, inclined at 30° to the horizontal, by a force of 100 N applied to the axle and parallel to the plane. After what distance up the plane will it acquire a translational velocity of $5.0\,\text{m s}^{-1}$, assuming it starts from rest? What is its angular velocity at this point? Assume $g = 10\,\text{m s}^{-2}$.

Conservation of angular momentum

Provided that no external torque acts, angular momentum is conserved. This is the case, for example, when a spinning skater draws her arms in.

Example 10

A skater is turning at $3.0\,\text{rad s}^{-1}$ with both arms outstretched, so that she has a moment of inertia of $4.0\,\text{kg m}^2$. Her arms are now drawn in, so that her new moment of inertia is $1.8\,\text{kg m}^2$. Calculate (a) her final angular velocity, (b) the increase in her angular KE.

Method

(a) Angular momentum is conserved, so

$$I_0\omega_0 = I\omega \qquad (11.10)$$

We have $I_0 = 4.0$, $\omega_0 = 3.0$ and $I = 1.8$. Equation 11.10 gives $\omega = 6.67$.

(b) Original KE $= \frac{1}{2}I_0\omega_0^2 = \frac{1}{2} \times 4 \times 3^2 = 18$ J

Final KE $= \frac{1}{2}I\omega^2 = \frac{1}{2} \times 1.8 \times 6.67^2 = 40.0$ J

$\therefore$ Increase in angular KE $= 22$ J

This arises because of the work done by the skater as she pulls her arms in.

Answer

(a) $6.7\,\text{rad s}^{-1}$, (b) 22 J.

Exercise 39

1 A skater is rotating at 2.0 rad s^{-1} with both arms outstretched, when her moment of inertia is 4.5 kg m^2. She now pulls in her arms and rotates at 8.0 rad s^{-1}. Calculate (a) her new moment of inertia, (b) the increase in her angular KE.

2 A turntable is rotating freely about an axis with an angular velocity of 6.0 rad s^{-1} and has a moment of inertia of 1.5 kg m^2. A rough disc is gently dropped on to the turntable so that the centres coincide. Eventually the combined turntable and disc rotate at 4.5 rad s^{-1}.* Calculate (a) the moment of inertia of the disc about the rotation axis, (b) the original angular KE of the turntable, (c) the final angular KE of the combination, (d) the angular KE 'lost'. (Note: this is due to work done against friction as the angular velocity of the disc increases from zero to 4.5 rad s^{-1}.)

Exercise 40: Questions of A-level standard

(Assume g = 10 m s^{-2}.)

1 Airline baggage is delivered to a carousel, which is a horizontal circular platform that rotates about a vertical axis through its centre. When the motor is switched on, it takes 0.50 s for the unloaded carousel, of moment of inertia 9000 kg m^2, to acquire its steady angular velocity of 0.20 rad s^{-1}.

(a) Find the average angular acceleration of the carousel during this period.

(b) Hence calculate the average torque causing this acceleration.

Luggage of mass 50 kg is loaded on to the carousel at a point 3.0 m from the axis. If the motor were disengaged and the platform could revolve freely, explain why this addition of luggage would cause the carousel to rotate at a slower rate; find this new angular velocity. [C 82, p]

Combined moment of inertia = Sum of separate moments of inertia.

2 A flywheel of moment of inertia 80 kg m^2 and rotating at 6000 rad s^{-1} was braked uniformly; it stopped in 30 s.

(a) What was its angular acceleration while the brake was applied?

(b) How many revolutions did it make in this time?

(c) What torque did the brake supply? [C 81]

3 A flywheel has moment of inertia 0.40 kg m^2 about an axis through its centre and perpendicular to the plane of the wheel, and is spinning at 50 revolutions per minute about this axis. What constant braking torque is required to bring it to rest in 20 revolutions? [AEB 80, p]

4 A flywheel of radius 0.250 m is mounted on a horizontal axle of radius 0.015 m. The moment of inertia of the system about its axis of rotation is 0.225 kg m^2, and the frictional couple at the bearings is negligible. A constant force of 60 N is applied tangentially to the axle for 4.0 s starting from rest. Calculate

(a) the angular acceleration during the first 4.0 s;

(b) the angular velocity and KE after 4.0 s;

(c) the constant tangential braking force which must then be applied to the flywheel to bring it to rest in 10 revolutions. [JMB 82, p]

5 A flywheel of mass M and moment of inertia I is initially rotating at n revolutions per second. In coming to rest the amount of work done is

A $\frac{1}{2}Mn^2$ **B** $\frac{1}{2}In^2$ **C** $\frac{1}{2}\pi^2In^2$
D $2\pi^2\dfrac{In^2}{M}$ **E** $2\pi^2In^2$ [AEB 80]

6 The moment of inertia of a sphere about an axis through its centre is $2mr^2/5$, where m is its mass and r is its radius. When a sphere rolls on a surface with velocity v, without sliding, the ratio of its translational KE to its rotational KE is

A $2r:5$ **B** $5r:2$ **C** $2m:5$
D $5:2$ **E** $2:5$ [L 80]

7

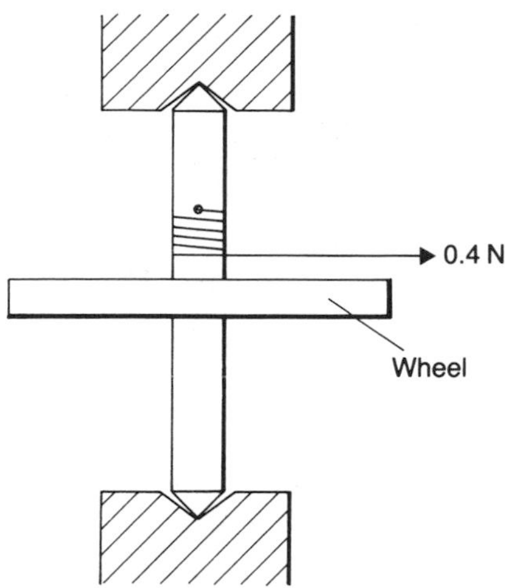

Fig 11.5 Diagram for Question 7

Fig 11.5 represents the rotor and bearings of a small gyroscope. The wheel is set rotating by pulling a fine cord which is extended at right angles to the axle with a steady tension of 0.4 N; the cord detaches itself from the axle when it has been pulled through a distance of 0.6 m. The moment of inertia of the wheel and axle about its axis of rotation is 5×10^{-4} kg m^2. Assuming that friction is negligible, calculate
(a) the kinetic energy gained by the wheel and axle;
(b) its final angular velocity;
(c) its final angular momentum.
Supposing that the wheel is now retarded by a constant couple, and that it continues to spin for a time of 100 s before coming to rest, find
(d) the magnitude of this couple;
(e) the number of revolutions the wheel makes during this stage. [O 80, p]

8 A flywheel of moment of inertia 5.0 kg m^2 is rotated at a steady speed of 20 revolutions per second by an electric motor which develops 50 W of mechanical power. What is
(a) the kinetic energy of the flywheel;
(b) the frictional couple that the motor has to work against to maintain a steady speed?
[SUJB 81]

9 A long but light string is attached to, and wrapped many times round, a stationary flywheel mounted with its axis horizontal. A mass of 20 kg is attached to the free end of the taut string, and is released from rest to fall 50 m down a pit shaft. When the mass arrives at the bottom of the pit with velocity 30 m s^{-1}, the flywheel is found to have acquired an angular velocity of 4π rad s^{-1}. Assuming no energy losses, find the moment of inertia of the flywheel about its axis. [WJEC 81]

10 The moment of inertia of the earth about its axis of rotation is 8.0×10^{37} kg m^2. Estimate (a) the earth's angular momentum and (b) the earth's angular kinetic energy.

Due to the frictional effect of tides on the ocean bed the length of the earth's day is very slowly increasing. What effect does this have on the kinetic energy and the angular momentum of the earth? [L 82]

11 A stationary horizontal hoop of mass 0.05 kg and mean radius of 0.2 m is dropped from a small height centrally and symmetrically on to a gramophone turntable which was originally rotating freely with an angular velocity of 3.5 rad s^{-1}. Eventually the combined turntable and hoop rotate together with an angular velocity of 1.5 rad s^{-1}. Find the moment of inertia of the turntable about its rotation axis. What was the original rotational kinetic energy of the turntable? Account for any loss of kinetic energy that has occurred.
[WJEC 79, p]

12 GRAVITATION

Gravitational force

Bodies attract each other solely as a result of the matter they contain. The gravitational force F (N) between two particles m_1 (kg) and m_2 (kg) placed distance r (m) apart is given by

$$F = \frac{Gm_1m_2}{r^2} \qquad (12.1)$$

where G is the universal gravitational constant and has value $6.7 \times 10^{-11}\,\mathrm{N\,m^2\,kg^{-2}}$.

Example 1

Calculate the gravitational attraction force between bodies of mass 3.0 kg and 2.0 kg placed with their centres 50 cm apart.

Method

We assume that the bodies are uniform spheres so they act, for this purpose, as if they are point masses (particles) located at their centres. We have $G = 6.7 \times 10^{-11}$, $m_1 = 3.0$, $m_2 = 2.0$ and $r = 0.50$. From Equation 12.1

$$F = \frac{Gm_1m_2}{r^2} = \frac{6.7 \times 10^{-11} \times 3 \times 2}{(0.5)^2}$$

$$= 1.6 \times 10^{-9}\,\mathrm{N}$$

This is a very small force. To get an appreciable force one or both of the objects must be very large. Our weight* is the result of the gravitational attraction force from the earth.

Answer

1.6×10^{-9} N.

**We neglect effects due to the earth's rotation which make a difference of about 0.3%.*

Example 2

Two 'particles' of mass 0.20 kg and 0.30 kg are placed 0.15 m apart. A third particle of mass 0.050 kg is placed between them on the line joining the first two particles. Calculate (a) the gravitational force acting on the third particle if it is placed 0.050 m from the 0.30 kg mass and (b) where along the line it should be placed for no gravitational force to be exerted on it.

Method

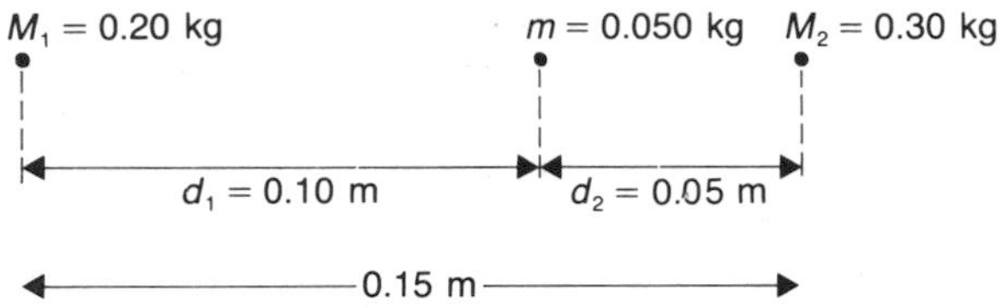

Fig 12.1 Solution to Example 2

Refer to Fig 12.1

(a) Both masses M_1 and M_2 attract m. Using Equation 12.1, we have for mass M_1 an attractive force F_1 (towards M_1) given by

$$F_1 = \frac{GM_1m}{d_1^{\,2}} = \frac{6.7 \times 10^{-11} \times 0.2 \times 0.05}{(0.1)^2}$$

$$= 6.7 \times 10^{-11}\,\mathrm{N}$$

For mass M_2 an attractive force F_2 (towards M_2) exists given by

$$F_2 = \frac{GM_2m}{d_2^2} = \frac{6.7 \times 10^{-11} \times 0.3 \times 0.05}{(0.05)^2}$$

$$= 40.2 \times 10^{-11}\,\mathrm{N}$$

Thus the net force F (towards M_2) is given by

$$F = F_2 - F_1 = 33.5 \times 10^{-11}\,\mathrm{N}$$

(b) Suppose mass m is x from M_1 and thus $(0.15 - x)$ from M_2. Then

$$F_1 = \frac{GM_1m}{x^2} \quad \text{and} \quad F_2 = \frac{GM_2m}{(0.15 - x)^2}$$

For no gravitational force to act on mass m, $F_1 = F_2$. Thus

$$\frac{GM_1m}{x^2} = \frac{GM_2m}{(0.15 - x)^2} \tag{12.2}$$

Note that G and m cancel out, so that d is independent of m. Substituting $M_1 = 0.2$ and $M_2 = 0.3$ into Equation 12.2 gives

$$\frac{0.2}{x^2} = \frac{0.3}{(0.15 - x)^2}$$

Taking square roots and cross multiplying gives

$$\sqrt{2} \times (0.15 - x) = \sqrt{3} \times x$$

This gives $x = 0.067$ m.

Answer

(a) 34×10^{-11} N, (b) 0.067 m from M_1 (0.20 kg).

Exercise 41

(Assume $G = 6.7 \times 10^{-11}\,\text{N m}^2\,\text{kg}^{-2}$.)

1 Calculate the gravitational attraction force between two 'particles', each of mass 20 kg, placed 1.0 m apart.

2 Consider the earth as a uniform sphere of radius 6.4×10^6 m and mass 6.0×10^{24} kg. Find the gravitational force on a mass of 5.0 kg placed on the surface of the earth. (Assume the earth can be replaced by a point mass acting at its centre.) Compare this with the weight of a 5.0 kg mass on earth.

3 Two small spheres of mass 4.0 kg and M kg are placed 80 cm apart. If the gravitational force is zero at a point 20 cm from the 4 kg mass along the line between the two masses, calculate the value of M.

4 The mass of the earth is 6.0×10^{24} kg and that of the moon is 7.4×10^{22} kg. If the distance between their centres is 3.8×10^8 m, calculate at what point on the line joining their centres there is no gravitational force. Neglect the effect of other planets and the sun.

Gravitational field strength

The gravitational field strength g (N kg^{-1}) is defined as the gravitational force acting on unit mass placed at the point in question. It equals the acceleration due to gravity g (m s^{-2}) at this point.

Example 3

Assuming that the earth is a uniform sphere of radius 6.4×10^6 m and mass 6.0×10^{24} kg, find the gravitational field strength g at a point (a) on the surface, (b) at height 0.50 times its radius above the earth's surface.

Method

(a) We assume that the earth can be replaced by a point mass acting at its centre. Then in Equation 12.1 $F = g$ if $m_1 = 1$. If M is the mass of the planet,

$$g = \frac{GM}{r^2} \tag{12.3}$$

This is a general expression.
We have $G = 6.7 \times 10^{-11}$, $M = 6.0 \times 10^{24}$ and $r = 6.4 \times 10^6$.
Substituting in Equation 12.3 gives $g = 9.8\,\text{N kg}^{-1}$.
Note: this equals the acceleration due to gravity at the earth's surface.

(b) We now have distance $r_1 = 1.5r$. Equation 12.3 tells us $g \propto 1/(\text{distance})^2$. If g_1 is the new value, then

$$\frac{g_1}{g} = \frac{r^2}{r_1^2} = \frac{r^2}{(1.5r)^2} = 0.444$$

$$g_1 = 0.444g = 4.36\,\text{N kg}^{-1}$$

Answer

(a) $9.8\,\text{N kg}^{-1}$, (b) $4.4\,\text{N kg}^{-1}$.

Example 4

The acceleration due to gravity at the earth's surface is $9.8\,\text{m s}^{-2}$. Calculate the acceleration due to gravity on a planet which has (a) the same mass and twice the density, (b) the same density and twice the radius.

Method

Acceleration due to gravity equals the gravitational field strength g. Equation 12.3 tells us that g depends on mass M and radius r of the planet.

(a) In this case the radius r_1 of the planet differs from earth radius r. Let the density of earth be ρ and of the planet be 2ρ. Since both have the same mass M,

$$M = \underbrace{\tfrac{4}{3}\pi r^3\rho}_{\text{for earth}} = \underbrace{\tfrac{4}{3}\pi r_1^3 \times 2\rho}_{\text{for planet}}$$

or $$r^3 = 2r_1^3 \quad \text{giving} \quad \frac{r}{r_1} = 2^{1/3}$$

From Equation 12.3 we see that $g \propto 1/r^2$ for two planets of the same mass. So, if g_1 is the gravitational strength on the planet,

$$\frac{g_1}{g} = \frac{r^2}{r_1^2} = (2)^{2/3}$$

since $r = (2)^{1/3}r_1$. As $g = 9.8$,

$$g_1 = (2)^{2/3} \times 9.8 = 15.6$$

(b) The new planet has radius $2r$. Let its mass be M_2. It has density ρ, therefore

$$M_2 = \tfrac{4}{3}\pi(2r)^3\rho = 8M$$

since $M = \frac{4}{3}\pi r^3\rho$. From Equation 12.3 we see that $g \propto M/r^2$. If g_2 is the gravitational field strength on the planet,

$$\frac{g_2}{g} = \frac{M_2}{(2r)^2} \div \frac{M}{r^2} = 2$$

since $M_2 = 8M$. Thus $g_2 = 2g = 19.6$.

Answer

(a) $15.6\ \mathrm{m\,s^{-2}}$, (b) $19.6\ \mathrm{m\,s^{-2}}$.

Exercise 42

(Assume $G = 6.7 \times 10^{-11}\ \mathrm{N\,m^2\,kg^{-2}}$.)

1 The gravitational field strength on the surface of the moon is $1.7\ \mathrm{N\,kg^{-1}}$. Assuming that the moon is a uniform sphere of radius 1.7×10^6 m, calculate (a) the mass of the moon, (b) the gravitational field strength 1.0×10^6 m above its surface.

2 The acceleration due to gravity at the earth's surface is $9.8\ \mathrm{m\,s^{-2}}$. Calculate the acceleration due to gravity on a planet which has (a) the same mass and twice the radius, (b) the same radius and twice the density, (c) half the radius and twice the density.

3 If the earth has radius r and the acceleration due to gravity at its surface is $9.8\ \mathrm{m\,s^{-2}}$, calculate the acceleration due to gravity at a point that is distance r above the surface of a planet with half the radius and the same density as the earth.

Gravitational potential and escape speed

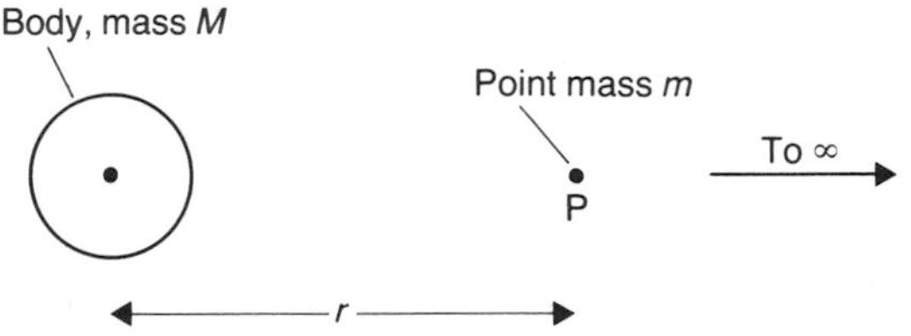

Fig 12.2 Gravitational potential at P

Refer to Fig 12.2. The gravitational potential U at point P due to the gravitational attractive force of mass M is given by

$$U = -\frac{GM}{r} \qquad (12.4)$$

The negative sign indicates that work must be done to take a mass from P to infinity (where the potential is zero). U is the work done per kg.

Example 5

Assuming that the earth is a uniform sphere of radius 6.4×10^6 m and mass 6.0×10^{24} kg, calculate (a) the gravitational potential at (i) the earth's surface and (ii) a point 6.0×10^5 m above the earth's surface, (b) the work done in taking a 5.0 kg mass from the earth's surface to a point 6.0×10^5 m above it, (c) the work done in taking a 5.0 kg mass from the earth's surface to a point where the earth's gravitational effect is negligible.

Method

(a) We use Equation 12.4 in which $G = 6.7 \times 10^{-11}$ and $M = 6.0 \times 10^{24}$.

(i) We have $r = r_1 = 6.4 \times 10^6$. So, if U_1 is the potential here,

$$U_1 = \frac{-GM}{r_1} = \frac{-6.7 \times 10^{-11} \times 6.0 \times 10^{24}}{6.4 \times 10^6}$$

$$= -6.28 \times 10^7\ \mathrm{J\,kg^{-1}}$$

(ii) We have $r = r_2 = (6.4 + 0.6) \times 10^6$ m. If U_2 is the potential at r_2, Equation 12.4 gives $U_2 = -5.74 \times 10^7$ J kg^{-1}.

(b) *The work required W, per kg,* is the difference in gravitational potential, so

$$W = U_2 - U_1 = 0.54 \times 10^7 \text{ J}$$

Note: we subtract U_1 from U_2 since there is an *increase* in gravitational potential as we move away from the earth. For a 5.0 kg mass we require $5.0 \times 0.54 \times 10^7 = 2.7 \times 10^7$ J. (We cannot use the simple form *mgh* to calculate work required, since *g* changes appreciably between the two points.)

(c) The work required W', per kg, is given by

$$W' = \text{Potential at } \infty - \text{Potential at earth's surface}$$
$$= 0 - (-6.28 \times 10^7)$$
$$= 6.28 \times 10^7 \text{ J}$$

For a 5.0 kg mass the work required is

$$5 \times 6.28 \times 10^7 = 31.4 \times 10^7 \text{ J}$$

Answer
(a) (i) -6.3×10^7 J kg^{-1} and (ii) -5.7×10^7 J kg^{-1}, (b) 2.7×10^7 J, (c) 31×10^7 J.

Example 6
Calculate the minimum speed which a body must have to escape from the moon's gravitational field, given that the moon has mass 7.7×10^{22} kg and radius 1.7×10^6 m.

Method
As the body moves away from the moon's surface, its kinetic energy decreases because its gravitational potential increases. Referring to Fig 12.2 we see that the work required to take a body of mass *m* from P to infinity is *GMm*/*r*. Suppose the body has speed *v* at point P, then it will have just enough kinetic energy to escape, provided that

$$\tfrac{1}{2}mv^2 = \frac{GMm}{r}$$

or

$$v = \sqrt{\frac{2GM}{r}} \qquad (12.5)$$

We have $G = 6.7 \times 10^{-11}$, $M = 7.7 \times 10^{22}$ and $r = 1.7 \times 10^6$.

Substituting into Equation 12.5 gives
$v = 2.46 \times 10^3$ m s^{-1}.

Answer
Escape speed $= 2.5 \times 10^3$ m s^{-1}.

Exercise 43

(Assume $G = 6.7 \times 10^{-11}$ N m^2 kg^{-2}.)

1 The gravitational potential difference between two points is 3.0×10^3 J kg^{-1}. Calculate the work done in moving a mass of 4.0 kg between the two points.

2 The moon has mass 7.7×10^{22} kg and radius 1.7×10^6 m. Calculate (a) the gravitational potential at its surface and (b) the work needed to completely remove a 1.5×10^3 kg space craft from its surface into outer space. Neglect the effect of the earth, planets, sun, etc.

3 A planet has radius 5.0×10^5 m and mean density 3.0×10^3 kg m^{-3}. Calculate the escape speed of bodies on its surface.

4 A neutron star has radius 10 km and mass 2.5×10^{29} kg. A meteorite is drawn into its gravitational field. Calculate the speed with which it will strike the surface of the star. Neglect the initial speed of the meteorite.

Exercise 44: Questions of A-level standard

(Assume $G = 6.7 \times 10^{-11}$ N m^2 kg^{-2}.)

1

P and Q represent the centres of two small spheres of masses *m* and 4*m* respectively. The gravitational field strengths due to the two spheres at R are equal in magnitude. The value of *x*/*y* is

A $\frac{1}{16}$ **B** $\frac{1}{4}$ **C** $\frac{1}{2}$ **D** 2 **E** 4

[AEB 82]

2 Two stationary particles of mass M_1 and M_2 are a distance *d* apart. A third particle experiences no gravitational force if it lies on the line joining M_1 and M_2 and at a distance from M_1 of

A $\dfrac{M_2}{M_1}d$ **B** $d\sqrt{\dfrac{M_1}{M_2}}$ **C** $d\sqrt{\dfrac{M_1}{M_1 + M_2}}$

D $\dfrac{M_1 d}{M_1 + M_2}$ **E** $\dfrac{\sqrt{(M_1)}d}{\sqrt{(M_1)} + \sqrt{(M_2)}}$

[O & C 81]

3 At a point outside the earth at a distance x from its centre the intensity of the earth's gravitational field is $5\,\mathrm{N\,kg^{-1}}$. At the earth's surface the intensity is $10\,\mathrm{N\,kg^{-1}}$. What is an approximate value for the radius of the earth?

A $\frac{1}{10}x$ **B** $\frac{1}{5}x$ **C** $\frac{1}{2}x$
D $\left(\frac{1}{\sqrt{2}}\right)x$ **E** $(\sqrt{2})x$ [O 82]

4 A neutron star has a mass of 2.0×10^{30} kg and a diameter of 20 km.
(a) What is the mean density of the star?
(b) Calculate the gravitational field strength at the surface. [O 82, p]

5 The mass and diameter of a planet are both four times those of the earth. Assuming that each body is a sphere of uniform density, then the ratio of the acceleration of free fall on the planet's surface to that on the earth's surface is

A 16:1 **B** 4:1 **C** 1:1
D 1:4 **E** 1:16 [AEB 82]

6 The acceleration due to gravity at the surface of the earth is g. At the surface of a planet of the same mass as the earth, but of twice the density, the acceleration due to gravity would be

A $2^{-2/3}g$ **B** $2^{-1/3}g$ **C** g
D $2^{1/3}g$ **E** $2^{2/3}g$ [O & C 82]

7

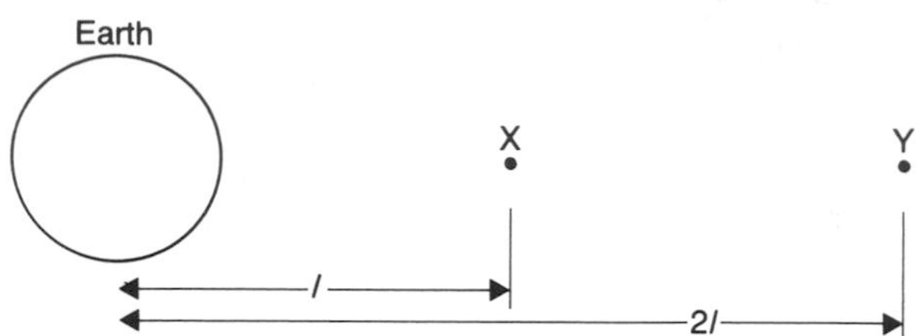

Fig 12.3 Diagram for Question 7

Fig 12.3 represents two points X and Y at distances l and $2l$ from the centre of the earth. The gravitational potential at X is $-8\,\mathrm{kJ\,kg^{-1}}$. When a 1 kg mass is taken from X to Y the work done on the mass is

A −4 kJ **B** −2 kJ **C** +2 kJ
D +4 kJ **E** +8 kJ [O 80]

Section C
Matter

13 ELASTICITY

Stress and strain

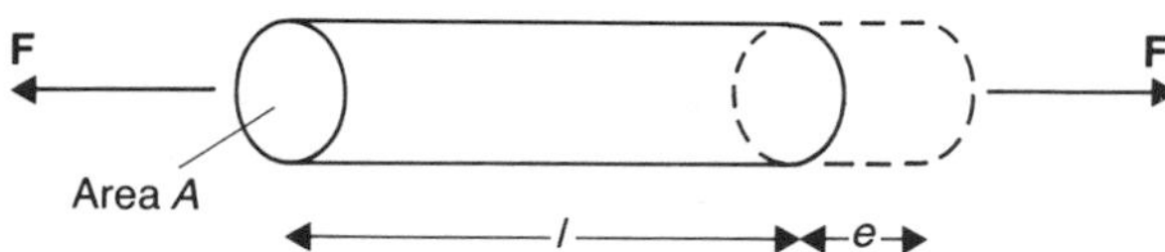

Fig 13.1 A solid specimen under tension

Refer to Fig 13.1 in which a specimen of original length l (m) and cross-sectional area A (m^2) is subjected to a tensile force F (N), so that its extension is e (m). We define

$$\text{Tensile stress } \sigma = \frac{F}{A} (\text{N m}^{-2} \text{ or Pa})^* \qquad (13.1)$$

$$\text{Tensile strain } \epsilon = \frac{e}{l} \text{ (no units)} \qquad (13.2)$$

Example 1

A metal bar is of length 2.0 m and has a square cross-section of side 40 mm. When a tensile force of 80 kN is applied, it extends by 0.046 mm. Calculate (a) the stress, (b) the strain in the specimen.

Method

We have $l = 2.0$, $A = (40 \times 10^{-3})^2 = 16 \times 10^{-4}$, $F = 80 \times 10^3$ and $e = 0.046 \times 10^{-3}$. So Equations 13.1 and 13.2 give

(a) $$\sigma = \frac{F}{A} = \frac{80 \times 10^3}{16 \times 10^{-4}} = 5.0 \times 10^7 \text{ N m}^{-2}$$

(b) $$\epsilon = \frac{e}{l} = \frac{0.046 \times 10^{-3}}{2.0} = 2.3 \times 10^{-5}$$

Answer

(a) 5.0×10^7 N m^{-2}, (b) 2.3×10^{-5}.

*$1 \text{ N m}^{-2} = 1$ Pa (pascal).

Exercise 45

1 A metal bar has circular cross-section of diameter 20 mm. If the maximum permissible tensile stress is 80 MN m^{-2} (80×10^6 N m^{-2}), calculate the maximum force which the bar can withstand.

2 A metal specimen has length 0.50 m. If the maximum permissible strain is not to exceed 0.10% (1.0×10^{-3}), calculate its maximum extension.

3 A metal bar of length 50 mm and square cross-section of side 20 mm is extended by 0.015 mm under a tensile load of 30 kN. Calculate (a) the stress, (b) the strain in the specimen.

Young's modulus

Up to a certain load, called the limit of proportionality,* extension is proportional to applied force, so that strain is proportional to stress. The slope of the stress–strain graph in the linear region is called Young's modulus E. So we define

$$E = \frac{\sigma}{\epsilon} \quad (\text{N m}^{-2}) \qquad (13.3)$$

Example 2

A steel bar is of length 0.50 m and has a rectangular cross-section 15 mm by 30 mm. If a tensile force of 36 kN produces an extension of 0.20 mm, calculate Young's modulus for steel. Assume that the limit of proportionality is not exceeded.

*Sometimes no distinction is made between this and the elastic limit.

Method

From Equations 13.3, 13.1 and 13.2

$$E = \frac{\text{Stress}}{\text{Strain}} = \frac{(\text{Force} \div \text{Area})}{(\text{Extension} \div \text{Original length})} \quad (13.4)$$

We have Force $= 36 \times 10^3$ N

Area $= 15 \times 30 = 450\ \text{mm}^2 = 450 \times 10^{-6}\ \text{m}^2$

Extension $= 0.20$ mm $= 0.20 \times 10^{-3}$ m

Original length $= 0.50$ m

So Equation 13.4 gives

$$E = \frac{(36 \times 10^3) \div (450 \times 10^{-6})}{(0.2 \times 10^{-3}) \div 0.5}$$
$$= 2.0 \times 10^{11}\ \text{N m}^{-2}$$

Answer

Young's modulus for steel $= 2.0 \times 10^{11}$ N m^{-2}.

Example 3

An aluminium alloy strut in the landing gear of an aircraft has a cross-sectional area of 60 mm^2 and a length of 0.45 m. During landing the strut is subjected to a compressive force of 3.6 kN. Calculate by how much the strut will shorten under this force. Assume that Young's modulus for the alloy is 90 GN m^{-2} and the proportional limit is not exceeded.

Method

Equation 13.4 gives

$$E = \frac{F \div A}{e \div l}$$

In this case the strut is compressed. Since materials in general have the same value for the elastic modulus in tension as in compression, it is necessary only to replace extension e in the above equation by compression c.

We have $A = 60\ \text{mm}^2 = 60 \times 10^{-6}\ \text{m}^2$, $l = 0.45$, $F = 3.6 \times 10^3$, $E = 90 \times 10^9$, and require the compression c. So

$$90 \times 10^9 = \frac{(3.6 \times 10^3) \div (60 \times 10^{-6})}{(c \div 0.45)}$$

Rearranging gives $c = 0.30 \times 10^{-3}$ m.

Answer

The strut shortens by 0.30 mm.

Exercise 46

(Assume that the proportional limit is not exceeded.)

1 A vertical copper wire is 1.0 m long and has radius 1.0 mm. A load of 180 N is attached to the bottom end and produces an extension of 0.45 mm. Calculate (a) the tensile stress, (b) the tensile strain, (c) the value of Young's modulus for copper.

2 A steel strut has a cross-sectional area of 25×10^3 mm^2 and is 2.0 m long. Calculate the magnitude of the compressive force which will cause it to shorten by 0.30 mm. Assume that E for steel is 200 GN m^{-2}.

3 A bronze wire of length 1.5 m and radius 1.0 mm is joined end-to-end to a steel wire of identical size to form a wire 3.0 m long. Calculate (a) the resultant extension if a force of 200 N is applied, (b) the force required to produce an extension of 0.30 mm. Assume that E for bronze is 1.0×10^{11} N m^{-2}; for steel 2.0×10^{11} N m^{-2}.
Hint: (a) total force acts on each wire, extension equals the sum of extensions, (b) $e \propto 1/E$ for each wire, or use $e \propto F$.

Work done in stretching a specimen

Work is done on a specimen when it is extended (or compressed). The work done is equal to the area under the force–extension (or compression) graph. Within the proportional limit

$$\text{Work done (J)} = \begin{pmatrix}\text{Area under force–} \\ \text{extension graph}\end{pmatrix}$$
$$= \tfrac{1}{2}F \times e \quad (13.5)$$

where F (N) is the force required to produce an extension e (m).

The work done becomes potential energy stored within the specimen, termed strain energy. Up to the elastic limit this energy is recoverable.

Example 4

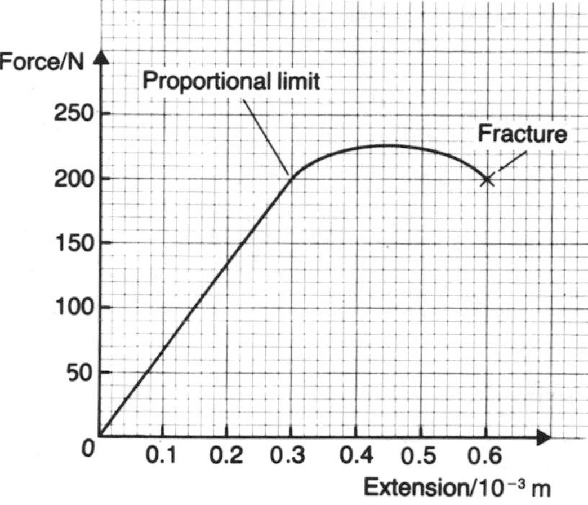

Fig 13.2 Information for Example 4

Fig 13.2 shows a force–extension graph for a metal specimen. Calculate the work done in stretching the specimen up to (a) the proportional limit, (b) fracture.

Method

Work done = Area under force–extension graph, where force is in newtons and extension in metres.

(a) Area under linear portion of graph

$$= \tfrac{1}{2}\text{Height} \times \text{Base}$$

$$= \tfrac{1}{2} \times 200 \times 0.3 \times 10^{-3}$$

$$= 3.0 \times 10^{-2}\,\text{J}$$

Note: we could have used Equation 13.5, with $F = 200\,\text{N}$ and $e = 0.3 \times 10^{-3}\,\text{m}$.

(b) We must add to (a) the area under the graph beyond the proportional limit and up to fracture. This is found by 'counting squares' on the graph paper and is approximately equal to

$$66 \times 10^{-3}\,\text{J} = 6.6 \times 10^{-2}\,\text{J}.$$

So, total work done up to fracture equals

$$6.6 \times 10^{-2} + 3.0 \times 10^{-2} = 9.6 \times 10^{-2}\,\text{J}$$

Answer

(a) 3.0×10^{-2} J, (b) 9.6×10^{-2} J.

Example 5

A mass of 3.5 kg is gradually applied to the lower end of a vertical wire and produces an extension of 0.80 mm. Calculate (a) the energy stored in the wire and (b) the loss in gravitational potential energy of the mass during loading. Account for the difference between the two answers. Assume that the proportional limit is not exceeded and $g = 10\,\text{m}\,\text{s}^{-2}$.

Method

(a) We have

$$F = 3.5 \times g = 35\,\text{N}$$

and

$$e = 0.80 \times 10^{-3}\,\text{m}$$

Equation 13.5 gives

$$\text{Work done} = \tfrac{1}{2}Fe = \tfrac{1}{2} \times 35 \times 0.8 \times 10^{-3}$$

$$= 14 \times 10^{-3}\,\text{J}$$

This is stored as elastic 'strain' energy.

(b) Loss in PE $= mgh$. We have $m = 3.5$, $g = 10$ and $h = 0.80 \times 10^{-3}$.

$$\therefore \quad \text{loss in PE} = 3.5 \times 10 \times 0.8 \times 10^{-3}$$

$$= 28 \times 10^{-3}\,\text{J}$$

The energy stored is only *half* the loss in gravitational PE because the wire needs a *gradually* increasing load, from zero to 35 N, to extend it. The remaining gravitational PE is given to the loading system (e,g, the hand as it gradually attaches the load to the wire). Note that if the load is *suddenly* applied the initial extension would be 1.6 mm; that is, *twice* the equilibrium extension.

Answer

(a) 14×10^{-3} J, (b) 28×10^{-3} J.

Example 6

A vertical steel wire of length 0.80 m and radius 1.0 mm has a mass of 20 kg applied to its lower end. Assuming that the proportional limit is not exceeded, calculate (a) the extension, (b) the energy stored in the wire. Take the Young modulus for steel as $2.0 \times 10^{11}\,\text{N}\,\text{m}^{-2}$ and g as $10\,\text{m}\,\text{s}^{-2}$.

Method

(a) Rearranging Equation 13.4 gives

$$e = \frac{F \div A}{E \div l} = \frac{Fl}{EA}$$

We have

$$F = 20 \times g = 200\,\text{N},$$

$$A = \pi \times (\text{radius})^2 = \pi \times (1.0 \times 10^{-3})^2$$

$$= \pi \times 10^{-6}\,\text{m}^2$$

$$l = 0.80$$

and $E = 2.0 \times 10^{11}$

So

$$e = \frac{Fl}{EA} = \frac{200 \times 0.80}{2.0 \times 10^{11} \times \pi \times 10^{-6}}$$
$$= 0.255 \times 10^{-3}\,\text{m}$$

(b) We have $F = 200\,\text{N}$ and $e = 0.255 \times 10^{-3}\,\text{m}$. From Equation 13.5

$$\text{Work done} = \text{Energy stored} = \tfrac{1}{2}Fe$$
$$= 25.5 \times 10^{-3}\,\text{J}$$

Answer

(a) 0.25 mm, (b) 25.5 mJ.

Exercise 47

1 The following tensile test data were obtained using a metal specimen:

Load/10^2 N	0	2.0	4.0	4.5	5.0	5.5
Extension/mm	0	0.10	0.20	0.24	0.30	0.40

Plot the load–extension graph and calculate the work done in stretching the specimen up to (a) the proportional limit (load = 4.0×10^2 N), (b) fracture (load = 5.5×10^2 N).

2 A metal column shortens by 0.25 mm when a load of 120 kN is placed upon it. Calculate (a) the energy stored in the column and (b) the loss in gravitational PE of the load. Explain why the values in (a) and (b) differ. Assume that the proportional limit is not exceeded.

3 A load of 120 N is gradually applied to a copper wire of length 1.5 m and area of cross-section 8.0 mm^2. Assuming that the proportional limit is not exceeded, calculate (a) the extension, (b) the energy stored in the wire. Take the Young modulus for copper as $1.2 \times 10^{11}\,\text{N m}^{-2}$.

4 A steel bar has a rectangular cross-section 50 mm by 40 mm and is 2.0 m long. Calculate the work done in extending it by 6.0 mm, assuming the proportional limit is not exceeded. Take E for steel as $2.0 \times 10^{11}\,\text{N m}^{-2}$.

Temperature effects

When the temperature of a rod changes then its length will, if unrestrained, change such that:

$$\Delta l = \alpha l \Delta T \qquad (13.6)$$

where Δl is the change in length, in metres, α the linear expansivity (unit = °C^{-1} or K^{-1}), l the original length in metres and ΔT the rise in temperature, in °C or K.

If, during a temperature change, the rod is to be prevented from changing in length, large forces are often required.

Example 7

A solid copper rod is of cross-sectional area 15 mm^2 and length 2.0 m. Calculate (a) its change in length when its temperature rises by 30 °C, (b) the force needed to prevent it from expanding by the amount in (a). Take the linear expansivity α for copper as $20 \times 10^{-6}\,\text{K}^{-1}$ and the Young modulus E for copper as $1.2 \times 10^{11}\,\text{N m}^{-2}$. Assume that the proportional limit is not exceeded.

Method

(a) We have $l = 2.0$, $\Delta T = +30\,°\text{C}$ (+ sign for temperature rise) and $\alpha = 20 \times 10^{-6}$. Equation 13.6 gives

$$\Delta = \alpha l \Delta T = 20 \times 10^{-6} \times 2 \times 30$$
$$= 12 \times 10^{-4}\,\text{m}$$

(b) A compressive force F (N) must be supplied which is sufficient to decrease the length by

$$\Delta l = 12 \times 10^{-4}\,\text{m}$$

Rearranging Equation 13.4 gives

$$F = \frac{EeA}{l}$$

We have $E = 1.2 \times 10^{11}$, $e = \Delta l = 12 \times 10^{-4}$, $A = 15\,\text{mm}^2 = 15 \times 10^{-6}\,\text{m}^2$ and $l = 2.0$.

$$\therefore \quad F = \frac{EeA}{l}$$
$$= \frac{1.2 \times 10^{11} \times 12 \times 10^{-4} \times 15 \times 10^{-6}}{2.0}$$
$$= 1080\,\text{N}$$

Answer

(a) 1.2 mm, (b) 1.1 kN.

Exercise 48

(For steel, take $\alpha = 12 \times 10^{-6}\,K^{-1}$ and $E = 2.0 \times 10^{11}\,N\,m^{-2}$. Assume that the proportional limit is not exceeded.)

1 Calculate the force required to extend a steel rod of cross-sectional area $4.0\,mm^2$ by the same amount as would occur due to a temperature rise of 60 K. Hint: let length = l; this cancels out.

2 A section of railway track consists of a steel bar of length 15 m and cross-sectional area $80\,cm^2$. It is rigidly clamped at its ends on a day when the temperature is 20 °C. If the temperature falls to 0 °C, calculate (a) the force the clamps must exert to stop the bar contracting and (b) the strain energy stored in the bar.

Exercise 49: Questions of A-level standard

(Assume $g = 10\,m\,s^{-2}$.)

1

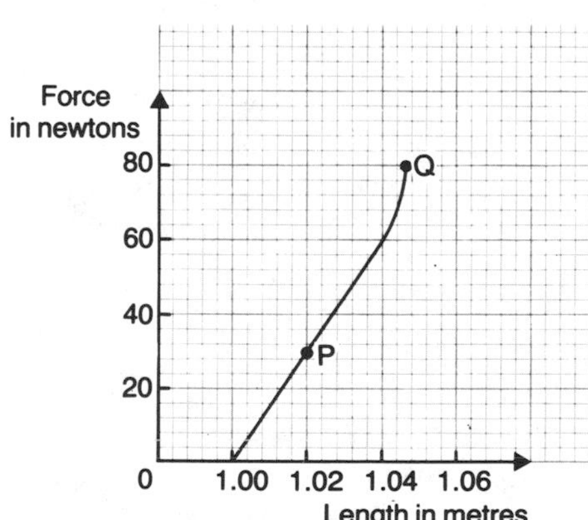

Fig 13.3 Graph for Question 1

Fig 13.3 shows how the length of an acrylic material varies with the tension in the specimen. The specimen breaks at Q, at which point its cross-section is $4\,mm^2$.

Which of the following can correctly be deduced from this information?

1. The strain at P is 2%.
2. The breaking stress is $20 \times 10^6\,N\,m^{-2}$.
3. Hooke's law is obeyed up to a force of 60 N.

[O & C Nuff 82]

2 Two wires each 2.0 m long, one of steel of diameter 0.80 mm and the other of brass of diameter 0.68 mm, are suspended vertically from two points in the same horizontal plane 0.10 m apart. Their lower ends are fixed to a light horizontal bar at points 0.10 m apart. A force of 100 N is applied vertically downwards to the centre of the bar, which tilts at 1° to the horizontal, the brass wire stretching more than the steel. Assuming that the wires remain vertical

(a) calculate the difference between the extensions of the two wires;
(b) calculate the tension in each wire;
(c) show that the extension of the steel wire is 1.0 mm;
(d) calculate the Young modulus for the brass wire;
(e) calculate the energy stored in the steel wire due to its extension.

(The Young modulus for steel = $2.0 \times 10^{11}\,N\,m^{-2}$.)

[AEB 80, p]

3 A 1 m length of copper wire of diameter 2 mm is joined end-to-end to a 1 m length of steel wire of diameter 1 mm. The resulting 2 m length of wire is now stretched by 6 mm. Given that the Young modulus of steel is twice that of copper, what are the extensions produced in the copper and steel wires?

[SUJB 81, p]

4 An elastic string of cross-sectional area $4\,mm^2$ requires a force of 2.8 N to increase its length by one-tenth. Find Young's modulus for the string. If the original length of the string was 1 m, find the energy stored in the string when it is extended.

[WJEC 79]

5 Explain the term, *Young modulus*.
A nylon guitar string 62.8 cm long and 1 mm diameter is tuned by stretching it 2.0 cm. Calculate (a) the tension, (b) the elastic energy stored in the string.
(Young's modulus of nylon = 2×10^9 Pa.)

[SUJB 80]

6 (a) Define the Young modulus. Describe how you would make an accurate determination of this quantity for material in the form of a wire.

(b) A wire of length 3.0 m and cross-sectional area $1.0 \times 10^{-6}\,m^2$ has a mass of 15 kg hung on it. What is the stress produced in the wire? If the Young modulus for the material is $2.0 \times 10^{11}\,N\,m^{-2}$, what is the extension x produced? When extended, how much energy is stored in the wire?

If the mass of 15 kg were allowed to fall through a distance x, what would be the change in its gravitational potential energy? Why is this change not equal to the final energy stored in the wire when extended by the mass? [L 80]

7 The table shows typical working stresses and strains for five different materials.

	Strain%	Stress/N m^{-2}
Steel	0.3	700×10^6
Bronze	0.3	400×10^6
Wood (yew)	0.9	120×10^6
Tendon (e.g. in ankle)	8.0	70×10^6
Rubber	300	7×10^6

(a) Which material is *stiffest*?

(b) The energy stored per unit volume in a stretched material is $\frac{1}{2}$(stress) $\times$ (strain). Which material stores the greatest energy per unit volume, in these typical working conditions?

[O & C Nuff 81]

8 It is modern practice for railways to use long welded steel rails. If these are laid without any stress in the rails on a day when the temperature is 15 °C, estimate the internal force when the temperature of the rails is 35 °C. You may assume that the rails are so held that there is no longitudinal movement, and that the cross-sectional area of the rails is $7.0 \times 10^{-3}\,m^2$.

(The Young modulus for steel = $2.0 \times 10^{11}\,N\,m^{-2}$, linear expansivity of steel = $1.0 \times 10^{-5}\,K^{-1}$.)

[AEB 79]

9 A solid copper wire of cross-sectional area 8 mm^2 and original length 110 m is set up as a telephone line with a uniform tension of 3.6×10^3 N. Assuming that the wire stretches elastically, calculate

(a) the extension of the wire;

(b) the elastic energy stored in the wire.

During cold weather the temperature of the wire falls by 15 K. Calculate

(c) the heat lost by the wire during this cooling;

(d) the change in elastic energy, assuming that the elastic properties of copper are unaffected by the temperature change.

(Data for copper:
Young's modulus = 1.2×10^{11} Pa;
density = $8.9 \times 10^3\,kg\,m^{-3}$;
specific heat capacity = $400\,J\,kg^{-1}\,K^{-1}$;
linear expansivity = $20 \times 10^{-6}\,K^{-1}$.)

[O 82, p]

Section D Oscillations and Waves

14 SIMPLE HARMONIC MOTION

Definition of SHM

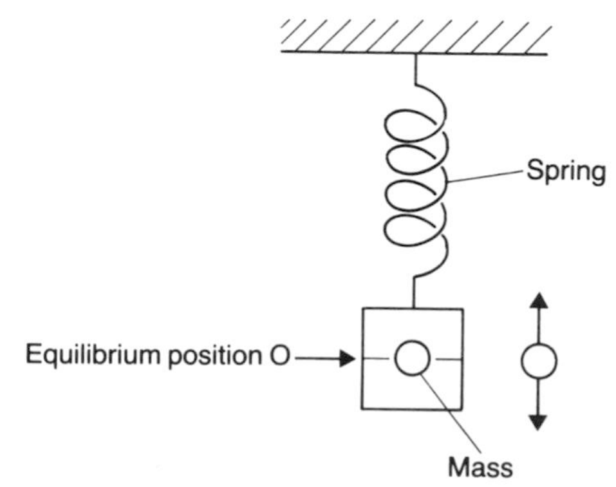

Fig 14.1 Vertical oscillations

Fig 14.1 shows a mass on the end of a spring. When displaced vertically it will perform simple harmonic motion because a restoring force acts which is proportional to the displacement of the mass from its equilibrium position O. Thus *its acceleration is always directed towards the point O and is proportional to the displacement from that point.*

Fig 14.2 illustrates some characteristics of the motion. Fig 14.2c is the displacement–time graph of the vertical SHM shown in Fig 14.2a. Fig 14.2b shows the rotating radius or 'phasor' representation of SHM — the point R moves in uniform circular motion with angular velocity ω. It can be shown that the motion of R projected on to the vertical diameter XY is the same as the SHM shown in Fig 14.2a and c. Note that the amplitude of SHM is the radius OR, and the 'phase angle' $\theta = \omega t$.

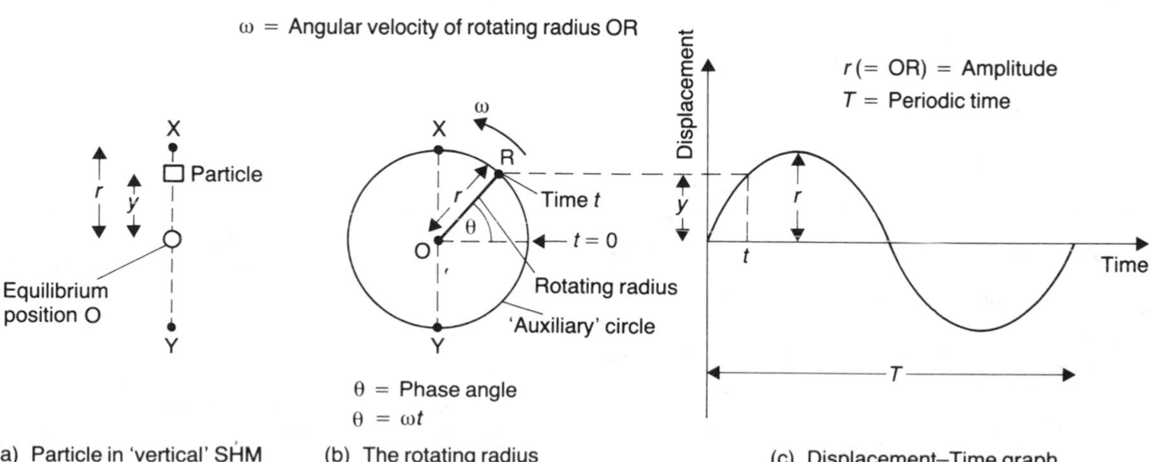

Fig 14.2 SHM and phasor representation

From the definition of SHM, the acceleration a of the particle is related to its displacement y from the equilibrium position by

$$a = -\text{Constant} \times y$$

where the negative sign indicates that the acceleration is in the opposite direction to the displacement. Also, it can be shown that

$$a = -\omega^2 y \qquad (14.1)$$

where ω is the angular velocity of the rotating radius in Fig 14.2b.

Example 1

A body oscillates vertically in SHM with an amplitude of 30 mm and a frequency of 5.0 Hz. Calculate the acceleration of the particle (a) at the extremities of the motion, (b) at the centre of the motion, (c) at a position midway between the centre and the extremity.

Method

We have frequency $f = 5.0$ Hz. Thus the angular velocity ω of the rotating radius in Fig 14.2b is, from Chapter 10, given by

$$\omega = 2\pi f = 10\pi \text{ rad s}^{-1}$$

We use Equation 14.1 to find the acceleration a.

(a) At the top of the motion we ascribe y a positive value, so $y = +0.030$ m. Thus

$$a = -\omega^2 y = -\ 100\pi^2 \times 0.03$$
$$= -3\pi^2 \text{ m s}^{-2}$$

Note that acceleration a is negative (downwards) when displacement y is positive (upwards).

Similarly at the bottom of the oscillation $y = -0.030$ m, so $a = +3\pi^2 \text{ m s}^{-2}$.

Note: a is positive (upwards) when y is negative (downwards).

(b) We have $y = 0$, so $a = -\omega^2 y = 0$.

(c) At a position halfway upwards $y = +0.015$ m. So

$$a = -\omega^2 y = -100\pi^2 \times 0.015$$
$$= -1.5\pi^2 \text{ m s}^{-2}$$

At a position halfway downwards $y = -0.015$ m and $a = +1.5\pi^2 \text{ m s}^{-2}$.

Answer

(a) $\mp 3.0\pi^2 \text{ m s}^{-2}$, (b) 0, (c) $\mp 1.5\pi^2 \text{ m s}^{-2}$.

Example 2

A horizontal platform vibrates vertically in SHM with a period of 0.20 s and with slowly increasing amplitude. What is the maximum value of the amplitude which will allow a mass, resting on the platform, to remain in contact with the platform? Assume acceleration due to gravity $g = 10 \text{ m s}^{-2}$.

Method

When the platform moves downwards the mass will remain in contact with it only so long as the platform accelerates downwards with value less than or equal to g. The maximum downwards acceleration of the platform is at the top of its motion. If the amplitude is r, then, using Equation 14.1,

$$a = -\omega^2 y = -\omega^2 r$$

at the top of the motion. Now $\omega = 2\pi/T$ where T, the period of the motion, equals 0.20 s.

When the mass is on the point of leaving the platform $a = -g$ (negative indicates downwards), so

$$-g = -\omega^2 r = -\left(\frac{2\pi}{T}\right)^2 r \qquad (14.2)$$

We have $g = 10$, $T = 0.2$ and require r.

Rearranging Equation 14.2 gives

$$r = g\left(\frac{T}{2\pi}\right)^2 = 10 \times \left(\frac{0.2}{2\pi}\right)^2$$
$$= 0.010 \text{ m}$$

Answer

Maximum amplitude = 10 mm.

Exercise 50

1 A body oscillates in SHM with an amplitude of 2.0 cm and a periodic time of 0.25 s. Calculate (a) its frequency, (b) the acceleration at the extremities and at the centre of the oscillation, (c) the acceleration when it is displaced 0.5 cm above the centre of the oscillation. Note: $f = 1/T$.

2 The piston in a particular car engine moves in approximately SHM with an amplitude of 8.0 cm. The mass of the piston is 0.80 kg and the piston makes 100 oscillations per second. Calculate (a) the maximum value of the acceleration of the piston, (b) the force needed to produce this acceleration.

3 A body of mass 0.40 kg has a maximum force of 1.2 N acting on it when it moves in SHM with an amplitude of 30 mm. Calculate (a) the frequency, (b) the periodic time of the motion.

4 A small mass rests on a horizontal platform which vibrates vertically in SHM with a constant amplitude of 30 mm and with a slowly increasing frequency. Find the maximum value of the frequency which will allow the mass to remain in contact with the platform. Assume $g = 10\,\text{m s}^{-2}$.

Mass on a spring

When a mass m (kg) is attached to the end of a spring of force constant k (N m^{-1}), the periodic time T (s) of oscillations is given by

$$T = 2\pi\sqrt{\frac{m}{k}} \qquad (14.3)$$

Since $T = 2\pi/\omega$ we have

$$\omega^2 = \frac{k}{m} \qquad (14.4)$$

Example 3

A mass of 0.2 kg is attached to the lower end of a light helical spring and produces an extension of 5.0 cm. Calculate (a) the force constant of the spring.

The mass is now pulled down a further distance of 2.0 cm and released. Calculate (b) the time period of subsequent oscillations, (c) the maximum value of the acceleration during the motion. Assume $g = 10\,\text{m s}^{-2}$.

Method

(a) We assume that the spring obeys Hooke's law — i.e. the extension (or compression) is directly proportional to the applied force. In mathematical terms, an applied force F (N) produces a change in length e (m) given by

$$F = ke \qquad (14.5)$$

where k (N m^{-1}) is the force constant of the spring.

We have $F = mg$ where $m = 0.2$ kg and $g = 10$. Since $e = 5.0 \times 10^{-2}$ m, Equation 14.5 gives

$$0.2 \times 10 = k \times 5.0 \times 10^{-2}$$

$$\therefore \quad k = 40\,\text{N m}^{-1}$$

(b) We use Equation 14.3 with $m = 0.2$ and $k = 40$.

$$T = 2\pi\sqrt{\frac{m}{k}} = 2\pi\sqrt{\frac{0.2}{40}} = 0.44\,\text{s}$$

Note that T is independent of the initial displacement (2.0 cm in this case).

(c) From Equation 14.4

$$\omega^2 = \frac{k}{m} = \frac{40}{0.2} = 200\,\text{rad}^2\,\text{s}^{-2}$$

For the maximum acceleration we use Equation 14.1, with displacement y at its maximum value of 2.0×10^{-2} m.

$$a = -\omega^2 y = -200 \times 2.0 \times 10^{-2} = -4.0\,\text{m s}^{-2}$$

The negative sign indicates direction.

Note an alternative way to find a. At maximum displacement, the net force acting on the mass is

$$F = k \times \text{Displacement} = 40 \times 2.0 \times 10^{-2} = 0.80\,\text{N}$$

Thus the maximum acceleration a is given by

$$a = \frac{\text{Force}}{\text{Mass}} = \frac{0.80}{0.20} = 4.0\,\text{m s}^{-2}$$

Answer

(a) $40\,\text{N m}^{-1}$, (b) 0.44 s, (c) $4.0\,\text{m s}^{-2}$.

The simple pendulum

The periodic time T (s) of 'small angle' oscillations of a simple pendulum of length l (m) is given by

$$T = 2\pi\sqrt{\frac{l}{g}} \qquad (14.6)$$

where g is the acceleration due to gravity.

Example 4

Calculate the frequency of oscillation of a simple pendulum of length 80 cm. Assume $g = 10\,\text{m s}^{-2}$.

Method

We use Equation 14.6 with $l = 0.80$ and $g = 10$.

$$T = 2\pi\sqrt{\frac{l}{g}} = 2\pi\sqrt{\frac{0.80}{10}} = 1.777$$

Now frequency $f = \dfrac{1}{T} = \dfrac{1}{1.777} = 0.56\,\text{Hz}$

Answer

0.56 Hz.

Example 5

Two simple pendulums of length 0.40 m and 0.60 m are set off oscillating in step. Calculate (a) after what further time the two pendulums will once again be in step, (b) the number of oscillations made by each pendulum during this time. (Assume $g = 10\,\mathrm{m\,s^{-2}}$.)

Method

(a) The two pendulums become out of step since they have different periodic times. Let T_1 be the periodic time of the pendulum of length $l_1 = 0.40$ m and T_2 that of the pendulum of length $l_2 = 0.60$ m. Using Equation 14.6

$$T_1 = 2\pi\sqrt{\frac{l_1}{g}} = 2\pi\sqrt{\frac{0.4}{10}} = 1.257\,\mathrm{s}$$

$$T_2 = 2\pi\sqrt{\frac{l_2}{g}} = 2\pi\sqrt{\frac{0.6}{10}} = 1.539\,\mathrm{s}$$

During the required time interval the shorter pendulum will complete one more oscillation than the longer pendulum. Let t be the time interval between the pendulums falling in step. If n equals the number of oscillations of the shorter pendulum, $(n-1)$ equals the number of oscillations of the longer pendulum. Thus

$$t = nT_1 = (n-1)T_2$$

So, since $T_1 = 1.257$ s and $T_2 = 1.539$ s,

$$n \times 1.257 = (n-1) \times 1.539$$

$$\therefore \quad n = \frac{1.539}{0.282} = 5.46$$

But $t = nT_1$, so

$$t = 5.46 \times 1.257 = 6.86\,\mathrm{s}$$

(b) The shorter pendulum makes $n = 5.5$ oscillations and the longer pendulum $(n-1) = 4.5$ oscillations.

Answer

(a) 6.9 s, (b) 5.5 and 4.5 oscillations.

Exercise 51

(Assume $g = 10\,\mathrm{m\,s^{-2}}$.)

1 A mass of 0.60 kg is hung on the end of a vertical light spring of force constant $30\,\mathrm{N\,m^{-1}}$. Calculate (a) the extension produced, (b) the time period of any subsequent oscillations, (c) the number of oscillations in 1 minute.

2

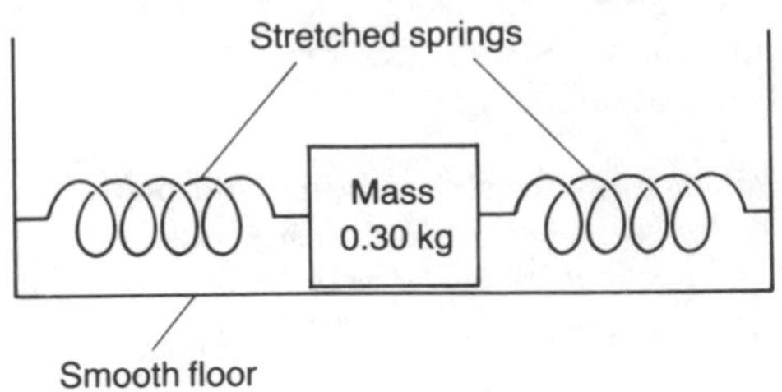

Fig 14.3 Diagram for Question 2

Refer to Fig 14.3, in which the 0.30 kg mass is tethered by two identical springs of force constant $2.5\,\mathrm{N\,m^{-1}}$. If the mass is now displaced by 20 mm to the left of its equilibrium position and released, calculate (a) the time period and frequency of subsequent oscillations, (b) the acceleration at the centre and extremities of the oscillation.
Note: effective force constant is twice that for one spring.

3 Calculate the length of a simple pendulum of periodic time (a) 1.0 s, (b) 0.5 s. If the two are set off oscillating in step, calculate (c) the number of times they will be in step over a 60 s period.

4 Two simple pendulums, of slightly different length, are set off oscillating in step. The next time they are in step is after a time of 20 s has elapsed, during which time the longer pendulum has completed exactly 10 oscillations. Find the length of each pendulum.

Displacement, velocity and acceleration variation

Refer to Fig 14.2. The following relationships apply in SHM:

(1) The displacement y is related to time t by

$$y = r\sin\theta = r\sin\omega t \qquad (14.7)$$

This assumes $y = 0$ when $t = 0$. The maximum displacement equals the amplitude r.

(2) The instantaneous velocity v is related to displacement y by

$$v = \omega\sqrt{(r^2 - y^2)} \qquad (14.8)$$

Note that $v = 0$ at the extremities of the oscillation, when $y = r$. Also v has maximum value $\pm\omega r$ when $y = 0$, at the centre of the oscillation.

(3) The instantaneous acceleration a is related to the displacement y by Equation 14.1:

$$a = -\omega^2 y \qquad (14.1)$$

Example 6

A body vibrates in SHM in a vertical direction with an amplitude of 50 mm and a periodic time of 4.0 s. Calculate the displacement after (a) 2.5 s, (b) 5.0 s, assuming that the displacement is zero at time zero.

Method

The angular velocity ω of the motion is given by $\omega = 2\pi/T$ where $T = 4.0$ s. So $\omega = 0.5\pi \text{ rad s}^{-1}$. We use Equation 14.7 with $r = 50 \times 10^{-3}$ m to find displacement.

(a) We have $t = 2.5$, so

$$y = r \sin \omega t = 50 \times 10^{-3} \sin (0.5\pi \times 2.5)$$
$$= 50 \times 10^{-3} \sin 1.25\pi$$

Now π rad = 180°, so $1.25\pi = 225°$, and

$$y = 50 \times 10^{-3} \sin 225° = -35 \times 10^{-3} \text{ m}$$

Note: we assumed that the body was initially moving in a positive direction. The negative sign indicates a displacement in the opposite direction to this.

(b) We have $t = 5.0$ s, so

$$y = r \sin \omega t = 50 \times 10^{-3} \sin (0.5\pi \times 5)$$
$$= 50 \times 10^{-3} \sin 450°$$

We subtract multiples of 360°, which means that previous whole oscillations are ignored. Subtracting 360° from 450°, we have

$$y = 50 \times 10^{-3} \sin 90°$$
$$= 50 \times 10^{-3} \text{ m}$$

Answer

(a) −35 mm, (b) 50 mm.

Example 7

A body moves in SHM with an amplitude of 30 mm and a frequency of 2.0 Hz. Calculate the values of (a) acceleration at the centre and extremities of the oscillation, (b) velocity at these positions, (c) velocity and acceleration at a point midway between the centre and extremity of the oscillation.

Method

We have $\omega = 2\pi f$ and $f = 2.0$. So $\omega = 4.0\pi \text{ rad s}^{-1}$.

(a) We use Equation 14.1. At the centre $y = 0$ so $a = 0$.

At the extremities the displacement equals 30×10^{-3} m. So

$$a = -\omega^2 y = -(4\pi)^2 \times 30 \times 10^{-3}$$
$$= -0.48\pi^2 \text{ m s}^{-2}$$

When y is positive a is negative and vice versa.

(b) We use Equation 14.8. At the centre $y = 0$ and $v = \pm\omega r$, depending on whether the body is moving upwards (+) or downwards (−) at that instant. Since $r = 30 \times 10^{-3}$ and $\omega = 4.0\pi$,

$$v = \pm\omega r = \pm 0.12\pi \text{ m s}^{-1}$$

At the extremities $v = 0$.

(c) At the midway point $y = 15 \times 10^{-3}$ m. Since $r = 30 \times 10^{-3}$, Equation 14.8 gives

$$v = \omega\sqrt{(r^2 - y^2)} = 4\pi\sqrt{(30^2 - 15^2)} \times 10^{-3}$$
$$= 0.33 \text{ m s}^{-1}$$

This can be positive or negative depending on which way the body is moving.

Equation 14.1 gives

$$a = -\omega^2 y = -(16\pi^2) \times 15 \times 10^{-3}$$
$$= -0.24\pi^2 \text{ m s}^{-2}$$

When y is positive a is negative and vice versa.

Answer

(a) 0, $\pm 0.48\pi^2 \text{ m s}^{-2}$, (b) $\pm 0.12\pi \text{ m s}^{-1}$, 0
(c) $\pm 0.33 \text{ m s}^{-1}$, $\pm 0.24\pi^2 \text{ m s}^{-2}$.

Exercise 52

1 A body is vibrating in SHM in a vertical direction with an amplitude of 40 mm and a frequency of 0.50 Hz. Assume at $t = 0$ the displacement is zero and it is moving upwards. Calculate the values of displacement, velocity and acceleration at each of the following times (in seconds): 0.00, 0.25, 0.50, 0.75, 1.00, 1.25, 1.50, 1.75, 2.00. Sketch the graphs of displacement, velocity and acceleration against time.

2 A body vibrates in SHM with an amplitude of 30 mm and frequency of 0.50 Hz. Calculate (a) the maximum acceleration, (b) the maximum velocity, (c) the magnitude of acceleration and velocity when the body is displaced 10 mm from its equilibrium position.

State the value of the constants r (in metres) and ω (in rad s^{-1}) in the equation $y = r \sin \omega t$ which describes the motion of the body.

Energy in SHM

There is a continuous interchange between kinetic energy (KE) and potential energy (PE) during vibration. Assuming no energy losses, the total energy is constant. At the centre of the oscillation we take PE as zero, so all the energy here is KE. Thus at the centre of the oscillation

$$\text{Total energy} = \text{KE} = \frac{1}{2}mv^2$$

Now $v = (\pm)\omega r$ at the centre, so

$$\text{Total energy} = \text{KE} = \frac{1}{2}m\omega^2 r^2 \qquad (14.9)$$

Example 8

A body of mass 0.10 kg oscillates in SHM with an amplitude of 5.0 cm and with a frequency of 0.50 Hz. Calculate (a) the maximum value and (b) the minimum value of its kinetic energy. State where these occur.

Method

(a) The maximum KE is at the centre of the motion. We use Equation 14.9 in which $m = 0.10$ kg, $\omega = 2\pi f = 2\pi \times 0.50 = \pi$ rad s^{-1} and amplitude $r = 5.0 \times 10^{-2}$ m.

$$\begin{aligned}\text{KE} &= \tfrac{1}{2}m\omega^2 r^2 \\ &= \tfrac{1}{2} \times 0.1 \times \pi^2 \times (5.0 \times 10^{-2})^2 \\ &= 12 \times 10^{-4}\ \text{J}\end{aligned}$$

(b) The minimum value of KE is at the extremities of the motion. Since velocity v is zero here KE is zero.

Answer

(a) 12×10^{-4} J, at centre, (b) zero, at extremities.

Exercise 53

1 A mass of 0.50 kg vibrates in SHM with a maximum KE of 3.0 mJ. If its amplitude is 20 mm, calculate the frequency of the motion.

2 A mass oscillates in SHM on the end of a spring of force constant 40 N m^{-1}. If the amplitude of the motion is 30 mm, calculate the maximum KE of the mass. (Hint: $\omega^2 = k/m$.)

3 A body oscillates in SHM with a total energy of 2.0 mJ. Calculate the total energy if (separately)
(a) the amplitude is doubled (frequency being constant);
(b) the frequency is halved (amplitude being constant);
(c) the amplitude and frequency are both doubled.

Exercise 54: Questions of A-level standard

1 A horizontal platform is made to perform vertical SHM of constant amplitude 5 cm but of slowly increasing frequency. At what value of the frequency will a mass resting on the platform first lose contact with the platform, and at what vertical position of the platform does this occur? (Assume $g = 9.8$ m s^{-2}.) [WJEC 80]

2 A trolley of mass 2 kg is secured by taut springs (as shown in Fig 14.4) and is found to need a force of 10 N to displace it by a horizontal distance of 0.05 m from its rest position.

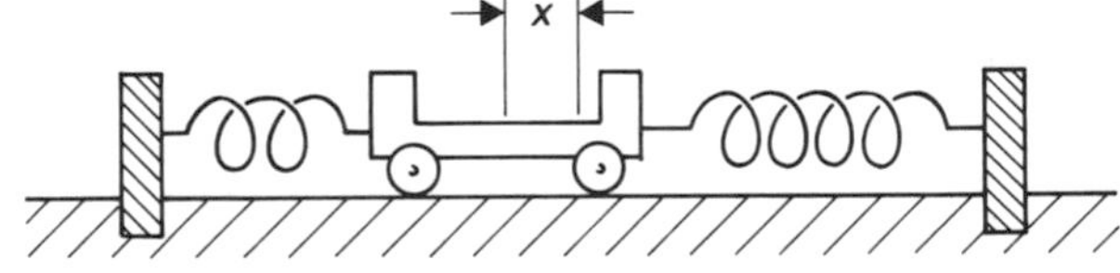

Fig 14.4 Diagram for Question 2

If the trolley is then released it oscillates, with its motion given by the equation

$$\frac{d^2x}{dt^2} = -\omega^2 x$$

(Note: ω is more usually given as rad s^{-1} but the radian is dimensionless.)

What is the value of ω^2 in this equation?

A $0.25\ s^{-2}$ **B** $1.0\ s^{-2}$ **C** $4.0\ s^{-2}$
D $100\ s^{-2}$ **E** $400\ s^{-2}$ [O 82]

3 A light car with very soft springs bounces up and down on its springs after hitting a bump, with a period of roughly $\sqrt{2}$ s. The mass of car and driver is 300 kg.

If the driver now packs in several friends, so that the mass of car and occupants is 600 kg, which one of the following is the best estimate of the new period of oscillation of the car on its springs?

A $2\sqrt{2}$ s **B** 2 s **C** $\sqrt{2}$ s

D 1 s **E** $\frac{1}{\sqrt{2}}$ s [NUFF 82]

4 Calculate the period of oscillation of a pendulum of length 1.8 m with a bob of mass 2.2 kg. What assumption is made in this calculation?

If the bob of this pendulum is pulled aside a horizontal distance of 20 cm and released, what will be the values of (a) the kinetic energy, (b) the velocity of the bob at the lowest point of its swing? Assume $g = 9.8\ m\ s^{-2}$. [L 81, p]

5 Two simple pendulums of slightly different lengths are set off oscillating in phase. The time periods are 1.00 s and 0.98 s. Calculate the number of oscillations made by the *shorter* pendulum during the time interval it takes for the two pendulums to be once again moving in phase.

6 A particle executes simple harmonic motion of amplitude 0.02 m and frequency 2.5 Hz. Its maximum velocity, in $m\ s^{-1}$, is

A 0.008 **B** 0.050 **C** 0.125
D 0.157 **E** 0.314 [O & C 80]

7 When a mass of 0.20 kg is attached to the lower end of a light helical spring it produces an extension of 0.20 m. The mass is now raised vertically by 0.04 m and released:

(a) calculate the time period of subsequent oscillations;

(b) derive the equation relating the displacement y (in metres) of the mass from its equilibrium position to the time t (in seconds) after release. Assume $g = 10\ m\ s^{-2}$.

8 A spring rests on a frictionless, massive, horizontal board. One end of the spring is fixed to the board and a mass of 0.15 kg is fixed to the other end. When the mass is suitably displaced a distance 0.02 m from its equilibrium position A and released, it vibrates in simple harmonic motion with a period of 4 s.

Determine (a) the velocity and acceleration of the mass when it is 0.015 m from A, (b) the shortest time it takes to travel from a point 0.01 m on one side of A to a point 0.01 m from A on the other side and (c) the total energy of the system. Does this energy vary with time? [WJEC 77, p]

9

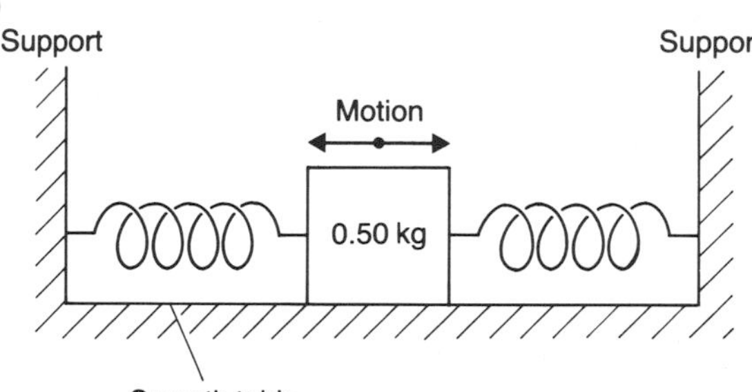

Fig 14.5 Diagram for Question 9

Fig 14.5 shows a mass of 0.50 kg which is in contact with a smooth horizontal table. It is attached by two light springs to two fixed supports as shown. If the mass moves in linear simple harmonic motion with a period of 2.0 s and an amplitude of 4.0 cm, calculate the energy associated with this motion.

10 A particle is oscillating with simple harmonic motion. If the amplitude of the oscillation is doubled but the frequency stays the same, which of the following quantities is/are doubled?

1. The maximum velocity of the particle.
2. The total energy of the system.
3. The maximum acceleration of the particle.

[NUFF 79]

15 PROGRESSIVE WAVES

Wave relationships

A progressive wave transfers energy from its source with speed c ($\mathrm{m\,s^{-1}}$). If the wave has wavelength λ (m) and frequency f (Hz), then

$$c = f\lambda \tag{15.1}$$

The periodic time T (s) of the wave motion is related to frequency f by

$$f = \frac{1}{T} \tag{15.2}$$

Equations 15.1 and 15.2 apply to longitudinal and transverse waves.

Example 1

A progressive wave travels a distance of 18 cm in 1.5 s. If the distance between successive crests is 60 mm, calculate (a) the frequency, (b) the periodic time of the wave motion.

Method

The speed c is given by

$$c = \frac{\text{Distance travelled (m)}}{\text{Time taken (s)}} = \frac{18 \times 10^{-2}}{1.5}$$

$$= 0.12\ \mathrm{m\,s^{-1}}$$

Now wavelength $\lambda = 60\ \mathrm{mm} = 0.060\ \mathrm{m}$. Rearranging Equation 15.1 gives

$$f = \frac{c}{\lambda} = \frac{0.12}{0.06} = 2.0\ \mathrm{Hz}$$

Rearranging Equation 15.2 gives

$$T = \frac{1}{f} = \frac{1}{2} = 0.50\ \mathrm{s}$$

Answer

(a) 2.0 Hz, (b) 0.50 s.

Factors affecting speed

The speed of sound waves in a gas is given by

$$c = \sqrt{\frac{\gamma P}{\rho}} = \sqrt{\frac{\gamma RT}{M}} \tag{15.3}$$

where γ is the ratio of principal heat capacities (see Chapter 26), P the gas pressure ($\mathrm{N\,m^{-2}}$), ρ the density of the gas ($\mathrm{kg\,m^{-3}}$), R the universal molar gas constant ($\mathrm{J\,mol^{-1}\,K^{-1}}$), M the mass of one mole of the gas ($\mathrm{kg\,mol^{-1}}$) and T the absolute temperature (K).

The speed of propagation of transverse waves along a string or wire is given by

$$c = \sqrt{\frac{T}{m}} \tag{15.4}$$

where T is the tension in the string, in newtons, m the mass per unit length of the string, in $\mathrm{kg\,m^{-1}}$.

Example 2

The speed of sound in dry air at 0 °C is $330\ \mathrm{m\,s^{-1}}$. Calculate the speed of sound in air at 30 °C.

Method

Equation 15.3 tells us that, since γ, R and M are constants, $c \propto \sqrt{T}$. Now speed $c_1 = 330$ at temperature $T_1 = 0\ °\mathrm{C} = 273\ \mathrm{K}$. Let c_2 be the unknown speed at temperature $T_2 = 30\ °\mathrm{C} = 303\ \mathrm{K}$. Since $c \propto \sqrt{T}$

$$\frac{c_1}{c_2} = \frac{\sqrt{T_1}}{\sqrt{T_2}}$$

Rearranging gives

$$c_2 = c_1\frac{\sqrt{T_2}}{\sqrt{T_1}} = 330 \times \frac{\sqrt{303}}{\sqrt{273}}$$

$$= 348\ \mathrm{m\,s^{-1}}$$

Answer

$348\ \mathrm{m\,s^{-1}}$.

Example 3

A horizontal stretched elastic string has length 3.0 m and mass 12 g. It is subject to a tension of 1.6 N. Transverse waves of frequency 40 Hz are propagated down the string. Calculate the distance between successive crests of this wave motion.

Method

We use Equation 15.4, with mass per unit length of string $m = (12 \times 10^{-3}) \div 3.0 = 4.0 \times 10^{-3}\,\mathrm{kg\,m^{-1}}$. Since $T = 1.6\,\mathrm{N}$,

$$c = \sqrt{\frac{T}{m}} = \sqrt{\frac{1.6}{4.0 \times 10^{-3}}}$$

$$= 20\,\mathrm{m\,s^{-1}}$$

The distance between successive crests is the wavelength λ. We have frequency $f = 40\,\mathrm{Hz}$. Rearranging Equation 15.1 gives

$$\lambda = \frac{c}{f} = \frac{20}{40} = 0.50\,\mathrm{m}$$

Answer

0.50 m.

Exercise 55

1 The speed of electromagnetic waves in air is $3.0 \times 10^{8}\,\mathrm{m\,s^{-1}}$. Calculate (a) the frequency of yellow light of wavelength $0.60 \times 10^{-6}\,\mathrm{m}$, (b) the wavelength of radio waves of frequency $2.0 \times 10^{5}\,\mathrm{Hz}$.

2 If the speed of sound at 0 °C is $330\,\mathrm{m\,s^{-1}}$, at what temperature (°C) will its speed be $340\,\mathrm{m\,s^{-1}}$?

3 Calculate the speed of sound in air at a pressure of $1.00 \times 10^{5}\,\mathrm{N\,m^{-2}}$ when its density is $1.40\,\mathrm{kg\,m^{-3}}$. Take the ratio of principal heat capacities as 1.40.

4 The speed of transverse waves along a stretched wire is $50\,\mathrm{m\,s^{-1}}$. What is the speed when the tension in the wire is doubled?

5 A horizontal stretched elastic string is subject to a tension of 2.5 N. Transverse waves of frequency 50 Hz and wavelength 2.0 m are propagated down the string. Calculate (a) the speed of the waves, (b) the mass per unit length of the string.

Phase angle

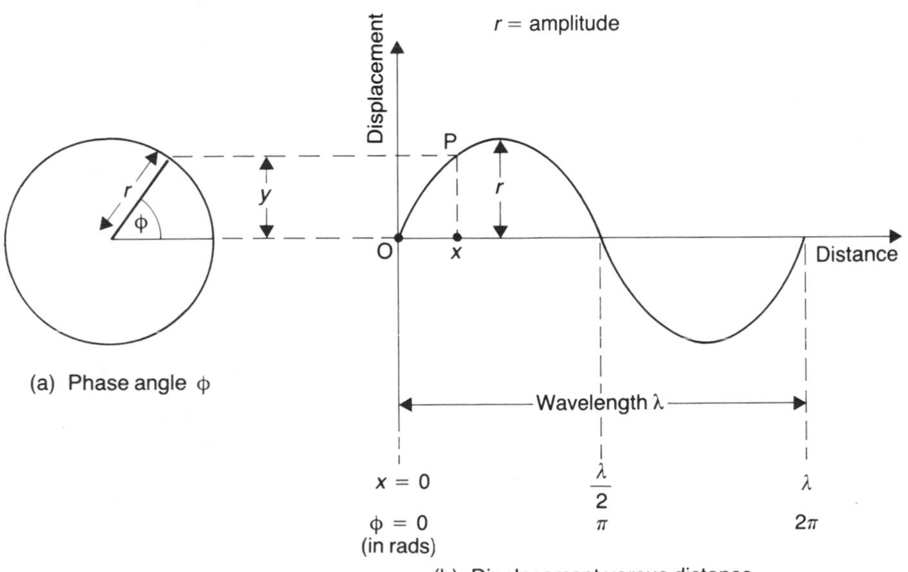

Fig 15.1 Displacement at a fixed time

Fig 15.1 shows the displacement y at all points on a sine wave, at a fixed time, over a single wavelength. It shows how displacement* y varies with distance x. The particle at P lags behind the particle at O by phase angle ϕ (in radians) given by

$$\phi = \frac{2\pi}{\lambda} \times x \qquad (15.5)$$

The relation between y and x in Fig 15.1b

$$y = r\sin\phi = r\sin\frac{2\pi}{\lambda}x = r\sin kx \qquad (15.6)$$

where $k = 2\pi/\lambda$ is called the wave number.

Example 4

A progressive wave has amplitude 0.40 m and wavelength 2.0 m. At a given time the displacement $y = 0$ at $x = 0$. Calculate

(a) the displacement at $x = 0.50$ m and 1.4 m;
(b) the phase angles at $x = 0.50$ m and 0.80 m;
(c) the phase difference between any two points which are 0.30 m apart on the wave.

Method

We have amplitude $r = 0.40$ and wavelength $\lambda = 2.0$.

(a) Using Equation 15.6, with

$$k = 2\pi/\lambda = 2\pi/2 = \pi\ \mathrm{m}^{-1},$$

we have:

for $x = 0.5$,

$$y = r\sin kx = 0.4\sin(\pi \times 0.5)$$
$$= 0.4\sin 90° = 0.40\ \mathrm{m}$$

(Note here that $y = r$, since $x = \frac{1}{4}\lambda$.)

for $x = 1.4$,

$$y = r\sin kx = 0.4\sin\pi \times 1.4$$
$$= 0.4\sin 252° = -0.38\ \mathrm{m}$$

Note the negative sign which indicates a *downwards* displacement, assuming upwards is positive.

(b) Using Equation 15.5:

for $x = 0.5$, $\quad \phi = \frac{2\pi x}{\lambda} = 0.5\pi$

for $x = 0.8$, $\quad \phi = \frac{2\pi x}{\lambda} = 0.8\pi$

(c) We can replace Equation 15.5 by

$$\Delta\phi = \frac{2\pi}{\lambda} \times \Delta x$$

where $\Delta\phi$ is the phase difference in radians between two points spaced Δx (m) apart on the wave. We have $\Delta x = 0.30$ and $\lambda = 2.0$, so

$$\Delta\phi = \frac{2\pi}{\lambda} \times \Delta x = \frac{2\pi}{2} \times 0.3 = 0.3\pi$$

Note that this agrees with part (b), since two points at 0.5 m and 0.8 m have phase angles 0.5π and 0.8π.

Answer

(a) 0.40 m, −0.38 m; (b) 0.5π rad, 0.8π rad;
(c) 0.3π rad.

**Displacement variation with time, at a given point on a sine wave, is dealt with in Chapter 14 on simple harmonic motion — see Fig 14.2 and Equation 14.7.*

Exercise 56

1 A wave on a stretched string has amplitude 5.0 cm and wavelength 30 cm. At a given time the displacement $y = 0$ at $x = 0$. Calculate (a) the wave displacements at $x = 10$ cm and $x = 50$ cm, (b) the phase angles at $x = 10$ cm and $x = 50$ cm.

2 A progressive wave has wavelength 20 cm. Calculate the minimum distance between two points which differ in phase by 60° ($\pi/3$ rad).

3 A transverse wave travels along a horizontal stretched string. In front of the string is a screen with two slots in it so that all an observer can see is the motion of two points on the string placed 3.0 m apart. The observer notes that the two points perform SHM with a period of 2.0 s, and that one point lags in phase by 90° compared with the other. Calculate (a) the frequency of the wave, (b) *two* possible values for the wavelength of the wave.

Exercise 57: Questions of A-level standard

1 A monochromatic source of light emits a continuous burst of waves lasting 0.020 μs. If the light is of wavelength 0.50 μm, calculate the number of complete waves emitted. Assume the speed of light is $3.0 \times 10^8\ \mathrm{m\,s^{-1}}$.

2 Water waves moving across the surface of a pond travel a distance of 14 cm in 0.70 s. The horizontal distance between a crest and a neighbouring trough is 2.0 mm. Calculate the frequency of the waves.

3 The bow of a ship steaming at 5 m s^{-1} into ocean rollers travelling in the opposite direction rises and falls with a period of 10 s. It is noted that when the crest of a wave is under the bow, an adjacent trough is under the stern. If the length of the ship is 60 m what is the speed of the ocean waves relative to the sea-bed?

A 1 m s^{-1} **B** 7 m s^{-1} **C** 11 m s^{-1}
D 12 m s^{-1} **E** 17 m s^{-1} [O 82]

4 A wave of frequency 2.5 Hz travels in the x direction with a speed of 20 m s^{-1}. The phase difference between the oscillations at points 2 m apart in the x direction is

A $\frac{\pi}{4}$ **B** $\frac{\pi}{2}$ **C** π
D 2π **E** 4π [O & C 82]

5 A sinusoidal transverse wave in a stretched elastic string has amplitude 2 mm, frequency 50 Hz and wavelength 3 m. It is travelling in the negative direction of the x axis. Insert the appropriate values for a, ω and k in the expression

$$y = a \sin (kx \pm \omega t)$$

which describes the wave, and state whether the sign should be + or −. What is the speed of the wave? If the frequency is doubled without altering the tension, what is the effect, if any, on ω and k?
[WJEC 80]

16 INTERFERENCE AND BEATS

Interference

This phenomenon occurs for all types of waves — for example sound, water waves and electromagnetic waves (light, microwaves and so on). To simplify the situation our initial treatment considers continuous waves, like sound or water waves.

Interference occurs due to the superposition of waves — the resultant displacement being the sum of the separate displacements of the individual wave motions. Fig 16.1 shows two sources S_1 and S_2 which emit waves of the same frequency and wavelength λ and of approximately the same amplitude. Regions of constructive and destructive interference exist. At a given point Q in the

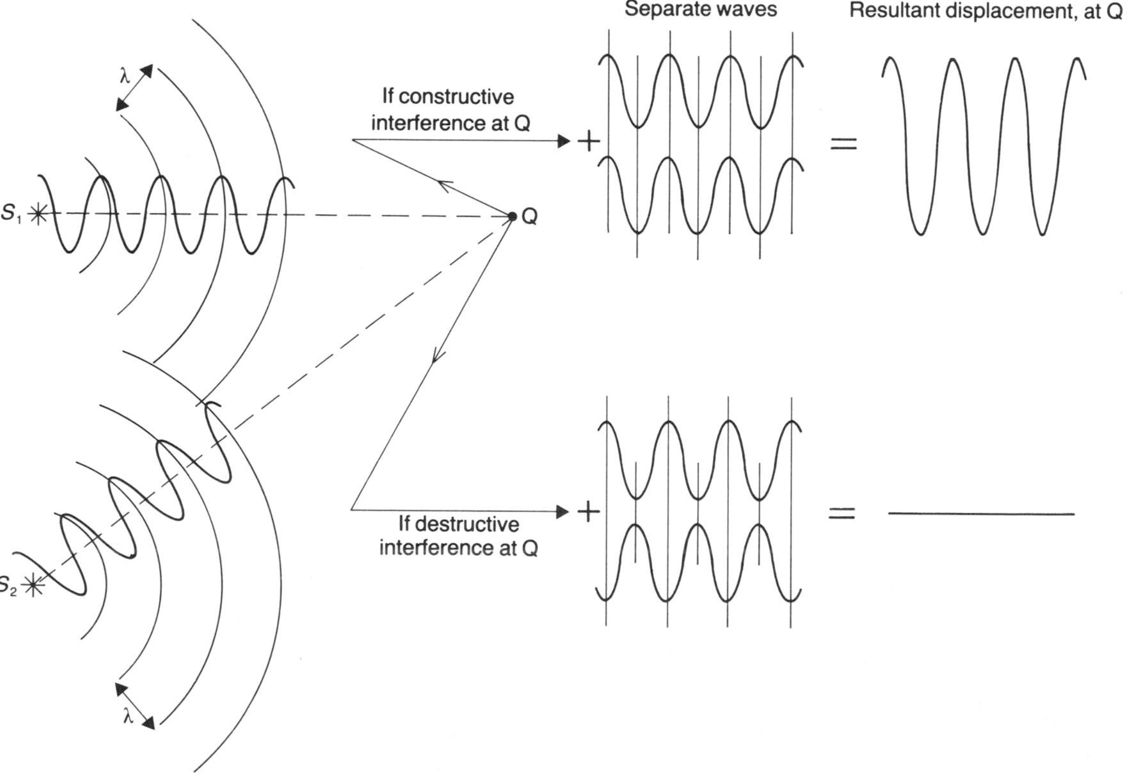

Fig 16.1 Interference at Q between waves from two sources

interference pattern

$S_2Q - S_1Q = n\lambda$ for constructive interference

$S_2Q - S_1Q = (n + \frac{1}{2})\lambda$ for destructive interference

where $n = 0, 1, 2, 3 \ldots$ This assumes the waves from S_1 and S_2 set off in phase.

When waves from two sources arrive at a point in phase there is constructive interference. If the waves arrive out of phase there is destructive interference.

Example 1

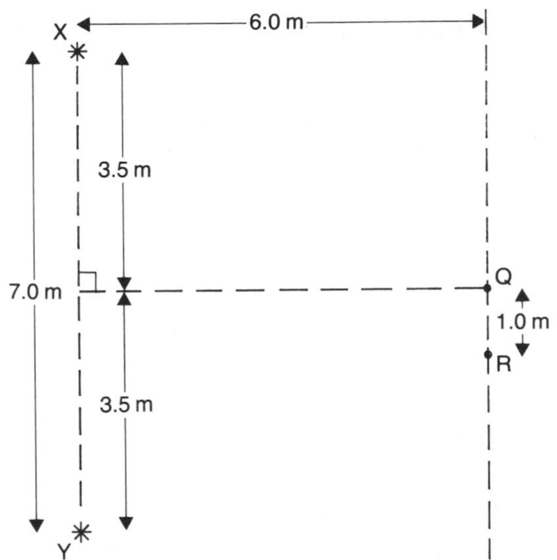

Fig 16.2 Information for Example 1

Fig 16.2 shows two sources X and Y which emit sound of wavelength 2.0 m. The two sources emit in phase, and emit waves of equal amplitude. What does an observer hear (a) at Q, (b) at R.

Method

(a) Q is equidistant from X and Y, so XQ = YQ. Thus

$$XQ - YQ = 0$$

There is constructive interference at Q, since the two sets of waves arrive in phase. The resultant amplitude of the sound at Q is twice that due to each source acting individually.

(b) We must find the path difference XR − YR.

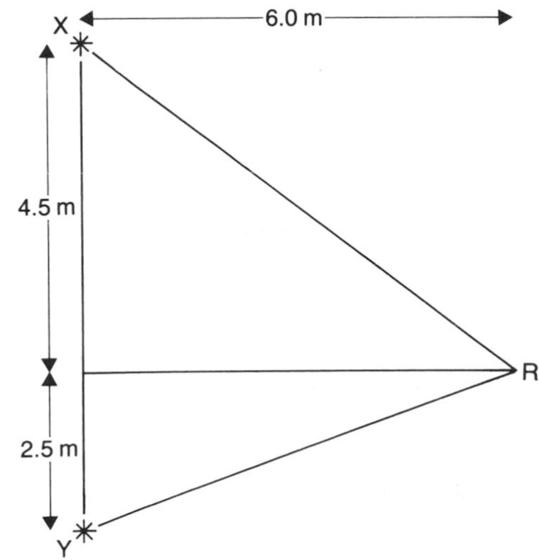

Fig 16.3 Solution to Example 1

Refer to Fig 16.3. Using Pythagoras' theorem, we see that

$$XR^2 = 4.5^2 + 6.0^2 = 56.25$$

$$\therefore \quad XR = 7.5\,\text{m}$$

Also $YR^2 = 2.5^2 + 6.0^2 = 42.25$

$$\therefore \quad YR = 6.5\,\text{m}$$

So $XR - YR = 1.0\,\text{m} = \frac{1}{2}\lambda$

since wavelength $\lambda = 2.0$ m. There is destructive interference at R because the two sets of waves arrive with a path difference of $\frac{1}{2}\lambda$, i.e. 180° out of phase. The resultant amplitude at R will be zero,* so that an observer will hear nothing at R.

Answer

(a) A sound of double amplitude, (b) nothing.

**This ignores any difference in amplitude of the waves which may occur because R is further from X than Y.*

Example 2

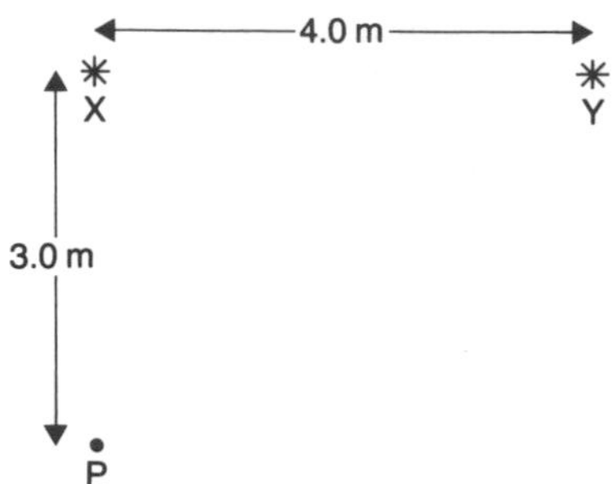

Fig 16.4 Information for Example 2

Fig 16.4 shows two sources X and Y which are identical and emit in phase. Calculate *two* possible values of wavelength for which (a) constructive interference, (b) destructive interference would occur at point P.

Method

We must calculate the path difference YP − XP. Using Pythagoras' theorem,

$$YP^2 = 3^2 + 4^2 = 25$$

$$\therefore \quad YP = 5.0\,\text{m}$$

$$\therefore \quad YP - XP = 2.0\,\text{m}$$

(a) For constructive interference $YP - XP = n\lambda$. Thus wavelength λ is given by $n\lambda = 2.0$ m

$$\therefore \quad \lambda = \frac{2.0}{n} \quad \text{where} \quad n = 0, 1, 2, 3, \ldots$$

For $n = 0$, $\lambda = \infty$ which is not practical. For $n = 1$, $\lambda = 2.0$ m.

For $n = 2$, $\lambda = 1.0$ m. Clearly other (smaller) values of λ are also suitable.

(b) For destructive interference $YP - XP = (n + \frac{1}{2})\lambda$. Thus wavelength λ is given by $(n + \frac{1}{2})\lambda = 2.0$.

$$\therefore \quad \lambda = \frac{2.0}{(n + \frac{1}{2})} \quad \text{where} \quad n = 0, 1, 2, 3, \ldots$$

For $n = 0$, $\lambda = 4.0$ m. For $n = 1$, $\lambda = 4/3$ m. Other (smaller) values of λ are also suitable.

Answer

(a) 2.0 m, 1.0 m, (b) 4.0 m, $\frac{4}{3}$ m.

Exercise 58

1 Referring to Fig 16.2, suppose that source X is 180° out of phase with source Y. What does an observer hear (a) at Q, (b) at R?

2

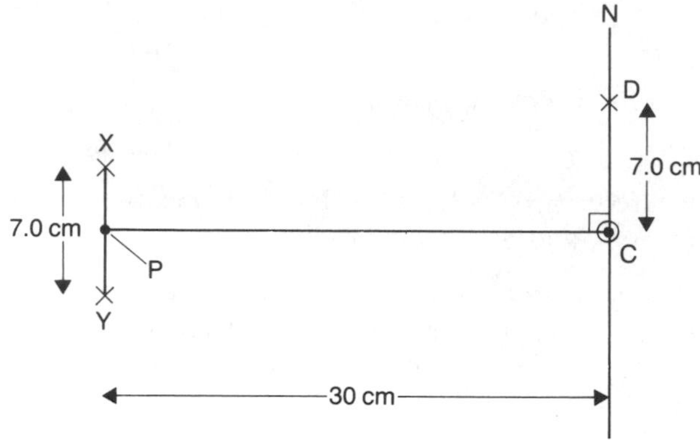

Fig 16.5 Information for Question 2

Fig 16.5 shows two identical microwave sources X and Y which emit in phase. There is constructive interference at C, which is on the perpendicular bisector of the line XY and 30 cm from P, the midpoint of XY. A detector moved from C towards N locates the first minimum at D. If CD = 7.0 cm calculate the wavelength of the microwaves emitted by X and Y.

3

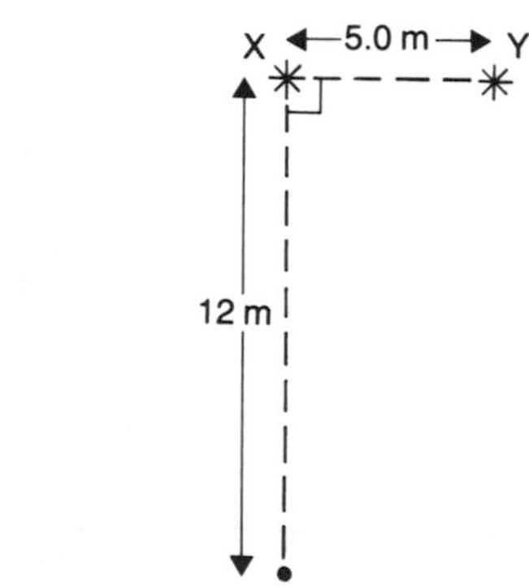

Fig 16.6 Diagram for Question 3

X and Y in Fig 16.6 are two identical sources of sound which emit in phase. Calculate the largest two values of wavelength (excluding $\lambda = \infty$) for which (a) constructive, (b) destructive interference will occur at Q. If the velocity of sound in air is $340\,\text{m}\,\text{s}^{-1}$, calculate the frequencies to which these wavelengths correspond.

4

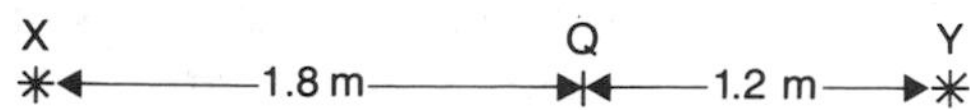

Fig 16.7 Diagram for Question 4

X and Y in Fig 16.7 are two identical sources of sound which emit in phase. Calculate the lowest possible value of frequency of the sources for there to be (a) constructive, (b) destructive interference at Q. (Velocity of sound = $340\,\mathrm{m\,s^{-1}}$.)

Young's double-slit arrangement

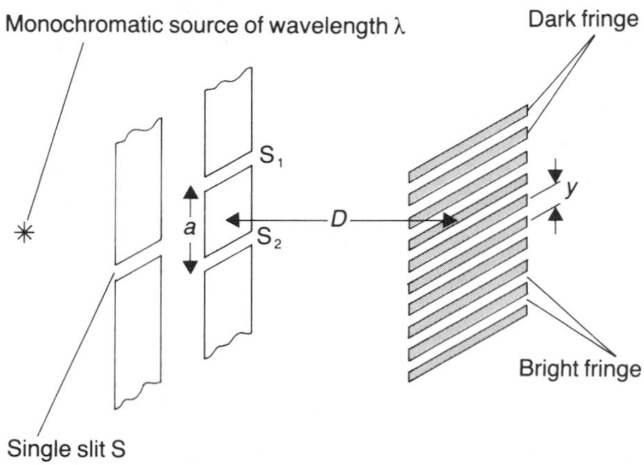

Fig 16.8 Young's double-slit arrangement

Fig.16.8 shows the set-up. The dark and bright fringes arise due to the interference of light emerging from two slits S_1 and S_2. In order that the sources S_1 and S_2 are coherent (i.e. phase-linked and of the same frequency) they must receive light from the same point on the source — this is ensured by diffraction of light at the single slit S.

The fringe separation y, in metres, is given by

$$y = \frac{\lambda D}{a} \tag{16.1}$$

where λ is the wavelength of source, in metres, D the distance, in metres, from slits to fringes and a the slit separation, in metres.

Example 3

In a Young's double-slit experiment, mercury green light of wavelength $0.54\,\mu\mathrm{m}$ $(0.54 \times 10^{-6}\,\mathrm{m})$ was used with a pair of parallel slits of separation 0.60 mm. The fringes were observed at a distance of 40 cm from the slits, Calculate the fringe separation.

Method

We have $\lambda = 0.54 \times 10^{-6}$, $a = 0.60 \times 10^{-3}$ and $D = 0.40$. Using Equation 16.1

$$y = \frac{\lambda D}{a} = \frac{0.54 \times 10^{-6} \times 0.40}{0.60 \times 10^{-3}}$$

$$= 0.36 \times 10^{-3}\,\mathrm{m}$$

Answer

Fringe separation = 0.36 mm.

Example 4

In a Young's double-slit arrangement green monochromatic light of wavelength $0.50\,\mu\mathrm{m}$ was used. Five fringes were found to occupy a distance of 4.0 mm on the screen. Calculate the fringe separation if (independently) (a) red light of wavelength $0.65\,\mu\mathrm{m}$ was used, (b) the slit separation was doubled, (c) the slits–screen distance was doubled.

Method

Five fringes occupy 4.0 mm. So the fringe separation is 4.0/5 = 0.80 mm.

(a) We see from Equation 16.1 that, for fixed D and a value, $y \propto \lambda$. If λ increases by a factor of $(0.65 \times 10^{-6}) \div (0.50 \times 10^{-6}) = 1.3$, then y will increase by a factor of 1.3. Thus y becomes

$$1.3 \times 0.80 = 1.04\,\mathrm{mm}$$

(b) For given λ and D values, $y \propto 1/a$. So if a is doubled, y becomes halved. Thus y becomes

$$0.50 \times 0.80 = 0.40\,\mathrm{mm}$$

(c) For given λ and a values $y \propto D$. So if D is doubled, y is doubled. Thus y becomes 1.6 mm.

Answer

(a) 1.0 mm, (b) 0.40 mm, (c) 1.6 mm.

Exercise 59

1 In a Young's double-slit experiment, sodium light of wavelength $0.59 \times 10^{-6}\,\mathrm{m}$ was used to illuminate a double slit with separation 0.36 mm. If the fringes are observed at a distance of 30 cm from the double slits, calculate the fringe separation.

2 In an experiment using Young's slits, six fringes* were found to occupy 3.0 mm when viewed at a distance of 36 cm from the double slits. If the wavelength of the light used is 0.59 μm, calculate the separation of the double slits.

3 When red monochromatic light of wavelength 0.70 μm is used in a Young's double-slit arrangement, fringes with separation 0.60 mm are observed. The slit separation is 0.40 mm. Find the fringe spacing if (independently)
(a) yellow light of wavelength 0.60 μm is used;
(b) the slit separation becomes 0.30 mm;
(c) the slit separation is 0.30 mm and the slits–fringe distance is doubled.

The air wedge

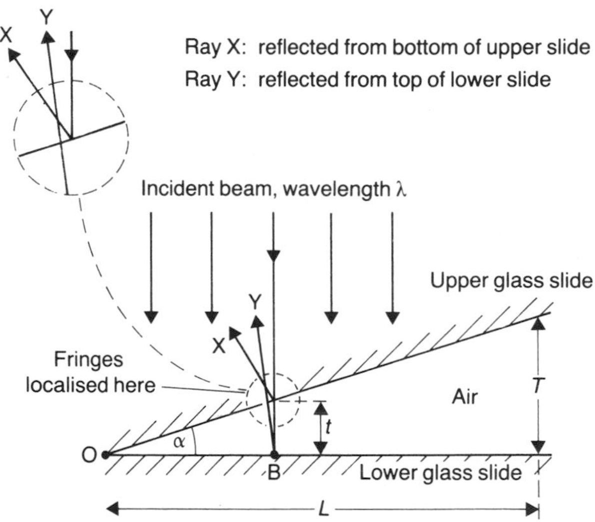

(a) The arrangement

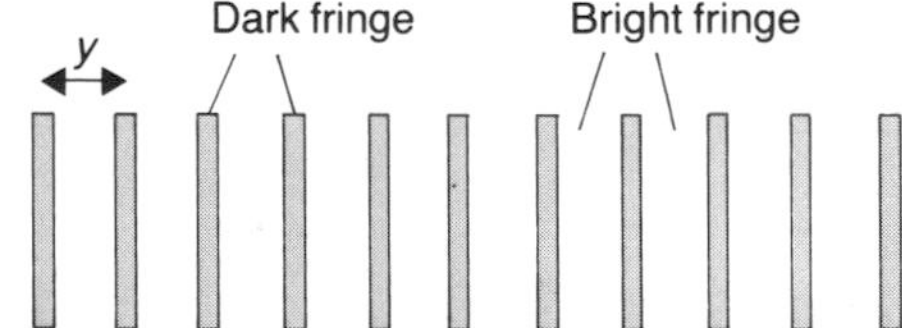

(b) Appearance of the fringes from above the wedge

Fig 16.9 The air wedge arrangement

One fringe means one fringe separation.

An air wedge is shown in Fig 16.9. Interference occurs between reflected rays X and Y resulting in fringes 'localised' where X and Y cross. The path difference $pd_{X\to Y}$ between rays X and Y is given by

$$pd_{X\to Y} = 2t + \frac{\lambda}{2} \qquad (16.2)$$

where t is the thickness of the wedge as shown, and $\lambda/2$ is the extra path difference introduced due to reflection at the less dense (air) to more dense (glass) reflection at B. The fringes follow contours of constant air-film thickness, and for plane surfaces are parallel straight fringes as shown in Fig 16.9b. The fringe separation y, in metres, is given by

$$y = \frac{\lambda}{2\alpha} \qquad (16.3)$$

where λ, in metres, is the wavelength of the light illuminating the wedge and α is the angle of the wedge, in radians. Also

$$\alpha = \frac{T}{L} \qquad (16.4)$$

where T, in metres, is the thickness of the wedge of length L, in metres, from the contact point O between upper and lower surfaces of the wedge.

Example 5

Interference fringes are formed in an air wedge using monochromatic light of wavelength 0.60×10^{-6} m. The fringes are formed parallel to the line of contact, and a dark fringe is observed along the line of contact. Calculate the thickness of the air wedge at positions where (a) the twentieth dark fringe and (b) the thirtieth bright fringe *from the line of contact* are observed.

Method

The path difference $pd_{X\to Y}$ is given by Equation 16.2.

(a) For a dark fringe the $pd_{X\to Y}$ must be an odd number of half wavelengths, i.e.

$$pd_{X\to Y} = \left(2t + \frac{\lambda}{2}\right) = (m + \tfrac{1}{2})\lambda \qquad (16.5)$$

where m, called the order of the fringe, has value 0, 1, 2, 3, . . . The first dark fringe, along the line of contact, must correspond to $pd_{X\to Y}$ equal to $\frac{1}{2}\lambda$, assuming 'perfect' (optical) contact at O in Fig 16.9. The dark fringe along the line of contact thus corresponds to $m = 0$ in Equation 16.5. We shall take the twentieth dark fringe as corresponding to $m = 20$. So, from Equation 16.5

$$(2t + \tfrac{1}{2}\lambda) = (m + \tfrac{1}{2})\lambda$$

or $$2t = m\lambda \qquad (16.6)$$

We have $m = 20$ and $\lambda = 0.60 \times 10^{-6}$. Rearranging Equation 16.6

$$t = \frac{m\lambda}{2} = \frac{20 \times 0.6 \times 10^{-6}}{2}$$

$$= 6.0 \times 10^{-6}\,\text{m}$$

(b) For a bright fringe the $pd_{X \to Y}$ must be a whole number of wavelengths. That is

$$pd_{X \to Y} = 2t + \tfrac{1}{2}\lambda = m\lambda \qquad (16.7)$$

where $m = 1, 2, 3, \ldots$ Note that the first bright fringe corresponds to a path difference of λ, so $m = 1$. We shall take the thirtieth bright fringe as corresponding to $m = 30$. Rearranging Equation 16.7

$$t = \tfrac{1}{2}(m - \tfrac{1}{2})\lambda$$

where $m = 30$ and $\lambda = 0.60 \times 10^{-6}$. So

$$t = \tfrac{1}{2}(30 - \tfrac{1}{2}) \times 0.6 \times 10^{-6}$$

$$= 8.85 \times 10^{-6}\,\text{m}$$

Answer

(a) 6.0 μm, (b) 8.9 μm.

Example 6

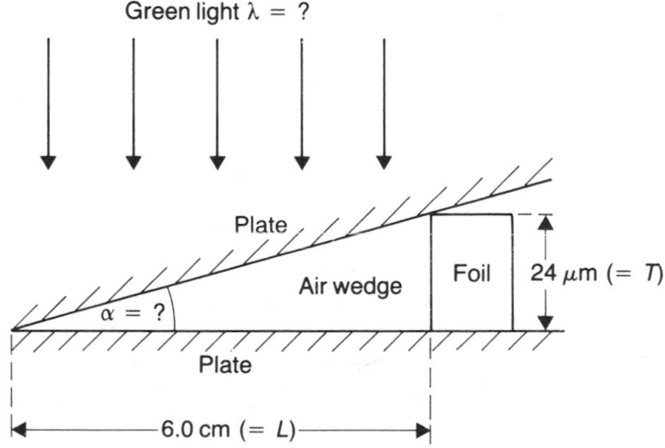

Fig 16.10 Information for Example 6

An air wedge is formed using two plane glass plates which are in contact along one edge. At a distance of 6.0 cm from this contact line the plates are separated by a piece of foil which is 24×10^{-6} m thick, as shown in Fig 16.10. Interference fringes, parallel to the line of contact, are observed using monochromatic green light. If the fringes are of spacing 0.70 mm, calculate (a) the angle of the wedge, (b) the wavelength of the light used, (c) the fringe spacing if red light of wavelength 0.65 μm is used, the angle of the wedge being unchanged.

Method

(a) To find α we use Equation 16.4 in which T, which equals 24×10^{-6} m, corresponds to the thickness of the foil and L, which equals 6.0×10^{-2} m, corresponds to the distance from the foil to the line of contact. Thus

$$\alpha = \frac{T}{L} = \frac{24 \times 10^{-6}}{6 \times 10^{-2}} = 4.0 \times 10^{-4}\,\text{rad}$$

(b) We have $\alpha = 4 \times 10^{-4}$, fringe separation $y = 0.70 \times 10^{-3}$ m and require wavelength λ. Rearranging Equation 16.3 gives

$$\lambda = 2\alpha y = 2 \times 4 \times 10^{-4} \times 0.7 \times 10^{-3}$$

$$= 0.56 \times 10^{-6}\,\text{m}$$

(c) Let y' be the fringe spacing using red light. To find y' we use Equation 16.3 with $\lambda' = 0.65 \times 10^{-6}$ and $\alpha = 4.0 \times 10^{-4}$ rad. Thus

$$y' = \frac{\lambda'}{2\alpha} = \frac{0.65 \times 10^{-6}}{2 \times 4.0 \times 10^{-4}}$$

$$= 0.81 \times 10^{-3}\,\text{m}$$

Alternatively we note that y is directly proportional to λ for a given α. So

$$\frac{y'}{y} = \frac{\lambda'}{\lambda}$$

where $y = 0.70$, $\lambda = 0.56\,\mu$m and $\lambda' = 0.65\,\mu$m. Thus

$$\frac{y'}{0.70} = \frac{0.65}{0.56}$$

$$\therefore \quad y' = 0.81\,\text{mm}$$

Answer

(a) 4.0×10^{-4} rad, (b) 0.56 μm, (c) 0.81 mm.

Exercise 60

1 When interference fringes are formed using an air wedge, it is found that the twentieth bright fringe is formed at an air thickness of 6.8 μm. Calculate (a) the wavelength of the light used, (b) the thickness of the air wedge for the tenth dark fringe.

2 Two plane glass plates are in contact at one edge and are separated by a piece of thin wire 12 cm from that edge. When illuminated with yellow light of wavelength 0.60×10^{-6} m interference fringes are observed parallel to the line of contact, with separation 0.15 mm. Calculate (a) the angle of the air wedge, (b) the diameter of the wire.

3 Interference fringes of separation 0.40 mm are observed when an air wedge is illuminated with yellow light of wavelength 0.60 μm. Calculate the fringe spacing if blue light of wavelength 0.45 μm is used. Hint: y is directly proportional to λ for a given angle of the wedge.

4 Interference fringes of spacing 1.0 mm are obtained using light of wavelength λ incident on an air wedge of angle α. The angle of the wedge is now doubled and the light replaced by one of wavelength 1.5λ. Calculate the new fringe separation.

5 An air wedge is illuminated normally with yellow light of wavelength 0.60×10^{-6} m and blue light of wavelength 0.45×10^{-6} m. This produces two fringe patterns of different spacing. Use Equation 16.6 to show that the third-order dark fringe in the yellow fringe system coincides with the fourth-order dark fringe of the blue fringe system. Calculate (a) the thickness of the air wedge at this position, (b) the angle of the wedge if this occurs at 9.0 mm from the line of contact of the plates. Explain why the pattern immediately on each side of this dark fringe appears white.

Effect of refractive index

Fig 16.11 indicates the effect of introducing a material of refractive index n instead of air. The wavelength of the waves inside the medium becomes $1/n$ times that in air. This means that if the air in Young's slits or the air wedge is replaced by such a medium, then the fringe spacing would become $1/n$ of its original value using air. In

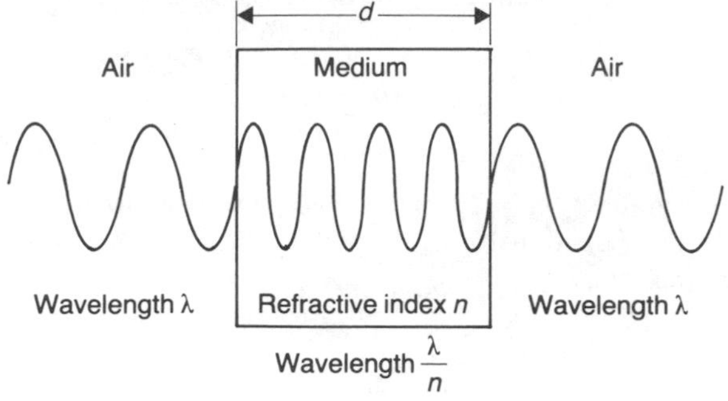

Fig 16.11 Change of refractive index and wavelength

addition a given thickness of material contains n times as many waves as the same thickness of air. Thus a thickness d of material in one of the beam paths (say in front of one of the slits in Young's slits) is equivalent to an *extra* thickness $(n - 1)d$ of air.

Example 7

Interference fringes of separation 0.90 mm are formed in an air wedge. If the air is replaced by a liquid of refractive index 1.3, all other factors remaining unchanged, calculate the new fringe separation.

Method

The new fringe spacing y' is $1/n$ times that using air. Thus, since $n = 1.3$.

$$y' = \frac{1}{1.3} \times 0.90 \times 10^{-3}$$

$$= 0.69 \times 10^{-3}\ \text{m}$$

Answer

The new fringe separation is 0.69 mm.

Example 8

In a Young's-slits experiment, light of wavelength 0.55 μm is used to illuminate the double slits. When *one* of these has a thin slice of material of refractive index 1.5 placed over it, the fringe pattern is displaced to one side by a distance which corresponds to 40 fringe separations. Calculate the thickness of the slice of material.

Method

Since the fringe pattern is displaced by 40 fringes, this means that an extra 40 wavelengths path difference has been introduced by inserting the slice of material. But (see previous page) a thickness d of material is equivalent to an *extra* thickness $(n - 1)d$ of air. Thus

$$40\lambda = (n - 1)d$$

where $\lambda = 0.55 \times 10^{-6}$ m, $n = 1.5$ and we require d. Rearranging gives

$$d = \frac{40\lambda}{(n-1)} = \frac{40 \times 0.55 \times 10^{-6}}{(1.5 - 1)}$$
$$= 44 \times 10^{-6}\ \text{m}$$

Answer
The material is 44 μm thick.

Exercise 61

1 Interference fringes of separation 0.72 mm are obtained using an air wedge. When the air is replaced by a liquid the new fringe spacing is 0.54 mm. Calculate the refractive index of the liquid.

2 In an air wedge the air has been replaced by a liquid of refractive index 1.4. The interference fringes have separation 0.50 mm. What would be the separation if the liquid were replaced by one of refractive index 1.3?

3 In a Young's-slits experiment light of wavelength 0.60 μm illuminates the double slits and produces fringes of spacing 0.50 mm. When one of the slits has a film of material of thickness 50 μm placed over it, the fringe pattern is displaced to one side by 22 mm. Calculate the refractive index of the material.

Beats

Beats occur during the superposition of waves with different frequencies. Fig 16.12a and b show the displacement–time graphs of two waves of frequency f_1 and f_2. Here $f_2 > f_1$ and the amplitudes

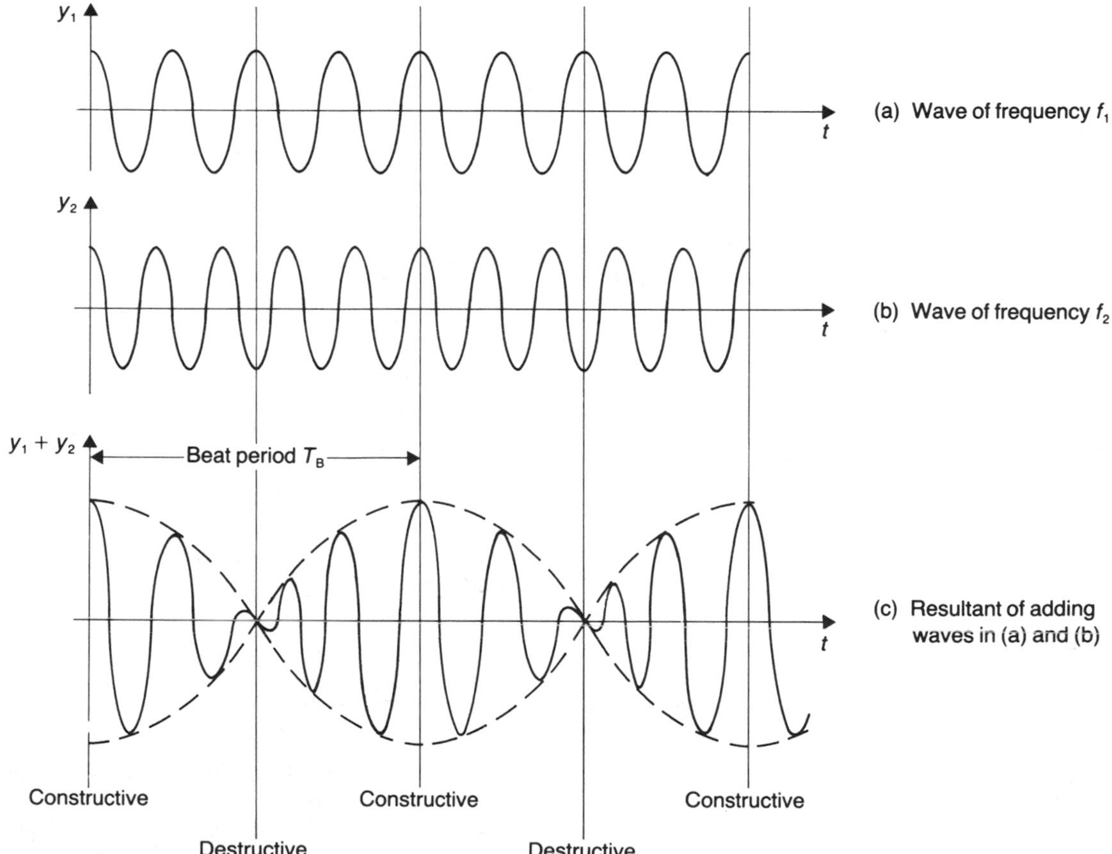

Fig 16.12 Superposition of two waves of different frequency

are the same. Fig 16.12c shows the resultant displacement, which has positions of maximum and zero disturbance. This gives rise to 'beats'. If the beat period is T_B seconds and the beat frequency f_B hertz, then

$$f_B = \frac{1}{T_B} = (f_2 - f_1) \qquad (16.8)$$

Example 9

A tuning fork of frequency 256 Hz and one of frequency 264 Hz are sounded together. Calculate (a) the beat frequency, (b) the beat period.

Method

We have $f_1 = 256$ and $f_2 = 264$.

(a) To find f_B we use Equation 16.8:

$$f_B = f_2 - f_1 = 264 - 256$$
$$= 8\text{ Hz}$$

(b) Rearranging Equation 16.8,

$$T_B = \frac{1}{f_B} = \tfrac{1}{8}\text{ s}$$

Answer

(a) 8 Hz, (b) $\frac{1}{8}$ s.

Example 10

A loudspeaker emits a note which gives a beat frequency of 4 Hz when sounded with a standard tuning fork of frequency 280 Hz. The beat frequency *decreases* when the fork is 'loaded' by adding a small piece of plasticine to its prongs. Calculate the frequency of the note emitted by the loudspeaker.

Method

Let the loudspeaker note be of frequency f_{unknown}. We know that f_2 differs from the fork frequency f_{known} (= 280 Hz) by 4 Hz. However *we do not know if f_{unknown} is greater than or less than f_{known}*. This is decided by noting that *when a tuning fork is loaded, its frequency decreases*. Now in this case the beat frequency also decreases on loading, so that f_{unknown} must be less than f_{known} (a beat frequency decrease implies the frequency of the loaded fork is closer to that of the loudspeaker note than that of the unloaded fork). We thus use Equation 16.8 with the lower frequency $f_1 = f_{\text{unknown}}$ and the higher frequency $f_2 = f_{\text{known}} = 280$. Because we have $f_B = 4$,

$$280 - f_{\text{unknown}} = 4$$
$$\therefore \quad f_{\text{unknown}} = 276\text{ Hz}$$

Answer

The loudspeaker note has frequency 276 Hz.

Example 11

Two simple pendulums, one of length 0.600 m and one of length 0.500 m are made to oscillate and commence in step. Calculate how long it will be before one pendulum has made exactly one more oscillation than the other. (Assume $g = 10\text{ m s}^{-2}$.)

Method

Problems similar to this were solved in Chapter 14 on simple harmonic motion. If the numbers are not simple, this type of problem can be solved more easily using the ideas of beat frequency and beat period. We call the time required for one pendulum to make one oscillation more than the other the beat period T_B. To find T_B we first calculate f_B, the beat frequency, by calculating the difference between the frequencies of both pendulums.

The periodic time T of a simple pendulum of length l is, from Equation 14.6, given by

$$T = 2\pi\sqrt{\frac{l}{g}}$$

Since $f = 1/T$, the frequency f of oscillation is given by

$$f = \frac{1}{2\pi}\sqrt{\frac{g}{l}}$$

where $g = 10\text{ m s}^{-2}$. For the pendulum of length $l_1 = 0.600$ m the frequency f_1 is

$$f_1 = \frac{1}{2\pi}\sqrt{\frac{g}{l_1}} = \frac{1}{2\pi}\sqrt{\frac{10}{0.60}} = 0.650\text{ Hz}$$

For the pendulum of length $l_2 = 0.500$ m we have

$$f_2 = \frac{1}{2\pi}\sqrt{\frac{g}{l_2}} = \frac{1}{2\pi}\sqrt{\frac{10}{0.50}} = 0.712\text{ Hz}$$

Thus, from Equation 16.8,

$$f_B = f_2 - f_1 = 0.712 - 0.650$$
$$= 0.062\text{ Hz}$$

and $$T_B = \frac{1}{f_B} = \frac{1}{0.062} = 16.1\text{ s}$$

Answer

16 s elapses before one pendulum makes one oscillation more than the other.

Exercise 62

1 A note from a loudspeaker gives a beat frequency of 10 Hz when sounded with a tuning fork of frequency 440 Hz. Calculate (a) the beat period, (b) two possible values for the frequency of the note emitted by the loudspeaker.

2 A vibrating sonometer wire emits a note which gives a beat frequency of 6.0 Hz when sounded in unison with a standard tuning fork of frequency 256 Hz. When the fork is loaded the beat frequency increases. What is the frequency of the note emitted by the sonometer?

3 Two identical springs of force constant 20 N m^{-1} are hung side by side. One has a mass of 0.200 kg attached to its lower end and the other a mass of 0.250 kg. Calculate (a) the frequency of oscillation of each mass–spring system, (b) the time period which elapses during which one system performs exactly one more oscillation than the other. Note: from Chapter 14, $T = 2\pi\sqrt{(m/k)}$ for a spring–mass system.

4 A simple pendulum is set up to swing in front of a clock pendulum with a periodic time of 2.00 s. The simple pendulum 'gains' on the clock, and the two swing in phase at intervals of 24 s. Which has the higher frequency? Calculate (a) the 'beat frequency', (b) the frequency of the clock pendulum, (c) the frequency of the simple pendulum.

Exercise 63: Questions of A-level standard

1 In Fig 16.13 X and Y are two generators of water waves of wavelength 0.50 m. Each of the generators, when operating on its own, produces waves of amplitude 60 mm at P, which is 2.00 m from X.

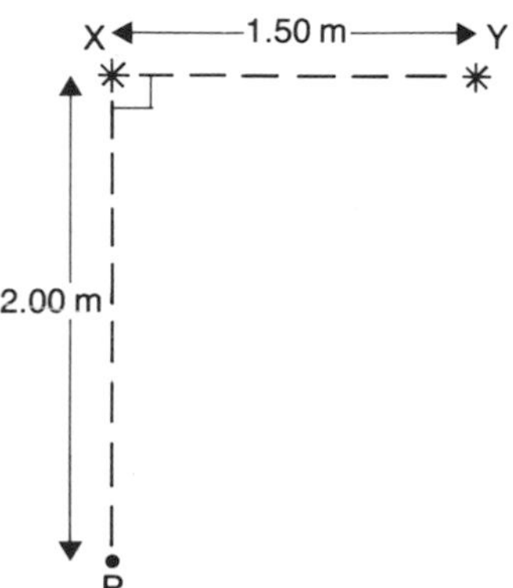

Fig 16.13 Diagram for Question 1

Find the amplitude of the resultant disturbance at P when the generators X and Y are operating (a) in phase, (b) 180° out of phase.

2

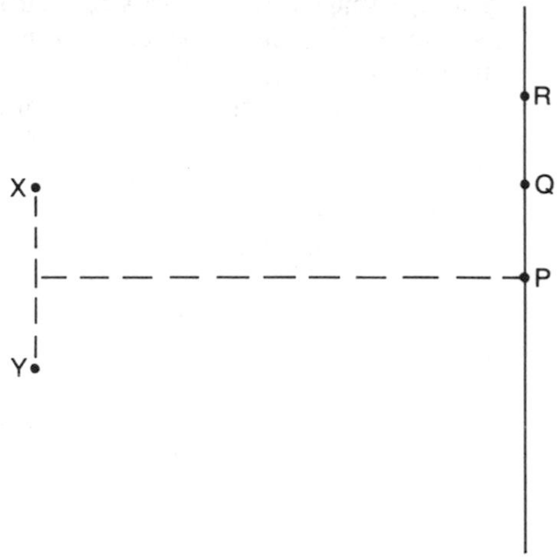

Fig 16.14 Diagram for Question 2

Two sources of waves of the same wavelength λ, phase and amplitude are situated at X and Y (Fig 16.14). At P there is a maximum, at Q a minimum and at R a maximum disturbance. Such an observation may be obtained if

1. XP = YP
2. YQ − XQ = λ
3. YR − XR = $\lambda/2$ [AEB 82]

3 Coherent light of wavelength 590 nm illuminates double slits of separation 10^{-4} m. A graticule is placed 0.2 m from the plane of the slits, and an eyepiece, focused on the graticule, is used to observe the interference fringes. Calculate the spacing of the fringes as measured by using the graticule. [O & C 82, p]

4 A Young's-slits experiment is intially performed in air. If the space between the double slit and the screen where the fringes are observed is filled with water of refractive index 4/3, the distance between the fringes

A remains the same

B is decreased to $\frac{1}{4}$ of its previous value

C is decreased to $\frac{1}{3}$ of its previous value

D is decreased to $\frac{3}{4}$ of its previous value

E is increased to $\frac{4}{3}$ of its previous value

[O & C 80]

5 (a) A wedge-shaped film of air is formed between two thin parallel-sided glass plates by means of a straight piece of wire. The two plates are in contact along one edge of the film and the wire is parallel to this edge.

(i) Draw and label a diagram of the experimental arrangement you would use to observe and make measurements on interference fringes produced with light normally incident on the film.

(ii) Explain the function of each part of the apparatus.

(b) In such an experiment, using light of wavelength 589 nm, the distance between the seventh and one hundred and sixty-seventh dark fringes was 26.3 mm, and the distance between the junction of the glass plates and the wire was 35.6 mm. Calculate the angle of the wedge and the diameter of the wire. [JMB 79]

6 A tuning fork of frequency 512 Hz has one of its prongs loaded with a small piece of plasticine. When sounded together with an unloaded fork of frequency 512 Hz, a beat frequency of 4 Hz is heard. The frequency of the loaded fork is therefore

A 504 Hz **B** 508 Hz **C** 510 Hz
D 514 Hz **E** 516 Hz [L 81]

17 DIFFRACTION AND THE DIFFRACTION GRATING

Diffraction

When waves pass through an aperture or meet an obstacle, the waves spread to some extent into a region of geometrical shadow. This effect is called diffraction. Calculations are usually restricted to the transmission grating, designed for visible light (i.e. the optical grating), and its use in a spectrometer.

The optical diffraction grating

A transmission grating consists of many parallel equidistant slits of width and spacing of the order of the wavelength of light. If plane waves (parallel light) are incident on it, then, by superposition of the secondary wavelets from each slit, it can be shown that a transmitted wavefront is formed only along a few specified directions.

If the incident parallel beam is at normal incidence (see Figs 17.1 and 17.2), then emergent parallel beams are seen only in directions such that

$$d \sin \theta = n\lambda \qquad (17.1)$$

where d is the spacing of the slits, n $(= 0, 1, 2, \ldots)$ the order of diffracted beam, λ the wavelength of incident light and θ the angle of diffracted beam to the normal.

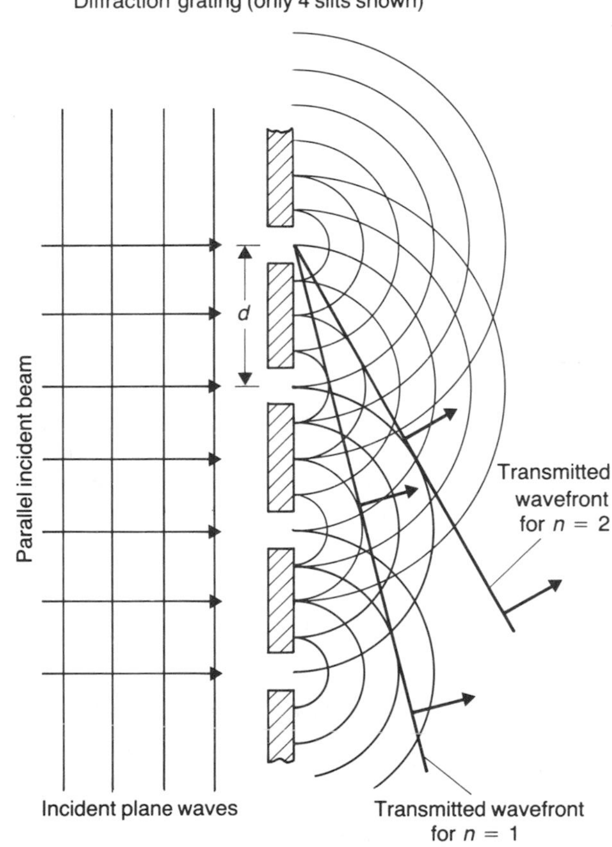

Fig 17.1 Action of the diffraction grating: formation of transmitted wavefronts

For constructive interference between beams from adjacent slits, path difference must be $n\lambda$

So: $d \sin \theta = n\lambda$

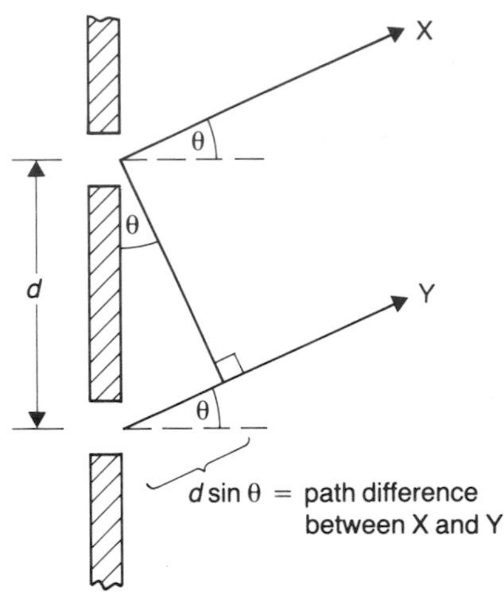

Fig 17.2 Action of the diffraction grating showing $d \sin \theta = n\lambda$

(iv) $n = 4$ gives $\sin \theta_3 = 1.2$, which is impossible (see Chapter 2). Thus the fourth order is not observed.

(b) Fig 17.3 is a schematic diagram showing the positions of the various maxima. Note that for $n = 0$ (the zeroth order) $\theta = 0$. Thus seven diffracted maxima are observed.

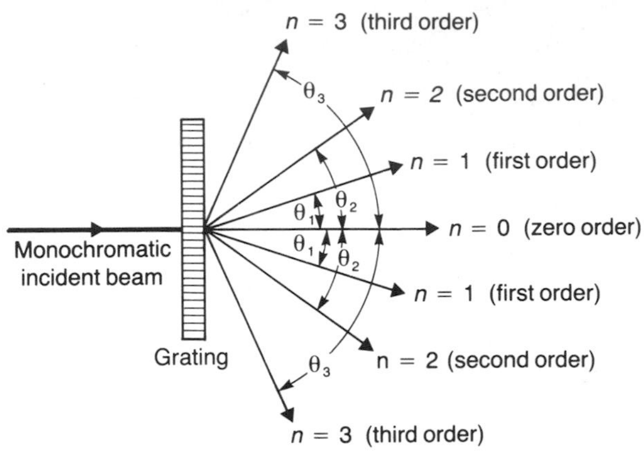

Fig 17.3 Angular distribution of diffracted beams

The following examples involve use of Equation 17.1.

Example 1

Monochromatic light of wavelength 600 nm is incident normally on an optical transmission grating of spacing 2.00 μm. Calculate (a) the angular positions of the maxima; (b) the number of diffracted beams which can be observed; (c) the maximum order possible.

Method

We are given

$$\lambda = 600 \times 10^{-9}\,\text{m}$$

$$d = 2.00 \times 10^{-6}\,\text{m}$$

(a) We substitute into Equation 17.1 as follows:

(i) for $n = 1$

$$2.00 \times 10^{-6} \times \sin \theta_1 = 1 \times 600 \times 10^{-9}$$

This gives $\sin \theta_1 = 0.3$ or $\theta_1 = 17.5°$

(ii) for $n = 2$

$$2.00 \times 10^{-6} \times \sin \theta_2 = 2 \times 600 \times 10^{-9}$$

This gives $\sin \theta_2 = 0.6$, or $\theta_2 = 36.9°$

(iii) $n = 3$ gives $\sin \theta_3 = 0.9$, or $\theta_3 = 64.2°$

(c) This has been covered in part (a), which shows that since $n = 4$ is impossible, the maximum order is 3. A quicker way to do this is as follows:

$$\sin \theta \leqslant 1$$

From Equation 17.1, $\sin \theta = n\lambda/d$. So

$$\frac{n\lambda}{d} \leqslant 1$$

or $$n \leqslant \frac{d}{\lambda} = \frac{2.00 \times 10^{-6}}{600 \times 10^{-9}} = 3.33$$

$$\therefore \quad n \leqslant 3.33$$

Since n must be an integer its maximum value is 3.

Answer

(a) The angular positions are

$n = 1$	$\theta_1 = 17.5°$
$n = 2$	$\theta_2 = 36.9°$
$n = 3$	$\theta_3 = 64.2°$

Note the trivial case of $\theta = 0$ for $n = 0$.

(b) There are seven diffracted beams.

(c) The maximum order is $n = 3$.

Example 2

Light consisting of wavelengths 420 nm and 650 nm is incident normally on a transmission grating of 6.00×10^5 lines m^{-1}. Calculate the angular separation of the wavelengths in the second-order spectrum.

Method

There are 6.00×10^5 lines per metre of grating. So the grating spacing d is given by

$$d = \frac{1}{6.00 \times 10^5} = 1.666 \times 10^{-6}\,\text{m}$$

Using Equation 17.1, for the second-order spectrum ($n = 2$) we have

(a) for $\lambda = 420 \times 10^{-9}$ m

$$1.666 \times 10^{-6} \times \sin\theta_2 = 2 \times 420 \times 10^{-9}$$

This gives $\sin\theta_2 = 0.504$ and $\theta_2 = 30.3°$.

(b) for $\lambda' = 650 \times 10^{-9}$ m

$$1.666 \times 10^{-6} \times \sin\theta'_2 = 2 \times 650 \times 10^{-9}$$

This gives $\sin\theta'_2{}' = 0.780$ and $\theta'_2 = 51.3°$.

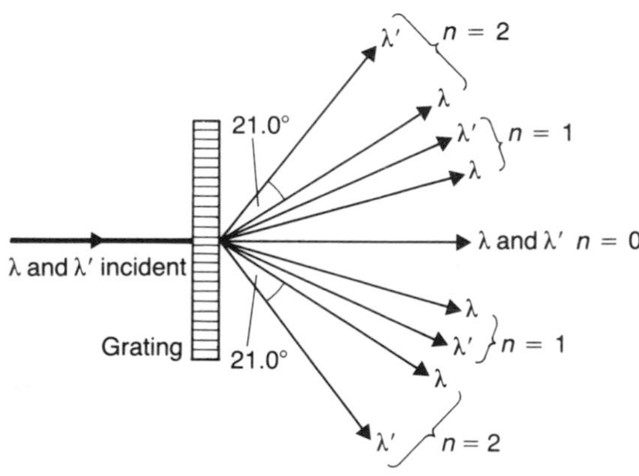

Fig 17.4 Angular separation as in Example 2

A schematic diagram of the situation is given in Fig 17.4. The angular separation

$$\theta'_2 - \theta_2 = 51.3 - 30.3° = 21.0°$$

Answer

The angular separation in the second-order spectrum is 21.0°.

Example 3

White light which has been passed through a certain filter has a range of wavelengths from 450 nm to 700 nm. It is incident normally on a diffraction grating. Show that if there are second- and third-order spectra, they will overlap.

Method

For any particular grating the angle of diffraction, for a given order, is greater for the longer wavelengths. This is seen by rearranging Equation 17.1:

$$\sin\theta = \frac{n\lambda}{d} \qquad (17.2)$$

Thus, for a given d and n value $\sin\theta \propto \lambda$.

We must therefore show that the second-order red (700 nm) has a higher θ value than the third-order blue (450 nm). For the given grating the d value is constant, so Equation 17.2 becomes

$$\sin\theta = \text{Constant} \times n\lambda$$

For $\lambda_1 = 700$ nm in the second order ($n = 2$)

$$\sin\theta_1 = \text{Constant} \times 2 \times 700 \times 10^{-9}$$
$$= \text{Constant} \times 1.40 \times 10^{-6}$$

For $\lambda_2 = 450$ nm in the third order ($n = 3$)

$$\sin\theta_2 = \text{Constant} \times 3 \times 450 \times 10^{-9}$$
$$= \text{Constant} \times 1.35 \times 10^{-6}$$

Since $\sin\theta_1 > \sin\theta_2$, then $\theta_1 > \theta_2$. So the second order at the red end overlaps with the third order at the blue end.

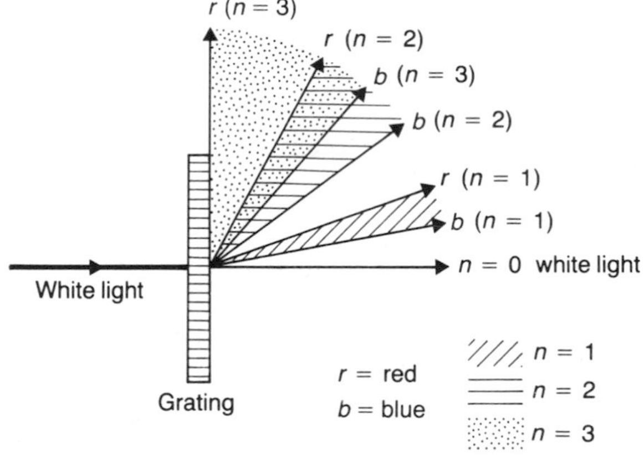

Fig 17.5 Appearance of diffraction spectra using white light (upper half only is shown)

Fig 17.5 is a schematic diagram of the white light diffraction spectra using a typical grating. The angular spread in a given order, and the maximum order, depend upon the grating spacing. However, the second- and third- and higher-order spectra (if present) will always overlap with each other as discussed above.

Exercise 64

1 What is the wavelength of light which gives a first-order maximum at an angle of 22° 30′ when incident normally on a grating with 600 lines mm^{-1}?

2 Light of wavelength 600 nm is incident normally on a diffraction grating of width 20.0 mm, on which 10.0×10^3 lines have been ruled. Calculate the angular positions of the various orders.

3 A source emits spectral lines of wavelength 589 nm and 615 nm. This light is incident normally on a diffraction grating having 600 lines per mm. Calculate the angular separation between the first-order diffracted waves. Find the maximum order for each of the wavelengths.

4 When a certain grating is illuminated normally by monochromatic light of wavelength 600 nm, the first-order maximum is observed at an angle of 21.1°. If the same grating is now illuminated with light with wavelength from 500 nm to 700 nm, find the angular spread of the first-order spectrum.

Exercise 65: Questions of A-level standard

1 A spectral line of known wavelength (5.792×10^{-7} m) emitted from a mercury vapour lamp is used to determine the spacing between the lines ruled on a plane diffraction grating. When the light is incident normally on the grating, the third-order spectrum, measured using a spectrometer, occurs at an angle of 60° 19′ to the normal. Calculate the grating spacing.

Why is the value obtained using the third-order spectrum likely to be more accurate than if the first-order were used? [L 81]

2 (a) A spectrometer and diffraction grating are adjusted to view the spectrum of a source of light, the plane of the grating being normal to the incident parallel beam. The source emits four discrete wavelengths which are listed in the table below together with the setting of the telescope crosswires on the first-order diffraction maxima.

Wavelength/nm	*Telescope setting*	
	Left	*Right*
448	165.7°	194.4°
501	164.0°	196.0°
588	161.1°	198.9°
668	158.4°	201.6°

If the setting of the crosswires on the central diffraction maximum is 180.0°, use the data to draw a straight-line graph, and use the graph to determine the number of lines per metre ruled on the grating.

(b) (i) The source is replaced by a monochromatic one. The crosswire settings for the second-order diffraction maxima are 136.8° (left), 223.2° (right). Calculate the wavelength of the light emitted by the source.

(ii) Excluding the zero-order, what is the total number of diffraction maxima produced? Give the reason for your answer. [JMB 82, p]

3 Monochromatic radiation of wavelength 600 nm, after passing through the slit and collimator of a spectrometer, falls normally on a transmission diffraction grating with 300 lines per mm. When the telescope is moved from the straight-through position through 90°, the number of diffraction maxima (excluding the undeviated beam) which can be observed is

A 1 **B** 2 **C** 5 **D** 6 **E** 11 [AEB 82]

4 Light from a mercury lamp is incident normally on a plane diffraction grating ruled with 6000 lines per cm. The spectrum contains two strong yellow lines of wavelengths 577 nm and 579 nm. What is the angular separation of the second-order diffracted beams corresponding to these two wavelengths? [C 80]

5 A source emits light of two distinct wavelengths, one of which is 600 nm. When the light falls normally on a diffraction grating, it is found that the third-order image formed by the light of wavelength 600 nm coincides with the fourth-order image for the other wavelength. The angle of diffraction for each image is 46°. Calculate the second wavelength emitted by the source and the number of lines per metre of the grating. [AEB 80]

6 Two monochromatic radiations X and Y fall normally on a diffraction grating, and the second-order intensity maximum for X emerges at the same angle as the third-order intensity maximum for Y. The ratio of the wavelength of X to that of Y is

A $\frac{2}{3}$ **B** $\frac{3}{2}$ **C** 2 **D** $\frac{1}{2}$ **E** $\sin^{-1}(\frac{2}{3})$

[O & C 81]

7 (a) The line spectrum of a certain substance consists of three prominent lines: blue (B), yellow (Y) and red (R). When the spectrum is examined with a diffraction grating of spacing 4×10^{-6} m, it is found that the sequence of lines, moving from the centre, is B Y R B Y B R. Give a brief explanation for this.

(b) Further, it is found that the diffraction angles θ of the fifth and seventh lines are 17.46° and 20.49° respectively, and that the sixth line is at an angular position exactly halfway between them. Find the wavelength of the blue line.

[WJEC 82, p]

18 STATIONARY WAVES

Formation of stationary waves

Stationary waves occur as a result of interference between progressive waves (see Chapter 16) of the same frequency and wavelength travelling along the same line. They may be formed due to interference between waves from two separate sources, as shown in Fig 18.1, or alternatively, due to interference between incident and reflected waves (see below).

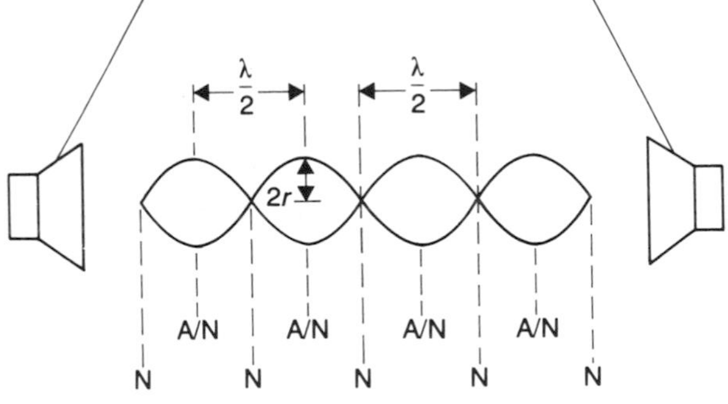

Fig 18.1 Formation of a stationary wave in the region between two sources of progressive waves.

If the two progressive waves which form the stationary wave have equal amplitude r, then the nodes, which are positions of permanent destructive interference, have zero amplitude. The antinodes, which are positions of maximum constructive interference, have amplitude $2r$. As shown in Fig 18.1, the separation of adjacent nodes and of adjacent antinodes is $\lambda/2$, where λ is the wavelength of the progressive waves from which the stationary wave is formed.

Example 1

Two loudspeakers which are connected to the same oscillator face each other and are separated by a distance of about 3 m. A small microphone, placed approximately midway along the line between the two loudspeakers, records positions of minimum intensity, which are separated by 4.2 cm. If the oscillator is set at a frequency 4.0 kHz, calculate the speed of sound in air.

Method

The loudspeaker separation of about 3 m happens to be a convenient distance, but is irrelevant in so far as calculation of the speed of sound is concerned. The microphone is moved around the midway position, since here the amplitude of the two waves arriving from the two sources will be about the same, so the nodes can be more accurately located.

The nodes are 4.2 cm apart. So $\lambda/2 = 4.2$ cm, hence wavelength $\lambda = 8.4\text{ cm} = 8.4 \times 10^{-2}$ m. Also we know frequency $f = 4.0 \times 10^3$ Hz. To find the speed of sound c we use Equation 15.1, i.e.

$$c = f\lambda = 4.0 \times 10^3 \times 8.4 \times 10^{-2}$$
$$= 336 \text{ m s}^{-1}$$

Answer

Speed of sound $= 0.34 \text{ km s}^{-1}$.

Note that the wavelength λ and speed c relate to the progressive waves which make up the stationary wave.

Example 2

A microwave transmitter is aimed at a metal plate, as shown in Fig 18.2.

(a) A small detector, moved along the line XY, travels 14 cm in moving from the first to the eleventh consecutive nodal position, Calculate the frequency of the microwaves emitted.

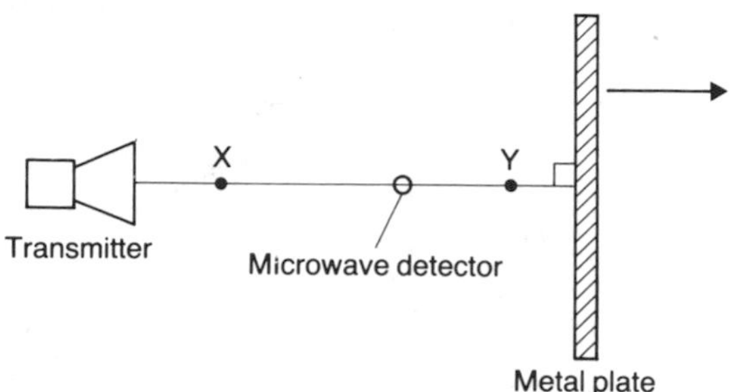

Fig 18.2 Information for Example 2

(b) The detector is now fixed in position and the metal plate is moved to the right, along the direction XY at a speed of $28\,\text{cm s}^{-1}$. Explain what the detector observes.
Assume that the speed of electromagnetic waves is $3.0 \times 10^8\,\text{m s}^{-1}$.

Method

(a) Between the first and eleventh nodes there are ten half-wavelengths. Thus $10 \times \lambda/2 = 14\,\text{cm}$, so wavelength $\lambda = 2.8\,\text{cm} = 2.8 \times 10^{-2}\,\text{m}$. We are given speed $c = 3.0 \times 10^8$; to find the frequency f we rearrange Equation 15.1:

$$f = \frac{c}{\lambda} = \frac{3 \times 10^8}{2.8 \times 10^{-2}}$$
$$= 1.07 \times 10^{10}\,\text{Hz}$$

(b) As the detector moves to the right, the stationary wave pattern moves also — and at the same speed, since there must always be a node* at the metal plate (it is a 'perfect' reflector). In 1 second a 28 cm 'length' of stationary wave will pass the detector, which will thus observe $(28 \div 1.4) = 20$ nodes and 20 antinodes.

Answer

(a) 1.1×10^{10} Hz, (b) the detector observes 20 success ive maxima, followed by minima, each second.

Exercise 66

1 Two loudspeakers face each other and are separated by a distance of about 20 m. They are connected to the same oscillator, which gives a signal frequency of 800 Hz.

(a) Calculate the separation of adjacent nodes along the line joining the two loudspeakers.

(b) A small microphone, moved at constant speed along this line, records a signal which varies periodically at 5.0 Hz. Calculate the speed at which the microphone moves.
Assume that the speed of sound is $340\,\text{m s}^{-1}$.

2 A source S of microwaves faces a detector D. A metal reflecting screen is now placed beyond D with its plane perpendicular to the line from S to D. As the screen is moved slowly away from D, the detector registers a series of maximum and minimum readings, the screen being displaced a distance of 5.6 cm between the first and fifth minimum. Calculate the wavelengths and frequency of the microwaves. Assume $c = 3.0 \times 10^8\,\text{m s}^{-1}$.

Electric field node.

Stationary waves in strings and wires

When a string or wire which is fixed at both ends is plucked, progressive transverse waves travel along the string or wire and are reflected at its ends. This results in the formation of stationary waves with certain allowed wavelengths and frequencies. Fig 18.3 shows the fundamental, which has the largest wavelength and hence the smallest frequency, and the first two overtones.

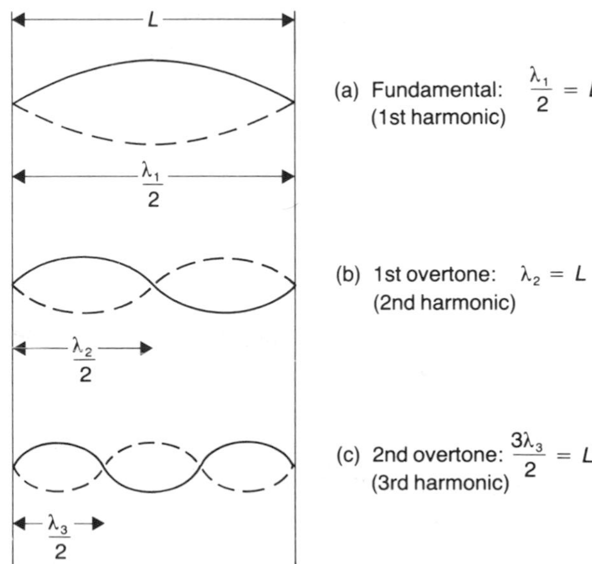

Fig 18.3 Stationary waves in a string or wire fixed at both ends

Now the speed of transverse waves along a stretched string or wire is given by Equation 15.4

$$c = \sqrt{\frac{T}{m}}$$

where T is the tension and m the mass per unit length. Thus the wavelengths and frequencies of the stationary waves in Fig 18.3 are as follows:

Table 18.1

Mode	*Wavelength*	*Frequency*
Fundamental	$\lambda_1 = 2L$	$f_1 = \dfrac{c}{\lambda_1} = \dfrac{1}{2L}\sqrt{\dfrac{T}{m}}$
1st overtone	$\lambda_2 = L$	$f_2 = \dfrac{c}{\lambda_2} = \dfrac{1}{L}\sqrt{\dfrac{T}{m}}$
2nd overtone	$\lambda_3 = \frac{2}{3}L$	$f_3 = \dfrac{c}{\lambda_3} = \dfrac{3}{2L}\sqrt{\dfrac{T}{m}}$

Note that $f_2 = 2f_1$, so the first overtone is the second harmonic, and $f_3 = 3f_1$, so the second overtone is the third harmonic. If the string or wire is held at the centre, only even harmonics (2nd, 4th and so on) can occur.

Example 3

A horizontal string is stretched between two points a distance 0.80 m apart. The tension in the string is 90 N and its mass is 4.5 g. Calculate (a) the speed of transverse waves along the string and (b) the wavelengths and frequencies of the three lowest frequency modes of vibration of the string. (c) Explain how your answer to (b) would differ if the string is held lightly at its centre position.

Method

(a) To find the speed c we use Equation 15.4, with $T = 90$ and $m = (4.5 \times 10^{-3}) \div 0.80$:

$$c = \sqrt{\frac{T}{m}} = \sqrt{\frac{90}{(4.5 \times 10^{-3}) \div 0.80}}$$

$$= 126 \text{ m s}^{-1}$$

(b) The fundamental has wavelength $\lambda_1 = 2L = 1.6$ m. Its frequency f_1 is given by

$$f_1 = \frac{c}{\lambda_1} = \frac{126}{1.6} = 78.8 \text{ Hz}$$

The first overtone has wavelength

$$\lambda_2 = L = 0.80 \text{ m}$$

Its frequency f_2 is given by

$$f_2 = \frac{c}{\lambda_2} = \frac{126}{0.8} = 158 \text{ Hz}$$

Alternatively, we could use $f_2 = 2f_1$ (see Table 18.1).

The second overtone has wavelength

$$\lambda_3 = \tfrac{2}{3}L = 0.533 \text{ m}$$

Its frequency f_3 is given by

$$f_3 = \frac{c}{\lambda_3} = \frac{126}{0.533} = 236 \text{ Hz}$$

Alternatively we could use $f_3 = 3f_1$ (see Table 18.1).

(c) If the string is held lightly at the centre, then only even harmonics are possible, i.e. those with the following wavelengths and frequencies:

2nd harmonic $\lambda_2 = L = 0.80$ m
$f_2 = 2f_1 = 158$ Hz

4th harmonic $\lambda_4 = \dfrac{L}{2} = 0.40$ m
$f_4 = 4f_1 = 316$ Hz

6th harmonic $\lambda_6 = \dfrac{L}{3} = 0.27$ m
$f_6 = 6f_1 = 474$ Hz

and so on.

Answer

(a) 126 m s^{-1}.
(b) Wavelengths: 1.6 m, 0.80 m, 0.53 m.
Frequencies: 79 Hz, 0.16 kHz, 0.24 kHz.
(c) Even harmonics only, as detailed above.

Example 4

The fundamental frequency of vibration of a stretched wire is 120 Hz. Calculate the new fundamental frequency if (a) the tension in the wire is doubled, the length remaining constant, (b) the length of the wire is doubled, the tension remaining constant, (c) the tension is doubled and the length of the wire is doubled.

Method

In Table 18.1 we see that the fundamental frequency f_1 is given by

$$f_1 = \frac{1}{2L}\sqrt{\frac{T}{m}} \qquad (18.1)$$

For a particular wire the mass per unit length m is constant.

(a) For a constant length L and for a constant m we see from Equation 18.1 that $f_1 \propto \sqrt{T}$. Since the tension doubles, the new fundamental frequency f'_1 is $\sqrt{2}$ times the original. Thus

$$f'_1 = \sqrt{2} \times 120 = 170 \text{ Hz}$$

(b) For a constant tension T and for a constant m we see from Equation 18.1 that $f_1 \propto 1/L$. Since the length doubles, the new fundamental frequency f''_1 is half the original, i.e. 60 Hz.

(c) For a constant m Equation 18.1 tells us that $f_1 \propto \sqrt{T}/L$. If the tension doubles and the length doubles, then the new fundamental frequency f'''_1 is $\sqrt{(2)}/2$ times the original. Thus

$$f'''_1 = \frac{\sqrt{2}}{2} \times 120 = 84.9\,\text{Hz}$$

Answer

(a) 170 Hz, (b) 60.0 Hz, (c) 84.9 Hz.

Example 5

A wire of cross-sectional area $0.50\,\text{mm}^2$ is fixed at two points and is subjected to a tension of 80 N, so that its fundamental frequency is 100 Hz. The wire is now changed for one of the same material but of cross-sectional area $0.80\,\text{mm}^2$. Calculate (a) the new fundamental frequency if the tension remains at 80 N, (b) the tension required to restore the fundamental frequency to 100 Hz.

Method

For a particular material the mass per unit length is directly proportional to the cross-sectional area. Let m and m' be the original and new mass per unit length respectively. Then

$$\frac{m}{m'} = \frac{0.50}{0.80}$$

(a) We see from Equation 18.1 that $f_1 \propto 1/\sqrt{m}$ if we have a constant T and a fixed L. Thus if f_1 is the original fundamental frequency and f'_1 the new fundamental frequency, then

$$\frac{f'_1}{f_1} = \sqrt{\frac{m}{m'}}$$

Now $f_1 = 100$ and $m/m' = 0.625$, so

$$f'_1 = f_1 \times \sqrt{\frac{m}{m'}} = 100 \times \sqrt{0.625} = 79.1\,\text{Hz}$$

(b) The mass per unit length m now remains constant. Its length L also remains constant. Thus Equation 18.1 tells us that $f_1 \propto \sqrt{T}$. Let f'_1 (= 79.1 Hz) be the fundamental frequency at tension T' (= 80 N). Let f_1 (= 100 Hz) be the fundamental frequency at the tension T which we require. Now

$$\frac{f_1}{f'_1} = \sqrt{\frac{T}{T'}}$$

Squaring and rearranging for T gives

$$T = \left(\frac{f_1}{f'_1}\right)^2 \times T' = \left(\frac{100}{79.1}\right)^2 \times 80 = 128\,\text{N}$$

Answer

(a) 79.1 Hz, (b) 128 N.

Exercise 67

1 A horizontal wire of fixed length 0.90 m and mass $4.5 \times 10^{-3}\,\text{kg m}^{-1}$ is subject to a fixed tension of 50 N. Find the wavelengths and frequencies of the three lowest frequency modes of vibration when the wire is (a) free to vibrate at its midpoint, (b) lightly held at its midpoint.

2 A wire of cross-sectional area $0.20\,\text{mm}^2$ and made of steel of density $8.0 \times 10^3\,\text{kg m}^{-3}$ is subject to a tension of 60 N. Calculate (a) the mass per unit length of the wire, (b) the speed of transverse waves propagated down the wire, (c) the wavelength of waves with frequency 120 Hz, (d) the length of wire which, when fixed at its ends, gives a fundamental frequency of 120 Hz.
Note: Mass = Length × Area × Density.

3 The fundamental frequency of vibration of a stretched wire is 150 Hz. Calculate the new fundamental frequency if (a) the tension in the wire is tripled, the length remaining constant, (b) the length of wire is halved, the tension remaining constant, (c) the tension is tripled and the length of wire is halved.

4 A wire having a diameter of 0.80 mm is fixed in a sonometer and has a fundamental frequency of 256 Hz. Alongside it is a wire made of the same material but of diameter 0.60 mm. Both wires are stretched over the same bridges on the sonometer but the thinner wire is subject to only half the tension of the thicker wire. Calculate the fundamental frequency of vibration of the thinner wire.

Stationary waves in pipes

When an air column is made to vibrate at one end, a progressive longitudinal (sound) wave travels along the air column and is reflected at its end so that a stationary longitudinal (sound) wave is formed.

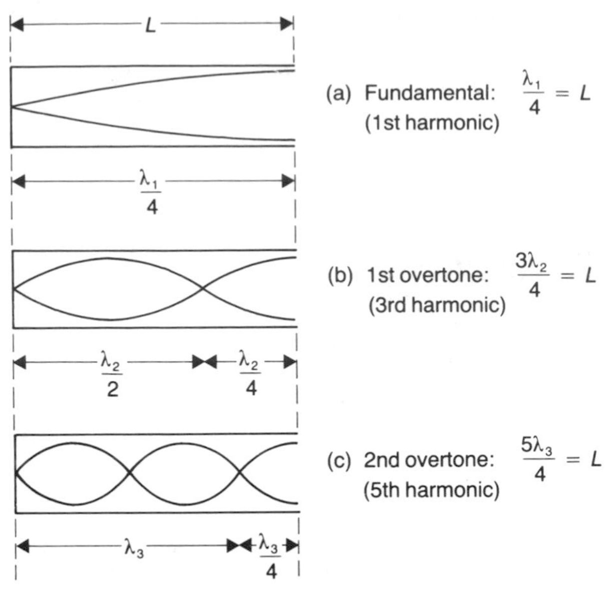

Fig 18.4 Stationary waves in a 'closed' pipe

Fig 18.4 shows a 'closed' or 'stopped' pipe, which means it is closed at one end. The fundamental and the first two overtones are shown. Let c be the speed of progressive sound waves in air at the particular temperature. The wavelengths and frequencies of the stationary waves in Fig 18.4 are as follows:

Table 18.2 Closed pipe

Mode	*Wavelength*	*Frequency*
Fundamental	$\lambda_1 = 4L$	$f_1 = \frac{c}{\lambda_1} = \frac{c}{4L}$
1st overtone	$\lambda_2 = \frac{4}{3}L$	$f_2 = \frac{c}{\lambda_2} = \frac{3c}{4L}$
2nd overtone	$\lambda_3 = \frac{4}{5}L$	$f_3 = \frac{c}{\lambda_3} = \frac{5c}{4L}$

Note that $f_2 = 3f_1$, so that the first overtone is the third harmonic, and $f_3 = 5f_1$, so that the second overtone is the fifth harmonic.

Fig 18.5 shows the fundamental and first two overtones for an 'open' pipe, i.e. a pipe which is open at both ends. The wavelengths and frequencies of the stationary waves shown in Fig 18.5 are as follows:

Table 18.3 Open pipe

Mode	*Wavelength*	*Frequency*
Fundamental	$\lambda_1 = 2L$	$f_1 = \frac{c}{\lambda_1} = \frac{c}{2L}$
1st overtone	$\lambda_2 = L$	$f_2 = \frac{c}{\lambda_2} = \frac{c}{L}$
2nd overtone	$\lambda_3 = \frac{2}{3}L$	$f_3 = \frac{c}{\lambda_3} = \frac{3c}{2L}$

Note that $f_2 = 2f_1$, so that the first overtone is the second harmonic, and $f_3 = 3f_1$, so that the second overtone is the third harmonic.

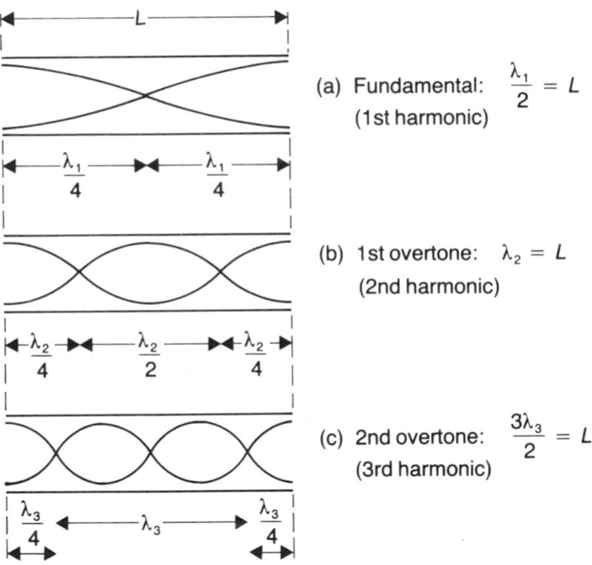

Fig 18.5 Stationary waves in an 'open' pipe

End correction

The antinode may not coincide exactly with the open end of the pipe. Usually it will be at a (short) distance e, called the end correction, beyond the open end as shown in Fig 18.6. Note that there are two end corrections for an 'open' pipe. In A-level calculations the end correction will often be taken to be negligible, but you should not assume that this is always the case.

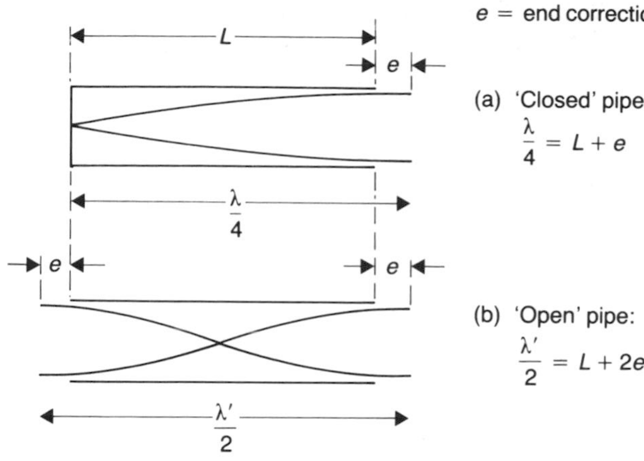

Fig 18.6 End correction

Example 6

A closed organ pipe is of length 0.680 m. Calculate the wavelengths and frequencies of the three lowest frequency modes of vibration. Take the speed of sound to be 340 m s^{-1} and neglect any end correction of the pipe.

Method

The pipe has length $L = 0.680$ m, and the speed of sound $c = 340\ \mathrm{m\,s^{-1}}$.

According to Table 18.2 the fundamental has wavelength $\lambda_1 = 4L = 2.72$ m. Its frequency f_1 is given by

$$f_1 = \frac{c}{\lambda_1} = \frac{340}{2.72} = 125\ \mathrm{Hz}$$

Similarly the first overtone has wavelength

$$\lambda_2 = 4L/3 = 0.907\ \mathrm{m}$$

and frequency f_2 given by

$$f_2 = \frac{c}{\lambda_2} = \frac{340}{0.907} = 375\ \mathrm{Hz}.$$

Alternatively we could use $f_2 = 3f_1$ (see Table 18.2).

The second overtone has wavelength

$$\lambda_3 = 4L/5 = 0.544\ \mathrm{m}$$

and frequency f_3 given by

$$f_3 = \frac{c}{\lambda_3} = \frac{340}{0.544} = 625\ \mathrm{Hz}$$

Alternatively we could use $f_3 = 5f_1$ (see Table 18.2).

Answer

The wavelengths are 2.72 m, 0.907 m and 0.544 m with frequencies 125 Hz, 375 Hz and 625 Hz respectively.

Example 7

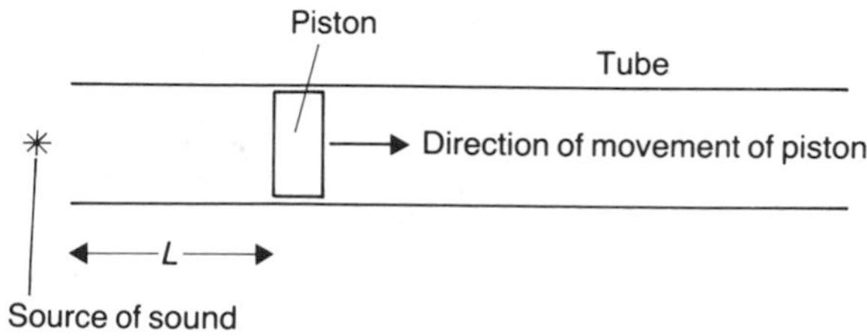

Fig 18.7 Information for Example 7

A small source of sound of frequency 400 Hz is placed near the open end of a tube as shown in Fig 18.7. The tube is closed by a movable piston, which is slowly moved in the direction shown so that the length L of the closed tube is gradually increased from a small value. The system produces a loud sound (resonance) first when $L = 0.194$ m and again when $L = 0.618$ m. Calculate (a) the speed of sound in air, (b) the end correction of the tube, (c) the next value of L at which resonance will occur.

Method

Resonance occurs when a stationary wave with *frequency equal to that of the sound source* is set up in the tube. Fig 18.8 shows that this *corresponds to tube lengths into which appropriate numbers of wavelengths fit exactly*.

Let the wavelength of the source be λ and the end correction be e, as shown in Fig 18.8.

(a) We know that the frequency f equals 400 Hz, so that in order to find the speed of sound c we must find the wavelength λ (which of course is constant in all three resonance positions). Since there is an end correction, we see from Fig 18.8a and b that

$$L_1 + e = \frac{\lambda}{4}$$

$$L_2 + e = \frac{3\lambda}{4}$$

Subtracting the top from the bottom equation we obtain

$$L_2 - L_1 = \frac{\lambda}{2}$$

i.e. the difference in the first two resonance lengths equals half a wavelength. This is readily seen by comparing Fig 18.8a and b.

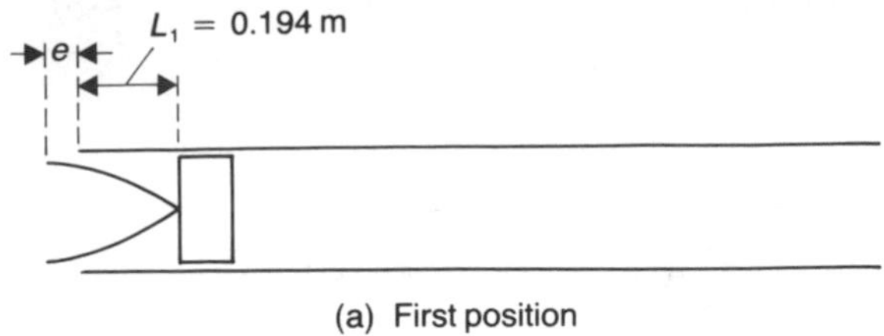

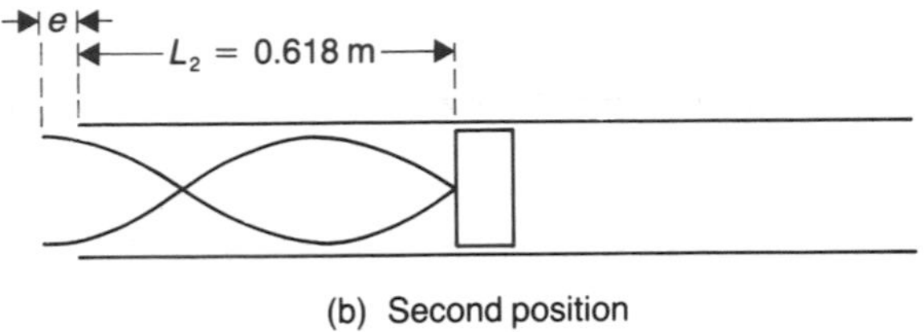

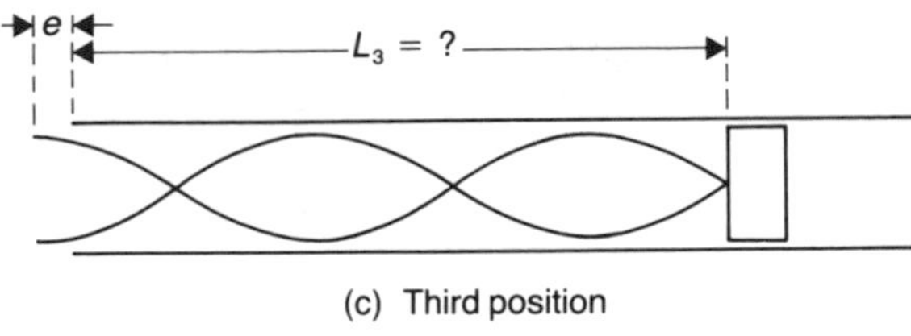

e = end correction

Fig 18.8 Resonance lengths for the tube in Example 7

Now $L_2 - L_1 = 0.618 - 0.194 = 0.424$ m. Thus

$$\lambda = 2(L_2 - L_1) = 2 \times 0.424$$
$$= 0.848 \text{ m}$$

We find the speed of sound c using Equation 15.1:

$$c = f\lambda = 400 \times 0.848$$
$$= 339 \text{ m s}^{-1}$$

(b) From the above we know that

$$L_1 + e = \frac{\lambda}{4}$$

We have $L_1 = 0.194$ and $\lambda = 0.848$. Rearranging gives

$$e = \frac{\lambda}{4} - L = 0.212 - 0.194$$
$$= 0.018 \text{ m}$$

(c) To find the third resonance length L_3 we note from Fig 18.8 that

$$L_3 = L_1 + \lambda = 0.194 + 0.848$$
$$= 1.042 \text{ m}$$

Alternatively we could use $L_3 = L_2 + \frac{\lambda}{2}$ or $L_3 + e = \frac{5\lambda}{4}$ to give the same answer.

Answer

(a) 339 m s^{-1}, (b) 0.018 m, (c) 1.04 m.

Example 8

Two open organ pipes are sounding together and produce a beat frequency of 12.0 Hz. If the longer pipe has a length of 0.400 m, calculate the length of the other pipe. Take the speed of sound as 340 m s^{-1} and ignore end corrections.

Method

We assume that the pipes are sounding their fundamental frequencies. We must first find the fundamental frequency f of the longer pipe. The wavelength λ of the fundamental note from an open pipe is, from Table 18.3, given by $\lambda/2 = L$ and since length $L = 0.400$ then $\lambda = 0.800$ m. Since the speed of sound $c = 340$,

$$f = \frac{c}{\lambda} = \frac{340}{0.8} = 425 \text{ Hz}$$

This is the fundamental frequency of the longer pipe. The shorter pipe must have a shorter wavelength wave, which means it must have a higher frequency (since $f \propto 1/\lambda$). This higher frequency f' is given by

$$f' = f + f_B = 425 + 12$$
$$= 437 \text{ Hz}$$

since the beat frequency $f_B = 12$ Hz (see Chapter 16). The wavelength λ' of the fundamental mode in the shorter pipe is thus given by

$$\lambda' = \frac{c}{f'} = \frac{340}{437} = 0.778 \text{ m}$$

Thus the length L' of the shorter pipe is given by

$$L' = \frac{\lambda'}{2} = \frac{0.778}{2} = 0.389 \text{ m}$$

Answer

The length of the shorter pipe is 0.389 m.

Exercise 68

(Assume that the speed of sound is 340 m s^{-1}.)

1 An open organ pipe is of length 0.664 m. Calculate the wavelengths and frequencies of the three lowest frequency modes of vibration. Ignore end corrections.

2 A piece of glass tubing is closed at one end by covering it with a sheet of metal. The fundamental frequency is found to be 280 Hz. If the metal sheet is now removed, calculate (a) what length the tube is, (b) the wavelengths and frequencies of the fundamental and the first overtone of the resulting open pipe. Ignore end corrections.

3 Calculate the length of a closed pipe with a fundamental frequency of 512 Hz. Ignore end corrections.

4 A tall vertical cylinder is filled with water, and a tuning fork of frequency 512 Hz is held over its open end. The water is slowly run out and the first resonance of the air column is heard when the water level is 15.6 cm below the open end. Calculate (a) the end correction of the tube and (b) the position of the water level when the second resonance is heard.

5 An open tube of length 30.0 cm has an end correction of 0.60 cm. Calculate its fundamental frequency.

6 Two open pipes of length 0.700 m and 0.750 m are sounded together and vibrate in their fundamental frequencies. Find the beat frequency, assuming that end corrections can be ignored.

Temperature effects

The frequency of a given mode of vibration depends on wavelength and the speed of sound ($f = c/\lambda$). The wavelength is determined solely by the length of the pipe and any end correction. The speed of sound depends upon temperature (see Chapter 15) and does not depend upon pressure. This means that the frequency f of vibration of a given pipe is dependent on temperature T. We shall see below that $f \propto \sqrt{T}$, where T is in kelvin.

Example 9

Two identical closed pipes of length 0.322 m are each vibrating with their fundamental frequency. If one pipe is held at 0 °C and the other at 17 °C, calculate the beat frequency which is observed. Take the speed of sound at 0 °C to be 331 m s^{-1} and ignore end corrections.

Method

The *wavelength* λ of the stationary waves in each pipe *is the same* and, from Table 18.2 is given by

$$\lambda = 4L = 4 \times 0.322 = 1.288 \text{ m}$$

The *frequency* of the two waves is different because the speed of sound is different as a result of the unequal temperatures. In Chapter 15 we see that the speed of sound $c \propto \sqrt{T}$, where T is the temperature in kelvin. If we let c and c' be the speed (of sound), in m s^{-1}, at temperatures T and T' respectively, then, from Equation 15.3,

$$\frac{c}{c'} = \frac{\sqrt{T}}{\sqrt{T'}}$$

Now the frequency f of the pipe at 0 °C is given by

$$f = \frac{c}{\lambda} = \frac{331}{1.288} = 257 \text{ Hz}$$

since the speed of sound $c = 331$ m s^{-1} at 0 °C.

The frequency f' of the pipe at 17 °C is given by $f' = c'/\lambda$ where c' is the speed of sound at 17 °C. Also we see that

$$\frac{f}{f'} = \frac{c/\lambda}{c'/\lambda} = \frac{c}{c'}$$

Now $$\frac{c}{c'} = \frac{\sqrt{T}}{\sqrt{T'}}$$

so $$\frac{f}{f'} = \frac{\sqrt{T}}{\sqrt{T'}} \qquad (18.2)^*$$

We have
$f = 257$, $T = 0\,°\text{C} = 273$ K, $T' = 17\,°\text{C} = 290$ K
and require f'. Rearranging the above gives us

$$f' = f \times \frac{\sqrt{T'}}{\sqrt{T}} = 257 \times \frac{\sqrt{290}}{\sqrt{273}}$$

$$= 265 \text{ Hz}$$

The beat frequency f_B is given by

$$f_B = f' - f = 265 - 257$$

$$= 8 \text{ Hz}$$

Answer

The beat frequency is 8 Hz.

**Equation 18.2 tells us that $f \propto \sqrt{T}$.*

Exercise 69

(Ignore end corrections.)

1 A closed pipe is of length 0.300 m. Calculate (a) its fundamental frequency at 0 °C, given that the speed of sound at 0 °C is 331 m s^{-1}, (b) the temperature, in °C, at which it will be in unison with a tuning fork of frequency 288 Hz.

2 Two open pipes of length 0.500 m and 0.550 m are sounded together and vibrate in their fundamental frequencies at 7 °C. Calculate (a) the beat frequency, given that the speed of sound at 7 °C is 335 m s^{-1}.

If the temperature of the longer pipe is now allowed to change whilst the shorter pipe stays at 7 °C, calculate (b) the value of the temperature of the air in the longer pipe at which the two pipes will be in unison.

Exercise 70: Questions of A-level standard

1 Stationary waves are set up in the space between a microwave transmitter and a plane reflector. Successive minima are spaced 15 mm apart. What is the frequency of the microwave oscillator? Take the speed of electromagnetic waves as 3.0×10^8 m s^{-1}. [C 81]

2 A stretched wire of length 0.7 m vibrates in its fundamental mode with a frequency of 320 Hz. Calculate the velocity of waves along the wire. Why does such a vibration not continue indefinitely? [WJEC 77, p]

3 State the formula for the fundamental frequency of the note emitted by a stretched string which involves the tension in the string, giving the meanings of the symbols used.

A stretched string, of length l and under tension T, emits a note of fundamental frequency f. The tension is now reduced to half of its original value and the vibrating length changed so that the frequency of the second harmonic is equal to f. What is the new length of the string in terms of the original length? [AEB 79]

4 A string of unstretched length 2.0 m and mass 0.15 kg has a force constant of 25 N m^{-1}. For the experiment, the string is stretched to a total length of 3.0 m. Calculate the velocity of propagation of transverse waves along the string.

The same string is clamped between two rigid supports 3.0 m apart, and set in vibration. Calculate the wavelengths and frequencies of the five lowest frequency modes of vibration which can be excited on the string, When the vibrating string is held lightly at the centre, in which of these modes does the string continue to vibrate? Explain your reasoning. [O & C 81, p]

5 A vertical steel wire is kept in tension by a piece of iron attached to one end. The wire is set in transverse vibration and emits a note of frequency 200 Hz. The iron is now completely immersed in water and the frequency of the note changes to 187 Hz. If the density of the water is 1000 kg m^{-3}, calculate the density of the iron. [AEB 82]

6 An organ pipe, of length 0.500 m, is closed at one end. What are the two lowest resonant frequencies of the pipe if the speed of sound in air is 330 m s^{-1}. Neglect any end corrections of the tube.

7 An air column, contained in a pipe closed at one end but open at the other, when excited has a fundamental frequency of 600 Hz. Another frequency at which the air column can be excited is

A 200 Hz **B** 300 Hz **C** 900 Hz
D 1200 Hz **E** 1800 Hz [O 80]

8 A tuning fork is sounded at the open end of a long cylindrical tube which is closed at the other end. Two successive positions of resonance are obtained when the length L of the vibrating air column has the following values: (i) L = 48.0 cm and (ii) L = 81.0 cm.
The wavelength, in cm, of the sound waves in the tube is

A 192 **B** 108 **C** 96 **D** 66 **E** 33 [AEB 82]

9 If the speed of sound in air is 336 m s^{-1}, what would be the length of an open organ pipe giving a fundamental frequency of 96 Hz? If this pipe were sounded together with another open pipe of length 2.10 m, what would be the beat frequency? Ignore end corrections. [AEB 79, p]

10 An open-ended organ pipe, of length 0.25 m, is sounding at 0 °C together with a tuning fork of slightly higher frequency, and ten beats per second are heard. Neglecting any end correction of the tube, and assuming that a change of temperature has no effect on the length of the pipe or frequency of the fork, calculate the change in temperature required to bring the pipe and fork into unison. (Speed of sound in air at 0 °C = 330 $m\,s^{-1}$.)
[AEB 80, p]

19 THE DOPPLER EFFECT

Introduction

The Doppler effect is a change in observed frequency due to the motion of the source of waves or the observer, or both.

The Doppler effect in sound

(1) Stationary source, moving observer:

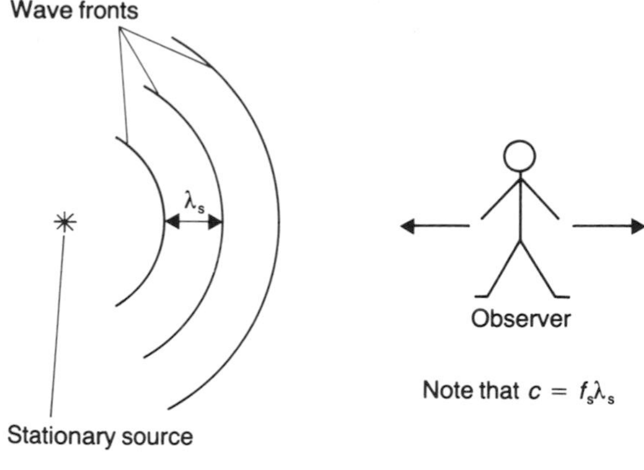

Fig 19.1 Stationary source and moving observer

Fig 19.1 shows a stationary source of sound waves of frequency f_s which travel towards an observer with speed c. If the observer moves towards the source, he receives more waves per second than if he is stationary, so that the frequency observed f_0 is greater than f_s. If he moves away from the source then f_0 is less than f_s. It can be shown that

$$f_0 = f_s\left(\frac{c \pm u_0}{c}\right) \qquad (19.1)$$

where u_0 is the speed of the observer along the line between source and observer. The *plus sign* applies when the observer moves *towards the source* and the *negative sign* when the observer moves *away from the source.*

(2) Moving source, stationary observer:

Note that a stationary source has wavelength λ_s as in Fig 19.1

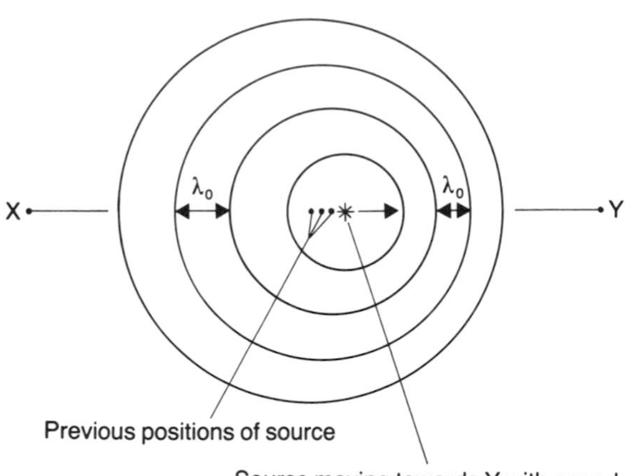

Fig 19.2 A moving source, showing change in wavelength

Fig 19.2 shows a source of waves similar to that in Fig 19.1, but which is now moving along the line XY, with speed u_s (assumed less than c), towards Y. The waves in front of the source have wavelength λ_0, which is smaller than λ_s, and those behind the source have wavelength λ_0, which is greater than λ_s.

It can be shown that

$$\lambda_0 = \left(\frac{c \mp u_s}{f_s}\right) \qquad (19.2)$$

The *negative sign* refers to waves received by an observer at Y, where the *source moves towards the observer*. The *positive sign* refers to waves received by an observer at X, where the *source moves away from the observer*.

An observer at Y thus receives waves of shorter wavelength, and hence observes a frequency f_0 which is higher than f_s, the frequency for a stationary source. For an observer at X, f_0 is less than f_s. Since $f_0 = c/\lambda_0$, it follows from Equation 19.2 that

$$f_0 = f_s\left(\frac{c}{c \mp u_s}\right) \tag{19.3}$$

(3) Moving source and moving observer:

The observed frequency f_0 arises due to combination of the effects described in parts (1) and (2). By combining Equations 19.1 and 19.3 we can show

$$f_0 = f_s\frac{(c \pm u_0)}{(c \mp u_s)} \tag{19.4}$$

where the *upper* signs refer to observer/source moving *towards* each other, and the *lower* signs refer to observer/source moving *away from* each other. Equation 19.4 can be used instead of remembering both of equations 19.1 and 19.3.

Example 1

A source of sound, when stationary, generates waves of frequency 500 Hz. The speed of sound is 340 m s^{-1}. Find from first principles the wavelength of the waves detected by the observer and the frequency observed when (a) the source is stationary and the observer moves towards it with speed 20.0 m s^{-1}, (b) the source moves away from the stationary observer with speed 30.0 m s^{-1}, (c) the source moves with speed 30.0 m s^{-1} in a direction away from the observer and the observer moves with speed 20.0 m s^{-1} towards the source.

In each case check your answers using the relevant formula.

Method

(a) The stationary source emits waves of frequency $f_s = 500$ Hz and wavelength λ_s, which travel with speed $c = 340$ m s^{-1}. Now $c = f_s\lambda_s$, so

$$\lambda_s = \frac{c}{f_s} = \frac{340}{500} = 0.680 \text{ m}$$

In 1 s a stationary observer will receive 500 waves since $f_s = 500$ Hz. If the observer moves *towards* the source he meets an *extra* number n of waves given by

$$n = \frac{\text{Distance travelled}}{\text{Wavelength of waves}}$$

We have distance travelled equal to 20 m in 1 s and wavelength $\lambda_s = 0.680$ m. So

$$n = \frac{20}{0.68} = 29.4$$

The total number of waves received per second by an observer moving towards the source is thus $500 + 29.4 = 529.4$, so $f_0 = 529.4$ Hz.

(Note that an observer moving *away* from the source at 20 m s^{-1} will receive 29.4 fewer waves per second, so $f_0 = 500 - 29.4 = 470.6$ Hz.)

To check our answer for f_0 we use Equation 19.1,* with $f_s = 500$, $c = 340$ and $u_0 = 20$. Using the plus sign, we have

$$f_0 = f_s\left(\frac{c + u_0}{c}\right) = 500 \times \left(\frac{340 + 20}{340}\right)$$
$$= 529.4 \text{ Hz}$$

(b) In 1 s the source emits 500 waves which, for a stationary source, occupy a distance of 340 m. Since the source moves *away* from the observer at 30 m s^{-1}, the 500 waves now occupy a distance of $340 + 30 = 370$ m. So wavelength λ_0 received by the observer is given by

$$\lambda_0 = \frac{\text{Distance occupied}}{\text{Number of waves}} = \frac{370}{500}$$
$$= 0.740 \text{ m}$$

To check our answer we use Equation 19.2, with $c = 340$, $u_s = 30$ and $f_s = 500$. Using the plus sign, we have

$$\lambda_0 = \left(\frac{c + u_s}{f_s}\right) = \left(\frac{340 + 30}{500}\right)$$
$$= 0.740 \text{ m}$$

To find the frequency observed f_0 we note that the speed with which the waves travel *relative to the ground* remains unchanged, and in this case equals 340 m s^{-1}. The observed frequency f_0 is thus given by

$$f_0 = \frac{\text{Speed of sound}}{\text{Wavelength } \lambda_0} = \frac{340}{0.740}$$
$$= 459 \text{ Hz}$$

Or Equation 19.4 with $u_s = 0$.

To check our answer we use Equation 19.3,* with $f_s = 500$, $c = 340$ and $u_s = 30$. So, using the positive sign, we have

$$f_0 = f_s\left(\frac{c}{c + u_s}\right) = 500 \times \left(\frac{340}{340 + 30}\right)$$
$$= 459 \text{ Hz}$$

(c) We combine parts (a) and (b). If the observer were stationary he would receive, according to part (b), 459 waves each second. Since he moves with speed 20 m s^{-1} towards the source, and since the waves he receives have wavelength $\lambda_0 = 0.740$ m, he intercepts an extra number n' of waves given by

$$n' = \frac{\text{Distance travelled}}{\text{Wavelength } \lambda_0} = \frac{20}{0.74}$$
$$= 27.0$$

Thus he receives $459 + 27 = 486$ waves each second, so frequency observed f_0 equals 486 Hz.

We can check this using Equation 19.4 with $f_s = 500$, $c = 340$, $u_0 = 20$ and $u_s = 30$. We use the upper (positive) sign in the top bracket and the lower (positive) sign in the lower bracket. Thus

$$f_0 = f_s\frac{(c + u_0)}{(c + u_s)} = 500\frac{(340 + 20)}{(340 + 30)}$$
$$= 486 \text{ Hz}$$

Answer

(a) 0.680 m, 529 Hz, (b) 0.740 m, 459 Hz,
(c) 0.740 m, 486 Hz.

Example 2

A train sounding its whistle (frequency 580 Hz) is travelling at 40.0 m s^{-1} along a straight section of track, and passes an observer standing close to the track. Calculate the maximum change in frequency which the observer will hear. Take the speed of sound in air as 340 m s^{-1}.

Method

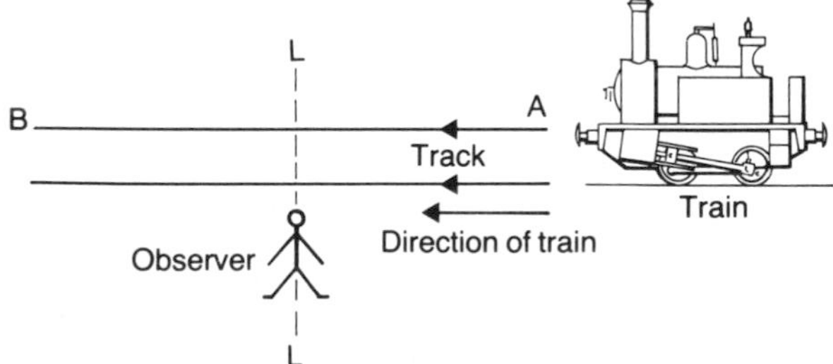

Fig 19.3 Solution to Example 2

*Or Equation 19.4 with $u_0 = 0$

Fig 19.3 shows the situation. It is the component of velocity of the train towards the observer which produces the change in frequency. If A and B are both a long way from the observer, then the component of velocity will be 40 m s^{-1} towards the observer for A, and 40 m s^{-1} away from the observer for B. Thus the frequency observed will be a maximum when the train is at A and a minimum when it is at B.

The observed frequency f_0 is found using Equation 19.3 in which $f_s = 580$, $c = 340$ and $u_s = 40$.

When the train is at A we use the negative sign in Equation 19.3, so

$$f_0 = f_s\left(\frac{c}{c - u_s}\right) = 580 \times \left(\frac{340}{340 - 40}\right)$$
$$= 657 \text{ Hz}$$

When the train is at B we use the positive sign in Equation 19.3, so

$$f_0 = f_s\left(\frac{c}{c + u_s}\right) = 580 \times \left(\frac{340}{340 + 40}\right)$$
$$= 519 \text{ Hz}$$

The maximum change in frequency is thus $657 - 519 = 138$ Hz.

Note that when the train is level with the observer at LL, it has no component of velocity towards the observer. The observed frequency at LL is thus 580 Hz.

Answer

The maximum frequency change is 138 Hz.

Exercise 71

(Take the speed of sound as 340 m s^{-1}.)

1 A motorist approaches a road junction at a constant speed of 15 m s^{-1}. A policeman standing at the junction blows on a whistle with frequency 680 Hz. (a) Find from first principles the frequency observed by the motorist. The motorist now reduces his speed at a rate of 5.0 m s^{-2}. Calculate (b) the frequency he observes at subsequent 1 s intervals until he stops.

2 A train sounding its whistle moves at a constant speed of 20 m s^{-1} along a long straight section of track. The train passes under a low bridge on which stands an observer. If the observer records a maximum frequency of 638 Hz, calculate (a) the frequency of the whistle, (b) the minimum frequency the observer hears.

3 A source of sound of frequency 400 Hz moves at a steady speed of 15 m s^{-1} towards an observer. If the observer moves at a steady speed of 25 m s^{-1} towards the source, calculate the frequency he observes.

4 A loudspeaker which emits a note of frequency 250 Hz is attached to a wire and whirled in a vertical circle of radius 1.00 m at a steady rate of 20.0 revolutions per minute. Calculate (a) the speed of rotation of the loudspeaker in m s^{-1}, (b) the maximum and minimum frequency detected by a stationary observer.

5 A train sounding its whistle travels at constant speed on a long straight section of track. An observer standing close to the track records a range of frequencies between 551 Hz and 658 Hz. Calculate (a) the speed of the train, (b) the frequency of its whistle.

Reflection of waves

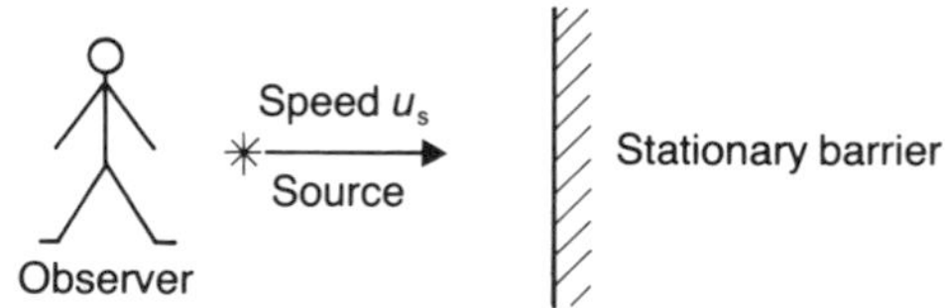

(a) The moving source and observer

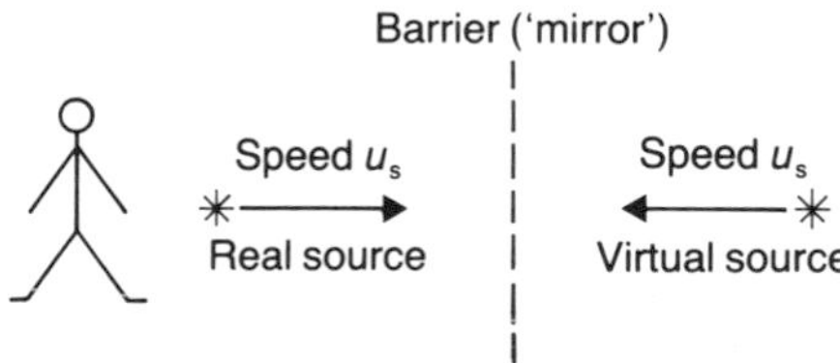

(b) The equivalent situation for reflected waves

Fig 19.4 Reflection of waves from a stationary barrier

Fig 19.4a shows a source of sound approaching a vertical stationary reflector. An observer situated as shown receives two sets of waves: (1) direct from the source which is moving away from him, (2) an echo due to waves reflected from the barrier.

The reflected waves have the same wavelength and frequency as the waves incident on the barrier and thus are like the waves which come from a source *moving towards* the observer. We can think of the barrier acting like a mirror and producing a 'virtual' source, as shown in Fig 19.4b.

Example 3

A hooter of frequency 360 Hz is sounded on a train approaching a tunnel in a cliff-face at 25 m s^{-1}, normal to the cliff. Calculate the observed frequency of the echo from the cliff-face, as heard by the train driver. Assume that the speed of sound in air is 330 m s^{-1}.

Method

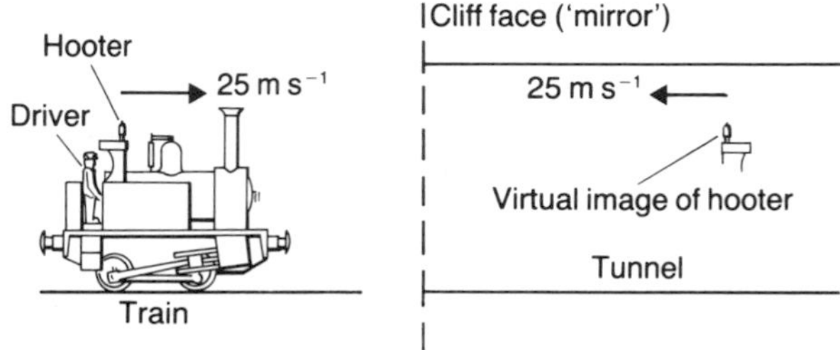

Fig 19.5 Solution to Example 3

The driver hears the echo which, as described above, can be considered to arise from a source moving at 25 m s^{-1} towards the train, as shown in Fig 19.5. Now the driver is also moving at 25 m s^{-1} towards the source. We use Equation 19.4 with the upper sign in both bracketed terms. We have $f_s = 360$, $c = 330$ and $u_0 = u_s = 25$. So

$$f_0 = f_s\frac{(c + u_0)}{(c - u_s)} = 360 \times \frac{(330 + 25)}{(330 - 25)}$$

$$= 419 \text{ Hz}$$

Answer

The echo has an observed frequency of 419 Hz.

Exercise 72

(Assume that the speed of sound is 340 m s^{-1}.)

1 A car travels at a constant speed of 30 m s^{-1} towards a tunnel and sounds its horn, which has a frequency of 200 Hz. The sound is reflected from the tunnel entrance. Calculate the frequency of the echo observed by (a) the driver of the car, (b) a stationary observer standing close to the road, (c) the driver of a car travelling at 20 m s^{-1} which is following the first car.

2 A train emerges from a tunnel at a speed of 20 m s^{-1} and sounds its whistle, which has a frequency of 450 Hz. Calculate the frequency of the echo from the tunnel entrance as observed by the train driver.

Exercise 73: Questions of A-level standard

1 In conditions in which the velocity of sound in air is 330 m s^{-1} a source of sound which is stationary with respect to the air is emitting a note of frequency 330 Hz. An observer receding from the source at a uniform speed of 30 m s^{-1} hears a note whose frequency, in Hz, is

A 363 **B** 360 **C** 310
D 300 **E** 275 [O & C 81]

2 A ship is travelling at 3 m s^{-1} towards a cliff in still air and is sounding its siren at 1 kHz. Find from first principles the frequency of the echo as measured by an observer on the ship. Give sufficient detail for your reasoning to be followed. The speed of sound in air is 330 m s^{-1}. [WJEC 82]

3 A railway engine travelling at constant speed emits a whistle of constant frequency. When the engine passes a stationary observer close to the track, the frequency of the sound heard by the observer changes from 600 Hz whilst approaching to 500 Hz whilst receding. Assuming the speed of sound is 340 m s^{-1}, calculate the speed of the engine.

Calculate the frequencies heard if the same engine passes an observer who is travelling at 10 m s^{-1} in the same direction as the engine and close to the track. [JMB 82, p]

Section E
Geometrical Optics

20 THIN LENSES

Single lenses

As shown in Fig 20.1, a lens acts to produce an image I from an object O. The lens formula is

$$\frac{1}{v} + \frac{1}{u} = \frac{1}{f} \qquad (20.1)$$

where, as shown in Fig 20.1, u is the object distance, v the image distance, and f the focal length of the lens.

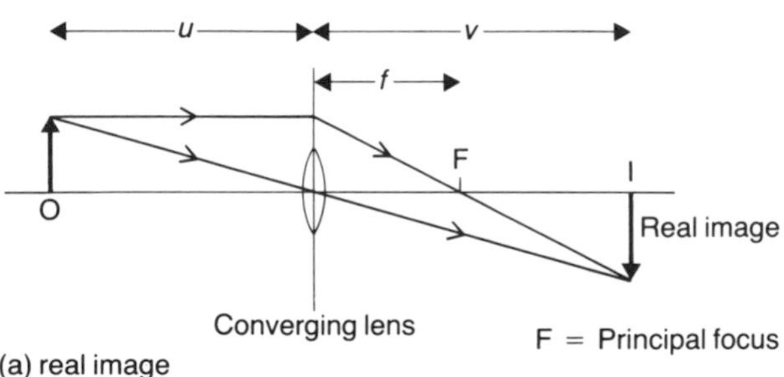

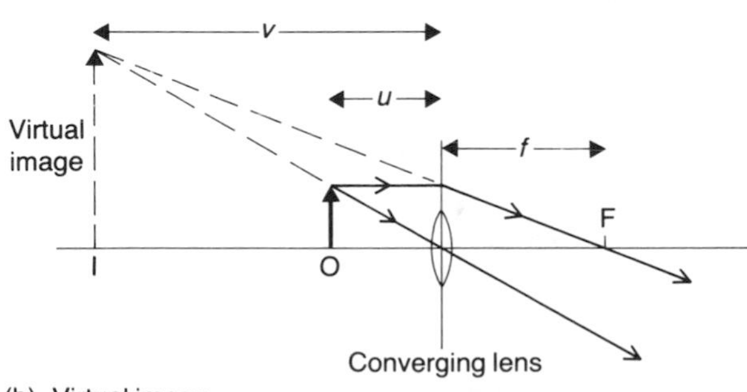

Fig 20.1 Formation of images by a converging lens

We shall use the real is positive, virtual is negative sign convention. This means that the focal length of a converging lens is positive and that of a diverging lens (see Fig 20.2) is negative.
Correct signs *must* be used in the lens formula.

Example 1

An object is placed (a) 25.0 cm, (b) 10.0 cm from a converging lens of focal length 15.0 cm. Calculate the image distance and lateral magnification produced in each case, and state the type of image produced.

Method

We have a converging lens so the focal length $f = +15$.

(a) This is a real object, so $u = +25$. We rearrange Equation 20.1 to find v:

$$\frac{1}{v} = \frac{1}{f} - \frac{1}{u} = \frac{1}{15} - \frac{1}{25} = \frac{2}{75}$$

$$\therefore \quad v = 37.5 \text{ cm}$$

Since v is positive the image is real. The situation is similar to that shown in Fig 20.1a.

The lateral magnification m is defined by

$$m = \frac{\text{Height of image}}{\text{Height of object}} \qquad (20.2)$$

It can be shown that

$$m = \frac{\text{Image distance}}{\text{Object distance}} = \frac{v}{u} \qquad (20.3)$$

We have $v = +37.5$ and $u = +25$. Thus

$$m = \frac{v}{u} = \frac{37.5}{25}$$

$$= 1.50$$

The image is 1.50 times as long as the object.

(b) We have a real object so $u = +10$. Rearranging Equation 20.1 to find v:

$$\frac{1}{v} = \frac{1}{f} - \frac{1}{u} = \frac{1}{15} - \frac{1}{10} = -\frac{1}{30}$$

$$\therefore \quad v = -30.0\,\text{cm}$$

Note that v is negative, so that the image is virtual. The situation is similar to that shown in Fig 20.1b. The lens acts as a simple magnifying glass. To find the lateral magnification m we use Equation 20.3 with $v = -30$ and $u = +10$:

$$m = \frac{v}{u} = \frac{-30}{10}$$

$$= -3.00$$

The image is 3.00 times as long as the object. The significance of the negative sign is that the image is virtual.

Answer

(a) 37.5 cm, 1.50 times, real, (b) 30.0 cm, 3.00 times, virtual.

Example 2

When a real object is placed in front of a diverging lens of focal length 20.0 cm, an image is formed 12.0 cm from the lens. Calculate (a) the object distance, (b) the lateral magnification produced. Draw a sketch to show the arrangement.

Method

We have a diverging lens, so the focal length $f = -20$. A real object always produces a virtual image when using a diverging lens, so that $v = -12$.

(a) Rearrange Equation 20.1 to find u:

$$\frac{1}{u} = \frac{1}{f} - \frac{1}{v} = \frac{1}{-20} - \frac{1}{(-12)} = \frac{2}{60}$$

$$\therefore \quad u = 30.0\,\text{cm}$$

(b) To find the lateral magnification m we use Equation 20.3, with $v = -12$ and $u = 30$:

$$m = \frac{v}{u} = \frac{-12}{30}$$

$$= -0.40$$

The image is 0.40 times as long as the object. The negative sign shows the virtual nature of the image. A sketch of the arrangement is given in Fig 20.2. A diverging lens always produces a virtual, erect diminished image when viewing a real object.

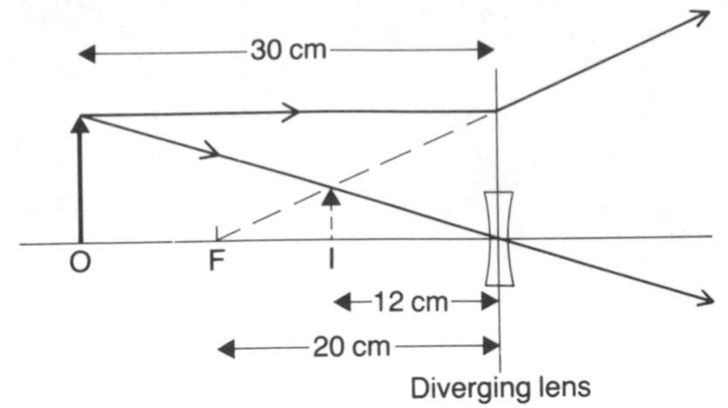

Fig 20.2 Solution to Example 2

Answer

(a) 30.0 cm, (b) (−)0.40 times.

Example 3

A camera has a lens of focal length 50.0 mm. If it can form images of objects from infinity down to 1.50 m from the lens, calculate the distance through which it must be possible to move the lens.

Method

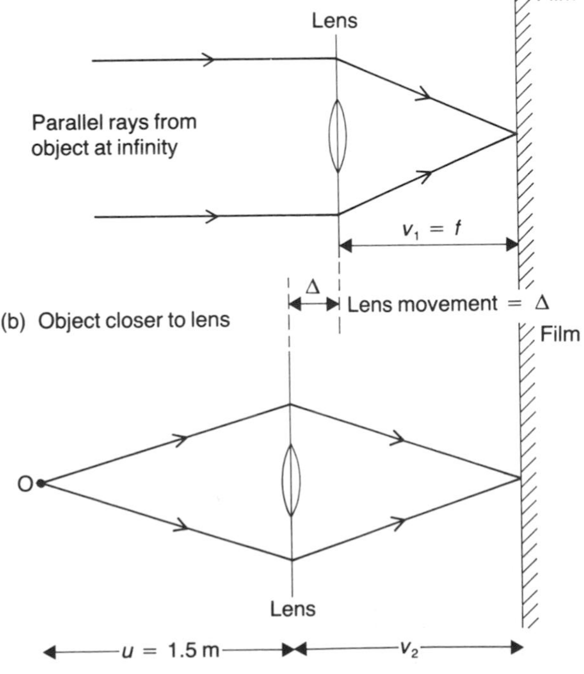

Fig 20.3 Formation of images using a camera lens

The lens in the camera forms real images on the film, as shown in Fig 20.3. The compound lens in the camera is thought of as a single thin, converging lens of focal length $f = 5.00$ cm.

As shown in Fig 20.3a, when the object is at infinity, the lens must be at a distance

$$v_1 = f = 5.00 \text{ cm}$$

from the lens. When the (real) object is at a distance

$$u = +1.50 \text{ m} = +150 \text{ cm}$$

from the lens, the image distance v_2, as shown in Fig 20.3b, is given by rearranging Equation 20.1:

$$\frac{1}{v_2} = \frac{1}{f} - \frac{1}{u} = \frac{1}{5} - \frac{1}{150} = \frac{29}{150}$$

$$\therefore \quad v_2 = 5.17 \text{ cm}$$

The required movement Δ of the lens is, as shown in Fig 20.3, given by

$$\Delta = v_2 - v_1 = 5.17 - 5.00$$
$$= 0.17 \text{ cm}$$

Answer

The lens must move by 0.17 cm.

Exercise 74

1 An object placed 20 cm from a converging lens results in a real image formed 30 cm from the lens. Calculate the focal length of the lens.

2 When an object is placed 10 cm from a converging lens, an erect image which is three times as long as the object is obtained. Calculate (a) the image distance, (b) the focal length of the lens.

3 An erect image, twice as long as the object, is obtained when using a simple magnifying glass of focal length 10 cm. Calculate (a) the object distance, (b) the image distance. Hint: $v/u = -2$.

4 When a real object is placed 12 cm in front of a diverging lens, a virtual image is formed 8.0 cm from the lens. Find the focal length of the lens.

5 The focal length of a camera lens is 100 mm. Calculate how far from the film the lens must be set in order to photograph an object which is (a) 100 cm, (b) 500 cm from the lens. Hence calculate (c) the movement of the lens between these two positions.

Lens combinations

When two lenses are used, the image produced by the first lens acts as an object for the second lens. This means that, in certain circumstances, we can have a virtual object for the second lens.

Lens combinations — separated lenses

In this case we apply the lens formula to each lens separately and in the correct order. Care must be taken to ensure that the distances and signs are correct for each lens.

Example 4

An object is placed at right angles to the axis of a converging lens of focal length 20.0 mm and 25.0 mm from it. A second converging lens of focal length 40.0 mm is placed 135 mm from the first lens and on the other side of it from the object. Calculate (a) the distance of the final image from the second lens, (b) the lateral magnification produced by the two-lens system. Draw a diagram to show the formation of the final image.

Method

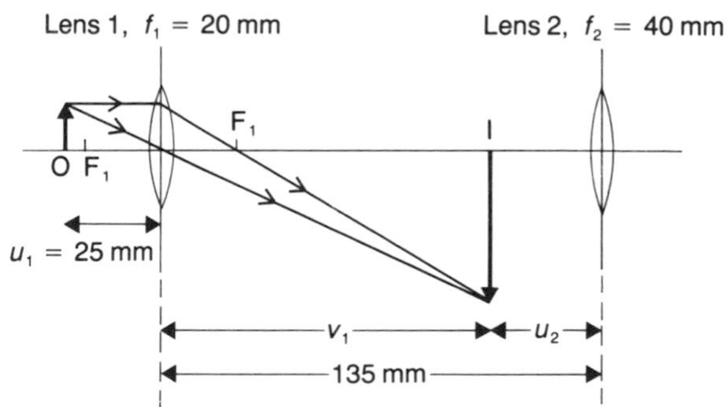

Image distance for the second lens is v_2

Fig 20.4 Solution to Example 4

We first draw a diagram of the arrangement — shown in Fig 20.4. Note that the image I produced by the first lens must be real, since the object O is placed beyond the focus of lens 1. This 'intermediate' *image I acts as a (real) object for the second lens.*

(a) To find the image distance v_2 from the second lens we must first find the object distance u_2 from lens 2. Thus we must find the image distance v_1 for lens 1 since, as shown in Fig 20.4, $v_1 + u_2 = 135$ mm. For convenience we shall work in millimetres throughout this calculation.

To find v_1 we apply Equation 20.1 *to lens 1*, with object distance $u_1 = +25$ and focal length $f_1 = +20$. Hence

$$\frac{1}{v_1} = \frac{1}{f_1} - \frac{1}{u_1} = \frac{1}{20} - \frac{1}{25} = \frac{1}{100}$$

$$\therefore \quad v_1 = 100 \text{ mm}$$

Now we know $v_1 + u_2 = 135$, so $u_2 = 35$ mm.

To find v_2 we now apply Equation 20.1 *to lens 2*, in which object distance $u_2 = +35$ and focal length $f_2 = 40$. Thus

$$\frac{1}{v_2} = \frac{1}{f_2} - \frac{1}{u_2} = \frac{1}{40} - \frac{1}{35} = \frac{-1}{280}$$

$$\therefore \quad v_2 = -280 \text{ mm}$$

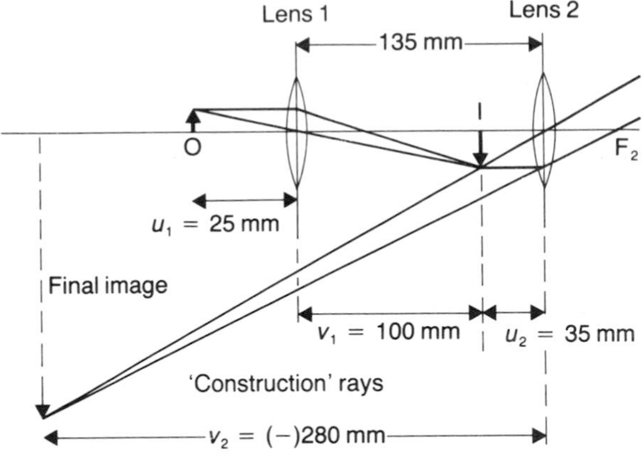

Fig 20.5 Formation of the final image in Example 4

Note that the final image is virtual, so that it must be to the left of lens 2. A diagram showing the formation of the final image is given in Fig 20.5. This arrangement is in fact that of the compound microscope. The first lens produces a magnified real image, which is further magnified by the second lens acting as a simple magnifying glass.

(b) We first find the lateral magnification produced by each lens separately by using Equation 20.3.

For lens 1

$$m_1 = \frac{v_1}{u_1} = \frac{100}{25} = 4.00$$

For lens 2

$$m_2 = \frac{v_2}{u_2} = \frac{-280}{35} = -8.00$$

The total lateral magnification m is the *product* of the separate magnifications, i.e.

$$m = m_1 \times m_2 = 4 \times -8 = -32$$

Note the negative sign, which tells us that the final image is virtual.

Answer

(a) 280 mm, (b) (−)32.0 times.

Example 5

A small illuminated object is placed on the axis of a converging lens of focal length 10.0 cm, and 12.0 cm from the lens. On the other side of the converging lens, and coaxial with it, is placed a diverging lens of focal length 25.0 cm. Find the position and nature of the final image when the lenses are (a) 50.0 cm, (b) 20.0 cm apart. In each case draw a diagram showing the formation of the final image.

Method

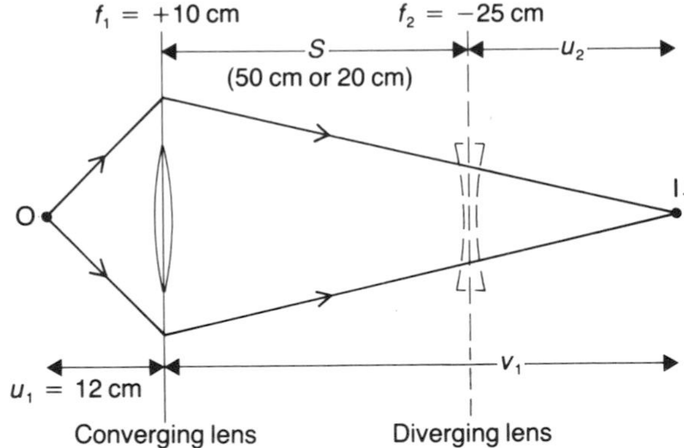

Fig 20.6 The lens arrangement in Example 5

The converging lens, *on its own*, would form an image as shown in Fig 20.6. To find the image distance v_1 we apply Equation 20.1, in which the object distance $u_1 = +12$ cm and the focal length $f_1 = +10$ cm. Thus

$$\frac{1}{v_1} = \frac{1}{f_1} - \frac{1}{u_1} = \frac{1}{10} - \frac{1}{12} = \frac{1}{60}$$

$$\therefore \quad v_1 = 60 \text{ cm}$$

When the diverging lens, shown dotted, is inserted then the image I_1 formed by the converging lens acts as a *virtual* object for the diverging lens. It is virtual, since the rays of light converging toward I_1 do not actually meet there once the diverging lens is inserted. The object distance u_2 is found from $u_2 = v_1 - S$, where S is the distance between the lenses.

(a) We have $S = 50$. Thus

$$u_2 = v_1 - S = 60 - 50 = 10 \text{ cm}$$

To denote the virtual nature of the object we put $u_2 = -10$ cm. To find the image distance v_2 from the diverging lens we apply Equation 20.1, with $u_2 = -10$ and focal length $f_2 = -25$. Thus

$$\frac{1}{v_2} = \frac{1}{f_2} - \frac{1}{u_2} = \frac{1}{-25} - \frac{1}{(-10)} = \frac{3}{50}$$

$$\therefore \quad v_2 = 16.7 \text{ cm}$$

Note that the *final image* I_2 *is real* and is 16.7 cm from the diverging lens. The arrangement is shown in Fig 20.7a.

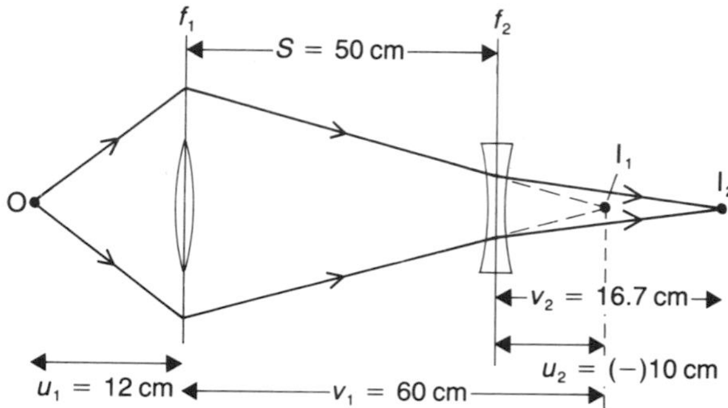

(a) For $S = 50$ cm

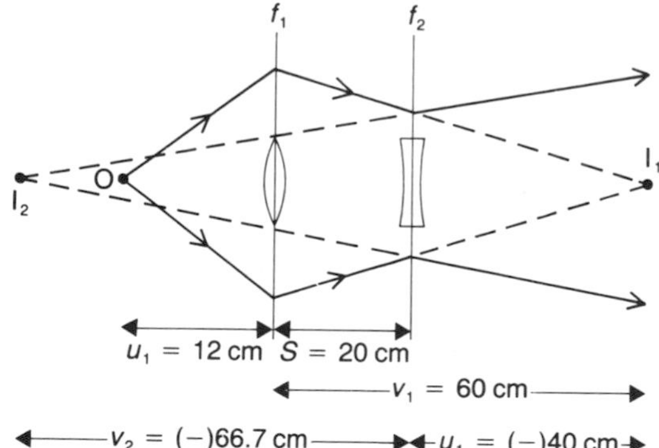

(b) For $S = 20$ cm

Fig 20.7 Solution to Example 5

(b) We have $S = 20$. Thus

$$u_2 = v_1 - S = 60 - 20 = 40 \text{ cm}$$

To denote the virtual nature of the object we put $u_2 = -40$ cm. To find the image distance v_2 we apply Equation 20.1 with $u_2 = -40$ and $f_2 = -25$. Thus

$$\frac{1}{v_2} = \frac{1}{f_2} - \frac{1}{u_2} = \frac{1}{-25} - \frac{1}{(-40)} = \frac{-3}{200}$$

$$\therefore \quad v_2 = -66.7 \text{ cm}$$

Note that the final image is virtual as shown in Fig 20.7b. The diverging beam of light *appears* to come from an image I_2 situated 66.7 cm to the left of the diverging lens.

A real image is produced in part (a) because the light from the first lens would converge to a point I_1 at a distance less than f_2 from the diverging lens. In part (b) the point I_2 is greater than f_2 from the diverging lens, so that the final image is virtual. Note that for the case of $u_2 = f_2$, or $S = 35$ cm, a parallel beam emerges to the right of the diverging lens.

Answer

(a) $v_2 = 16.7$ cm, real; (b) $v_2 = (-)66.7$ cm, virtual. See Fig 20.7

Exercise 75

1 An object 3.00 mm long is placed at right angles to the axis of and 40.0 mm away from a converging lens of focal length 30.0 mm. On the other side of this lens and 180 mm from it is placed a second converging lens of focal length 50.0 mm. Find the position, nature and size of (a) the image formed by the first lens, (b) the final image formed. Calculate (c) the total magnification produced. Draw a diagram to show the formation of the final image. Note: this shows the principle of the projection microscope.

2 A small illuminated object is placed on the axis of a converging lens of focal length 15 cm and 20 cm from it. On the other side of the converging lens, and 40 cm from it, is placed a diverging lens of focal length F. If the final real image is formed 60 cm from the diverging lens, calculate the value of F.

3 A telephoto lens in a camera consists of a converging lens of focal length 20 cm placed 8.0 cm in front of a diverging lens of focal length 15 cm. If the camera is used to focus a distant object, find the position and nature of (a) the image formed by the converging lens on its own, (b) the final image formed by the two lenses.

Lens combinations — lenses in contact

It is of course possible to treat this situation as a case of separated lenses in which the separation is zero. A more convenient method is to say that the combined lens system can be replaced by a single lens of focal length f given by

$$\frac{1}{f} = \frac{1}{f_1} + \frac{1}{f_2} \tag{20.4}$$

where, as shown in Fig 20.8, f_1 and f_2 are the individual focal lengths of the separate lenses. The thickness of the lenses is neglected.

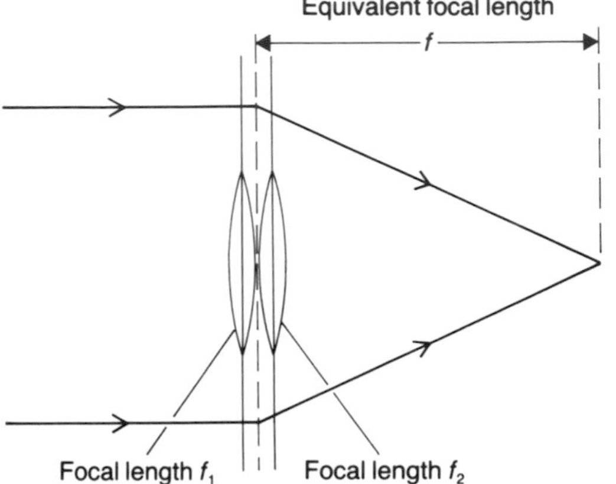

Fig 20.8 Lenses in contact

Example 6

A converging lens of focal length 30 cm is placed in contact with a diverging lens of focal length 20 cm. Calculate the focal length of the combination.

Method

We use Equation 20.4, in which for the converging lens $f_1 = +30$, and for the diverging lens $f_2 = -20$. Thus

$$\frac{1}{f} = \frac{1}{30} + \frac{1}{-20} = \frac{2-3}{60} = \frac{-1}{60}$$

$$\therefore \quad f = -60 \text{ cm}$$

The equivalent lens is diverging and has focal length 60 cm. Note that the combined lens must be diverging, since the single diverging lens is more powerful (i.e., has a shorter focal length) than the converging lens.

Answer

The combination is a diverging lens of focal length 60 cm.

Exercise 76

1 A converging achromatic doublet consists of a converging (crown glass) lens of focal length 20 cm and a diverging lens (made of flint glass). If the focal length of the doublet is 80 cm, calculate the focal length of the diverging lens.

Exercise 77: Questions of A-level standard

1 A lens is placed between an illuminated object and a screen which are 0.60 m apart. The lens required to produce an image the same size as the object is

A converging $f = 0.30$ m
B diverging $f = 0.30$ m
C converging $f = 0.15$ m
D diverging $f = 0.15$ m
E converging $f = 0.20$ m [AEB 81]

2 A camera has a lens, of focal length 120 mm, which can be moved along its principal axis towards and away from the film. If the camera is to be able to form perfectly focused images of objects from infinite distance down to 1.00 m from the camera, through what distance must it be possible to move the lens? [L 82]

3

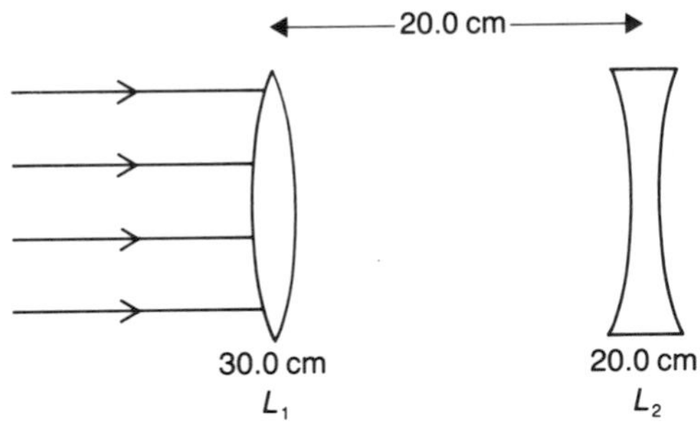

Fig 20.9 Diagram for Question 3

In Fig 20.9 parallel rays of light enter a thin converging lens L_1 of focal length 30.0 cm and then travel towards a thin diverging lens L_2 of focal length 20.0 cm, which is situated at a distance of 20.0 cm beyond L_1. The final image formed by this combination is

A at L_1 **B** 13.3 cm from L_1
C 40.0 cm from L_1 **D** at L_1
E at infinity [L 83]

4 An achromatic lens is constructed of two thin lenses in contact, a converging lens of focal length 150 mm and a diverging lens of focal length 100 mm. What is the numerical value of the focal length of the combination?

A 50 mm **B** 60 mm **C** 200 mm
D 250 mm **E** 300 mm [O 80]

21 OPTICAL INSTRUMENTS

Angular magnification

When an object is viewed, the *apparent* size of the object is determined by the length L of the image formed on the retina. As shown in Fig 21.1, L is determined by the *visual angle* θ which the object subtends at the eye. Throughout this chapter we assume θ to be small, in which case L is directly proportional to θ.

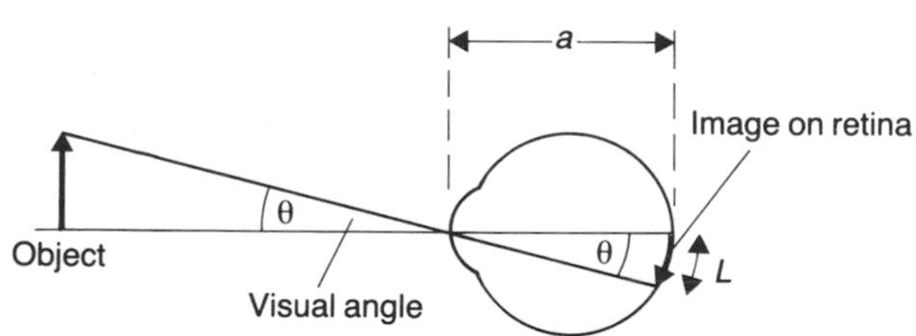

Note: $L = a\theta$
where L = length of image on retina (metres)
a = length of eyeball (metres)
θ = visual angle (radians); assumed small

Fig 21.1 Visual angle

The purpose of an optical instrument is to increase the size of the visual angle. In doing so the final image, when viewed through the instrument, appears to be larger than when the object is viewed using the unaided or 'naked' eye. We define the angular magnification (or magnifying power) M of an optical instrument by

$$M = \frac{\beta}{\alpha} \tag{21.1}$$

where β is the angle subtended at the eye by the *image* when using the instrument, and α is the angle subtended using the unaided eye by the *object* when at the appropriate distance.

The magnifying glass (simple microscope)

Using the unaided eye, the maximum apparent size of the object occurs when it is placed at the least distance of distinct vision D (typically 250 mm for adults) from the eye, as shown in Fig 21.2.

h = height of object O

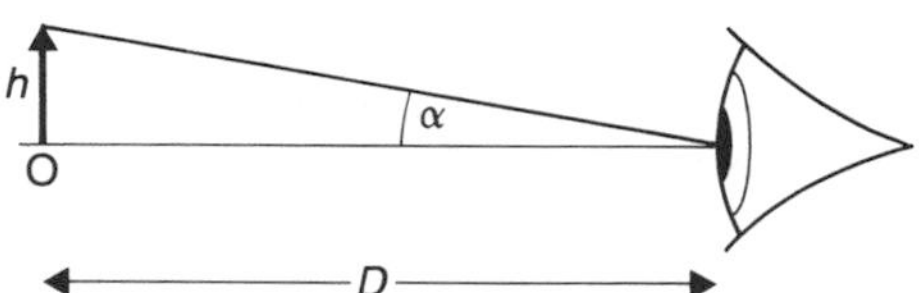

Fig 21.2 Visual angle α of an object at the least distance of distinct vision D

The angle subtended α, in radians (see Chapter 2), is given by

$$\alpha = \frac{h}{D} \tag{21.2}$$

where h is the height of the object O.

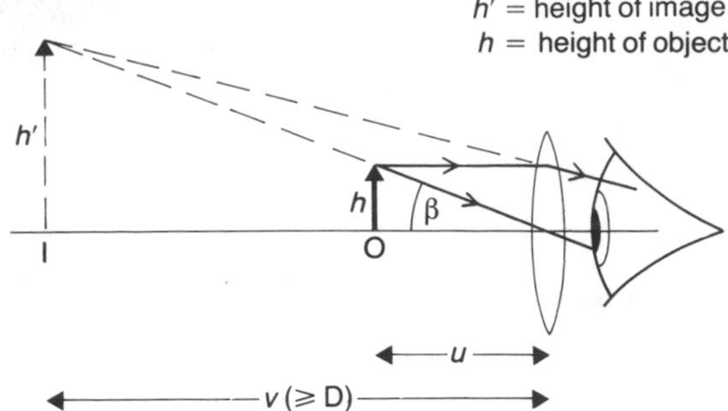

Fig 21.3 Visual angle β using a simple microscope

Fig 21.3 shows the formation of an image I when the object O is placed distance u from the magnifying glass. *Since u is less than D, β is greater than α.* Thus

$$\beta = \frac{h'}{v} = \frac{h}{u} \tag{21.3}$$

Combining Equations 21.1, 21.2 and 21.3 gives

$$M = \frac{\beta}{\alpha} = \frac{h/u}{h/D}$$

$$\therefore \quad M = \frac{D}{u} \tag{21.4}$$

This is a *general expression* and is true for whatever value of v (and u) we have. Note that *in normal adjustment the image distance v equals D.* It is convenient to use Equation 21.4, but you should also be able to work from first principles (see below).

Example 1

An object of height 2.00 mm is to be viewed using a simple magnifying glass of focal length 50.0 mm. If the final image is formed at the least distance of distinct vision (250 mm) from the eye, calculate the visual angle subtended (a) using the unaided eye, (b) using the magnifying glass. Hence calculate (c) the angular magnification achieved. Check your answer to part (c) using the appropriate formula.

Method

Using millimetres we have $h = 2.00$ and $D = 250$.

(a) From Equation 21.2

$$\alpha = \frac{h}{D} = \frac{2}{250}$$

$$= 8.00 \times 10^{-3}\ \text{rad}$$

(b) Referring to Fig 21.3 we have image distance equal to 250 and since the image is virtual, $v = -250$. To find β we require the object distance u. We can rearrange Equation 21.1, putting the focal length of the lens as $f = 50$:

$$\frac{1}{u} = \frac{1}{f} - \frac{1}{v} = \frac{1}{50} - \frac{1}{(-250)} = \frac{6}{250}$$

$$\therefore \quad u = \frac{250}{6}$$

From Equation 21.3

$$\beta = \frac{h}{u} = 2 \times \frac{6}{250}$$

$$= 48.0 \times 10^{-3}\ \text{rad}$$

(c) From Equation 21.1

$$M = \frac{\beta}{\alpha} = \frac{48 \times 10^{-3}}{8 \times 10^{-3}} = 6.00$$

Note that the image is virtual, so we should, strictly speaking, write $M = -6.00$. It is common practice, however, to write only the numerical value and to omit the sign. M does not depend on the height of the object, since h cancels in the derivation of Equation 21.4. To check our answer for M using Equation 21.4, we have $D = 250$ and $u = 250/6$, so

$$M = \frac{D}{u} = 250 \times \frac{6}{250} = 6.00$$

Answer

(a) 8.00×10^{-3} rad, (b) 48.0×10^{-3} rad, (c) 6.00 times.

Example 2

A man wishes to study a photograph in fine detail by using a lens as a simple magnifying glass in such a way that he sees an image magnified ten times and at a distance of 250 mm from the lens. What focal length lens should he use, and how far from the photograph should it be held? [L]

Method

We have $M = 10.0$. Refer to Fig 21.3. $v = -250$ and we have to *assume* that $D = -250$. We require u and f.

Rearranging Equation 21.4 gives us (without signs)

$$u = \frac{D}{M} = \frac{250}{10} = 25.0\ \text{mm}$$

The photograph is held 25 mm from the lens. From Equation 20.1

$$\frac{1}{f} = \frac{1}{v} + \frac{1}{u} = \frac{1}{-250} + \frac{1}{25} = \frac{9}{250}$$

$$\therefore \quad f = \frac{250}{9} = 27.8\,\text{mm}$$

Answer

A converging lens of focal length 27.8 mm is needed at 25.0 mm from the photograph.

Exercise 78

1 A man whose least distance of distinct vision is 250 mm views a stamp using a converging lens of focal length 30 mm. If the final image is located at the least distance of distinct vision, calculate (a) the distance of the stamp from the lens, (b) the angular magnification he achieves. Assume that the eye is close to the lens.

2 Repeat Question 1, but assume that the image is to be observed at infinity.

3 Repeat Question 1 for a man whose least distance of distinct vision is 180 mm.

The astronomical telescope

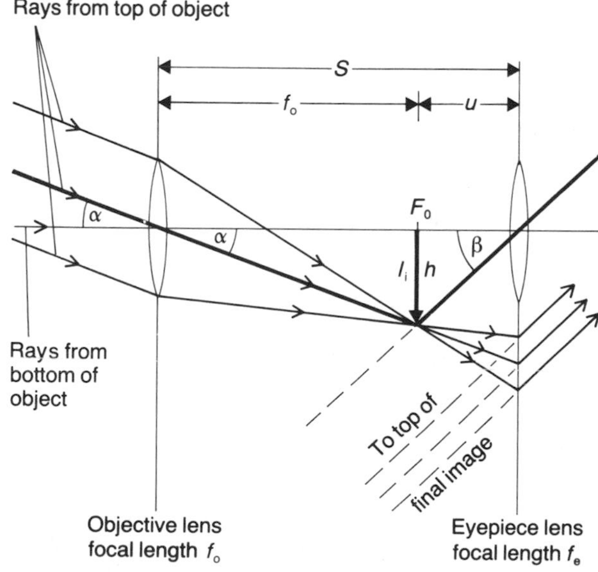

Fig 21.4 Visual angle in the astronomical telescope

Fig 21.4 shows how an astronomical telescope, used to observe a distant object such as a star, increases the visual angle. The object would subtend an angle α when viewed with the unaided eye. Use of the telescope leads to the formation of a final image which subtends an angle β at the eye (assuming that the eye is close to the eyepiece lens). From Fig 21.4, in which the intermediate image I_i is of height h and is formed at distance u from the eyepiece lens,

$$\alpha = \frac{h}{f_o} \qquad (21.5)$$

and

$$\beta = \frac{h}{u} \qquad (21.6)$$

where α and β, in radians, are small angles. Combining Equations 21.1, 21.5 and 21.6 gives the angular magnification (or magnifying power) M:

$$M = \frac{\beta}{\alpha} = \frac{h/u}{h/f_o}$$

or

$$M = \frac{f_o}{u} \qquad (21.7)$$

Equation 21.7 is a *general expression* and can be used at whatever distance the final virtual image is formed from the eye. In *normal* adjustment, in which the final image is formed at infinity, we note two special characteristics:

(1) $u = f_e$, so that $M = f_o/f_e$.

(2) The objective and eyepiece lenses are separated by a distance $(f_o + f_e)$.

Example 3

An astronomical telescope has an objective lens of focal length 100 cm and an eyepiece lens of focal length 5.00 cm. Calculate the angular magnification and the separation of the lenses when (a) the telescope is in normal adjustment, (b) the final virtual image is formed 25.0 cm from the eyepiece lens.

Method

Referring to Fig 21.4 we have $f_o = +100$ and $f_e = +5.00$.

(a) In normal adjustment $u = f_e = +5.00$. Thus from Equation 21.7

$$M = \frac{f_o}{u} = \frac{f_o}{f_e} = \frac{100}{5}$$

$$= 20.0$$

The separation S of the lenses is given by

$$S = (f_o + u) = (f_o + f_e) = 100 + 5$$
$$= 105\text{ cm}$$

(b) We must find u. With reference to the eyepiece lens, the final image is virtual, so $v = -25.0$ cm.

We apply Equation 21.1 to the *eyepiece lens*:

$$\frac{1}{u} = \frac{1}{f_e} - \frac{1}{v} = \frac{1}{5} - \frac{1}{(-25)} = \frac{6}{25}$$

$$\therefore \quad u = \frac{25}{6} = 4.17\text{ cm}$$

From Equation 21.7

$$M = \frac{f_o}{u} = 100 \times \frac{6}{25} = 24.0$$

The separation S of the lens is given by

$$S = (f_o + u) = 100 + 4.17 = 104\text{ cm}$$

Note that the angular magnification is *greater* when the image is formed closer to the eyepiece lens. This is because the eyepiece lens is closer to the intermediate image, and thus the visual angle β is larger.

Answer
(a) $M = 20.0$, lens separation is 105 cm,
(b) $M = 24.0$, lens separation is 104 cm.

Example 4

An astronomical telescope consists of two thin converging lenses. When it is in normal adjustment the lenses are 650 mm apart and the angular magnification is 12.0. Calculate the focal length of the objective lens and the eyepiece lens.

Method
We have $M = 12$ and lens separation $S = 650$ mm. Now, in normal adjustment (see p.135)

$$M = \frac{f_o}{f_e} \qquad \therefore \quad f_o = 12f_e \qquad \text{(i)}$$

and $$S = f_o + f_e \qquad \therefore \quad f_o + f_e = 650 \qquad \text{(ii)}$$

We have two simultaneous equations (see Chapter 2) so we substitute for f_o from (i) into (ii):

$$12f_e + f_e = 650$$
$$\therefore \quad f_e = 50\text{ mm}$$
$$\therefore \quad f_o = 12f_e = 600\text{ mm}$$

Answer
The objective has focal length 600 mm, the eyepiece 50 mm.

Example 5

A telescope has an objective of focal length 80 cm and an eyepiece of focal length 2.0 cm. It is focused on the moon, whose diameter subtends an angle of 8.0×10^{-3} rad at the objective. The eyepiece lens is adjusted so as to project a sharp image of the moon on to a screen placed 20 cm from the eyepiece lens. Calculate the diameter of (a) the intermediate image formed by the objective lens, (b) the image on the screen. Calculate (c) the lens separation.

Method

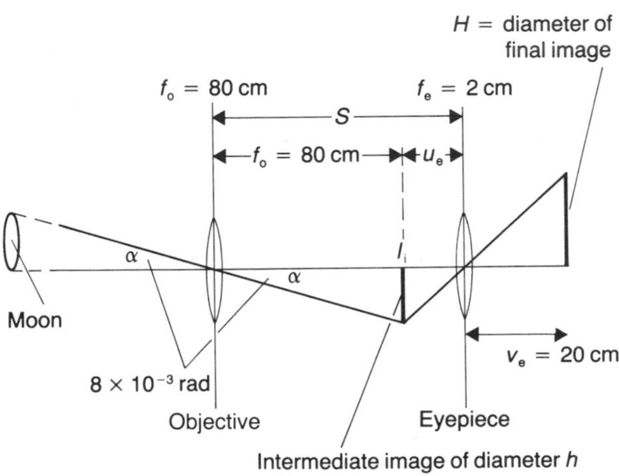

Fig 21.5 Image projection using a telescope (for Example 5)

The situation is shown in Fig 21.5 (two separate construction rays are shown in order to simplify the diagram). The intermediate real image I_i acts as the object for the eyepiece lens, which further magnifies it to produce a *magnified real image* on the screen.

(a) We have $f_o = 80$ cm and $\alpha = 8.0 \times 10^{-3}$. To find h we rearrange Equation 21.5:

$$h = \alpha f_o = 8.0 \times 10^{-3} \times 80$$
$$= 0.64\text{ cm}$$

(b) To find the diameter H of the final image we use the idea of lateral magnification m (see Chapter 20) by the eyepiece lens. From Equations 20.2 and 20.3

$$m = \frac{\text{Height of image } H}{\text{Height of object } h} = \frac{\text{Image distance } v_e}{\text{Object distance } u_e}$$

Rearranging gives

$$H = \frac{v_e}{u_e} \times h \qquad (21.8)$$

We have $v_e = +20$ cm and $h = 0.64$ cm. To find u_e we apply Equation 20.1 to the eyepiece lens, of focal length $f_e = 2.0$ cm:

$$\frac{1}{u_e} = \frac{1}{f_e} - \frac{1}{v_e} = \frac{1}{2} - \frac{1}{20} = \frac{9}{20}$$

$$\therefore \quad u_e = \frac{20}{9} = 2.2 \text{ cm}$$

Thus, from Equation 21.8,

$$H = \frac{v_e}{u_e} \times h = 20 \times \frac{9}{20} \times 0.64$$

$$= 5.8 \text{ cm}$$

(c) The lens separation S is given by

$$S = f_o + u_e = 80 + 2.2$$

$$= 82 \text{ cm}$$

Answer

(a) 0.64 cm, (b) 5.8 cm, (c) 82 cm.

Exercise 79

1 An astronomical telescope which is in normal adjustment consists of two thin converging lenses of focal length 60.0 cm and 3.00 cm. It is focused on a distant object which subtends an angle of 2.00×10^{-3} rad when viewed directly. Calculate (a) the angular magnification achieved, (b) the separation of the lenses, (c) the angle subtended by the final image.

2 Repeat Question 1, given that the final virtual image is formed 25 cm from the eyepiece lens.

3 An astronomical telescope has an objective of focal length 90 cm and an eyepiece of focal length 5.0 cm. When, in normal adjustment, it is used to view a full moon, the final image subtends an angle of 0.10 radian at the eye lens. Calculate (a) the angular magnification, (b) the angle subtended by the moon when viewed directly.

Given that the distance between the moon and the earth is 3.8×10^5 km, calculate (c) the diameter of the moon.

Exercise 80: Questions of A-level standard

1 Using a 5.0 cm focal length lens held close to the eye to read some small print, you find that the image is blurred unless the print is placed not less than a certain distance x away from the lens. Explain why this is so. Supposing that the near point of your own eye is at 15 cm, calculate x and the angular magnification that would be achieved.

What is the maximum separation of the lens and print at which an upright image of the print will be clearly visible? Explain this. [L 80]

2 A camera with a lens of focal length 75 mm is used to take a photograph of a distant scene. The negative is viewed with the unaided eye from the least distance of distinct vision of 0.25 m. What overall magnifying power is obtained?

By examining the negative through a magnifying lens it is possible to achieve an overall magnifying power of unity. Calculate the required focal length if the final image is to be (a) very distant and (b) at the least distance of distinct vision, 0.25 m, from the eye. [WJEC 77, p]

3 An astronomical telescope in normal adjustment consists of two thin converging lenses placed 550 mm apart. If the magnifying power of the telescope is 10.0, calculate the focal length of the eyepiece lens.

4 (a) Explain what is meant by the magnifying power of a telescope. Show that, if f_o and f_e are the focal lengths of the objective and eyepiece respectively, the magnifying power is f_o/f_e when the telescope is in normal adjustment.

(b) For an astronomical telescope, f_o and f_e are 0.90 m and 0.10 m respectively. When, in normal adjustment, it is used to view a full moon the image subtends an angle of 0.051 radian at the eye lens. If the distance between the moon and the earth is 3.8×10^5 km, calculate a value for the diameter of the moon. [JMB 82]

5 A refracting telescope has an objective of focal length 900 mm and an eyepiece of focal length 12 mm.

(a) What is the distance apart of the lenses when the telescope is in normal adjustment, and what is its magnifying power?

(b) If the telescope is used to project a sharp image of the sun on to a screen that is at a fixed distance of 150 mm from the eyepiece, what adjustments to the telescope will be needed?

(c) Given that the sun's diameter subtends an angle of 9×10^{-3} rad, calculate the diameter of the image on the screen. [O 81, p]

6 A refracting telescope has an objective of focal length 1.0 m and an eyepiece of focal length 2.0 cm. A real image of the sun, 10 cm in diameter, is formed on a screen 24 cm from the eyepiece. What angle does the sun subtend at the objective? [L 79]

Section F
Heat

22 TEMPERATURE MEASUREMENT

The Celsius scale

This method of measuring temperature (often called the Centigrade scale) describes the ice point as zero (0 °C) and the steam point as 100 degrees Celsius (100 °C), i.e. the fundamental interval is 100 degrees.

To describe the number of degrees θ for any temperature we use some temperature-dependent property such as the length of a mercury thread (the mercury-in-glass scale) or the resistance of a coil of pure platinum wire (the platinum resistance scale). The size of the property at 100 °C is denoted by X_{100}, and at 0 °C by X_0. Each degree then corresponds to a property change of $(X_{100} - X_0)/100$. A temperature change from 0 to θ gives a property change from X_0 to X, so that

$$X - X_0 = \theta \times \frac{X_{100} - X_0}{100}$$

or

$$\theta = \frac{X - X_0}{X_{100} - X_0} \times 100 \qquad (22.1)$$

Example 1

A certain platinum resistance thermometer has a resistance of 2.40 Ω at 0 °C, 3.34 Ω at 100 °C and 2.87 Ω at an unknown temperature. (a) Evaluate this temperature on the platinum resistance scale. (b) What value of resistance would be expected at −2 °C on this scale?

Method

(a)
$$\theta = \frac{X - X_0}{X_{100} - X_0} \times 100$$
$$= \frac{2.87 - 2.40}{3.34 - 2.40} \times 100 = 50\,°\text{C}$$

(b)
$$-2 = \frac{X - 2.40}{3.34 - 2.40} \times 100$$
$$\therefore \quad X - 2.40 = \frac{-2 \times 0.94}{100}$$
$$= -\,0.0188$$
$$\therefore \quad X = -0.0188 + 2.40$$
$$= 2.38\,\Omega$$

Answer

(a) 50 °C, (b) 2.38 Ω.

Agreement of scales

The resistance R of a pure metal, perhaps a coil of pure copper wire or platinum, conforms approximately to the formula $R = R_0(1 + \alpha\theta)$, where α is a constant for the given metal and θ is the Celsius temperature on the mercury scale. Consequently Equation 22.1 becomes

$$\underset{\text{(Resistance scale)}}{\theta\,°\text{C}} = \frac{[R_0(1 + \alpha\theta) - R_0]}{R_0(1 + 100\alpha) - R_0} \times 100$$

$$= \frac{\alpha\theta}{100\alpha} \times 100 = \underset{\text{(Mercury scale)}}{\theta\,°\text{C}}$$

We see that the resistance scale and mercury scale agree for most ordinary purposes because resistance increases linearly with mercury scale temperature. Differences will show up in accurate work. This is true also of other scales in common use.

The gas thermometer scale

The variation of gas pressure with temperature at constant volume, or the variation of volume with temperature (pressure being kept constant), or even the product of pressure times volume varying with temperature could each be used for temperature measurement — but the first of these possibilities is preferred.

$$\underset{\text{(Constant volume)}}{\theta\,°\mathrm{C}} = \frac{P - P_0}{P_{100} - P_0} \times 100 \qquad (22.2)$$

The Kelvin absolute thermodynamic scale of temperature

On this scale the zero is the lowest temperature that in theory could be reached; the ice point is 273.16 degrees and the steam point is 100 degrees higher. These degrees are called kelvins (K) and are equal in size to the Celsius degree. For ordinary purposes we can say absolute temperature $T = 273 +$ Celsius temperature. The temperature-dependent property used for this scale is the efficiency of an ideal heat engine. The temperature measured by this 'thermometer' is absolute in the sense of not being dependent on the behaviour of a particular material and it is superior to all others. It can be shown that a perfect-gas thermometer would give identical answers.

A nearly perfect gas can be used in a gas thermometer and corrections can be applied to allow for its not being perfect. Thus it is possible to realise the Kelvin thermodynamic scale.

For Kelvin temperatures

$$\frac{T}{T_f} = \frac{P}{P_f} \qquad (22.3)$$

where P and P_f are pressures at constant volume for temperatures T and T_f. (T_f could be a fixed point, e.g. ice point, triple or steam point.)

Customary temperature

All absolute temperatures must be recorded in kelvins and for any temperature *change* we should use the SI unit, which is the kelvin, but for stating a temperature (customary temperature) it is normal to write °C. So a rise from 10 °C to 15 °C is a change of 5 K, from 283 K to 288 K.

Example 2

The resistance of a certain coil of metal wire used as a resistance thermometer was 20.0 Ω, 20.7 Ω and 24.0 Ω at temperatures of 0, 20 and 100 °C respectively on a mercury-in-glass thermometer.

Calculate the temperature on the scale of the resistance thermometer when the mercury scale temperature is 20.0 °C.

Method

$$\underset{\text{(Resistance scale)}}{\theta} = \frac{X - X_0}{X_{100} - X_0} \times 100$$

with X representing the coil resistance. Writing $X = 20.7$ we get

$$\theta = \frac{20.7 - 20.0}{24.0 - 20.0} \times 100$$

$$= \frac{0.7}{4.0} \times 100$$

$$= 17.5\,°\mathrm{C}$$

Answer

17.5 °C.

Exercise 81

1 The mercury level in an uncalibrated mercury-in-glass thermometer was found to be 2.0 cm from the bulb when the thermometer was placed in melting ice, and 18 cm from the bulb when it was placed in pure, boiling water. Where will the mercury level be found at (a) 25 °C, (b) −5 °C?

2 A resistance thermometer has resistance of 75.0 Ω and 96.0 Ω respectively at the ice and steam points. What are the temperatures indicated by this thermometer corresponding to (a) 82.0 Ω, (b) 72.0 Ω?

3 The volume of some mercury increased by $12.0 \times 10^{-4}\,cm^3$ when the temperature as recorded by a constant-volume air thermometer rose from 0 °C to 65 °C. For a rise from 65 °C to 100 °C, still on this scale, the volume changed by $6.0 \times 10^{-4}\,cm^3$. Calculate the mercury scale temperature corresponding to 65 °C on the gas scale.

Practical realisation of the Kelvin scale

Resistance thermometers and thermocouples are particularly convenient for practical thermometry. A temperature close to the Kelvin scale temperature (or the Celsius equivalent) corresponding to a resistance R can be obtained by use of a suitable formula. An example is $R = R_0(1 + \alpha\theta + \beta\theta^2)$.

An example of a formula for use with a thermocouple is $E = a + b\theta + c\theta^2$ where E is the EMF at Celsius temperature θ, and a, b and c are constants for the thermocouple.

Thermocouples are also used for less accurate work because they have the advantage of being small. The formula $E = a\theta + b\theta^2$ is then satisfactory.

Example 3

The volume V of a fixed mass of mercury at a temperature t °C measured on the perfect gas scale is given by

$$V = V_0(1 + 1.818 \times 10^{-4}t + 0.8 \times 10^{-8}t^2)$$

where V_0 is the volume at 0 °C on the gas scale.

Calculate the temperature expected on a mercury thermometer when the gas scale temperature is 40.0 °C.

Method

For $t = 40.0$ °C the equation given becomes

$$V = V_0(1 + 0.007\,272 + 0.000\,012\,80)$$

so that

$$V = V_0 \times 1.007\,285$$

For $t = 100$ °C the equation becomes

$$V_{100} = V_0(1 + 0.018\,18 + 0.000\,08) = 1.018\,26V_0$$

The required temperature on the mercury scale is given (see Equation 22.1) by

$$\underset{\text{(Mercury scale)}}{\theta} = \frac{V - V_0}{V_{100} - V_0} \times 100$$

$$= \frac{1.007\,285V_0 - V_0}{1.018\,26V_0 - V_0} \times 100$$

$$= \frac{0.007\,285}{0.018\,26} \times 100$$

$$= 39.896 \quad \text{or} \quad 39.9\,°\text{C}$$

(Three significant figures were given for the temperature 40.0 °C in the question.)

Answer

39.9 °C.

Example 4

A copper–constantan thermocouple with its cold junction at 0 °C had an EMF of 4.28 mV with its other junction at 100 °C. The EMF became 9.29 mV when the temperature difference was 200 °C. If the EMF E is related to the temperature difference θ by the equation $E = A\theta + B\theta^2$, (a) calculate the values of A and B. (b) Up to what temperature may E be assumed proportional to θ without incurring an error of more than 1%?

Method

(a) $E = A\theta + B\theta^2$

$$\therefore \quad 4.28 = 100A + 10^4B \qquad \text{(i)}$$

$$\text{and} \quad 9.29 = 200A + 4 \times 10^4B \qquad \text{(ii)}$$

Multiplying (i) by two and subtracting (i) from (ii) will eliminate A and allow B to be evaluated. Thus

$$9.29 - 8.56 = 2 \times 10^4B$$

$$\therefore \quad B = \frac{0.73}{2 \times 10^4}$$

$$\therefore \quad B = 0.365 \times 10^{-4}\,\text{mV K}^{-2} \text{ or } 0.0365\,\mu\text{V K}^{-2}$$

Substituting this value for B into (i) gives

$$4.28 = 100A + 0.365$$

$$\therefore \quad A = \frac{4.28 - 0.365}{100}$$

$$\therefore \quad A = 3.915 \times 10^{-2} \quad \text{or} \quad 39.15 \times 10^{-3}\,\text{mV K}^{-1}$$

$$\text{or} \quad 39.15\,\mu\text{V K}^{-1}$$

(b) If $B\theta^2 = 1\%$ of E, then $B\theta^2/E = 1/100$

$$\therefore \quad \frac{B\theta^2}{A\theta + B\theta^2} = \frac{1}{100}$$

$$\therefore \quad 100B\theta^2 = A\theta + B\theta^2$$

$$\therefore \quad 99B\theta = A$$

$$\therefore \quad \theta = \frac{A}{99B}$$

$$= \frac{39.15}{99 \times 0.0365}$$

$$= 10.83\ ^\circ\text{C}$$

Answer

(a) $A = 39.2\ \mu\text{V K}^{-1}$, $B = 0.0365\ \mu\text{V K}^{-2}$;
(b) 10.8 °C.

Exercise 82

1 A certain resistance thermometer has a resistance of 5.00 Ω at 0 °C, and its resistance at any other Celsius temperature θ is given approximately by $R = R_0(1 + 4 \times 10^{-3}\theta)$. (a) Calculate the resistance at 40 °C. (b) Calculate the temperature at which the resistance is 5.50 Ω.

2 The resistance of a certain wire at a Celsius temperature t measured on the ideal gas scale is given by

$$R_t = R_0(1 + 4 \times 10^{-3}t + 10^{-5}t^2)$$

where R_0 is the resistance at 0 °C. What is the temperature on the resistance scale using this wire, when the temperature is 50 °C on the ideal gas scale?

3 The EMF E obtained from a chromel-constantan thermocouple is 6.32 mV at 100 °C and 24.91 mV at 600 °C. If E is related to the Celsius temperature by $E = A\theta + B\theta^2$, show that the temperature when $E = 16.4$ mV is approximately 300 °C.

Exercise 83: Questions of A-level standard

1 A coil of wire has resistance of 2.00 Ω, 2.80 Ω and 3.00 Ω at temperatures of 0 °C, 100 °C and t respectively. What is the value of t on the scale defined by this resistor?

A 20 °C **B** 25 °C **C** 80 °C
D 105 °C **E** 125 °C [L 80]

2 The volume of a certain liquid varies with temperature according to the relation $V = V_0(1 + at + bt^2)$, where V_0 is the volume at 0 °C, and V the volume at t °C on the scale of a gas thermometer. a and b are constants with $b = a/1000$. What will be the liquid thermometer reading if the gas thermometer reads 50.0 °C?

3 A temperature T can be defined by $T = T_f(X/X_f)$, where T_f is the assigned temperature of a fixed point and X and X_f are the values of a thermometric property of a substance at T and T_f respectively. On the ideal gas scale the fixed point is the triple point of water, and $T_f = 273.16$ K.

Part (f) of this question says:
The pressures recorded in a certain constant-volume gas thermometer at the triple point of water and at the boiling point of a liquid were 600 mm of Hg and 800 mm of Hg respectively. What is the apparent temperature of the boiling point? However, it was found that the volume of the thermometer increased by 1% between the two temperatures. Obtain a more accurate value of the boiling point. [WJEC 79, p]

4 A chromel–alumel thermocouple is connected across a galvanometer having a resistance 24 Ω. One junction of the thermocouple is kept at 0 °C and the other is at 200 °C. If the EMF of the thermocouple is 40 μV for each 1.0 K of temperature difference between the junctions, and the thermocouple has a resistance of 6.0 Ω, what size of current is expected?

23 SPECIFIC HEAT CAPACITY AND LATENT HEAT

Specific heat capacity of a substance

This quantity is usually denoted by c or s and is the quantity of heat required to raise the temperature of 1 kg of substance by 1 K.

The SI unit for s is J kg^{-1} K^{-1} or J/kg K. Removing s joule of heat from 1 kg causes a fall in temperature of 1 K.

The quantity of heat (H joule) needed to raise m kilogram of substance by θ kelvin is

$$H = ms\theta \qquad (23.1)$$

The (total) heat capacity of an object

This is the heat needed to raise the temperature of the object by 1 K. For a mass m of material whose specific heat capacity is s, the heat capacity is ms.

Water equivalent

The water equivalent of a mass m of substance is the mass of water that has equal heat capacity. It is given by $m \times$ SHC of substance $\div$ SHC of water.

Calorimeters

Vessels used for heat measurements are called calorimeters and are of two kinds. The first kind is constructed of a good heat conductor such as copper, and the temperature of the calorimeter is assumed always to be the same as its contents. The second kind is made of a heat-insulating material so that negligible heat enters or leaves the calorimeter during an experiment if the time concerned is short.

Conservation of energy

In most calorimetry calculations we apply the law of conservation of energy to the heat exchanges that occur. Heat that is lost by one thing cooling down is the heat gained by something else. (The term 'heat' is best reserved for energy that is moving from a hotter place to a colder place. Energy settled inside a material is best called internal energy, and is the kinetic and potential energy of the microscopic movements of the material's atoms and molecules.)

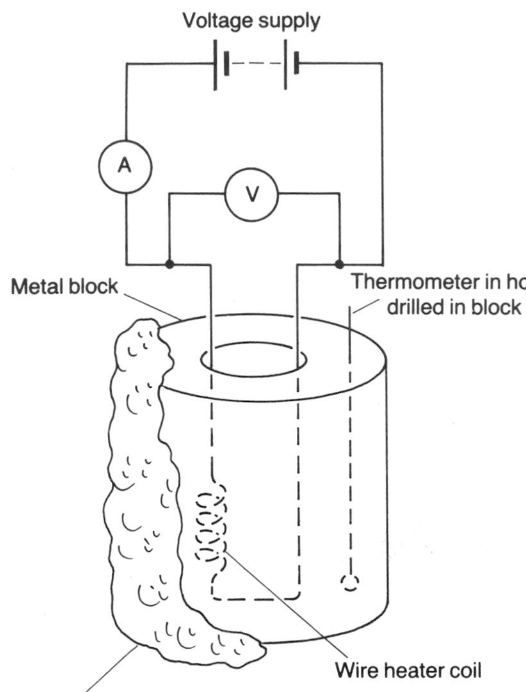

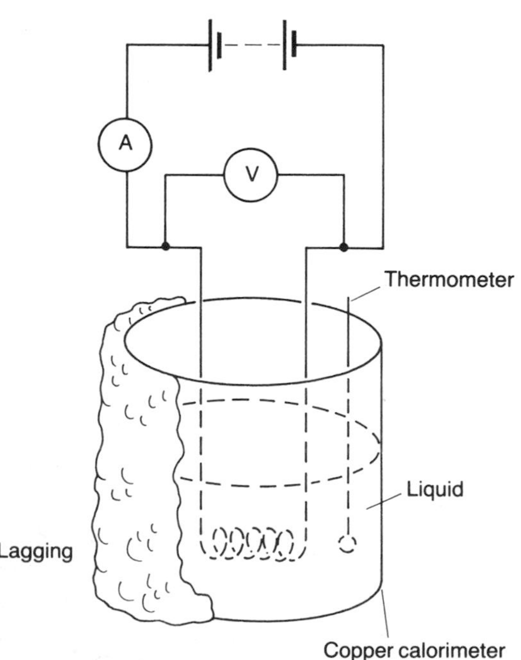

Fig 23.1 Apparatus for simple electrical calorimetry (not to scale)

Electrical heating

As pointed out in Chapter 27, the heat produced per second in an electric heater is given by VI where V is the potential difference ('voltage') across the heater wire and I is the current through it. This heat (per second) which spreads from the heater into the surrounding material is often described as the 'power dissipated'.

$$\text{Power} = V \times I \tag{23.2}$$

If the resistance of the heater wire is R (again see Chapter 27), we may use the relation $R = V/I$ to obtain

$$\text{Power } P = I^2R \quad \text{or} \quad V^2/R \tag{23.3}$$

In a time t seconds the heat produced by the heater is

$$\text{Heat } H = VIt \quad \text{or} \quad I^2Rt \quad \text{or} \quad \frac{V^2}{R}t \tag{23.4}$$

Example 1

A solid copper block of mass 5.0 kg is heated for 7 minutes exactly by an electric heater embedded in the block. A potential difference of 25 V is applied across the heater, and the current is recorded as 2.0 A. If the temperature of the block rises by 10 K, calculate the specific heat capacity of copper, assuming that no heat escapes from the apparatus and that the heat capacity of the heater itself is negligible (see Fig 23.1a).

Method

The heat given out by the heater is

$$H\ (= VIt) = 25 \times 2.0 \times 7 \times 60 = 21\,000\ \text{J}$$

The heat taken in that causes the temperature rise is

$$H\ (= ms\theta) = 5.0 \times s \times 10\ \text{J}$$

We can equate the heat given out and the heat taken in, as calculated above, because no significant amount of heat stays in the heater or escapes from the copper.

$$21\,000 = 50 \times s$$

$$\therefore \quad s = 420\ \text{J kg}^{-1}\,\text{K}^{-1} \quad \text{or} \quad 0.42\ \text{kJ kg}^{-1}\,\text{K}^{-1}$$

Answer

0.42 kJ $\text{kg}^{-1}\,\text{K}^{-1}$.

Example 2

50 g of water at 12 °C is placed in a copper calorimeter which weighs 0.10 kg. An electric heater coil of negligible thermal capacity is immersed in the water. With 7.0 V across the heater producing a steady current of 1.0 A for exactly 6 minutes, a final temperature of 22 °C was obtained. If the heat loss to the surroundings is negligible, what is the value of the specific heat capacity for water? (See Fig 23.1b.)
(SHC copper = 420 J kg^{-1} K^{-1}.)

Method

The heat given out by the heater is

$$H\ (= VIt) = 7.0 \times 1.0 \times 6 \times 60 = 2520\ \text{J}$$

The heat taken in by the water is

$$H_W(= ms\theta) = \frac{50}{1000} \times s \times (22 - 12) = 0.5 \times s$$

The heat taken in by the calorimeter (also initially at 12 °C and finally at 22 °C) is

$$H_c\ (= ms\theta) = \frac{100}{1000} \times 420 \times (22 - 12) = 420$$

Equating heat given out to heat gained ($H = H_W + H_c$), we get

$$2520 = 0.5s + 420$$

$$\therefore \quad s = \frac{2100}{0.5} = 4200\ \text{J kg}^{-1}\ \text{K}^{-1}$$

Answer

4.2 kJ kg^{-1} K^{-1}.

Heat losses

In practice there will always be some heat exchange between the calorimetry apparatus and the surroundings. One simple way of reducing the effect of this is to conduct experiments so that half the time the apparatus is warmer than the surroundings and half the time it is colder, so as to produce a net heat exchange of zero. In Example 2 this may have been done. If the room temperature was 17 °C, then one would hope that the apparatus would have temperatures equally above and below room temperature and for equal times. When an experiment like the one in Example 2 begins at room temperature θ_R, the apparatus will at all subsequent times exceed θ_R and there will be heat loss to the surroundings, which causes a smaller temperature rise. Heat losses will also cause a temperature fall when the heat supply is stopped. This is illustrated in Fig 23.2

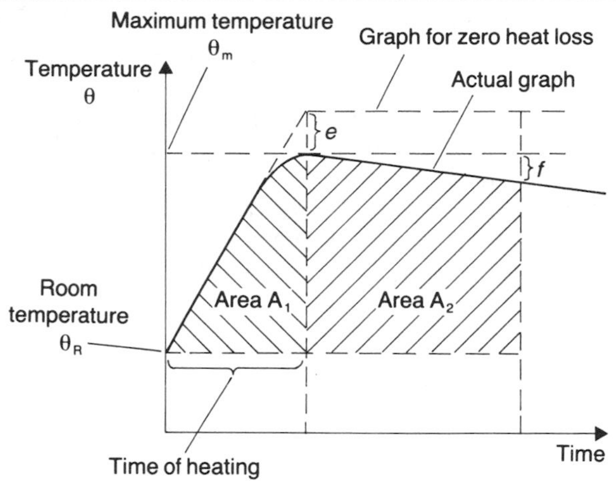

Fig 23.2 Consideration of heat loss to surroundings

The temperature shortfall, usually described as the temperature error or cooling correction, is denoted by e in the figure. Its value can be calculated from the formula

$$\frac{e}{f} = \frac{A_1}{A_2} \tag{23.5}$$

This formula is a consequence of Newton's law of cooling, according to which the heat loss per second is proportional to the excess temperature (actual temperature minus ambient temperature, θ_R).

Another consequence of Newton's law is that the average rate of heat loss and of 'temperature loss' during the time of heating is equal to half of the rate of loss that occurs in the vicinity of θ_m. So e can be deduced from

$$e = \frac{1}{2}\left(\begin{array}{c}\text{Rate of temperature}\\ \text{fall close to}\\ \text{maximum temperature}\end{array}\right) \times \left(\begin{array}{c}\text{Time of}\\ \text{heating}\end{array}\right) \tag{23.6}$$

If the heater stayed switched on indefinitely, the temperature would rise until the rate of heat escape equalled the rate of supply. The temperature then remains steady and a 'steady state' exists.

Example 3

75 g of liquid is placed in a copper calorimeter of mass 50 g. The initial temperature is 17.2 °C. A heater coil of negligible thermal capacity is immersed in the liquid and is operated at 1.8 A, 6.3 V, for exactly 4 minutes. After this time the temperature is recorded as 25.0 °C. Subsequently the temperature falls steadily to reach 24.7 °C after 2 minutes has elapsed from the time when the heating was stopped. Obtain a value for the specific heat capacity of the liquid.
(SHC copper = $0.42\ \mathrm{J\,g^{-1}\,K^{-1}}$.)

Method

The temperature correction is given by Equation 23.6 as

$$e = \frac{1}{2} \times \frac{25.0 - 24.7}{2} \times 4 = 0.3\ \mathrm{K}$$

The corrected temperature rise is $25.3 - 17.2 = 8.1$ K.

Heat supplied $= 6.3 \times 1.8 \times 4 \times 60 = 2721.6$

Heat gained by liquid $= 75 \times s \times 8.1$, where s is SHC of liquid

$$\begin{pmatrix}\text{Heat gained by}\\ \text{calorimeter}\end{pmatrix} = 50 \times 0.42 \times 8.1 = 170.1$$

$$\therefore \quad 2721.6 = (75 \times s \times 8.1) + 170.1$$

$$\therefore \quad s = \frac{2721.6 - 170.1}{75 \times 8.1}$$

$$= 4.2\ \mathrm{J\,g^{-1}\,K^{-1}}$$

Answer

$4.2\ \mathrm{J\,g^{-1}\,K^{-1}}$.

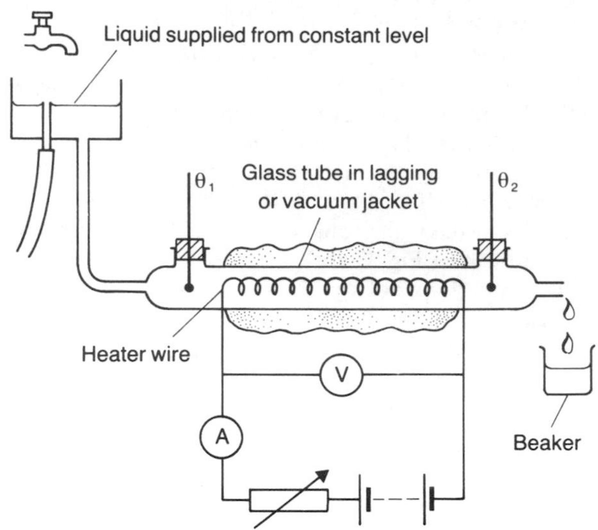

Fig 23.3 A continuous-flow calorimeter

The continuous-flow calorimeter

Fig 23.3 shows a continuous-flow calorimeter suitable for use with a liquid such as water.

When the temperature θ_2 is steady we know that the glassware has reached steady temperatures throughout.

The heat supplied per second is VI and equals $ms(\theta_2 - \theta_1)$ where m is the mass of liquid per second flowing through the apparatus and collected in the beaker. The specific heat capacity s can be calculated directly from this equation if heat loss to the surroundings is neglected. The heat loss, say h watt, which is usually quite small, can be allowed for as follows. The current is changed to a new value I', using a new voltage V', and the liquid flow is adjusted to m' to make up for this so that θ_2 is the same as before.

$$\therefore \quad VI = ms(\theta_2 - \theta_1) + h \qquad (23.7a)$$

$$\text{and} \quad V'I' = m's(\theta_2 - \theta_1) + h \qquad (23.7b)$$

The heat loss per second is the same for both situations because all parts of the apparatus have the same temperatures as before.

Subtracting these equations we get

$$V'I' - VI = s(\theta_2 - \theta_1)(m' - m)$$

from which an accurate value of s can be calculated.

Example 4

In a continuous-flow calorimeter the readings were: 6.0 V, 2.1 A, $\theta_1 = 17.0$ °C, $\theta_2 = 22.0$ °C, $35\ \mathrm{g\,min^{-1}}$ followed by 4.0 V, 1.4 A, $\theta_1 = 17.0$ °C, $\theta_2 = 22.0$ °C, $15\ \mathrm{g\,min^{-1}}$.

Obtain a value for the specific heat capacity of the liquid and the rate of loss of heat to the surroundings.

Method

Using $VI = ms(\theta_2 - \theta_1) + h$, we have, working in grams,

$$6.0 \times 2.1 = \left(\frac{35}{60} \times s \times 5.0\right) + h$$

and, using $V'I' = m's(\theta_2 - \theta_1) + h$, we have

$$4.0 \times 1.4 = \left(\frac{15}{60} \times s \times 5.0\right) + h$$

Subtracting these equations (see Chapter 2) gives

$$12.6 - 5.6 = \frac{20}{60} \times s \times 5.0$$

so that

$$s = \frac{7.0}{5.0} \times \frac{60}{20} = 4.2\ \mathrm{J\,g^{-1}\,K^{-1}}$$

Substituting this value for s in the first equation gives

$$12.6 = \left(\frac{35}{60} \times 4.2 \times 5.0\right) + h$$

from which

$$h = 12.6 - 12.25 = 0.35\ \mathrm{W}$$

Answer

$4.2\ \mathrm{J\,g^{-1}\,K^{-1}}$, 0.35 W.

Mixing hot with cold

If hot solid or liquid is introduced into cold liquid, the heat lost by the hot material cooling down is equal to the heat gained by the cold liquid and calorimeter plus any heat loss to the surroundings.

Example 5

21.0 g of liquid at 60.0 °C is mixed into 100 g of water at 12.5 °C which is already in a metal calorimeter of mass 70.0 g and specific heat capacity $400\ \mathrm{J\,kg^{-1}\,K^{-1}}$. If heat escape to the surroundings may be neglected, calculate the expected new temperature of the water, given that the specific heat capacity is $4200\ \mathrm{J\,kg^{-1}\,K^{-1}}$ for water and $4000\ \mathrm{J\,kg^{-1}\,K^{-1}}$ for the liquid.

Method

The liquid cools from 60 °C to the final temperature which can be called x °C. The calorimeter and water start at 12.5 °C and rise to x °C.

The heat given out is

$$\frac{21}{1000} \times 4000 \times (60 - x)$$

The heat taken in is

$$\frac{100}{1000} \times 4200 \times (x - 12.5) + \frac{70}{1000} \times 400 \times (x - 12.5)$$

Since these two heats are equal,

$$21 \times 4 \times (60 - x) = 420 \times (x - 12.5) + 7 \times 4 \times (x - 12.5)$$

$$\therefore \quad 5040 - 84x = 420x - 5250 + 28x - 350$$

$$\therefore \quad x = \frac{5040 + 5250 + 350}{420 + 28 + 84} = 20\ ^\circ\mathrm{C}$$

Answer

20 °C.

Exercise 84

1 A coil of wire of heat capacity $12\ \mathrm{J\,k^{-1}}$ has a PD of 5.0 V applied between its ends, so that a current of 0.20 A flows. Calculate the temperature rise produced in 1 minute. Assume the coil to be thermally insulated.

2 Using a constant-flow calorimeter for measuring the specific heat capacity of a liquid, a PD of 5.0 V was applied to the heating coil. The rate of flow of liquid was then doubled and, by adjusting the applied PD, the same inlet and outlet temperatures were obtained. Assuming heat losses to be negligible, calculate the new value of the applied PD. (It is necessary to assume that the resistance of the apparatus remains constant.)

3 380 g of a liquid at 12 °C in a copper calorimeter weighing 90 g is heated at a rate of 20 watt for exactly 3 minutes to produce a temperature of 17 °C. If the specific heat capacity of the calorimeter is $400\ \mathrm{J\,kg^{-1}\,K^{-1}}$, the thermal capacity of the heater is negligible, and there is negligible heat loss to the surroundings, obtain a value for the specific heat capacity of the liquid.

4 Calculate the final temperature when 200 g of water at 50 °C is mixed with 80 g of water at 10 °C. Heat losses to the calorimeter and surroundings may be neglected.

5 90 g of liquid were placed in a 40 g copper calorimeter (SHC $400\ \mathrm{J\,kg^{-1}\,K^{-1}}$) and heating was provided for 1 min 40 s using a PD of 14 V across an electric immersion heater. The heater current was constant at 3.0 A during this time. The heat capacity of the heater was negligible. The temperature, initially 10.0 °C, rose to 30.0 °C and then fell to 29.0 °C as the apparatus cooled for 50 s. Obtain a value for the specific heat capacity of the liquid.

6 (a) A metal block weighing 1.0 kg is heated by a 32 W electric heater inside it. If the rate of rise of temperature is $5.0\,\mathrm{K\,min^{-1}}$, what is the specific heat capacity of the metal? Assume the thermal capacity of the heater and the heat loss to the surroundings to be negligible.

(b) If a final temperature of 40 K greater than that of the surroundings is reached with the heater still on, what is the rate of escape of heat from the block (i) at this temperature, (ii) at an excess temperature of 10 K?

Specific latent heats

The heat required to melt unit mass of substance at its melting point is called its specific latent heat of fusion. It is usually measured in $\mathrm{J\,kg^{-1}}$. Similarly for evaporation at the boiling point we have the specific latent heat of vaporisation.

Measurement of specific latent heat of vaporisation

A suitable apparatus is illustrated in Fig 23.4.

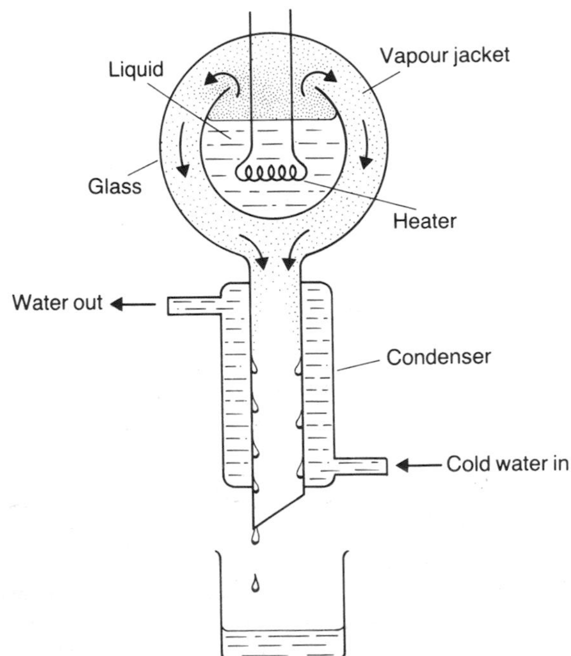

Fig 23.4 Apparatus for measuring the specific latent heat of vaporisation

The heat supplied is

$$V_1 I_1 t = m_1 l + Q \qquad (23.8)$$

where V_1 is the PD across the heater and I_1 the current through it. m_1 is the mass of substance evaporated in time t, l the specific latent heat and Q the heat loss in time t to the surroundings.

Using a different heater voltage V_2 we get a current I_2 and a mass m_2 is evaporated in time t. Thus

$$V_2 I_2 t = m_2 l + Q \qquad (23.9)$$

Eliminating Q from the above equations we get

$$V_1 I_1 t - m_1 l = V_2 I_2 t - m_2 l \qquad (23.10)$$

We have assumed Q to be the same for both rates of heating. This should be true enough since the liquid is at its boiling point each time and Q is already small due to the vapour jacket reducing heat loss.

Example 6

In an experiment to determine the specific latent heat of vaporisation of an alcohol using a self-jacketing vaporiser the following results were taken:

Experiment 1:
$V_1 = 7.40\,\mathrm{V}$, $I_1 = 2.60\,\mathrm{A}$, mass $m_1 = 5.80 \times 10^{-3}\,\mathrm{kg}$ collected in 300 s.

Experiment 2:
$V_2 = 10.0\,\mathrm{V}$, $I_2 = 3.60\,\mathrm{A}$, mass $m_2 = 11.3 \times 10^{-3}\,\mathrm{kg}$ collected in 300 s.

Calculate (a) the specific latent heat of vaporisation of the alcohol, (b) the average rate of heat loss to the surroundings, (c) the power of the heater required to produce a rate of evaporation of 1.50 g per minute.

Method

(a) We rearrange Equation 23.10 with $t = 300$:

$$l = \frac{(V_2 I_2 - V_1 I_1)t}{(m_2 - m_1)}$$

$$= \frac{(10 \times 3.6 - 7.4 \times 2.6) \times 300}{(11.3 - 5.8) \times 10^{-3}}$$

$$= 914 \times 10^3\,\mathrm{J\,kg^{-1}}$$

(b) To find Q we rearrange Equation 23.8 and use the above value for l. So

$$Q = V_1 I_1 t - m_1 l$$

$$= (7.4 \times 2.6 \times 300) - (5.8 \times 10^{-3} \times 914 \times 10^3)$$

$$= 471\,\mathrm{J}$$

The average rate of heat loss is

$$Q/t = 471/300 = 1.57\ \text{W}$$

(c) We can use Equation 23.8 or 23.9. Using Equation 23.8 we have power

$$P_1 = V_1 I_1 = 7.4 \times 2.6 = 19.2\ \text{W}$$

and $$\frac{m_1}{t} = \frac{5.8 \times 10^{-3}}{300} = 1.93 \times 10^{-5}\ \text{kg s}^{-1}$$

and require power P at which $m/t = 1.5$ g per minute or 2.50×10^{-5} kg s^{-1}. Rearranging Equation 23.10 gives

$$P = P_1 + \left(\frac{m}{t} - \frac{m_1}{t}\right)l$$

$$= 19.2 + (2.5 - 1.93) \times 10^{-5} \times 914 \times 10^3$$

$$= 24.4\ \text{W}$$

Alternatively, we write $V_3 I_3 = m_3 l + Q$. For $t = 300$ s (5 min), $Q = 471$ J, $m_3 = 3 \times 1.5 \times 10^{-3}$, $l = 914 \times 10^3$ and we calculate $V_3 I_3$.

Answer

(a) 914 kJ kg^{-1}, (b) 1.57 W, (c) 24.4 W.

Vapours

A vapour is a gas at a temperature below its critical temperature. This means that a vapour can be condensed without necessarily being cooled first. It condenses whenever its pressure exceeds a value called its saturation vapour pressure (SVP). Also, if liquid is present it will evaporate and the mass of vapour will increase until the SVP is reached. The SVP is decided by the particular gas concerned and by the temperature. In other respects the vapour behaves like other gases. For a worked example see Example 4 in Chapter 1.

Exercise 85

1 Assuming that heat losses can be neglected, calculate the power of a heater required to boil off water at a rate of 10.0 g per minute. Assume l for water = 2.26 MJ kg^{-1}.

2 An experiment was performed to determine the specific latent heat of vaporisation of a liquid at its boiling point. The following table summarises the results:

Voltage (V)	*Current* (A)	*Mass evaporated in* 400 s g^{-1}
10.0	2.00	14.6
15.0	2.50	30.6

Calculate (a) the specific latent heat of vaporisation of the liquid, (b) the heat loss to the surroundings in 400 s, (c) the rate of evaporation of the liquid when a 30.0 W rate of heating is used.

3 A closed vessel contains air which is just saturated with acetone vapour at 50 °C, at which temperature the total pressure is 1.50×10^5 N m^{-2}. Calculate the pressure when the vessel is cooled to 20 °C. (SVP of acetone at 50 °C = 86×10^3 N m^{-2}; at 20 °C = 26×10^3 N m^{-2}.)

Exercise 86: Questions of A-level standard

1 The dimensions of specific heat capacity are

A $[L][T]^{-2}[\theta]^{-1}$ **B** $[L][T]^{-1}[\theta]^{-1}$
C $[L]^2[T]^{-2}[\theta]^{-1}$ **D** $[L]^2[T]^{-1}[\theta]^{-1}$
E $[M]^{-1}[L]^2[\theta]^{-1}$ [AEB 81]

2 A piece of aluminium of mass 0.20 kg and specific heat capacity 1.2 kJ kg^{-1} K^{-1} is heated to a steady temperature t and is then quickly but carefully placed in 0.22 kg of water contained in a copper calorimeter of water equivalent 0.02 kg. The temperature of the water rises from 16 to 21 °C. Calculate the temperature t, given that the specific heat capacity of water is 4.2 kJ kg^{-1} K^{-1}.

3 A 12 W electric heater is used to supply heat energy to 200 g of liquid contained in a calorimeter of heat capacity 50 J K^{-1}. When the temperature has become steady, the heat supply is removed and the temperature of the liquid begins to fall at 1.2 K min^{-1}. What is the specific heat capacity of the liquid? [AEB 80, p]

4 In an experiment to determine the specific heat capacity of a liquid, the liquid flows past an electric heating coil, and in the steady state the inlet and outlet temperatures are 10.4 °C and 13.5 °C respectively. When the mass rate of flow of the liquid is 3.2×10^{-3} kg s^{-1}, the power supplied to the coil is 27.4 W. The flow rate is then changed to 2.2×10^{-3} kg s^{-1} and, in order to maintain the same inlet and outlet temperatures, the power supplied is adjusted to 19.3 W. Explain why two sets of data are obtained, and calculate the specific heat capacity of the liquid. [L 75, p]

5 In determining the specific latent heat of vaporisation of a liquid by an electrical method, the liquid evaporates at the rate of 1.0 g s^{-1} when the electrical power is P. When the power is $2P$, the rate of evaporation is 2.1 g s^{-1}. The specific latent heat of vaporisation of the liquid, in J g^{-1}, is

A $5P/21$ **B** $10P/11$ **C** P
D $2P$ **E** $21P/10$ [AEB 82]

6 A steel pressure vessel, partially filled with water, also contains air, the total pressure being 1.01×10^5 Pa and the temperature 300 K. It is sealed with a 2 cm diameter bung, and heated to temperature 400 K. Assuming that some of the water still remains liquid, calculate the pressure in the vessel and the force tending to push the bung out.

(Saturated vapour pressure of water at 300 K = 3.57×10^3 Pa; at 400 K = 2.47×10^5 Pa) [SUJB 82]

7 In an experiment to determine the specific heat capacity of aluminium, a cylindrical 1 kg block of aluminium was heated electrically by a 17.3 W immersion heater inserted into a hole in the centre of the block. The block was suspended in a draught-free room at 20 °C. The temperature of the block at first rose steadily (10 K in 10 min), then more slowly, finally stabilising at 85 °C.

(a) Explain, using the conservation-of-energy principle, why the temperature of the block stabilised although the heater was still switched on.

(b) Assuming that the rate of heat loss from the block was proportional to the excess temperature of the block above that of the room, calculate (i) the rate of heat loss from the block at 25 °C, and (ii) the specific heat capacity of aluminium (corrected for heat loss). [L 80, p]

24 THERMAL CONDUCTION

Thermal conductivity

The thermal conductivity k of a material describes how easy it is for heat to pass through it from a hotter place (temperature θ_1) to a cooler place (θ_2) separated by a distance l. It is defined by the equation

$$Q = \frac{kA(\theta_1 - \theta_2)}{l} \tag{24.1}$$

where (as shown in Fig 24.1) A is the area of cross-section perpendicular to the flow Q. Q is the heat energy per second passing through. The unit for Q is usually joule per second, i.e. watt.

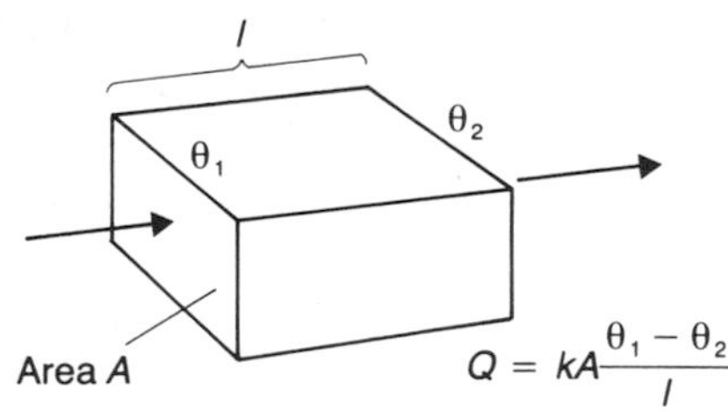

Fig 24.1 Flow of heat

Thermal conductivity k is analogous to electrical conductivity (see Chapter 27), kA/l is conductance, l/kA is resistance, Q is analogous to electric current and $\theta_1 - \theta_2$ is analogous to electrical potential difference. $(\theta_1 - \theta_2)/l$ may be called the temperature gradient and is constant if A, k and Q are constant (negligible heat loss from the sides). The unit for temperature gradient is $\mathrm{K\,m^{-1}}$ and the unit for k is $\mathrm{W\,m^{-1}\,K^{-1}}$.

Heat conductors in series

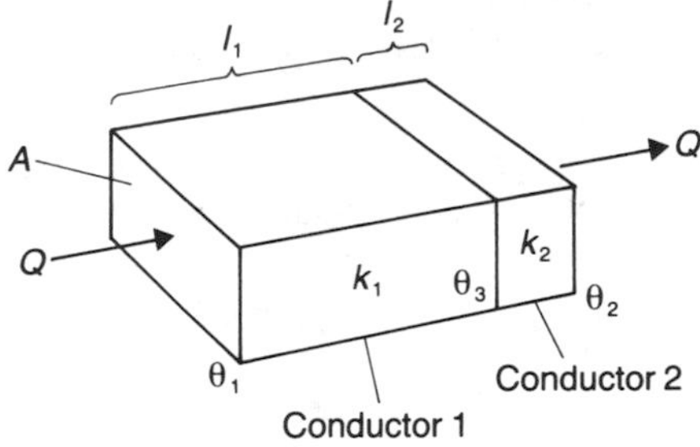

Fig 24.2 Conductors in series

In Fig 24.2, with no loss from the sides, $Q = k_1A(\theta_1 - \theta_3)/l_1$ and here, we emphasise, Q is the same for conductor 2. Therefore $Q = k_2A(\theta_3 - \theta_2)/l_2$.

Most problems can be solved by using these two equations.

From the two equations for Q, if we eliminate θ_3, we can get $Q = (\theta_1 - \theta_2)/R$. R is the thermal resistance given by $R = R_1 + R_2$ for the two conductors in series where $R_1 = l_1/k_1A$ and $R_2 = l_2/k_2A$.

Conduction of heat through walls

A wall may consist of (say) glass for part of its area and brick elsewhere. In this case the total heat flow through the wall (perhaps from a warm room to a cold exterior) equals $Q_{\text{glass}} + Q_{\text{brick}}$. Example 1 below is of this type.

If the wall consists of two layers then we have two conductors in series. Example 2 is of this kind.

Example 1

The cabin of a light aircraft can be regarded as an approximately rectangular box, sides 1.5 m by 1.5 m by 2 m. The windows are made of Perspex, 10 mm thick, and have a total area of 3 m^2; the remainder of the cabin wall is made of thin aluminium alloy sheet lined with insulating material, 20 mm thick. The cabin heater is able to maintain a temperature of 20 °C in the cabin when the outside temperature is −10 °C. Assuming the temperature difference across the aluminium alloy may be neglected, calculate (a) the power of the heater, (b) the percentage of the total energy which is lost through the Perspex.

Explain why, in practice, the power of the heater needed would be much less than that calculated. Take the thermal conductivity of Perspex to be 0.2 W m^{-1} K^{-1} and that of the insulating material to be 3.5×10^{-2} W m^{-1} K^{-1}. [O 81, p]

Method

Almost without exception a thermal conductivity question requires the use of

$$Q = \frac{kA(\theta_1 - \theta_2)}{l}$$

Using this for the windows, we have

$$Q_{\text{window}} = 0.2 \times 3 \times \frac{(20 - -10)}{10 \times 10^{-3}} = 1800\ \text{W}$$

If the heat flow Q_r through the remainder of the cabin surface is now found, the two Q values can be added to give the total heat escaping per second, which of course equals the power of the heater. (The presence of the aluminium alloy in series with the insulating material can be neglected.)

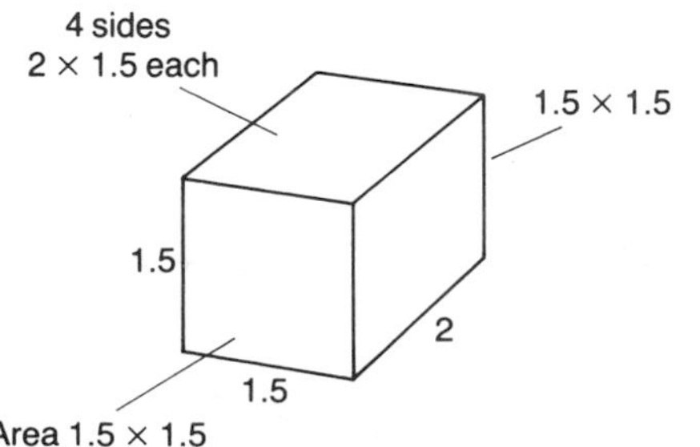

Fig 24.3 Diagram for Example 1

The area A of the box is, as seen in Fig 24.3, equal to $(1.5 \times 1.5) \times 2 + (1.5 \times 2) \times 4$ which amounts to 16.5 m^2. The area that is not window is 16.5 − 3, i.e. 13.5 m^2. So

$$Q_r = 3.5 \times 10^{-2} \times 13.5 \times \frac{(20 - -10)}{20 \times 10^{-3}}$$

which equals 709 W.

$$Q_{\text{total}} = 1800 + 709 = 2509\ \text{W} \quad \text{or} \quad 2.5\ \text{kW}$$

For (b), the percentage required is

$$\frac{Q_{\text{window}}}{Q_{\text{total}}} \times 100 \quad \text{or} \quad \frac{1800}{2509} \times 100 \quad \text{or} \quad 72\%$$

The heater power needed is less than that calculated because heat escape through windows is reduced by still air adjacent to the window surfaces.

Answer

(a) 2.5 kW, (b) 72%.

Searle's bar

To measure experimentally the thermal conductivity k of a good conductor of heat such as copper, an apparatus known as Searle's bar is suitable. Fig 24.4 shows a steam-heated design.

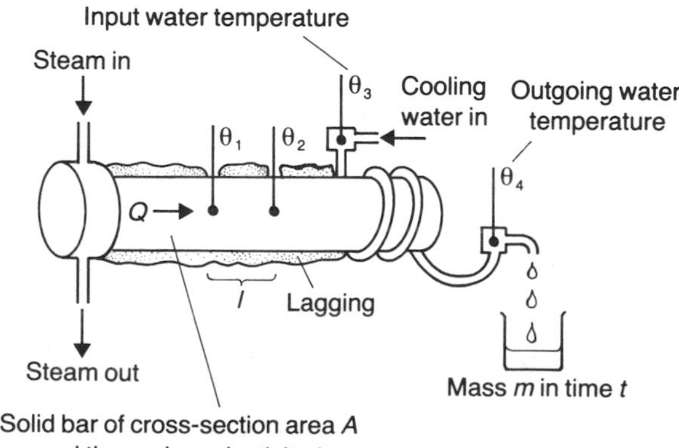

Fig 24.4 Searle's bar

The heat flow per second along the bar is

$$Q = \frac{kA(\theta_1 - \theta_2)}{l}$$

and Q is also equal to

$$\frac{ms}{t}(\theta_4 - \theta_3) \qquad (24.2)$$

which is the heat entering the cold water per second (s = specific heat capacity). The heat loss from the sides of the bar should be negligible.

Alternatively Q may be measured easily if in place of steam heating, an electric heater is used. Then $Q = VI$ (see p.179). At the cold end of the bar the water flow would still be needed to remove the heat, but measurements there would no longer be essential.

Example 2

In a Searle's bar experiment, heat is supplied at a rate of 80 W to one end of a well-lagged, uniform copper bar of cross-section area 10 cm^2 and total length 20 cm. The heat is removed by water cooling at the other end of the bar. Two thermometers, T_1 and T_2, are used to record the temperatures within the bar at distances of 5.0 cm and 15.0 cm from the hot end of the bar. These are 48 °C and 28 °C respectively. (a) Calculate a value for the thermal conductivity of copper. (b) Estimate the rate of flow (in g per minute) of cooling water sufficient for the water temperature rise not to exceed 5 K. (c) Estimate the temperature at the cold end of the bar. (The specific heat capacity of water is 4200 J kg^{-1} K^{-1}.)

Method

A suitable diagram is sketched (Fig 24.5).

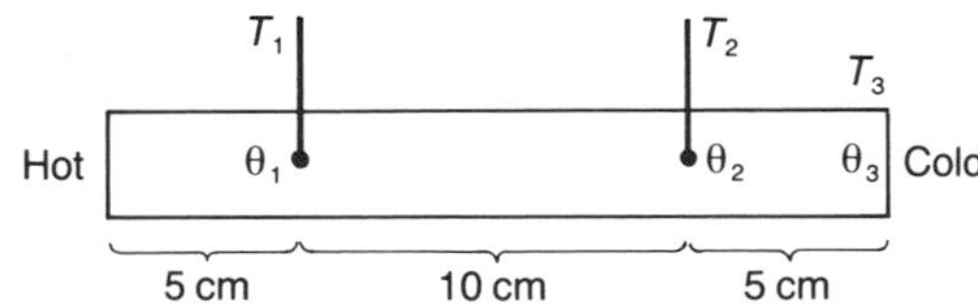

Fig 24.5 Diagram for Example 2

(a)
$$Q = \frac{kA(\theta_1 - \theta_2)}{l}$$

$$\therefore \quad 80 = \frac{k \times 10 \times 10^{-4} \times (48 - 28)}{10 \times 10^{-2}}$$

$$= k \times 10^{-2} \times 20$$

$$\therefore \quad k = \frac{80}{0.2} \text{ or } 400\ \text{W m}^{-1}\,\text{K}^{-1}$$

(b) To estimate the rate of flow of water (in kilogram per s), $ms\theta/t = Q$, where θ is the temperature rise of the water. This is obtained from equation 24.2 with $\theta = (\theta_4 - \theta_3)$.

$$\therefore \quad \frac{m}{t} = \frac{Q}{s\theta} = \frac{80}{4200 \times 5}\ \text{kg s}^{-1}$$

$$\text{or} \quad \frac{80 \times 10^3 \times 60}{4200 \times 5}\ \text{g min}^{-1} = 228.57\ \text{g min}^{-1}$$

(c) To find T_3, *either*:

$$Q = \frac{kA(\theta_2 - \theta_3)}{l}$$

$$\therefore \quad 80 = \frac{400 \times 10 \times 10^{-4}(28 - \theta_3)}{5 \times 10^{-2}}$$

$$\therefore \quad 28 - \theta_3 = 10$$

$$\therefore \quad \theta_3 = 18\ ^\circ\text{C}$$

or, using the fact that the temperature gradient is constant throughout the bar's length,

$$\frac{\theta_1 - \theta_2}{10} = \frac{\theta_2 - \theta_3}{5}$$

$$\therefore \quad \frac{48 - 28}{10} = \frac{28 - \theta_3}{5}$$

$$\therefore \quad T_3 = 18\ ^\circ\text{C}$$

Answer

(a) 400 W m^{-1} K^{-1}, (b) 229 or 2.3×10^2 g min^{-1}, (c) 18 °C.

Lees' disc

For measuring k for a poorly conducting material the Lees' disc apparatus may be used (Fig 24.6a).

In this apparatus Q is measurable when k is small because l is made small.

$$Q = kA\left(\frac{\theta_1 - \theta_2}{l}\right)$$

and θ_1 and θ_2 are recorded (Fig 24.6a). To evaluate Q the metal disc is heated to a temperature above the temperature θ_2 already recorded and is allowed to cool as in Fig 24.6b. During this cooling the temperature of the disc is recorded, and a graph is obtained as shown in Fig 24.6c. From this graph the temperature fall per second is deduced at θ_2. Then

$$Q = m \times s \times \text{temperature fall per s} \quad (24.3)$$

Like Searle's bar, it can have electric heating instead of steam. Q would then equal VI (see p.179), and the heat loss from the lower metal disc would not have to be measured.

(a) Typical apparatus

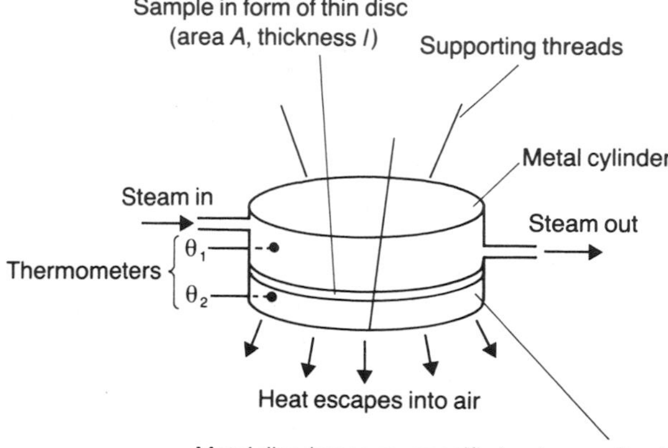

(b) Cooling of metal disc

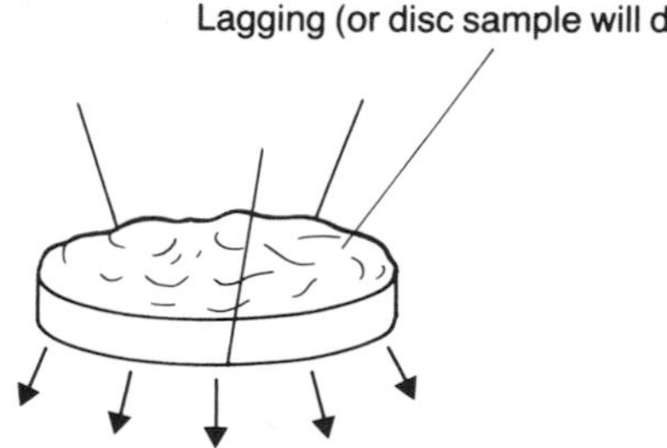

(c) Cooling curve

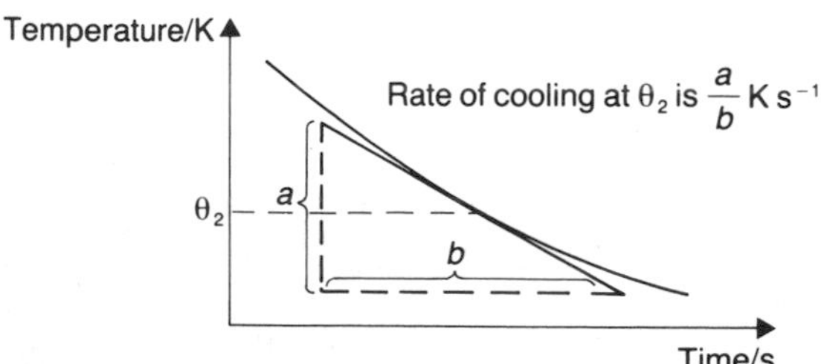

Fig 24.6 Lees' disc

Example 3

In a Lees' disc type experiment a sheet of glass was cut to form a disc of 5 cm radius and had a uniform thickness of 2 mm. One side was maintained at a steady temperature of 100 °C, while a copper block in good thermal contact with the glass was found to be at 70 °C. The block weighed 0.75 kg.

Following this, the cooling of the copper block was studied over a range of temperatures and the rate of cooling at 70 °C was found to be 16.5 K min^{-1}.

Calculate a value for the thermal conductivity of glass. (Specific heat capacity of copper = 400 J kg^{-1} K^{-1}.)

Method

$$Q = kA\frac{(\theta_1 - \theta_2)}{l}$$

and $Q = ms \times$ Rate of cooling at θ_2, see Equation 24.3. Area $A = \pi R^2$, $R = 5$ cm, $\theta_1 = 100$ °C, $\theta_2 = 70$ °C, $l = 2$ mm, $m = 0.75$ kg, $s = 400$ J kg^{-1} K^{-1}, rate of cooling at $\theta_2 = 16.5$ K min^{-1}.

From Equation 24.3,

$$Q = 0.75 \times 400 \times \frac{16.5}{60}$$

$$= 82.5 \text{ W}$$

Using this value for Q in the first equation,

$$82.5 = \frac{k \times \pi(5 \times 10^{-2})^2 \times (100 - 70)}{2 \times 10^{-3}}$$

$$\therefore \quad k = \frac{82.5 \times 2 \times 10^{-3}}{3.14 \times 25 \times 10^{-4} \times 30}$$

$$\therefore \quad k = 0.70 \text{ W m}^{-1}\text{ K}^{-1}$$

Answer

0.70 W m^{-1} K^{-1}.

Exercise 87

1 Calculate the heat flow through a layer of cork of 2 mm thickness and 24 cm^2 area when the temperature difference between its surfaces is 60 K. (k for cork = 0.05 W m^{-1} K^{-1}.)

2 A 10 cm long brass bar is joined end-on to a copper bar of equal length and diameter, so as to form a compound bar with a cross-section area of 6.0 cm^2. The join has negligible thermal resistance and the bar is well lagged. The free end of the brass bar is maintained at 100 °C and the far end of the compound bar is kept at 20 °C. Calculate the heat flow per second along the bar and also the temperature of the junction.
Assume k for copper = 400 W m^{-1} K^{-1} and for brass = 100 W m^{-1} K^{-1}.

3 Obtain a value for the heat flow through a window across which the temperature difference is 10 K, if the window consists of
(a) a single sheet of glass 6.0 mm thick and 2.5 m^2 in area,
(b) this sheet plus an identical sheet separated by a 10 mm air space.
Take the thermal conductivities of the glass and air to be 0.08 and 0.02 W m^{-1} K^{-1} respectively.

4 If heat flowing along a lagged copper bar at a rate of 60 W is removed steadily by water cooling, what temperature rise will occur in the water if the water flow is 100 g per minute?
(SHC water = 4200 J kg^{-1} K^{-1}.)

5 Heat is flowing along a uniform, lagged metal bar, and the temperature is 80 °C at 8.0 cm from the hot end, and 50 °C at 20 cm from this end. At what distance from the hot end is the temperature 60 °C?

6 In a Lees' disc experiment a cardboard specimen of area 25 cm^2 and 2 mm thickness was used, and in the steady state the temperatures on each side of the specimen were 99 °C and 59 °C. The heat passing through the cardboard enters a 600 g copper block, from which it escapes into the surrounding air. Subsequently this block, with the cardboard still in place, and covered with further insulation, was heated and then allowed to cool. The temperature fall from 60 °C to 58 °C took 48 s. Calculate the thermal conductivity of the cardboard, given that the specific heat capacity of copper is 420 J kg^{-1} K^{-1}.

Exercise 88: Questions of A-level standard

1 The SI unit for thermal conductivity is

A W m K^{-1} **B** J s m K^{-1} **C** W m^{-1} K^{-1}
D m K^{-1} **E** W^{-1} m K

2 A brick wall is 50 times thicker than a glass window and the thermal conductivity of brick is one-eighth that of glass. If the temperature difference across both materials is the same, the rate of heat flow per unit area through the glass is n times that through the brick, where n equals

A $\frac{1}{400}$ **B** $\left(\frac{8}{50}\right)^2$ **C** $\frac{50}{8}$
D $\left(\frac{50}{8}\right)^2$ **E** 400 [L 79]

3 The external wall of a brick house is of area 16 m^2 and thickness 0.3 m. The indoor and outdoor temperatures are 20 °C and 0 °C respectively. Find the rate at which heat is lost through the wall. What is the rate of loss when the internal surface of the wall is covered with expanded polystyrene tiles of thickness 20 mm, and what is the temperature of the brick–tile interface?
(Thermal conductivity of brick = 0.5 W m^{-1} K^{-1}; of expanded polystyrene = 0.03 W m^{-1} K^{-1}.)
[WJEC 80, p]

4

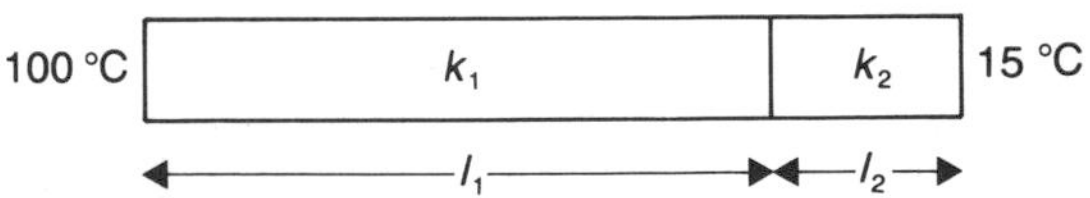

Fig 24.7 Diagram for Question 4

A lagged rod of length l_1 is made of metal whose thermal conductivity is k_1. It makes good thermal contact with a second lagged metal rod of equal cross-section area, which has a length l_2 and thermal conductivity k_2. The free ends of the rods are maintained at 100 °C and 15 °C, as shown in Fig 24.7. If $l_1 = 5l_2$ and $k_2 = 1.5k_1$, what is the temperature where the rods are joined?

5 A well-lagged copper bar having a uniform cross-section area of $2.0 \times 10^{-4}\,\mathrm{m}^2$ is heated at one end and is cooled by a flow of water at the other end. The temperature gradient along the bar is found to become steady at a value of $2.1\,\mathrm{K\,cm^{-1}}$ when the water flow is $1.0\,\mathrm{g\,s^{-1}}$. If the heat taken up by the water raises its temperature by 4.0 K, what is the thermal conductivity of the copper?
(The specific heat capacity of water is $4.2\,\mathrm{kJ\,kg^{-1}\,K^{-1}}$.)

25 THE IDEAL GAS LAWS AND KINETIC THEORY

The gas laws

The laws obeyed by a perfect or ideal gas are as follows (for a fixed mass of gas):

$$pV = \text{Constant, at constant } T \quad \text{(Boyle's law)}$$

$$\frac{V}{T} = \text{Constant, at constant } p \quad \text{(Charles' law)}$$

$$\frac{p}{T} = \text{Constant, at constant } V \quad \text{(Pressure law)}$$

where p is the pressure, V the volume and T the absolute temperature (K) of the gas.

The ideal gas equation

The three laws above are incorporated in the ideal gas equation:

$$\frac{pV}{T} = \text{Constant} \tag{25.1}$$

An alternative way of writing this is

$$\frac{p_1V_1}{T_1} = \frac{p_2V_2}{T_2} \tag{25.2}$$

where p_1, V_1, T_1 refer to the initial state and p_2, V_2, T_2 to the final state. Note that pressure and volume may be expressed in any suitable units (see Chapter 3) that we choose, but *temperature must be in kelvin.*

Example 1

A gas cylinder has a volume of $0.040\,\text{m}^3$ and contains air at a pressure of 2.0 MPa. Assuming that temperature remains constant calculate (a) the equivalent volume of air at atmospheric pressure ($1.0 \times 10^5\,\text{Pa}$), (b) the volume of air, at atmospheric pressure, which escapes from the cylinder when it is opened to the atmosphere.

Method

(a) If temperature is constant then $T_1 = T_2$, and Equation 25.2 reduces to the equation for Boyle's law and becomes

$$p_1V_1 = p_2V_2 \tag{25.3}$$

We have
$p_1 = 2.0 \times 10^6$, $V_1 = 0.040$, $p_2 = 1.0 \times 10^5$
and require V_2.

Rearranging Equation 25.3 gives

$$V_2 = \frac{p_1V_1}{p_2} = \frac{2 \times 10^6 \times 0.04}{1 \times 10^5}$$
$$= 0.80\,\text{m}^3$$

(b) Air escapes from the cylinder until it contains $0.04\,\text{m}^3$ of air at atmospheric pressure. It is then 'empty', so that a volume ΔV will escape where

$$\Delta V = 0.80 - 0.04 = 0.76\,\text{m}^3$$

Note that ΔV is the volume of air, at atmospheric pressure, which would have to be pumped into the 'empty' cylinder to raise its pressure to 2.0 MPa.

Answer

(a) $0.80\,\text{m}^3$, (b) $0.76\,\text{m}^3$.

Example 2

A flask containing air is corked when the atmospheric pressure is 750 mmHg and the temperature is 17 °C. The temperature of the flask is now raised gradually. The cork blows out when the pressure in the flask exceeds atmospheric pressure by 150 mmHg. Calculate the temperature of the flask when this happens.

Method

Note that we have to assume that corking the flask did not change the original pressure of the air inside it, that the atmospheric pressure remains unchanged and that the volume of the flask does not change appreciably during the change of temperature.

If the volume is constant then $V_1 = V_2$ and Equation 25.2 reduces to the equation for the Pressure law and becomes

$$\frac{p_1}{T_1} = \frac{p_2}{T_2} \qquad (25.4)$$

We have

$$p_1 = 750\,\text{mmHg}$$

$$T_1 = 273 + 17 = 290\,\text{K}$$

$$p_2 = \text{Atmospheric pressure} + \text{Excess pressure}$$

$$= 750 + 150 = 900\,\text{mmHg}$$

To find T_2 we rearrange Equation 25.4. This gives

$$T_2 = \frac{p_2 \times T_1}{p_1} = \frac{900 \times 290}{750}$$

$$= 348\,\text{K} = 75\,°\text{C}$$

Note again that the units can be mmHg for pressure *provided that both p_1 and p_2 are in the same units.*

Answer

The cork blows out at 75 °C.

Example 3

A gas cylinder of volume 4.0 litre ($4.0 \times 10^{-3}\,\text{m}^3$) contains oxygen at a temperature of 15 °C and a pressure of $2.5\,\text{MN}\,\text{m}^{-2}$. Calculate (a) the equivalent volume of oxygen at standard temperature and pressure (STP), (b) the mass of oxygen in the cylinder. The density of oxygen is $1.4\,\text{kg}\,\text{m}^{-3}$ at STP.

Method

Standard temperature and pressure (STP) are 0 °C and $1.0 \times 10^5\,\text{N}\,\text{m}^{-2}$ respectively.

(a) We use Equation 25.2 in which we have

$$p_1 = 2.5 \times 10^6, \quad V_1 = 4.0 \times 10^{-3},$$

$$T_1 = 273 + 15 = 288$$

$$p_2 = 1.0 \times 10^5, \quad V_2 = \text{unknown},$$

$$T_2 = 273 + 0 = 273$$

Rearranging Equation 25.2 gives

$$V_2 = \frac{p_1 V_1 T_2}{T_1 p_2}$$

$$= \frac{2.5 \times 10^6 \times 4 \times 10^{-3} \times 273}{288 \times 1 \times 10^5}$$

$$= 94.8 \times 10^{-3}\,\text{m}^3$$

(b) The density of oxygen is $1.4\,\text{kg}\,\text{m}^{-3}$. To find the mass of gas:

$$\text{Mass} = \text{Volume} \times \text{Density}$$

$$= 94.8 \times 10^{-3} \times 1.4 = 0.133\,\text{kg}$$

Note that since the density is quoted at STP we must use the volume of gas at STP.

Answer

(a) $95 \times 10^{-3}\,\text{m}^3$, (b) 0.13 kg.

Exercise 89

1 Change the following Celsius temperatures into degrees absolute:
(a) 7 °C, (b) 710 °C, (c) −80 °C, (d) −199 °C.

2 A fixed mass of gas is held at 27 °C. To what temperature must it be heated so that its volume doubles if its pressure remains constant?

3 A car tyre has a volume of $18 \times 10^{-3}\,\text{m}^3$ and contains air at an *excess* pressure of $2.5 \times 10^5\,\text{N}\,\text{m}^{-2}$ above atmospheric pressure ($1.0 \times 10^5\,\text{N}\,\text{m}^{-2}$). Calculate the volume which the air inside would occupy at atmospheric pressure, assuming that its temperature remains unchanged.

4 Inside a sealed container is a fixed mass of gas at a pressure of 1.5×10^5 Pa when the temperature is 17 °C. At what temperature will the pressure inside it be 2.5×10^5 Pa?

5 A fixed mass of gas has a volume of 200 cm^3 at a temperature of 57 °C and a pressure of 780 mm mercury. Find its volume at STP (0 °C and 760 mm mercury).

6 A gas cylinder has a volume of 20 litres ($20 \times 10^{-3}\,m^3$). It contains air at a temperature of 17 °C and an *excess* pressure of $3.0 \times 10^5\,N\,m^{-2}$ above atmospheric pressure ($1.0 \times 10^5\,N\,m^{-2}$). Calculate the mass of air in the cylinder, given that the density of air at STP is $1.3\,kg\,m^{-3}$.

The equation of state

For a *given amount* of an ideal gas, *the equation of state* is as follows:

$$pV = nRT \qquad (25.5)$$

where p is the pressure ($N\,m^{-2}$ or Pa), V the volume (m^3), n the number of moles of the gas (mol), R the *universal molar gas constant* (value $8.31\,J\,mol^{-1}\,K^{-1}$) and T the temperature (K). Note that one mole of a gas is the amount which contains Avogadro's number L ($= 6.02 \times 10^{23}$) of molecules.

Equation 25.5 can be rewritten to include the mass M_g (kg) of the gas involved. If M_m ($kg\,mol^{-1}$) is the molar mass (i.e. the mass of one mole), then the number of moles n is given by

$$n = \frac{\text{Mass of gas}}{\text{Molar mass}} = \frac{M_g}{M_m} \qquad (25.6)$$

Using Equation 25.6 to substitute for n in Equation 25.5 gives

$$pV = M_g\left(\frac{R}{M_m}\right)T \qquad (25.7)$$

Note that M_m depends on the particular gas. Also, if m is the mass of a molecule of the gas, then

$$M_m = \begin{pmatrix}\text{Avogadro's}\\ \text{number } L\end{pmatrix} \times \begin{pmatrix}\text{Mass of}\\ \text{molecule } m\end{pmatrix} \qquad (25.8)$$

Example 4

A cylinder of volume $2.00 \times 10^{-3}\,m^3$ contains a gas at a pressure of $1.50\,MN\,m^{-2}$ and at a temperature of 300 K. Calculate (a) the number of moles of the gas, (b) the number of molecules of the gas, (c) the mass of gas if its molar mass is 32.0×10^{-3} kg, (d) the mass of one molecule of the gas.

Assume that the universal gas consant R is $8.31\,J\,mol^{-1}\,K^{-1}$ and the Avogadro constant L is $6.02 \times 10^{23}\,mol^{-1}$.

Method

(a) We use Equation 25.5 in which $p = 1.5 \times 10^6$, $V = 2 \times 10^{-3}$, $R = 8.31$ and $T = 300$. Rearranging to find n gives us

$$n = \frac{pV}{RT} = \frac{1.5 \times 10^6 \times 2 \times 10^{-3}}{8.31 \times 300}$$
$$= 1.20$$

(b) One mole contains 6.02×10^{23} molecules, so that 1.20 mol contains $1.20 \times 6.02 \times 10^{23} = 7.22 \times 10^{23}$ molecules.

(c) We have $M_m = 32 \times 10^{-3}$, $n = 1.2$ and require the mass of gas M_g. Rearranging Equation 25.6 gives us

$$M_g = nM_m = 1.2 \times 32 \times 10^{-3}$$
$$= 38.4 \times 10^{-3}\,kg$$

(d) We use Equation 25.8, in which $M_m = 32 \times 10^{-3}$, $L = 6.02 \times 10^{23}$ and we require m. Thus

$$m = \frac{M_m}{L} = \frac{32 \times 10^{-3}}{6.02 \times 10^{23}}$$
$$= 5.32 \times 10^{-26}\,kg$$

Answer

(a) 1.20, (b) 7.22×10^{23}, (c) 38.4×10^{-3} kg, (d) 5.32×10^{-26} kg.

Example 5

A cylinder contains 2.0 kg of nitrogen at a pressure of $3.0 \times 10^6\,N\,m^{-2}$ and at a temperature of 17 °C. What mass of nitrogen would a cylinder of the same volume contain at STP (0 °C and $1.0 \times 10^5\,N\,m^{-2}$)?

Method

We use Equation 25.7, and note that V and M_m are constants for a given volume of a particular gas. At $p_1 = 3.0 \times 10^6$ and $T_1 = 17\,°C = 290$ K, $M_g = 2.0$. So Equation 25.7 gives

$$3.0 \times 10^6 \times V = 2 \times \left(\frac{R}{M_m}\right) \times 290 \qquad \text{(i)}$$

At STP we have $p_2 = 1.0 \times 10^5$, $T_2 = 273$ K and require the mass M_g in the cylinder. So

$$1.0 \times 10^5 \times V = M_g\left(\frac{R}{M_m}\right) \times 273 \qquad \text{(ii)}$$

Dividing (i) by (ii) to eliminate the constants gives

$$\frac{3.0 \times 10^6}{1.0 \times 10^5} = \frac{2 \times 290}{M_g \times 273}$$
$$\therefore\quad M_g = 7.08 \times 10^{-2}\,kg$$

Answer

7.1×10^{-2} kg at STP.

Example 6

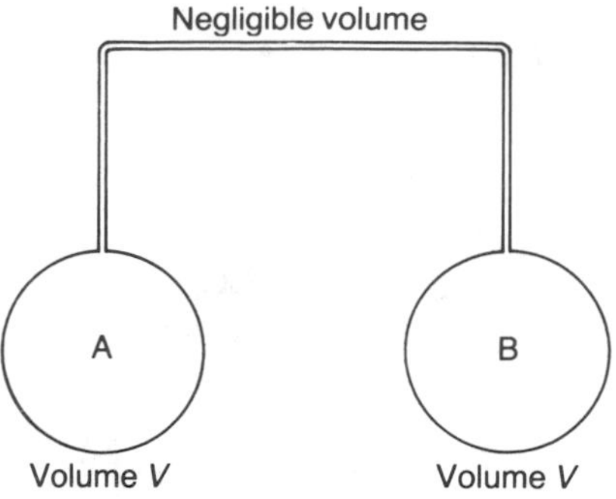

Fig 25.1 Information for Example 6

Two vessels A and B, of equal volume, are connected by a tube of negligible volume, as shown in Fig 25.1. The vessels contain a total mass of 2.50×10^{-3} kg of air and initially both vessels are at 27 °C when the pressure is $1.01 \times 10^5\,\text{N m}^{-2}$. Vessel A is now cooled to 0 °C and vessel B heated to 100 °C. Calculate (a) the mass of gas now in each vessel, (b) the pressure in the vessels.

Method

(a) Let the volume of each vessel be V (we assume this does not change). Note that *since the vessels are connected, the pressure is equal in the two vessels;* let the final pressure be p. We apply Equation 25.7 to each vessel separately:

Vessel A contains mass M_{gA} of gas at temperature 273 K, so

$$pV = M_{gA}\left(\frac{R}{M_m}\right) \times 273 \qquad \text{(i)}$$

Vessel B contains mass M_{gB} of gas at 373 K, so

$$pV = M_{gB}\left(\frac{R}{M_m}\right) \times 373 \qquad \text{(ii)}$$

Comparing (i) and (ii) we see that

$$M_{gA} \times 273 = M_{gB} \times 373 \qquad \text{(iii)}$$

Now the total mass of gas is 2.5×10^{-3} kg, so

$$M_{gA} + M_{gB} = 2.5 \times 10^{-3} \qquad \text{(iv)}$$

Substituting $M_{gA} = (373/273)M_{gB}$ from (iii) into (iv) we find

$M_{gA} = 1.44 \times 10^{-3}$ kg and $M_{gB} = 1.06 \times 10^{-3}$ kg.

(b) We apply Equation 25.7 to the original whole system at temperature $273 + 27 = 300$ K, pressure $1.01 \times 10^5\,\text{N m}^{-2}$, volume $2V$ (since A and B each have volume V) and mass $M_g = 2.5 \times 10^{-3}$ kg. Hence

$$1.01 \times 10^5 \times 2V = 2.5 \times 10^{-3}\left(\frac{R}{M_m}\right) \times 300 \qquad \text{(v)}$$

To find the final pressure p, we make use of (i), in which $M_{gA} = 1.44 \times 10^{-3}$, so

$$pV = 1.44 \times 10^{-3}\left(\frac{R}{M_m}\right) \times 273 \qquad \text{(i)}$$

Dividing (i) by (v) gives

$$\frac{p}{1.01 \times 10^5 \times 2} = \frac{1.44 \times 10^{-3} \times 273}{2.5 \times 10^{-3} \times 300}$$

or $p = 1.06 \times 10^5\,\text{N m}^{-2}$.

Using (ii) should give the same answer for p. Try this as a check.

Answer

(a) 1.44×10^{-3} kg (A), 1.06×10^{-3} kg (B),
(b) $1.06 \times 10^5\,\text{N m}^{-2}$.

Exercise 90

(Assume that the universal molar gas constant R is $8.31\,\text{J mol}^{-1}\,\text{K}^{-1}$ and Avogadro's number L is 6.02×10^{23}.)

1 Calculate the volume occupied by one mole of gas at standard temperature (0 °C) and standard pressure ($1.01 \times 10^5\,\text{N m}^{-2}$).

2 The molar mass of carbon dioxide is 44.0×10^{-3} kg. Calculate (a) the number of moles and (b) the number of molecules in 1.00 kg of the gas.

3 The molar mass of nitrogen is 28.0×10^{-3} kg. A sample of the gas contains 6.02×10^{22} molecules. Calculate (a) the number of moles of the gas, (b) the mass of the gas and (c) the volume occupied by the gas at a pressure of $0.110\,\text{MN m}^{-2}$ and a temperature of 290 K.

4 An oxygen cylinder contains 0.50 kg of gas at a pressure of $0.50\,\text{MN m}^{-2}$ and a temperature of 7 °C. What mass of oxygen must be pumped into the cylinder to raise its pressure to $3.0\,\text{MN m}^{-2}$ at a temperature of 27 °C. If the molar mass of oxygen is 32×10^{-3} kg calculate the volume of the cylinder.

5 Two vessels, one having three times the volume of the other, are connected by a narrow tube of negligible volume. Initially the whole system is filled with a gas at a pressure of 1.05×10^5 Pa and a temperature of 290 K. The smaller vessel is now cooled to 250 K and the larger heated to 400 K. Find the final pressure in the system.

Kinetic theory

The pressure exerted by a gas arises as a result of gas molecules bombarding the walls of the container. There are very many molecules in a typical sample of gas, and the molecules have a whole range of speeds. Fig 25.2 shows the number of molecules having speed c at a given temperature.

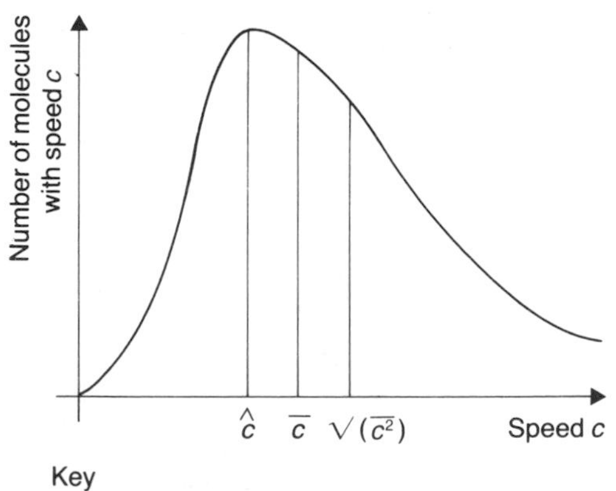

Fig 25.2 Distribution of molecular speeds in a gas

The laws of Newtonian mechanics are used to show that the pressure exerted by the gas is given by

$$p = \tfrac{1}{3}\rho\overline{c^2} \tag{25.9}$$

where ρ is the density of the gas and $\overline{c^2}$ the mean square speed of the molecules of the gas (i.e. the average of all the values of speed squared). Now

$$\rho = \frac{\text{Mass of gas}}{\text{Volume}} = \frac{M_g}{V}$$

Substituting ρ in Equation 25.9 gives

$$pV = \tfrac{1}{3}M_g\overline{c^2} \tag{25.10}$$

By comparing Equations 25.5 and 25.10, for one mole of a gas, we can show that the average *translational* kinetic energy ($\overline{\text{KE}}$) per molecule of a gas is given by

$$\overline{\text{KE}} = \tfrac{1}{2}m\overline{c^2} = \frac{3}{2}\frac{R}{L}T \tag{25.11}$$

where m is the mass of a molecule and $R/L = k$ is the Boltzmann constant.

The square root of $\overline{c^2}$ is called the root mean square (RMS) speed and has theoretical significance. Note from Equation 25.11 that, for a particular gas,

$$\sqrt{(\overline{c^2})} \propto \sqrt{T} \tag{25.12}$$

Example 7

At a certain time, the speeds of seven particles are as follows:

Speed/m s^{-1}	2.0	3.0	4.0	5.0	6.0
Number of particles	1	3	1	1	1

Calculate the root mean square speed of the particles.

Method

Table 25.1

Number of particles n	1	3	1	1	1
Speed c	2.0	3.0	4.0	5.0	6.0
c^2	4	9	16	25	36

We first square the speeds (see Table 25.1). The mean square speed $\overline{c^2}$ is the *average of the squares of the speeds*, as follows:

$$\overline{c^2} = \tfrac{1}{7}\{(1 \times 4) + (3 \times 9) + (1 \times 16) + (1 \times 25) + (1 \times 36)\}$$

$$= \tfrac{1}{7}\{4 + 27 + 16 + 25 + 36\}$$

$$= 15.4\ \text{m}^2\,\text{s}^{-2}$$

Note: This is done by adding up the *'speed squared' values for each particle* and dividing by the number of particles.

To find the RMS speed we take the square root of $\overline{c^2}$, hence

$$\text{RMS speed} = \sqrt{(\overline{c^2})} = \sqrt{15.4}$$

$$= 3.9\ \text{m}\,\text{s}^{-1}$$

Note that the *most probable speed* $\hat{c}$ is $3.0\ \text{m s}^{-1}$ since most (3) particles have this speed. The *average speed* c is found from the average of the speeds, as follows:

$$\bar{c} = \tfrac{1}{7}\{(1 \times 2) + (3 \times 3) + (1 \times 4) + (1 \times 5) + (1 \times 6)\}$$

$$= 3.7\ \text{m s}^{-1}$$

Answer

$3.9\ \text{m s}^{-1}$.

Example 8

Calculate the RMS speed of air molecules in a container in which the pressure is 1.0×10^5 Pa and the density of air is $1.3\ \text{kg m}^{-3}$.

Method

We have $p = 10^5$ and $\rho = 1.3$. Rearranging Equation 25.9 to find $\sqrt{(\overline{c^2})}$ gives

$$\sqrt{(\overline{c^2})} = \sqrt{\frac{3p}{\rho}} = \sqrt{\frac{3 \times 10^5}{1.3}}$$

$$= 480\ \text{m s}^{-1}$$

Answer

$0.48\ \text{km s}^{-1}$.

Example 9

Calculate the temperature at which the RMS speed of oxygen molecules is twice as great as their RMS speed at 27 °C.

Method

We use Equation 25.12. Thus, since 27 °C is 300 K,

$$\frac{\text{RMS at } T}{\text{RMS at } 300} = \frac{\sqrt{T}}{\sqrt{300}}$$

$$\therefore \quad 2 = \frac{\sqrt{T}}{\sqrt{300}}$$

Squaring both sides gives

$$T = 4 \times 300 = 1200\ \text{K} = 927\ °\text{C}$$

Answer

927 °C.

Exercise 91

1 Eight molecules have the following speeds: 300, 400, 400, 500, 600, 600, 700, 900 m s^{-1}. Calculate their RMS speed.

2 The following table shows the distribution of speed of 20 particles:

Speed/m s^{-1}	10	20	30	40	50	60
No. of particles	1	3	8	5	2	1

Find (a) the most probable speed, (b) the average speed, (c) the RMS speed.

3 The RMS speed of helium at STP is $1.30\ \text{km s}^{-1}$. If 1 standard atmosphere is $1.01 \times 10^5\ \text{N m}^{-2}$, calculate the density of helium at STP.

4 The RMS speed of nitrogen molecules at 127 °C is $600\ \text{m s}^{-1}$. Calculate the RMS speed at 1127 °C.

5 If the density of nitrogen at STP (1.01×10^5 Pa and 0 °C) is $1.25\ \text{kg m}^{-3}$, calculate the RMS speed of nitrogen at 227 °C.

Exercise 92: Questions of A-level standard

1 A diving bell of internal volume $6\ \text{m}^3$ is lowered into a fresh water lake until the volume of the contained air is $4\ \text{m}^3$. The height of the water barometer at the surface is 10 m. Assuming the temperature of the air in the bell does not change, the depth in metres below the surface of the lake of the surface of the water in the bell is

A 5 **B** $\frac{20}{3}$ **C** 15
D $\frac{50}{3}$ **E** 25 [O & C 81]

2 In the Boyle's law experiment (Fig 25.3) it was found that when $h = 50$ mm, $V = 18\ \text{cm}^3$ and when $h = 150$ mm, $V = 16\ \text{cm}^3$.
What was the atmospheric pressure in mm of mercury?

A 750 **B** 760 **C** 800
D 1200 **E** 1500 [AEB 81]

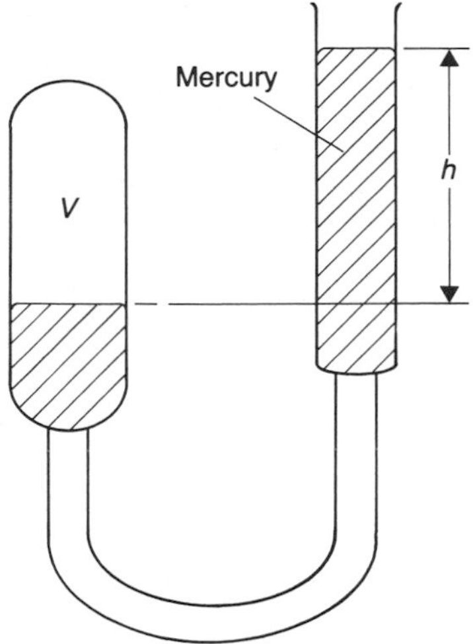

Fig 25.3 Diagram for Question 2

3 A bicycle tyre has a volume of $1.2 \times 10^{-3}\,\text{m}^3$ when fully inflated. The barrel of a bicycle pump has a working volume of $9 \times 10^{-5}\,\text{m}^3$. How many strokes of this pump are needed to inflate the completely flat tyre (i.e. zero air volume in it) to a total pressure of $3.0 \times 10^5\,\text{Pa}$, the atmospheric pressure being $1.0 \times 10^5\,\text{Pa}$? Assume the air is pumped in slowly, so that its temperature does not change.

Explain why the barrel of the bicycle pump becomes hot when the tyre is being inflated quickly. [SUJB 79]

4 If helium is an ideal gas of molar mass 0.004 kilogram, then in the equation $pV = nRT$ for 8 grams of helium the numerical value of n should be

A 0.004 **B** 0.008 **C** 2
D 4 **E** 8 [O 80]

5 Taking the molar gas constant to be $8\,\text{J}\,\text{K}^{-1}\,\text{mol}^{-1}$, estimate the molar volume of a gas at 300 K and at a pressure of $1 \times 10^5\,\text{Pa}$.

The volume of a single molecule of the gas is estimated to be $2 \times 10^{-29}\,\text{m}^3$. What fraction of the volume occupied by the gas is empty space?
Take the Avogadro constant L to be $6 \times 10^{23}\,\text{mol}^{-1}$. [C 80]

6 A cylinder containing 19 kg of compressed air at a pressure 9.5 times that of the atmosphere is kept in a store at 7 °C. When it is moved to the workshop, where the temperature is 27 °C, a safety valve on the cylinder operates, releasing some of the air. If the valve allows air to escape when its pressure exceeds 10 times that of the atmosphere, calculate the mass of air that escapes. [L 80]

7

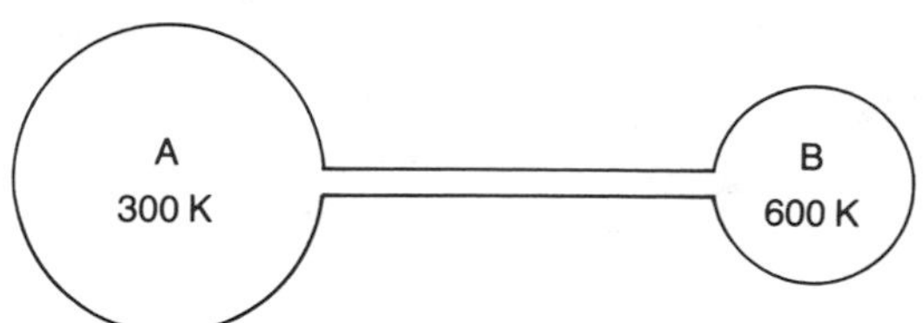

Fig 25.4 Diagram for Question 7

Fig 25.4 shows two flasks connected by an open pipe. Flask A has three times the volume of Flask B. The system is filled with an ideal gas and allowed to come to a steady state in which Flask A is at 300 K and Flask B at 600 K. If the mass of gas in A is m, calculate the mass of gas in B in terms of m.

8 Four gas particles have speeds of 1, 2, 3 and $4\,\text{km}\,\text{s}^{-1}$ respectively. Their RMS speed, in $\text{km}\,\text{s}^{-1}$, is

A 2.5 **B** $\sqrt{7.5}$ **C** $\sqrt{10}$
D $\sqrt{30}$ **E** 7.5 [AEB 82]

9 The density of hydrogen at a pressure of $2.0 \times 10^5\,\text{N}\,\text{m}^{-2}$ and a temperature of 50 °C is $0.15\,\text{kg}\,\text{m}^{-3}$. Under these conditions the RMS molecular speed, in $\text{m}\,\text{s}^{-1}$, is

A 6.7×10^2 **B** 2.0×10^3 **C** 4.0×10^3
D 1.2×10^4 **E** 4.0×10^6 [O & C 82]

10 The molecules of an ideal gas at thermodynamic (absolute) temperature T have a root mean square speed C_{RMS}. If the gas is heated to temperature $2T$, what is then the root mean square speed of the molecules?

A $(\sqrt{2})C_{RMS}$ **B** $2C_{RMS}$ **C** $(2\sqrt{2})C_{RMS}$
D $4C_{RMS}$ **E** $(4\sqrt{2})C_{RMS}$ [O 82]

11 At a certain temperature the root mean square speed of the molecules of an ideal gas is v. If the temperature of the gas is changed such that at constant pressure the volume of the gas has doubled, the root mean square speed of the molecules is

A $\frac{v}{2}$ **B** $\frac{v}{\sqrt{2}}$ **C** v
D $(\sqrt{2})v$ **E** $2v$ [AEB 81]

26 IDEAL GASES AND THERMODYNAMICS

The first law of thermodynamics

In mathematical terms the first law is written as

$$\Delta Q = \Delta U + \Delta W \qquad (26.1)$$

where ΔQ is the heat supplied *to* the system, ΔU the *increase* in internal energy of the system and ΔW the work done *by* the system on the surroundings.

Thus if 5 J (ΔQ) of energy was given to a sample of gas by heating it, and if the gas then expanded and did 3 J (ΔW) of work (e.g. by pushing a piston), Equation 26.1 tells us that 2 J (ΔU) of energy would remain inside the gas. For an ideal gas this would correspond to a rise in kinetic energy, only, of the molecules — so there would be an increase in RMS speed and temperature (see Equation 25.11). Note that no change in potential energy is possible since the interatomic forces are zero.

Work done by an expanding gas

Fig 26.1 shows a gas enclosed in a cylinder by a frictionless piston. If the gas expands and moves the piston outwards, the gas does work against the external force. The external work ΔW is given by

$$\Delta W = \int_{V_1}^{V_2} p \, dV \qquad (26.2)$$

(a) Before expansion

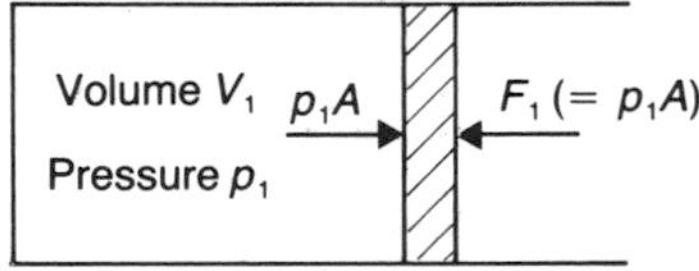

(b) After expansion

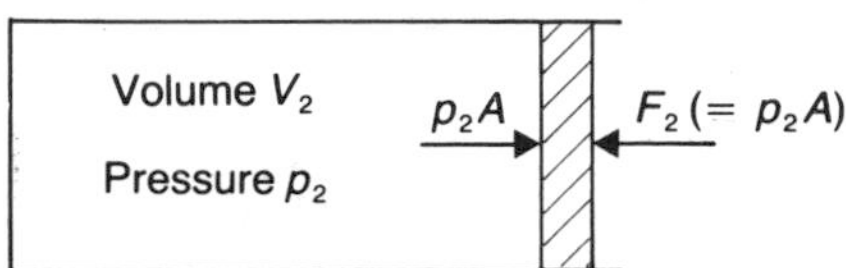

Fig 26.1 A gas expanding in a cylinder

This mathematical operation needs to be carried out if the pressure p of the gas changes as it expands. If the pressure remains constant, so that $p_1 = p_2 = p$, then Equation 26.2 becomes

$$\Delta W = p(V_2 - V_1) \qquad (26.3)$$

where ΔW is in joules when p is in pascals and $(V_2 - V_1)$ is in m^3.

Example 1

Fig 26.2 shows a sample of gas enclosed in a cylinder by a frictionless piston of area 100 cm^2. The cylinder is now heated, so that 250 J of energy is transferred to the gas, which then expands against atmospheric pressure (1.00×10^5 N m^{-2}) and pushes the piston 15.0 cm along the cylinder as shown. Calculate (a) the external work done by the gas, (b) the increase in internal energy of the gas.

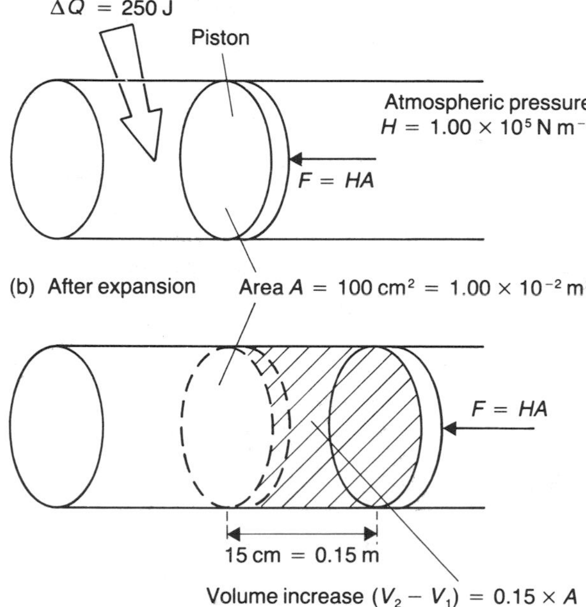

Fig 26.2 Information for Example 1

Method

(a) Referring to Fig 26.2, we see that the force F exerted by the atmosphere on the piston is given by

$$F = H \times A = 1 \times 10^5 \times 1 \times 10^{-2}$$
$$= 1 \times 10^3\,\text{N}$$

Thus the work done ΔW during expansion is

$$\Delta W = \text{Force } F \times \text{Distance moved by piston}$$
$$= 10^3 \times 0.15$$
$$= 150\,\text{J}$$

We could use Equation 26.3 to calculate ΔW to get the same answer, as follows. The pressure p of the gas is equal to atmospheric pressure during the expansion. Thus, since $(V_2 - V_1)$ is $0.15 \times A$,

$$\Delta W = p(V_2 - V_1)$$
$$= 1 \times 10^5 \times 0.15 \times 1 \times 10^{-2}$$
$$= 150\,\text{J}$$

(b) We have $\Delta Q = 250$ and $\Delta W = 150$. Rearranging Equation 26.1 gives

$$\Delta U = \Delta Q - \Delta W = 250 - 150$$
$$= 100\,\text{J}$$

Thus as heat is supplied to the gas the speed of the molecules increases. This would increase the pressure in the container, if it were not for the fact that the piston is pushed out. This decreases the density of the gas, and thus (see Equation 25.9) the pressure of the gas can remain at atmospheric pressure. The net effect is one of heat input being used to do work in pushing back the atmosphere, and to increase the internal energy (and so increase molecular speeds and temperature) of the gas.

Answer

(a) 150 J, (b) 100 J.

Example 2

When 1.50 kg of water is converted to steam (at 100 °C) at standard atmospheric pressure $(1.01 \times 10^5\,\text{N m}^{-2})$, 3.39 MJ of heat are required. During the transformation from liquid to vapour state, the increase in volume of the water is $2.50\,\text{m}^3$. Calculate the work done against the external pressure during the process of vaporisation. Explain what happens to the rest of the energy.

Method

When the liquid is converted into steam, the molecules have to push back the atmosphere during the accompanying increase in volume. We use Equation 26.3 with $p = 1.01 \times 10^5$ and $(V_2 - V_1) = 2.50$. So,

$$\Delta W = p(V_2 - V_1) = 1.01 \times 10^5 \times 2.50$$
$$= 0.253 \times 10^6\,\text{J}$$

The external work done $\Delta W = 0.253\,\text{MJ}$.

The rest of the energy goes to an increase in internal energy ΔU of the water molecules and is given by equation 26.1:

$$\Delta U = \Delta Q - \Delta W = 3.39 - 0.253$$
$$= 3.14\,\text{MJ}$$

This is needed to do work in separating the water molecules during the liquid–vapour transition. It thus becomes potential energy. No kinetic energy change occurs because there is no increase in temperature.

Answer

External work done is 0.253 MJ.

Exercise 93

1 A fixed mass of gas is cooled, so that its volume decreases from 4.0 litres to 2.5 litres at a constant pressure of 1.0×10^5 Pa. Calculate the external work done by the gas. Note: 1 litre = $10^{-3}\,\text{m}^3$.

2 Referring to Fig 26.2a, suppose that the sample of gas is cooled down so that 120 J of heat is extracted from it. If as a result the piston moves inwards 5.0 cm along the cylinder, calculate (a) the external work done by the gas, (b) the increase in internal energy of the gas.

3 The specific latent heat of vaporisation of steam is 2.26 MJ kg^{-1}. When 50 cm^3 of water is boiled at standard atmospheric pressure of 1.01×10^5 Pa, 83×10^3 cm^3 of steam are formed. Calculate (a) the mass of water boiled, (b) the heat energy input needed, (c) the external work done during vaporisation, (d) the increase in internal energy. (Density of water = 1000 kg m^{-3}; 1 cm^3 = 10^{-6} m^3.)

Principal heat capacities

The molar heat capacity of a gas is the amount of heat energy required to raise the temperature of 1 mol of the gas by 1 K. There are an infinite number of possible molar heat capacities, depending on how pressure and volume are allowed to vary. We choose two principal values as follows:

C_p: the molar heat capacity at *constant pressure*

C_v: the molar heat capacity at *constant volume*

The principal *specific* heat capacities, denoted c_p and c_v, are the principal heat capacities for *1 kg* of gas.

It can be shown that, for *all* gases,

$$C_p - C_v = R \tag{26.4}$$

where R is the universal molar gas constant (see Chapter 25). The ratio of the principal heat capacities is denoted by γ:

$$\gamma = \frac{C_p}{C_v} = \frac{c_p}{c_v} \tag{26.5}$$

The ratio γ has significance elsewhere (see Equations 15.3 and 26.9); it depends upon the atomicity of the gas, and has values at room temperature as follows:

$\gamma = 1.67$ for a monatomic gas

$\gamma = 1.40$ for a diatomic gas

$\gamma = 1.33$ for some polyatomic gases

Example 3

56.0×10^{-3} kg of nitrogen is to be heated from 270 K to 310 K. When this occurs in an insulated freely extensible container, 2.33 kJ of heat is required. When contained in an insulated rigid container, 1.66 kJ of heat is required. Calculate the principal molar heat capacities of nitrogen.
(Molar mass of nitrogen = 28.0×10^{-3} kg.)

Method

The number of moles n of gas is, from Equation 25.6,

$$n = \frac{\text{Mass of gas}}{\text{Molar mass}} = \frac{56}{28} = 2.00$$

Heat ΔQ required is given by

$$\Delta Q = \begin{pmatrix}\text{Number}\\ \text{of moles}\\ n\end{pmatrix} \times \begin{pmatrix}\text{Principal heat}\\ \text{capacity}\\ C\end{pmatrix} \times \begin{pmatrix}\text{Temperature}\\ \text{change}\\ \Delta T\end{pmatrix} \tag{26.6}$$

We have $n = 2$ and $\Delta T = 40$.

For the freely extensible container the gas is under constant pressure. Since $\Delta Q_p = 2.33 \times 10^3$, Equation 26.6 gives

$$2.33 \times 10^3 = 2 \times C_p \times 40$$

$$\therefore \quad C_p = 29.1\ \text{J mol}^{-1}\,\text{K}^{-1}$$

For the rigid container the gas remains at constant volume. Since $\Delta Q_v = 1.66 \times 10^3$,

$$1.66 \times 10^3 = 2 \times C_v \times 40$$

$$\therefore \quad C_v = 20.8\ \text{J mol}^{-1}\,\text{K}^{-1}$$

Note that

$$C_p - C_v = 29.1 - 20.8 = 8.3\ \text{J mol}^{-1}\,\text{K}^{-1}$$

The value of 8.3 corresponds to R, which agrees with Equation 26.4.

Answer

29.1 J mol^{-1} K^{-1}, 20.8 J mol^{-1} K^{-1}.

Example 4

The specific heat capacity of a diatomic gas at constant volume is 0.410 kJ kg^{-1} K^{-1}. Calculate (a) the specific heat capacity of the gas at constant pressure and (b) the specific gas constant for the gas.

Method

Note the use of 'specific': we refer to 1 kg of gas. For a diatomic gas $\gamma = 1.40$.

(a) Rearranging Equation 26.5, with $c_v = 0.41$,

$$c_p = \gamma c_v = 1.4 \times 0.41$$
$$= 0.574 \text{ kJ kg}^{-1}\text{ K}^{-1}$$

(b) Equation 26.4 refers to 1 mole in the case of all gases. For 1 kg of a particular gas we write

$$c_p - c_v = r \qquad (26.7)$$

where r is called the specific gas constant and is given by

$$r = \frac{R}{\text{Molar mass } M_m} \qquad (26.8)$$

Since R is a universal constant, r is a characteristic of the gas.

From Equation 26.7,

$$r = c_p - c_v = 0.574 - 0.410$$
$$= 0.164 \text{ kJ kg}^{-1}\text{ K}^{-1}$$

Answer

(a) $0.574 \text{ kJ kg}^{-1}\text{ K}^{-1}$, (b) $0.164 \text{ kJ kg}^{-1}\text{ K}^{-1}$.

Example 5

1 mol of oxygen at 280 K is in an insulated, infinitely flexible container. The atmospheric pressure outside the container is $5 \times 10^5 \text{ N m}^{-2}$. When 580 J of heat is supplied to the oxygen, the temperature increases to 300 K and the container increases in volume by $3.32 \times 10^{-4} \text{ m}^3$. Calculate the values of the two principal molar heat capacities and that of the universal molar gas constant. [WJEC 77, p]

Method

For 1 mol of gas, heat $\Delta Q_p = 580$ J is needed to increase its temperature by $\Delta T = 300 - 280 = 20$ K, at *constant pressure* (since the container is flexible). Rearranging Equation 26.6 gives

$$C_p = \frac{\Delta Q_p}{n\Delta T} = \frac{580}{1 \times 20}$$
$$= 29.0 \text{ J mol}^{-1}\text{ K}^{-1}$$

We now require to find the energy needed to increase its temperature at *constant volume*.

To find this we subtract the external work done ΔW when the gas expands.

Using Equation 26.3 with $p = 5 \times 10^5$ and $(V_2 - V_1) = 3.32 \times 10^{-4}$:

$$\Delta W = p(V_2 - V_1) = 5 \times 10^5 \times 3.32 \times 10^{-4}$$
$$= 166 \text{ J}$$

Thus 580 − 166 = 414 J are needed to raise the temperature of 1 mol of the gas by 20 K at *constant volume*. From Equation 26.6

$$C_v = \frac{\Delta Q_v}{n\Delta T} = \frac{414}{1 \times 20}$$
$$= 20.7 \text{ J mol}^{-1}\text{ K}^{-1}$$

To find R we use Equation 26.4:

$$R = C_p - C_v = 29.0 - 20.7$$
$$= 8.3 \text{ J mol}^{-1}\text{ K}^{-1}$$

Note that we have implicitly used Equation 26.1. In this equation we identify

$$\Delta Q = \Delta Q_p = C_p\Delta T = 580 \text{ J}$$
$$\Delta W = p(V_2 - V_1) = 166 \text{ J}$$
$$\therefore \quad \Delta U = \Delta Q - \Delta W = 414 \text{ J}$$

Note, then, that $\underline{\Delta U} = \Delta Q_v = \underline{C_v\Delta T}$ (for 1 mol) so that 414 J is needed to increase the internal energy (i.e. temperature only) of the gas when *no external work* is done: that is, at *constant volume*.

Answer

$C_p = 29.0 \text{ J mol}^{-1}\text{ K}^{-1}$, $C_v = 20.7 \text{ J mol}^{-1}\text{ K}^{-1}$, $R = 8.3 \text{ J mol}^{-1}\text{ K}^{-1}$.

Exercise 94

(Assume $R = 8.31 \text{ J mol}^{-1}\text{ K}^{-1}$.)

1 The molar heat capacity at constant volume of a certain gas is $5R/2$. Calculate γ for the gas.

2 The amount of heat required to raise the temperature of 3.00 mol of a polyatomic gas, at constant pressure, from 320 K to 370 K is 4.99 kJ. Calculate (a) C_p and C_v, (b) the value of γ, (c) the heat required to raise the temperature of 4.00 mol from 300 K to 400 K at constant volume.

3 Argon has a molar mass of 40×10^{-3} kg and a principal molar heat capacity, at constant volume, of 12.5 J mol^{-1} K^{-1}. Calculate (a) the value of γ, (b) the *specific* heat capacity at constant volume, (c) the amount of heat required to raise the temperature of 1.00 kg of argon by 80 K at constant volume.

4 2.00 mol of nitrogen, at 300 K, are in an insulated, freely extensible container, and the pressure outside the container is 1.00×10^5 N m^{-2}. The principal molar heat capacity of nitrogen at constant pressure is 29.0 J mol^{-1} K^{-1}. Calculate (a) the heat required to raise its temperature to 340 K, (b) the increase in volume of the gas during this process (use $pV = nRT$), (c) the external work done, (d) the internal energy change, (e) the heat required to effect the temperature change at constant volume. Compare (d) and (e) and comment.

Isothermal and adiabatic changes

An *isothermal change* is one which takes place in such a way that the *temperature remains constant*. Thus for an isothermal change, Equation 25.2 reduces to Equation 25.3:

$$p_1V_1 = p_2V_2 \qquad (25.3)$$

where p_1 and V_1 are the initial pressure and volume and p_2 and V_2 are pressure and volume after the isothermal change.

An *adiabatic change* is one which takes place in such a way that *no heat can enter or leave the system* during the process. This means that, from Equation 26.1, since $\Delta Q = 0$, any external work done by the gas must lead to a corresponding decrease in internal energy (and hence a temperature drop). Similarly an adiabatic compression leads to an increase in internal energy and hence a temperature rise. For an *adiabatic change* it can be shown that (for a fixed mass of gas)

$$p_1V_1^{\gamma} = p_2V_2^{\gamma} \qquad (26.9)$$

where p_1 and V_1 refer to initial pressure and volume, p_2 and V_2 to pressure and volume after the adiabatic change and γ is the ratio of the principal heat capacities. Any suitable units may be used for pressure and volume.

Note that *for any change, Equations 25.2 and 25.5 can be used.*

By combining Equations 26.9 and 25.2 we can eliminate pressure to get, for an adiabatic change,

$$T_1V_1^{(\gamma-1)} = T_2V_2^{(\gamma-1)} \qquad (26.10)$$

where T_1 and T_2 refer to initial and final temperature respectively.

Example 6

A gas at an initial pressure of 760 mm mercury is expanded adiabatically until its volume is doubled. Calculate the final pressure of the gas if the ratio of the principal specific heat capacities is 1.40.

Method

We have $p_1 = 760$ and $\gamma = 1.4$. Let $V_1 = V$, so $V_2 = 2V$.

Rearranging Equation 26.9 gives

$$p_2 = p_1\left(\frac{V_1}{V_2}\right)^{\gamma} = 760 \times \left(\frac{V}{2V}\right)^{1.4} = \frac{760}{2^{1.4}} = \frac{760}{2.64}$$

$$= 288$$

Answer

Final pressure is 288 mm mercury.

Example 7

The piston of a bicycle pump is slowly moved in until the volume of air enclosed is one-fifth of the total volume of the pump and is at room temperature (290 K). The outlet is then sealed and the piston suddenly drawn out to full extension. No air passes the piston. Find the temperature of the air in the pump immediately after withdrawing the piston, assuming that air is a perfect gas with $\gamma = 1.4$. [WJEC 80, p]

Method

The pushing-in of the piston results in some air remaining trapped in the body of the pump. Its initial temperature is $T_1 = 290$; let its initial volume $V_1 = V$. The act of *suddenly* drawing out the piston indicates an adiabatic expansion and, since no air passes the piston, a fixed mass of gas. The final volume $V_2 = 5V$ and we require the final temperature T_2. Rearranging Equation 26.10 with $\gamma - 1 = 0.4$:

$$T_2 = T_1\left(\frac{V_1}{V_2}\right)^{(\gamma-1)} = 290\left(\frac{V}{5V}\right)^{0.4}$$

$$= 152\ \text{K}$$

Note that we could have used Equation 26.9 to find p_2 in terms of p_1 and then used Equation 25.2 to find T_2. It is worth checking the answer using this method, which is equivalent to proving Equation 26.10.

The final temperature is less than the initial value because the external work done by the gas, on expansion, results in a corresponding decrease in internal energy, hence temperature.

Answer

Final temperature is 152 K.

Example 8

A fixed mass of gas, initially at 7 °C and a pressure of $1.00 \times 10^5\,\mathrm{N\,m^{-2}}$, is compressed isothermally to one-third of its original volume. It is then expanded adiabatically to its original volume. Calculate the final temperature and pressure, assuming $\gamma = 1.40$.

Method

We must treat the two processes *separately and in order.*

For the *isothermal* change we have $p_1 = 1 \times 10^5$; let $V_1 = V$ and $V_2 = V/3$. We rearrange Equation 25.3 to find p_2:

$$p_2 = \frac{p_1V_1}{V_2} = \frac{1 \times 10^5 \times V}{V/3}$$
$$= 3 \times 10^5\,\mathrm{N\,m^{-2}}$$

For the *adiabatic* change our initial temperature is still 7 °C, so

initial state: $p_2 = 3.00 \times 10^5$, $V_2 = V/3$,

$T_2 = 273 + 7 = 280\,\mathrm{K}$

final state: $p_3 = ?$, $V_3 = V$, $T_3 = ?$

Note that we have the initial state with suffix 2 and the final state with suffix 3, so Equation 26.9 becomes $p_2V_2^{\gamma} = p_3V_3^{\gamma}$. Rearranging to find p_3:

$$p_3 = p_2\left(\frac{V_2}{V_3}\right)^{\gamma} = 3 \times 10^5\left(\frac{V/3}{V}\right)^{1.4} = \frac{3 \times 10^5}{3^{1.4}}$$
$$= 0.644 \times 10^5\,\mathrm{N\,m^{-2}}$$

To find T_3 we can use Equation 25.2 (or, alternatively, Equation 26.10):

$$T_3 = \frac{p_3V_3T_2}{p_2V_2} = \frac{0.644 \times 10^5 \times V \times 280}{3 \times 10^5 \times V/3}$$
$$= 180\,\mathrm{K}$$

Answer

Final temperature is 180 K, final pressure is $0.644 \times 10^5\,\mathrm{N\,m^{-2}}$.

Exercise 95

(Assume $\gamma = 1.40$ for air.)

1 2.00 litre of air initially at a pressure of $1.01 \times 10^5\,\mathrm{N\,m^{-2}}$ and a temperature of 17 °C is compressed to a volume of 0.30 litre (a) under isothermal conditions, (b) under adiabatic conditions. Calculate in each case the final pressure and temperature.

2 $3.00 \times 10^{-4}\,\mathrm{m^3}$ of air at 7 °C and a pressure of $5.00 \times 10^5\,\mathrm{N\,m^{-2}}$ is allowed to expand until the pressure falls to $1.00 \times 10^5\,\mathrm{N\,m^{-2}}$. Calculate the final volume and temperature in each case if the expansion takes place under (a) isothermal, (b) adiabatic conditions.

3 A fixed mass of air at an initial pressure of 760 mm mercury and 0 °C is expanded adiabatically to 1.50 times its volume and then compressed isothermally to 0.50 times its *original* volume. Calculate its final temperature and pressure.

Exercise 96: Questions of A-level standard

1 A fixed mass of gas is heated at a constant pressure of 10^5 Pa ($\mathrm{N\,m^{-2}}$), so that its volume increases from $5\,\mathrm{m^3}$ to $8\,\mathrm{m^3}$. The external work done by the gas in expanding, expressed in joules, is

A 2.50×10^3 **B** 3.33×10^4 **C** 3.00×10^5
D 5.00×10^5 **E** 8.00×10^5 [AEB 82]

2 At a temperature of 100 °C and a pressure of 1.01×10^5 Pa, 1.00 kg of steam occupies $1.67\,\mathrm{m^3}$, but the same mass of water occupies only $1.04 \times 10^{-3}\,\mathrm{m^3}$. The specific latent heat of vaporisation of water at 100 °C is $2.26 \times 10^6\,\mathrm{J\,kg^{-1}}$. For a system consisting of 1.00 kg of water changing to steam at 100 °C and 1.01×10^5 Pa, find

(a) the heat supplied to the system;
(b) the work done by the system;
(c) the increase in internal energy of the system.

[C 79]

3 The ratio of the principal heat capacities of an ideal gas is γ, and the molar gas constant is R. The molar heat capacity at constant pressure of the gas is

A R **B** $\dfrac{R}{\gamma - 1}$ **C** $\dfrac{\gamma - 1}{R}$

D $\dfrac{(\gamma - 1)R}{\gamma}$ **E** $\dfrac{\gamma R}{\gamma - 1}$

4 The specific heat capacity at constant volume of a certain ideal gas is $6 \times 10^2\,\mathrm{J\,K^{-1}\,kg^{-1}}$ and is independent of temperature. Find the internal energy of $5.0 \times 10^{-3}\,\mathrm{kg}$ of the gas at 27 °C. [C 82]

5 An ideal gas at an initial temperature of 15 °C and pressure of $1.10 \times 10^5\,\mathrm{Pa}$ is compressed isothermally to one-quarter of its original volume. What will be its final pressure and temperature? What would have been the pressure and temperature if the compression had been adiabatic? (Ratio of the principal specific heat capacities of the gas = 1.40.) [AEB 82, p]

6 In the adiabatic compression of an ideal gas ($\gamma = 5/3$) from a volume V to a volume $V/8$ the temperature will increase by a factor

A 2 **B** 4 **C** 8 **D** 16 **E** 32 [AEB 82]

7 The molar heat capacity at constant volume of a certain ideal gas is $5R/2$. When the gas is compressed slowly and adiabatically to twice its initial pressure, the initial volume is multiplied by a factor of

A $2^{-3/5}$ **B** $2^{-5/7}$ **C** $2^{7/5}$

D $2^{5/3}$ **E** $2^{5/7}$ [O & C 80]

8 A cylinder of volume $5 \times 10^{-4}\,\mathrm{m^3}$ contains oxygen at a pressure of $2 \times 10^5\,\mathrm{Pa}$ and a temperature of 300 K. Find (a) the number of molecules and (b) their mean kinetic energy*.

When the gas is compressed adiabatically to a volume of $2 \times 10^{-4}\,\mathrm{m^3}$, the temperature rises to 434 K, (c) Find γ, the ratio of the principal heat capacities, for oxygen.
(Molar gas constant $R = 8\,\mathrm{J\,mol^{-1}\,K^{-1}}$, Avogadro's constant $N_A = 6 \times 10^{23}\,\mathrm{mol^{-1}}$.) [WJEC 82]

9 In a diesel engine, fuel oil is injected into a cylinder in which air has been heated by adiabatic compression to above the ignition temperature of the oil. The ignition temperature of a certain fuel is 630 °C, and the air enters the cylinder, which has an initial volume of $5.0 \times 10^{-4}\,\mathrm{m^3}$, at a pressure of $1.0 \times 10^5\,\mathrm{Pa}$ and a temperature of 28 °C.

(a) What minimum compression ratio (the ratio of the initial to the final volume of the cylinder) is required to heat the air to the fuel ignition temperature?

(b) How much work is done in compressing the air?†

Assume all adiabatic changes to be reversible, so that the usual equations apply.
(For air, $\gamma = 1.40$) [C 82, p]

* *Translational KE per molecule.*

† *For an adiabatic expansion*

$$\Delta W = \int_{V_1}^{V_2} p\,\mathrm{d}V = \frac{1}{(\gamma - 1)}(p_1V_1 - p_2V_2)$$

Section G
Electricity, Magnetism and Electronics

27 DIRECT CURRENT ELECTRICITY

Electric charge

Objects normally contain equal numbers of protons and electrons. A negatively charged object has a surplus of electrons and a positively charged object has more protons than electrons, i.e. a deficiency of electrons. The (net) charge on an object is measured in coulombs (abbreviation C). This is the SI unit for charge and it is defined below.

Electric current

A current is a flow of charge. In a metal wire many electrons are free to move, so that a current can flow in a metal wire as a flow of electrons, i.e. the current carriers are electrons.

The unit for current is the ampere (A), defined in Chapter 32. Current size I is related to charge Q moving through (entering and leaving) a wire in time t seconds by

$$I = \frac{Q}{t} \quad \text{and} \quad 1\,\text{A} = 1\,\text{C}\,\text{s}^{-1} \qquad (27.1)$$

Equation 27.1 defines the coulomb as 1 A s.

The direction of current flow is taken to be that of positive charge flow, i.e. opposite to that of electron flow.

Carrier velocity

If carriers, e.g. electrons in a metal wire, are moving with an average drift velocity along the wire of v metre per second, then

$$I = nAqv \qquad (27.2)$$

where n is the carrier density (number per m^3), A is the cross-section area of the wire (so that nA is the carriers per metre length of wire) and q is the charge of each carrier.

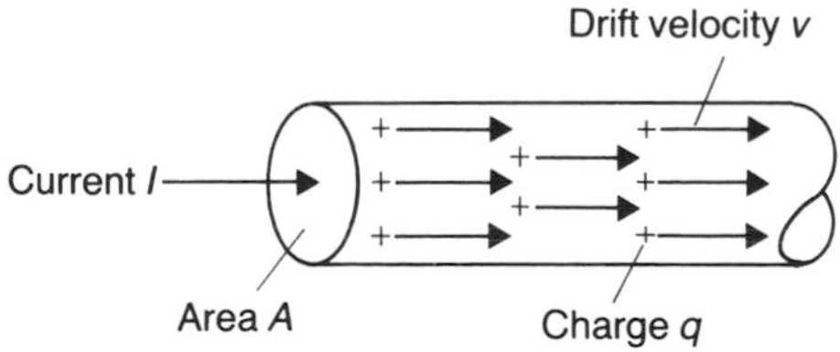

Fig 27.1 $I = nAqv$

Example 1

How many electrons are passing through a wire per second if the current is 1.00 mA, given that the charge carried by each electron is 1.6×10^{-19} C?

Method

$I = 10^{-3}$, $q = 1.6 \times 10^{-19}$ C; let time $t = 1$ s and the number of electrons be n. Using $I = Q/t$ (Equation 27.1) we have

$$10^{-3} = \frac{n \times 1.6 \times 10^{-19}}{1}$$

$$\therefore \quad n = \frac{10^{-3}}{1.6 \times 10^{-19}}$$

$$\therefore \quad n = 6.25 \times 10^{15}$$

Answer
6.2×10^{15}.

Example 2
Calculate the mean velocity of electron flow (the drift velocity) in a wire where the free electron density is $5 \times 10^{28}\,\text{m}^{-3}$ if the current is 1 A and the wire has a uniform cross-section area of $1\,\text{mm}^2$. (Electron charge $= -1.6 \times 10^{-19}\,\text{C}$.)

Method
$I = nqvA$ (Equation 27.2) and $I = 1\,\text{A}$, $n = 5 \times 10^{28}\,\text{m}^{-3}$, $q = 1.6 \times 10^{-19}\,\text{C}$ and $A = 10^{-6}\,\text{m}^2$

$$v = \frac{I}{nqA} = \frac{1}{5 \times 10^{28} \times 1.6 \times 10^{-19} \times 10^{-6}}$$

$$= \frac{1}{8} \times 10^{-3} = 0.125 \times 10^{-3}\,\text{m s}^{-1}$$

Answer
$1.2 \times 10^{-4}\,\text{m s}^{-1}$ (if we assume an accuracy of two significant figures).

Potential and potential difference

The potential of a place may be thought of as its attractiveness for electrons or unattractiveness for positive charges. A place where there is a high concentration of electrons or which has a lot of electrons near it will have a low potential.

The difference of potential (PD) V between two places is defined as the work done W per coulomb of charge moved from the one place to the other.

$$V = \frac{W}{Q} \qquad (27.3)$$

where W is the work done (e.g. if positive charge Q moves from lower potential (−) to higher potential (+)) or energy obtainable from the movement (e.g. if negative charge Q goes from − to + place).

The unit for PD is the volt (V).

The potential of a place measured in volts is the PD between the place concerned and some reference point, usually taken to be a place far away from any electric charges (i.e. at infinity), or otherwise the earth. In other words, either of these places may be taken as zero potential.

Electric current flows spontaneously from a higher potential place (+) to a lower potential place (−) if the two places are joined by a conducting path.

Ohm's law

This law states that the current I through a given conductor is proportional to the PD between its ends, provided that its temperature does not change.

$$I \propto V \quad \text{or} \quad \frac{V}{I} = \text{Constant} \qquad (27.4)$$

This law applies to metallic conductors and many others.

Resistance *R* of a conductor

This is the opposition of the conductor to current flow through it, and it is defined as the PD needed across it (between its ends) per ampere of current:

$$R = \frac{V}{I} \qquad (27.5)$$

The unit for resistance is the ohm (Ω).

Resistors

These are devices whose purpose is to provide resistance to the flow of current. Some variable resistors are called rheostats.

Electric circuits

Often a current is produced by use of a voltaic cell or battery (two or more cells joined together). The cell creates and maintains a PD between its terminals. A current is obtained if these two

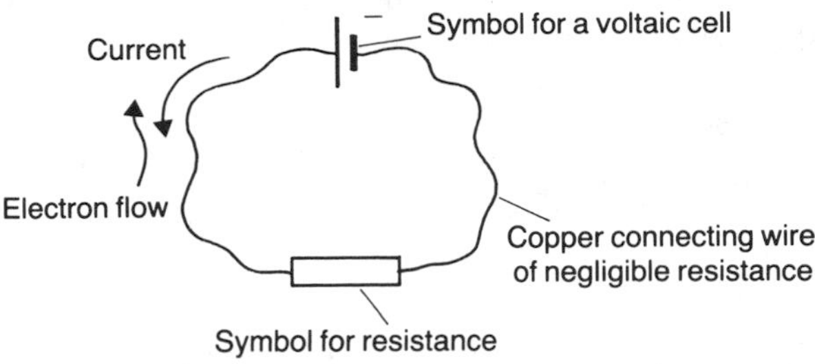

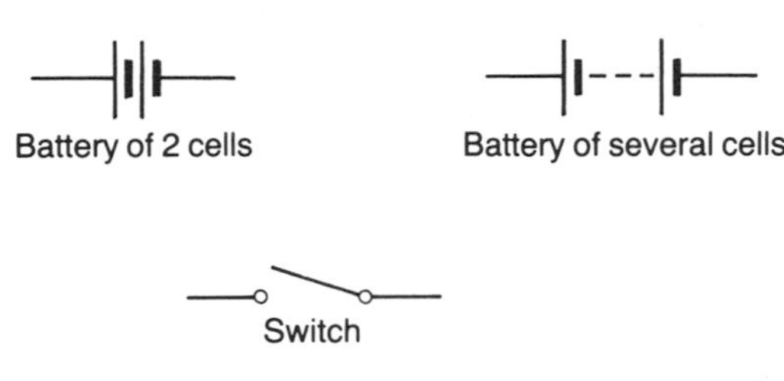

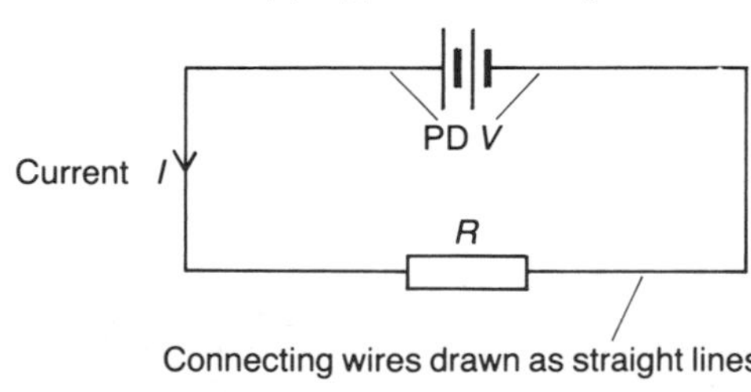

Fig 27.2

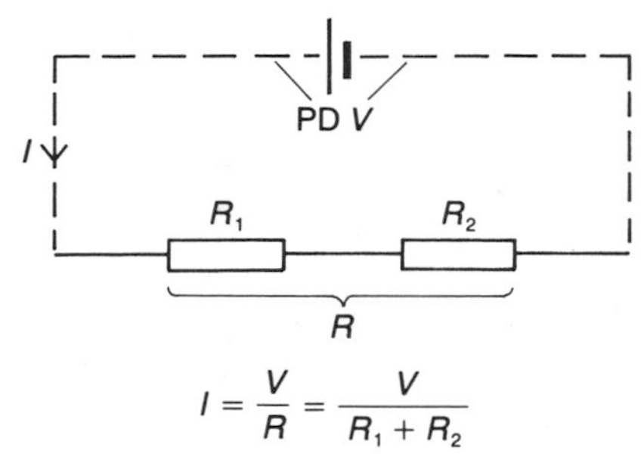

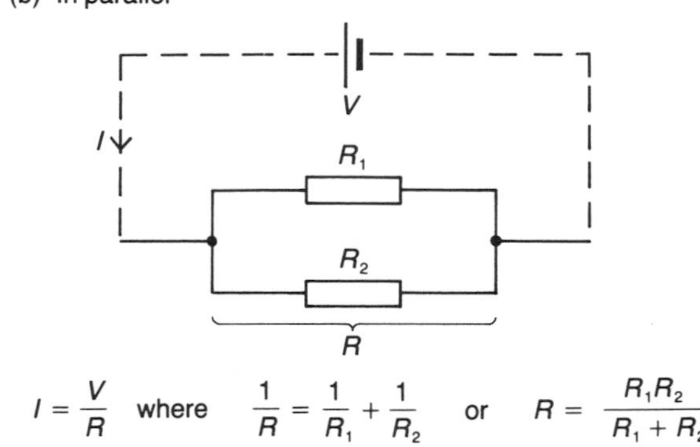

Fig 27.3 Resistors in series and parallel

terminals are joined by a conducting path, i.e. when a complete circuit is formed. (Fig 27.2)

The current obtained from a voltaic cell is direct current (DC) because its direction is constant.

Resistors in series

When two resistances R_1 and R_2 ohm are connected as shown in Fig 27.3a they are in series and the total resistance is R, where

$$R = R_1 + R_2 \qquad (27.6)$$

R_1 and R_2 carry the same current.

Resistances in parallel

In this arrangement the resistance of the combination is given by

$$\frac{1}{R} = \frac{1}{R_1} + \frac{1}{R_2} \quad \text{or} \quad R = \frac{R_1 R_2}{R_1 + R_2} \qquad (27.7)$$

In a parallel combination the PD across one resistor is the same as that across the other, but the total circuit current I in Fig 27.3b is shared between the resistors.

Example 3

Calculate the current through, and PD across, each of the resistors in the circuit shown (Fig 27.4, p.178).

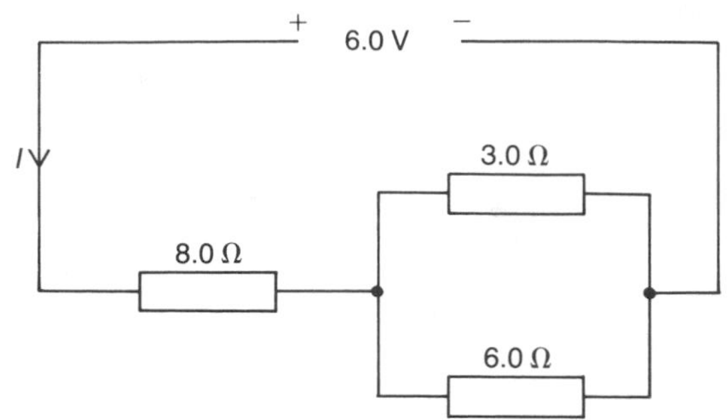

Fig 27.4 Circuit diagram for Example 3

Method

The resistance of 3 Ω in parallel with 6 Ω is

$$R = \frac{R_1R_2}{R_1 + R_2} = \frac{3.0 \times 6.0}{3.0 + 6.0} = 2.0\,\Omega$$

or $$\frac{1}{R} = \frac{1}{R_1} + \frac{1}{R_2} = \frac{1}{3.0} + \frac{1}{6.0} = 0.50^*$$

$$\therefore\ R = 2.0\,\Omega$$

We see that the circuit can be regarded as 8.0 Ω in series with 2.0 Ω. Circuit resistance is

$$R = R_1 + R_2 = 8.0 + 2.0 = 10\,\Omega$$

$$\therefore\quad I = \frac{V}{R} = \frac{6.0}{10} = 0.60\text{ A}$$

Note that we know only one PD, namely 6.0 V, and to use $I = V/R$ we must use $V = 6.0$ with the correct resistance. It is the 10 Ω across which the PD is 6.0 V.

The current through the 8.0 Ω resistor is I, which is 0.60 A.

PD across the 8.0 Ω (using $V = IR$ for this resistor now that its current is known) is given by

$$V = 0.6 \times 8.0 = 4.8\text{ V}$$

To obtain answers for the 3.0 Ω and 6.0 Ω we can say either:

PD across the 3.0 Ω and 6.0 Ω is 6.0 V − 4.8 V = 1.2 V. Then current I_3 through the 3.0 Ω is

$$I_3 = 1.2/3.0 = 0.40\text{ A}$$

and for the 6.0 Ω the current I_6 is 1.2/6.0 or 0.20 A; *or* (in view of the simple values of 3.0 and 6.0 for the parallel resistors) we can say:

**A common error is to forget that this is 1/R, not R.*

The 3.0 Ω and 6.0 Ω are in the ratio of 1 : 2, so that the easier route for the current (3.0 Ω) will carry two parts of the 0.60 A while the 6.0 Ω route will carry one part. The 6.0 Ω carries one-third of the 0.60 A, namely 0.20 A; the 3.0 Ω carries two-thirds, namely 0.40 A.

Answer

0.60 A, 4.8 V; 0.20 A, 1.2 V; 0.40 A, 1.2 V.

Resistivity ρ of a material

The resistance R of a conductor is proportional to its length l, inversely proportional to its area of cross-section A, and dependent upon the nature of the material, described by its resistivity ρ, which is defined by the following equation:

$$R = \rho\frac{l}{A} \qquad (27.8)$$

The unit for ρ (which is given by $\rho = RA/l$) is Ω m.

The term 'conductivity' σ of a material is used for the reciprocal of ρ so that

$$\sigma = \frac{1}{\rho} \qquad (27.9)$$

Temperature coefficient of resistance or resistivity

This quantity is denoted by α.

The resistance of many materials, e.g. a metal, increases steadily with increase of temperature in accordance with the equation

$$\alpha = \frac{R - R_0}{R_0\theta} \quad \text{or} \quad R = R_0(1 + \alpha\theta) \qquad (27.10)$$

where R is the resistance at Celsius temperature θ, R_0 is the resistance at 0 °C and α is the temperature coefficient of resistance of the material.

The unit for α is K^{-1}.

We can also write

$$\rho = \rho_0(1 + \alpha\theta)$$

where ρ and ρ_0 are resistivities at temperature θ and zero.

Example 4

Calculate the length of wire of 1.0 mm diameter and $5.0 \times 10^{-6}\,\Omega$ m resistivity that would have a resistance of $5.0\,\Omega$.

Method

$$R = \frac{\rho l}{A} \qquad A = \frac{\pi d^2}{4} \qquad R = 5.0,$$

$$\rho = 5.0 \times 10^{-6}, \quad d = 1.0 \times 10^{-3}$$

$$R = 5.0 = \frac{5.0 \times 10^{-6} \times l}{A}$$

$$\therefore \quad A = \frac{\pi \times 10^{-6}}{4} = 7.9 \times 10^{-7}$$

$$\therefore \quad l = \frac{RA}{\rho} = \frac{5 \times 7.9 \times 10^{-7}}{5 \times 10^{-6}} = 0.79\text{ m}$$

Answer

0.79 m

Example 5

A coil of wire has resistance $6.00\,\Omega$ at 60 °C and $5.25\,\Omega$ at 15 °C. What is its temperature coefficient of resistance?

Method

$$R = R_0(1 + \alpha\theta) \qquad \therefore \quad 6.00 = R_0(1 + \alpha 60) \quad \text{and}$$

$$5.25 = R_0(1 + \alpha 15)$$

$$\therefore \quad \frac{5.25}{6.00} = \frac{R_0(1 + 15\alpha)}{R_0(1 + 60\alpha)}$$

Cancelling the R_0 factor and cross multiplying gives

$$5.25 + 315\alpha = 6.00 + 90\alpha$$

$$\therefore \quad 315\alpha - 90\alpha = 6.00 - 5.25$$

$$\therefore \quad \alpha = \frac{0.75}{225} = 0.0033\text{ K}^{-1}$$

Answer

0.0033 K^{-1}.

Electrical heating in a resistance

When current flows through a resistance there is a PD V across the resistance and, for Q coulombs passing through, electrical potential energy is lost (work W is done), this becoming internal energy of the resisting material (its temperature has risen). Since $V = W/Q$ and $Q = It$ (Equations 27.3 and 27.1) we have

$$W = VIt \qquad (27.11)$$

i.e. the heat produced is VIt where t is the time for which current flows.

Using $R = V/I$ (Equation 27.5), we can also write

$$W = \frac{V^2 t}{R} \qquad (27.12)$$

or

$$W = I^2 Rt \qquad (27.13)$$

The work done per second or heat produced per second is the power P and

$$P = VI \quad \left(\text{or } \frac{V^2}{R} \text{ or } I^2 R\right) \qquad (27.14)$$

The unit for power is watt (W). $1\text{ W} = 1\text{ J s}^{-1}$.

The expression 'power dissipated' (in a resistance) is often used. It means 'heat produced (per second)', but reminds us that the heat normally spreads and escapes from the place where it is produced.

Electromotive force and internal resistance

The PD between the terminals of a cell is caused by a chemical action which stops when the PD reaches a value characteristic of the type of cell, called the EMF of the cell. EMF stands for electromotive force, although it is a voltage not a force. When the cell is producing no current, i.e. it is on open circuit, the terminal PD V equals the EMF E:

$$V = E, \quad \text{on open circuit} \qquad (27.15)$$

When a current is being produced, the PD falls from the EMF value E, the chemical action starts again and the terminal PD V that is maintained is less than E by an amount called the 'lost volts'. This drop $E - V$ is a consequence of internal resistance r in the cell that hinders the cell's working. The lost volts equals $I \times r$, so that

$$E - V = Ir \qquad (27.16)$$

Either of the statements $E = V$ when $I = 0$ or $E - V = Ir$ may be used to define E, but a more satisfactory definition is

$$E = \frac{P}{I} \qquad (27.17)$$

where P is the total power $(I^2R + I^2r)$ dissipated

in the circuit resistance R and the internal resistance r. This means that

$$E = \frac{I^2R + I^2r}{I} = IR + Ir$$

$$\therefore \quad E = V + Ir$$

which agrees with Equation 27.16 and gives $E = V$ when $I = 0$.

For calculations a cell or other voltage source can be regarded as a cell of zero internal resistance with a separate resistance r in series with it (Fig 27.5a).

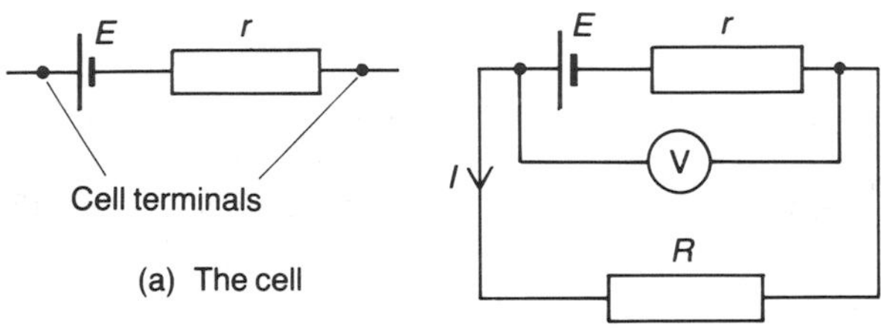

Fig 27.5 Cell with EMF *E* and internal resistance *r*

A cell represented in this way is seen in the circuit of Fig 27.5b. This circuit agrees with $E - V = Ir$ and $E = P/I$ and it is also seen that

$$I = \frac{E}{R + r} \qquad (27.18)$$

Example 6

A cell of EMF 1.5 V and internal resistance 1.0 Ω is connected to a 5.0 Ω resistor to form a complete circuit. Calculate the current expected, the terminal PD and the power dissipated in the external circuit and in the cell.

Method

A suitable diagram is shown in Fig 27.6.

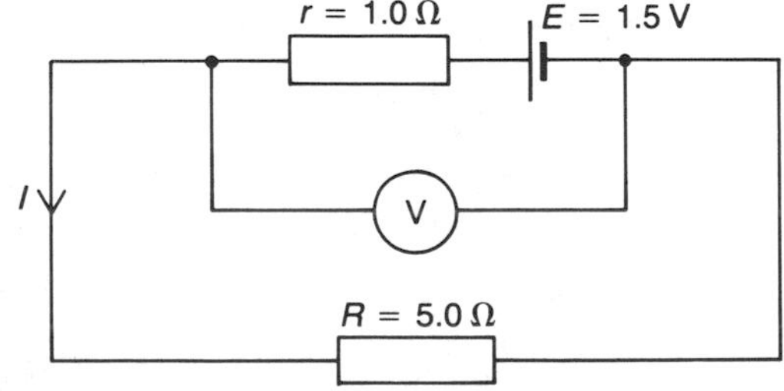

Fig 27.6 Circuit diagram for Example 6

$$I = \frac{E}{R + r} \text{ with } E = 1.5,\ R = 5.0 \text{ and } r = 1.0.$$

$$\therefore \quad I = \frac{1.5}{5.0 + 1.0} = 0.25\ \text{A}$$

Using $E - V = Ir$,

$$V = E - Ir = 1.5 - 0.25 \times 1.0 = 1.25\ \text{V}$$

The power in the 5.0 Ω is $I^2R = 0.25^2 \times 5.0 = 0.31$ W.

Alternatively, this power equals

PD across $R \times$ Current $= VI = 1.25 \times 0.25 = 0.31$ W

The power in the 1.0 Ω internal resistance is

$$I^2r = 0.25^2 \times 1.0$$

$$= 0.0625\ \text{W}$$

Alternatively, this power equals lost volts squared × internal resistance. Also the total power $I^2R + I^2r$ can be equated to $E \times I$.

Answer

0.25 A, 1.2 V, 0.31 W, 0.062 W.

Cells in series and parallel

When cells are joined in series, each cell adds its EMF to the total EMF if its + terminal connects to the − terminal of the next cell. It subtracts if − joins on to −. The internal resistances add.

For identical cells (same E and r) connected in parallel the EMF of the combination equals E, while the internal resistance of the combination is that of equal resistors in parallel (see Equation 27.7).

Maximum power

When a cell or other voltage source, having internal resistance r, is connected to a 'load' resistance (R in Fig 27.5b), the current through R is given by $E/(R + r)$, the PD across it is $ER/(R + r)$ and the power dissipated in it is equal to the product of these. The current is at its largest when $R \ll r$; the PD is large when $R \gg r$ and, it can be shown, the power is greatest when $R = r$.

Example 7

A cell of EMF 1.5 V and internal resistance 0.50 Ω is joined by its positive terminal to the positive terminal of a 1.0 V cell having internal resistance 1.0 Ω. The free terminals of the cells are then connected by a 3.5 Ω resistor. Calculate (a) the current through, and power dissipated in, the 3.5 Ω resistor and (b) the power dissipated in each cell.

Method

A suitable circuit diagram is shown in Fig 27.7.

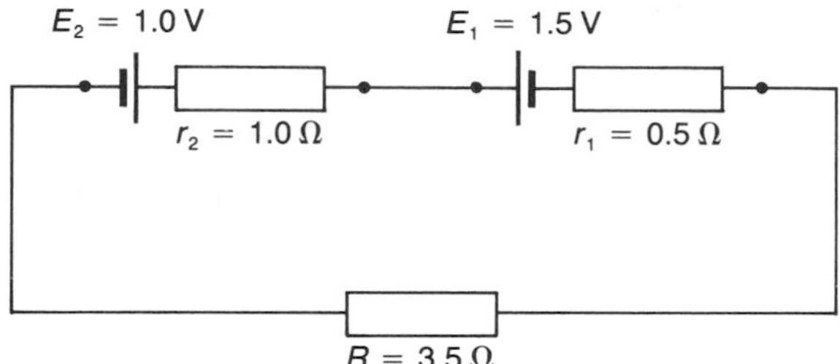

Fig 27.7 Diagram for Example 7

The total EMF is 1.5 − 1.0 = 0.5 V with polarity + to left.

The total internal resistance is 0.5 + 1.0 = 1.5 Ω.

$$\text{Current } I = \frac{E}{R + r} = \frac{0.5}{3.5 + 1.5} = 0.1 \text{ A}$$

The power dissipated in R is

$$I^2R = 0.1^2 \times 3.5 = 0.035 \text{ W}$$

The power dissipated in r_1 is

$$I^2r_1 = 0.1^2 \times 0.5 = 0.005 \text{ W}$$

The power dissipated in r_2 is

$$I^2r_2 = 0.1^2 \times 1.0 = 0.01 \text{ W}$$

(As a check the total power is EI (see Equation 27.17) and equals 0.5 × 0.1 or 0.05 W. This agrees with 0.035 + 0.005 + 0.01 as required.)

Answer

(a) 0.1 A, 0.035 W, (b) 0.005 W, 0.010 W.

Exercise 97

1 In a certain semiconducting material the current carriers each have a charge of 1.6×10^{-19} C. How many are entering the semiconductor per second when the current is 2 μA?

2 How many free electrons are there per metre length of wire if a current of 2.0 A requires the electron drift velocity to be 10^{-3} m s^{-1}?
(Electronic charge = 1.6×10^{-19} C.)

3 Calculate the current through each resistor and the PD across each in the circuit shown in Fig 27.8.

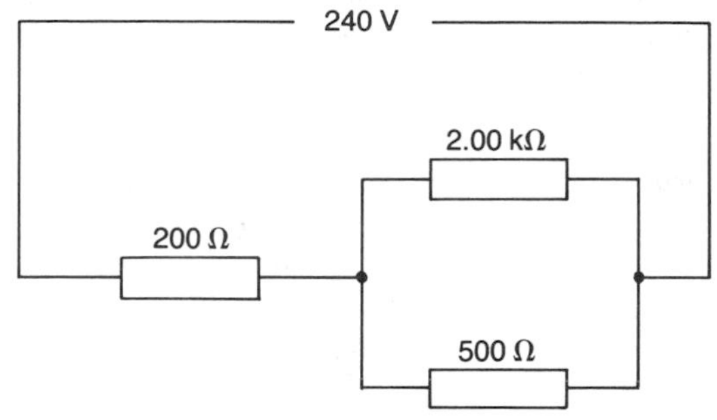

Fig 27.8 Circuit for Question 3

4 The electrical resistivity of manganin is 45×10^{-8} Ω m and is affected very little by temperature change. Calculate the resistance of 2.0 m of manganin wire of 1.0 mm diameter.

5 The resistivity of mild steel is 15×10^{-8} Ω m at 20 °C and its temperature coefficient is 50×10^{-4} K^{-1}. Calculate the resistivity at 60 °C.

6 Calculate the heat produced in 5 minutes in a pair of 10 Ω resistors connected in parallel with a PD of 2.0 V across the combination.

7 A certain large 6.0 V battery is used to produce a current of 60 A. (a) If this current is obtained when the load resistance is 0.08 Ω, what is the internal resistance of the battery? (b) What would the maximum current be that could be drawn from the battery? (c) How much heat would be produced per second in the battery when this maximum current is flowing, if the internal resistance is assumed to remain constant?

Kirchhoff's circuit laws

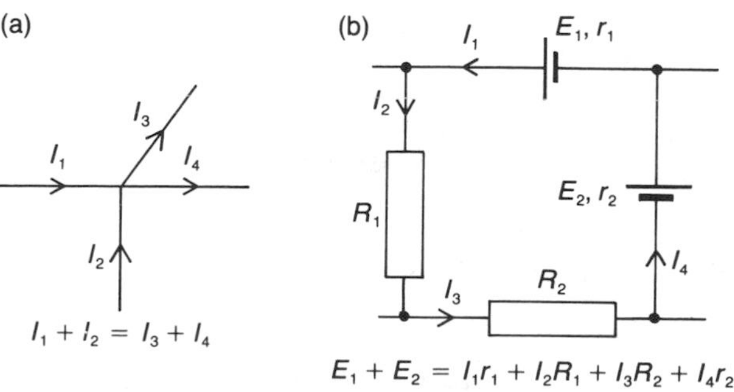

Fig 27.9 Kirchhoff's laws

The two Kirchhoff laws are:

(1) At any point in a circuit where conductors join, the total current towards the point equals the total current flowing away from it (Fig 27.9a).

(2) For any path that forms a complete loop the total of the EMFs equals the sum of the products of current and resistance, allowing for polarity: i.e. the algebraic sum of the EMFs equals the algebraic sum of the *IR* products ($\Sigma E = \Sigma IR$). This law is illustrated in Fig 27.9b.

Example 8

Calculate the currents passing through the two cells in the circuit shown (Fig 27.10a).

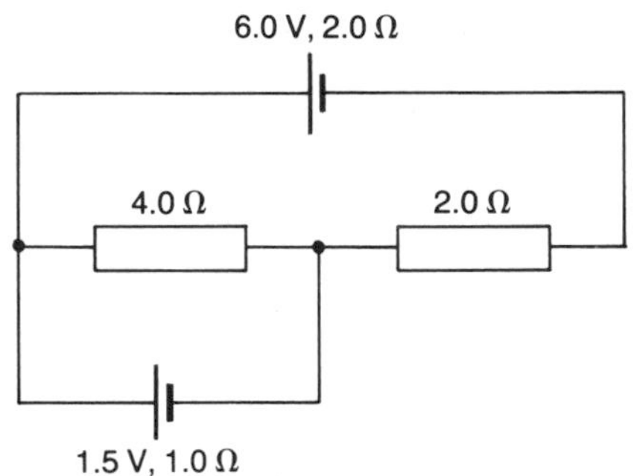

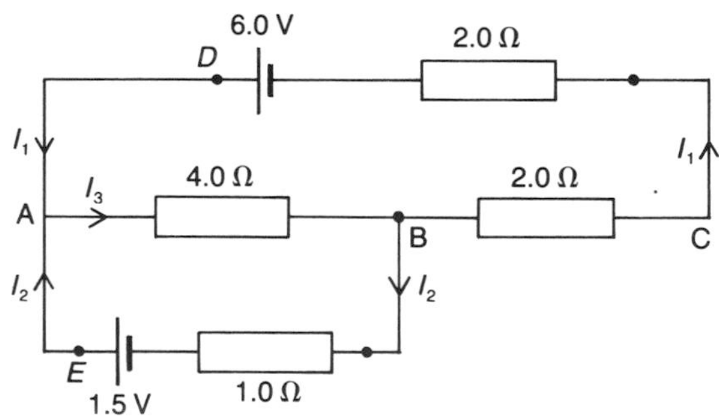

Fig 27.10 Circuits for Example 8

Method

Using Kirchhoff's first law for the point A in Fig 27.10b:

$$I_3 = I_1 + I_2 \quad \text{(i)}$$

Using Kirchhoff's second law for loop ABCD:

$$6 = 4I_3 + 2I_1 + 2I_1 \quad \text{(ii)}$$

Similarly for loop ABE:

$$1.5 = 4I_3 + 1.0I_2 \quad \text{(iii)}$$

Replacing I_3 in (ii) by $I_1 + I_2$ from (i) gives

$$6 = 4I_1 + 4I_2 + 2I_1 + 2I_1 = 8I_1 + 4I_2 \quad \text{(iv)}$$

Eliminating I_3 from (iii) in the same way gives

$$1.5 = 4I_1 + 4I_2 + I_2 = 4I_1 + 5I_2 \quad \text{(v)}$$

Replacing I_2 in (v) by $(6 - 8I_1)/4$ from (iv) we get

$$1.5 = 4I_1 + \frac{30 - 40I_1}{4}$$

from which I_1 is found to be 1.0 A.

Using this value for I_1 in (iv) ((v) could be used instead) gives

$$6 = 8 \times 1.0 + 4I_2 \quad \text{from which} \quad I_2 = -0.50\text{ A}$$

(Using these values for I_1 and I_2 in (i) gives $I_3 = 0.50$ A.)

Answer

Current through 6.0 V cell is 1.0 ampere.
Current through 1.5 V cell is −0.50 ampere (i.e. 0.50 A in direction opposite to that shown in Fig 27.10).

Exercise 98

1 Write down the equation that results from applying the second Kirchhoff law (a) to loop BCD in Fig 27.11, (b) to loop ABD and (c) to loop ABCD.

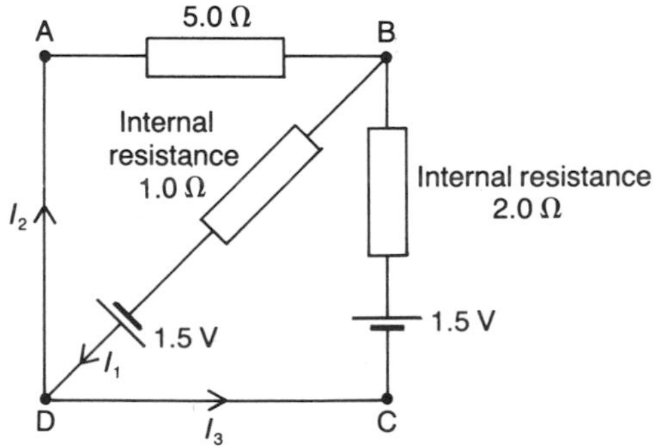

Fig 27.11 Circuit for Question 1

2 Use Kirchhoff's laws to carry out the calculation of Example 3 (p.177).

Note: The number of equations needed when using Kirchhoff's laws is of course equal to the number of unknown quantities. Note too that a loop only gives a helpful equation if part of the loop has not featured in other loops used.

Exercise 99: Questions of A-level standard

1 The resistivity of aluminium at room temperature is $3.2 \times 10^{-8}\,\Omega$ m. Assuming that it has 5×10^{28} free electrons per cubic metre, calculate the drift velocity of the electrons if a potential gradient of $1\,\text{V m}^{-1}$ were applied.
(Electron charge = -1.6×10^{-19} C.) [C 77, p]

2 A cable consists of 10 strands of copper wire each of cross-sectional area $1.1 \times 10^{-3}\,\text{cm}^2$. Calculate (a) the resistance per metre of the cable, (b) the minimum number of strands which would be required if the resistance per metre is not to exceed $0.010\,\Omega$. (Resistivity of copper = $1.8 \times 10^{-8}\,\Omega$ m.) [AEB 81, p]

3 A generator produces 100 kW of power at a potential difference of 10 kV. The power is transmitted through cables of total resistance 5 Ω. What is the power loss in the cables?

A 50 W **B** 250 W **C** 500 W
D 1000 W **E** 50 000 W [O 81]

4 A constant potential difference of 2.0 V is applied to the ends of a metal strip of length 0.30 m, width 4.0 mm and thickness 10^{-5} m. The metal has resistivity $1.5 \times 10^{-7}\,\Omega$ m at 0 °C and temperature coefficient of resistance $3.3 \times 10^{-3}\,\text{K}^{-1}$. The strip loses heat to its surroundings at a rate proportional to the surface area and to the excess temperature above the surroundings which are at 0 °C. Calculate (a) the resistance of the strip at 0 °C, (b) the electrical power dissipated when the strip is at a uniform temperature of 20 °C. [O & C 81, p]

5 In the circuit shown in Fig 27.12 what is the potential difference between the points B and D? What resistor could you add to the 12 Ω resistor in branch ADC in order to make the potential difference between B and D zero? [L 81, p]

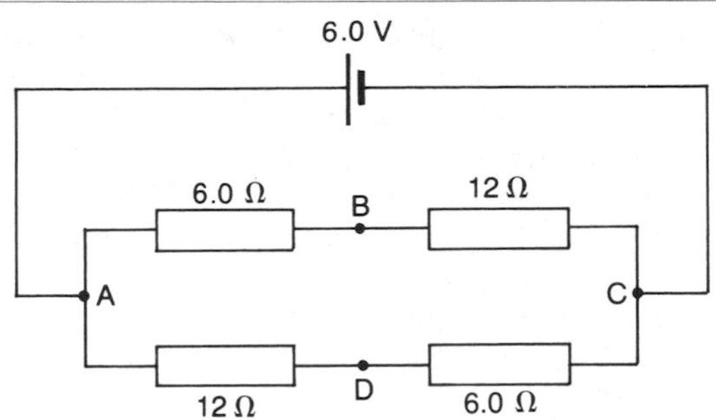

Fig 27.12 Circuit diagram for Question 5

6 In the circuit shown (Fig 27.13) a potential difference of 9.0 V is applied between A and B. What is (a) the current through the 10 Ω resistor and (b) the potential difference between C and D?

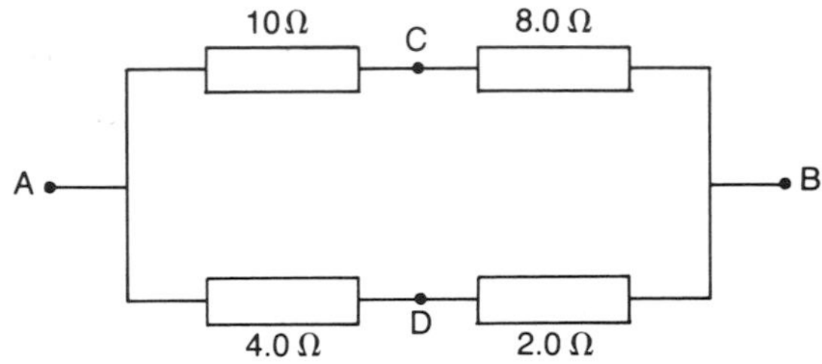

Fig 27.13 Circuit diagram for Question 6

7 The circuit diagram (Fig 27.14) shows a network of resistors each of resistance R.

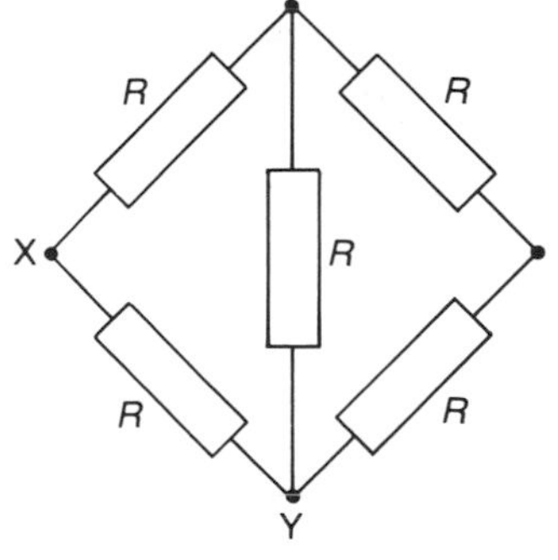

Fig 27.14 Circuit diagram for Question 7

What is the effective resistance between the points X and Y?

A $\frac{2}{7}R$ **B** $\frac{1}{2}R$ **C** $\frac{5}{8}R$
D $\frac{2}{3}R$ **E** $\frac{3}{4}R$ [O 80]

8 A battery is assembled from N cells, each of EMF E and internal resistance r, arranged in p rows each containing the same number of cells. The cells of each row are in series and the rows are in parallel with each other.

(a) Write down expressions for the EMF and internal resistance of the battery.

(b) Deduce an expression for the current flowing in a load resistance R connected to the battery.

(c) For a battery assembled from 12 cells, each of internal resistance 1.0 Ω, find an arrangement of cells to give the maximum current in a load resistor of 1.3 Ω.

(d) The EMF of each cell is 1.5 V. Calculate (i) the maximum current, and the power dissipated (ii) in the load resistor and (iii) in the internal resistance of the battery. [O & C 82, p]

9 The battery in the circuit shown in Fig 27.15 has an EMF E and an internal resistance r. If the potential difference across R is 2.0 V when R equals 4.0 Ω and 2.4 V when R equals 8.0 Ω, calculate the values of E and r.

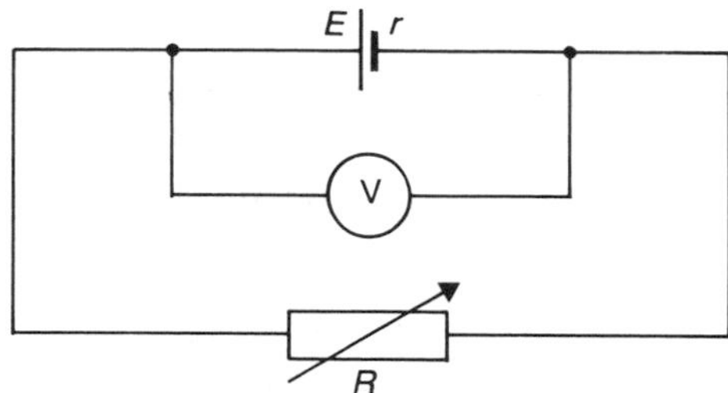

Fig 27.15 Circuit diagram for Question 9

10 A bulb is used in a torch which is powered by two identical cells in series each of EMF 1.5 V. The bulb then dissipates power at the rate of 625 mW and the PD across the bulb is 2.5 V. Calculate (a) the internal resistance of each cell and (b) the energy dissipated in each cell in 1 minute. [JMB 79, p]

11 State Kirchhoff's laws for circuit networks.

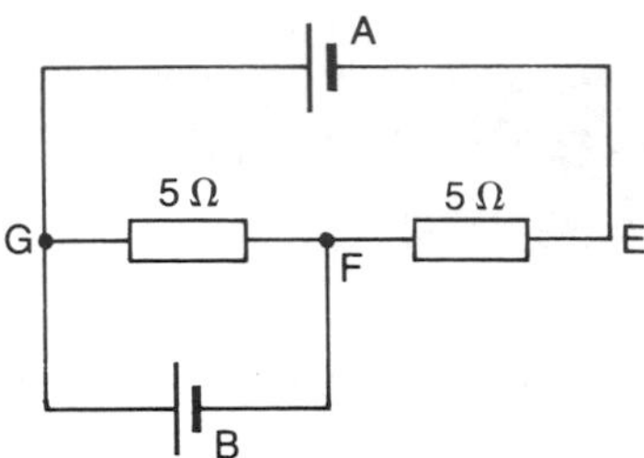

Fig 27.16 Circuit diagram for Question 11

In the circuit shown in Fig 27.16, cell A has an EMF of 10 V and an internal resistance of 2 Ω; cell B has an EMF of 3 V and an internal resistance of 3 Ω.

(a) Show that the currents through A and B are 65/71 A and 14/71 A respectively. What is the magnitude of the current through GF?

(b) Determine the power dissipated as heat in the resistor FE. If the circuit is switched on for 30 minutes, calculate the energy dissipated in FE in kilowatt-hours.

(c) What is the potential difference (i) across the terminals of cell A, (ii) across the terminals of cell B? [WJEC 78]

28 ELECTRICAL METERS

The moving-coil meter

The commonest type of meter is the 'moving-coil' design. Its action is explained in detail in Chapter 32. This kind of meter can be very sensitive and is made so that the pointer deflection is proportional to the current.

A galvanometer is an instrument that is suitable for detecting the presence of a current.

Conversion of a sensitive current-measuring meter to measure large currents

This range multiplication is common practice with sensitive moving-coil meters. A resistance of suitable value is fitted in parallel with the sensitive meter. This resistance is called a 'shunt'.

Only a fraction of the current to be measured passes through the sensitive meter. How the shunt achieves the required conversion is best explained by an example, as follows.

Example 1
Calculate the shunt resistance required to convert a 0–10 mA moving-coil meter whose resistance is 5.0 Ω into a 0–2.0 A meter.

Method
Fig 28.1 shows the position of the shunt and illustrates the situation when the current to be measured is at its highest value, namely 2.0 A (2000 mA). The meter must then give full scale deflection, i.e. *10 mA flows through it.*

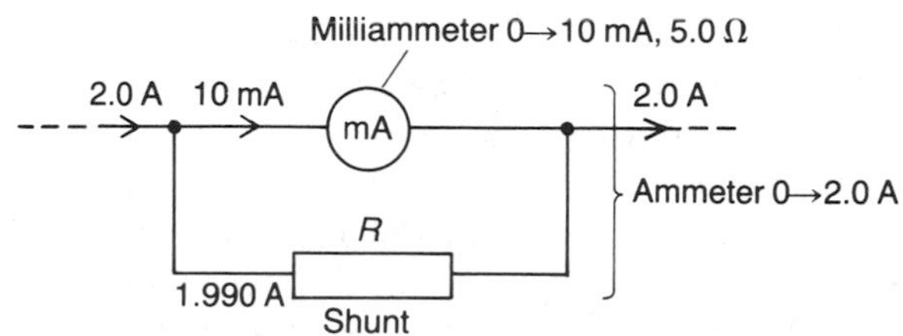

Fig 28.1 Circuit diagram for Example 1

The current through the shunt resistance R must be 2000 mA minus 10 mA, i.e. 1990 mA.

We know that the meter has resistance 5.0 Ω and has 10 mA through it. Therefore the PD across it is 5.0 × 10/1000 volt, i.e. 50 mV. Because this is also the PD across R and we know the current through R, we can deduce R from $R = V/I$.

$$R\left(=\frac{V}{I}\right) = \frac{50 \times 10^{-3}}{1990 \times 10^{-3}}$$

$$= \frac{50}{1990}$$

$$= 0.025\,125\ \Omega \quad \text{or} \quad 0.025\ \Omega$$

Answer
0.025 Ω or 25×10^{-3} Ω.

Meter resistance

The resistance of a current-measuring meter should be so small that the current to be measured is not changed when the meter is fitted into the circuit. A shunted milliammeter usually satisfies this requirement. In contrast a voltmeter should have as high a resistance as possible.

Voltmeters

The common type of voltmeter is the moving-coil design.

The moving-coil voltmeter works on the principle that a PD can be measured by allowing it to produce a current, which is measured. A larger PD gives a larger current. For example, the 0–10 mA meter mentioned in Example 1 could be used as a 0–50 mV meter but it would be a very poor voltmeter because its resistance is only 5 Ω. When it is connected to a circuit, perhaps to measure the PD between the ends of a certain resistor, the 5 Ω would be in parallel with the resistor and would completely change the current through, and therefore the PD across, the resistor before the measurement is made. A good voltmeter should have a resistance that is high compared with the circuitry under test. Satisfactory voltmeters can be obtained by having a high resistance fitted in series with a sensitive moving-coil meter. This is shown in Fig 28.2.

The series resistor is often called a *multiplier*.

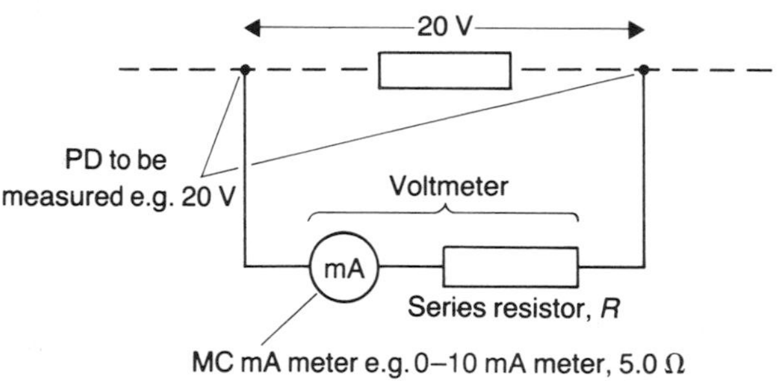

Fig 28.2 Using an MC meter to make a voltmeter

Example 2

Calculate the value required for the series resistor when a 0–10 mA, 5.0 Ω moving-coil meter is converted into a 0–20 V meter.

Method

The diagram required is shown in Fig 28.2.

We consider the situation where the meter reading is at its maximum, i.e. 20 V is being measured and the current is then 10 mA. Using the fact that $I = V/R$ we have

$$10 \times 10^{-3} = \frac{20}{R + 5}$$

$$\therefore \quad R + 5 = \frac{20}{10 \times 10^{-3}} = 2000\ \Omega$$

$$\therefore \quad R = 1995\ \Omega$$

Answer

2.0 kΩ.

Exercise 100

1 A microammeter gives full scale deflection (FSD) with 100 μA and it has a resistance of 100 Ω.
(a) What PD is needed across it for FSD?
(b) If a 1.00 Ω shunt were fitted across the meter, what current would flow through the shunt when the meter gives FSD?

2 How exactly could a 10 Ω, 10 mA FSD moving-coil meter be converted to read (a) 0 to 2.5 A, (b) 0 to 20 V? (c) If this meter were converted into a 0–200 mV meter by fitting a series resistor, what would be the resistance of the voltmeter?

3 A milliammeter has a resistance of 10 Ω and requires a PD of 10 mV between its terminals to give full scale deflection (FSD). How could it be adapted to produce (a) an ammeter with 1.0 A FSD, (b) a voltmeter with 20 V FSD?

4 The resistor fitted as a shunt across a moving-coil meter has a resistance which is 1/99 of the meter resistance. By what factor is the current range multiplied?

A 99 **B** $\frac{1}{99}$ **C** 100 **D** 0.01 **E** 1000

Voltmeter of insufficient resistance

Two examples in which the resistance of a voltmeter is not sufficiently high to be neglected are given below (Examples 3 and 4). Another problem of this kind may be found in Chapter 29, Exercise 105, Question 3.

Example 3

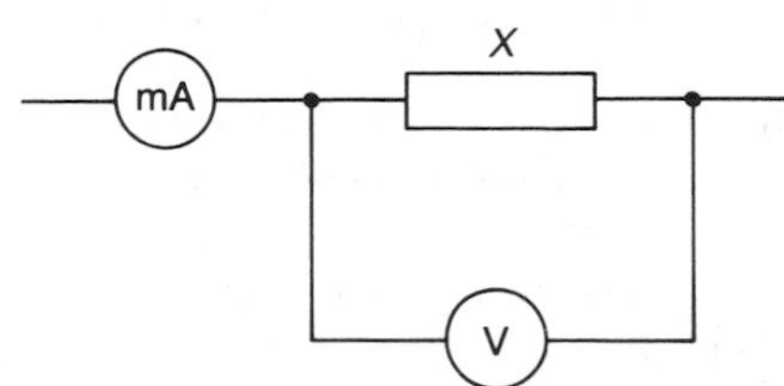

Fig 28.3 Diagram for Example 3

In Fig 28.3 the voltmeter reads 9.0 V when the milliammeter reads 1.0 mA. Calculate the value of the resistance X, (a) assuming the voltmeter resistance to be very high, (b) given that the resistance of the voltmeter is 10 000 Ω.

Method

(a) $R = \frac{V}{I} = \frac{9.0}{1.0 \times 10^{-3}} = 9000\,\Omega$ or $9.0\,\text{k}\Omega$

$$\therefore \quad X = 9.0\,\text{k}\Omega$$

(b) The value of V/I, namely 9.0 kΩ, is the resistance of X in parallel with the voltmeter.

$$\therefore \quad \frac{1}{9000} = \frac{1}{X} + \frac{1}{10\,000}$$

$$\therefore \quad \frac{1}{X} = \frac{1}{9000} - \frac{1}{10\,000} = \frac{10}{90\,000} - \frac{9}{90\,000}$$

$$\therefore \quad \frac{1}{X} = \frac{1}{90\,000}$$

$$\therefore \quad X = 90\,000\,\Omega \quad \text{or} \quad 90\,\text{k}\Omega$$

Answer

(a) 9.0 kΩ, (b) 90 kΩ.

Example 4

A DC voltage supply gives 50 V on open circuit and has an internal resistance (or 'output resistance') of 1.0 kΩ. What reading will a voltmeter give when connected to the terminals of this supply if the resistance of the voltmeter is only 4.0 kΩ?

Method

The current I which flows through the voltmeter is given by Equation 27.18 (p.180):

$$I = \frac{E}{R + r}$$

where $E = 50\,\text{V}$, $r = 1000\,\Omega$ and $R = 4000\,\Omega$.

$$\therefore \quad I = \frac{50}{4000 + 1000} = \frac{1}{100}\,\text{A}$$

The PD V between the supply's terminals is given by Equation 27.16:

$$E - V = Ir$$

$$\therefore \quad 50 - V = \frac{1}{100} \times 1000$$

$$\therefore \quad V = 40\,\text{V}$$

Answer

40 V.

Effect of temperature on a moving-coil meter

The coil of a moving-coil meter is usually made of copper wire; the temperature coefficient of copper is 0.0043 K^{-1}. If the meter's temperature increases, it would cause an increase in the coil resistance and this might produce a significant increase in the proportion of the measured current flowing through the meter's shunt. This is illustrated in Example 5.

Example 5

A moving-coil meter gives full scale deflection (FSD) with 1.5 mA and its resistance when calibrated at 10 °C was 20.00 Ω.

(a) Calculate the resistance of the shunt needed to convert it to a 1.0 A FSD meter.

(b) If the temperature coefficient for the meter coil is 0.0043 K^{-1} while the shunt has effectively zero temperature coefficient, what percentage error will occur when the ammeter is used at 20 °C?

Method

(a) At full scale deflection the meter current is 1.5 mA, so that the PD across the coil (and across the shunt, see Fig 28.1) is given by

$$V = IR = 1.5 \times 10^{-3} \times 20.00 = 0.030\,\text{V}$$

A shunt resistance R_S is needed so that with 0.030 V across it a current of 1.000 A minus 1.5 mA flows through it.

$$\text{Current} = 1.0000 - 0.0015 = \frac{V}{R_S} = \frac{0.030}{R_S}$$

$$\therefore \quad R_S = \frac{0.030}{0.9985} = 0.030\,04\,\Omega$$

or 0.030 Ω to two significant figures.

(b) The new coil resistance at 20 °C (call it R') is given by (see Example 5 in Chapter 27)

$$\frac{R}{R'} = \frac{R_0(1 + 10\alpha)}{R_0(1 + 20\alpha)}$$

where R is the original resistance at 10 °C and α is the temperature coefficient.

$$\frac{R'}{20.00} = \frac{(1 + 0.0043 \times 20)}{(1 + 0.0043 \times 10)}$$

$$\therefore \quad R' = \frac{1.086}{1.043} \times 20.00$$

$$= 20.825\ \Omega$$

The coil current for FSD is still 1.5 mA but this will be obtained with a PD of

$$V' = IR' = 1.5\ \text{mA} \times R' = 1.5\ \text{mA} \times 20.825$$

The new shunt current will be V'/R_S which is

$$I' = \frac{1.5 \times 10^{-3} \times 20.825}{0.030\,04} = 1.0399\ \text{A}$$

So the FSD current for the warmer meter is 1.0399 plus the coil current of 0.0015 A which is 1.0414 A. The meter has become less sensitive. The error is 0.0414 A when the correct value of the measured current is 1.0414 A.

The percentage error is therefore

$$\frac{0.0414}{1.0414} \times 100 = 4.0\%$$

Answer

(a) 0.030 Ω, (b) 4%.

Alternating current meters

Some meters, the hot-wire ammeter for example, will work with either DC or AC.

Moving-coil meters can be fitted with a rectifier. See Chapter 36.

Exercise 101

1 In a circuit a 1.0 kΩ resistor is connected in series with a milliammeter of resistance 10 Ω. A 4000 Ω voltmeter is connected first across the 1.0 kΩ resistor and then across the milliammeter and 1.0 kΩ resistor together. If the reading of the milliammeter is 5.0 mA on both occasions, what are the two readings of the voltmeter?

2 A 10 Ω, 10 mA FSD (100 mV) moving-coil meter is fitted with a shunt which has a resistance of 0.020 Ω at 15 °C. Calculate the current that gives full scale deflection (a) at 15 °C, (b) at 30 °C, if the coil resistance remains unchanged but the temperature coefficient of the shunt material is 0.040 K^{-1}.

3 Two copper–constantan thermocouples give the same reading of 2.0 mV on a high-resistance voltmeter. What readings would they give on a 100-division, 100 Ω, 0–500 μA moving-coil meter if one of the thermocouples has a resistance of 1.1 Ω and the other 2.1 Ω?

Exercise 102: Questions of A-level standard

1 A moving-coil meter has a resistance of 5.0 Ω, and full scale deflection is produced by a current of 1.0 mA. How can this meter be adapted for use as (a) a voltmeter reading up to 10 V, (b) an ammeter reading up to 2 A? [SUJB 80]

2 A milliammeter of resistance 8.4 Ω is shunted so that it will read a full scale current seven times as great as before. The value of the shunt is

A 0.12 Ω **B** 1.2 Ω **C** 1.4 Ω
D 50.4Ω **E** 58.8 Ω [L 81]

3 A moving-coil ammeter at 0 °C has a resistance of 5 Ω and gives full scale deflection for a current of 15 mA. What value of shunt resistance is required so that it reads 1.0 A at 0 °C? What circuit current will give a full scale deflection of the instrument if the temperature of the shunt rises to 25 °C while the instrument remains at 0 °C?
(Temperature coefficient of resistance of material of shunt = $4.0 \times 10^{-3}\ K^{-1}$.) [AEB 81]

4 (a) A 12 V battery with internal resistance of 2 Ω is connected across a 10 Ω load resistance. Calculate the potential difference across the load resistance.
(b) A voltmeter of full scale deflection 10 V and resistance 10 Ω is now connected across the load resistance. What potential difference will it indicate? [C 82, p]

29 POTENTIOMETERS AND BRIDGES

The voltage divider

(a)

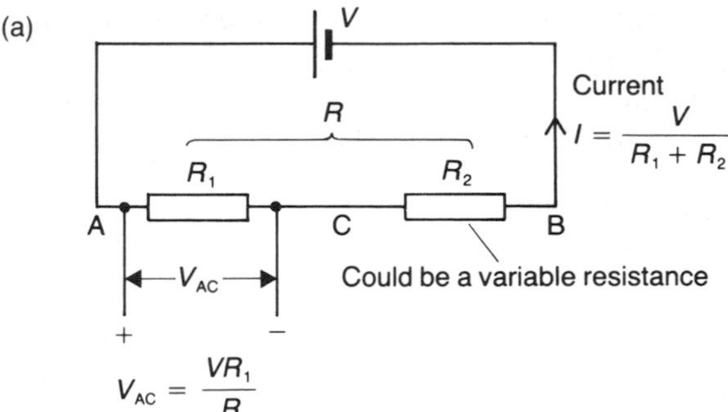

(b)

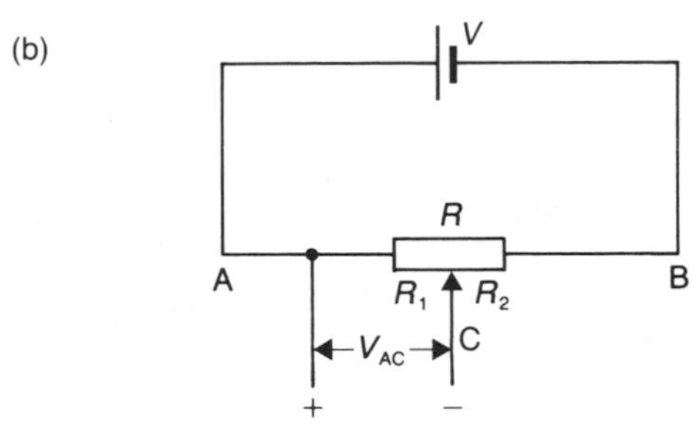

(c)

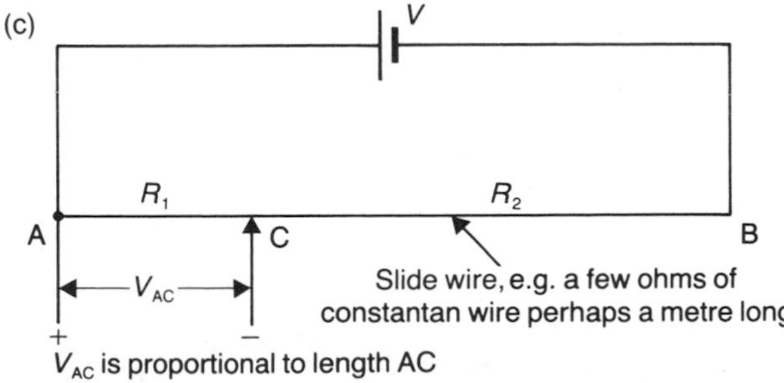

Fig 29.1 Voltage dividers

Fig 29.1 shows circuits each of which provides a PD V_{AC} which is a fraction $R_1/(R_1 + R_2)$ of the supply voltage V. These circuits are often called 'potential dividers' although 'potential difference dividers' would be better. 'Voltage dividers' avoids the difficulty.

Effect of load on PD obtained

The load (see Fig 29.2) is in parallel with part of the voltage divider.

Example 1

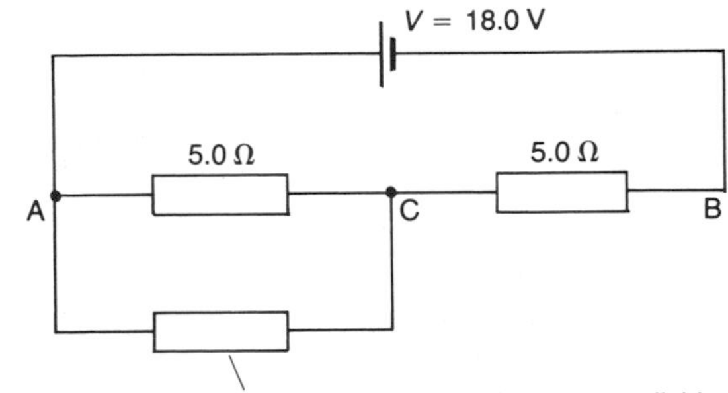

Fig 29.2 Circuit for Example 1

Calculate the PD between A and C in the circuit of Fig 29.2.

Method

We have 20 Ω in parallel with the 5.0 Ω. Using $R = R_1R_2/(R_1 + R_2)$ (Equation 27.7) we get

$$R_{AC} = \frac{20 \times 5.0}{20 + 5.0} = 4.0\,\Omega$$

$$V_{AC} = \frac{V \times R_1}{R_1 + R_2}$$

(see Fig 29.1) and in this equation $R_1 = R_{AC}$, $V = 18$ V and $R_2 = 5.0\,\Omega$.

$$V_{AC} = \frac{18 \times 4.0}{4.0 + 5.0} = 8.0\text{ V}$$

(This compares with 9.0 V when the load is a very high resistance or open circuit, i.e. when $R_{AC} = 5.0\,\Omega$.)

Answer
8.0 V.

The potentiometer

Potentiometer circuits are for measuring potential *differences*.

The PD to be measured is put into a circuit together with an opposing variable PD obtained from a voltage divider. The voltage divider is adjusted until its PD V_{AC} equals the PD being measured. A simple potentiometer is shown in Fig 29.3.

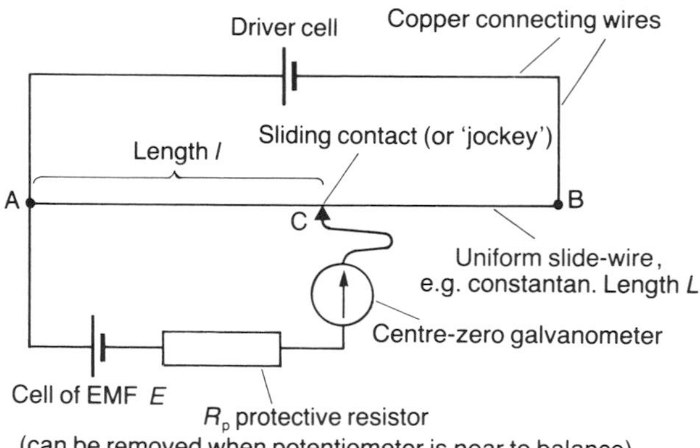

Fig 29.3 A simple potentiometer

The sliding contact in Fig 29.3 is moved until the galvanometer indicates zero, i.e. the potentiometer is 'balanced'. The current in the lower part of the circuit is zero because PD V_{AC} equals the PD E provided by the cell under test, and this is true regardless of the value of the resistor R_p (e.g. 1 or 2 kΩ usually), which serves only to protect the galvanometer from excessive current. E can be calculated *approximately* from

$$E = \frac{V_{\text{driver}} \times l}{L} \qquad (29.1)$$

This answer is approximate because V_{driver} is not accurately known, either because the driver cell EMF is not precisely known or because the terminal PD of the cell is less than its EMF. For an accurate evaluation of E, the cell under test that has required length l for balance is now removed and is replaced by a cell of accurately known EMF (E_S), i.e. a standard cell. A new balance is obtained with this cell and the balance length AC required is measured (l_s). Then

$$\frac{E}{E_S} = \frac{l \times V_{\text{driver}}/L}{l_s \times V_{\text{driver}}/L} = \frac{l}{l_S} \qquad (29.2)$$

Note that, because no current is flowing through the cell being tested when balance is obtained, the PD obtained from the cell under test is the EMF value as indicated above.

Example 2

A simple potentiometer consisting of a 2.0 V driver cell, slide wire and galvanometer with a 2.0 kΩ series resistor is used to compare the EMFs of a Leclanché cell and a Daniell cell. The balance lengths recorded are 75 cm and 54 cm.

(a) If the Daniell cell's EMF is known to be 1.08 V, what is the EMF of the Leclanché cell?

(b) When the Daniell cell is being used, approximately what current flows through the galvanometer if the protective resistor is in use and the sliding contact is moved to (i) one end, (ii) the other end of the slide wire? (The resistances of the galvanometer and of the cells may be neglected.)

Method

(a) $$\frac{E}{E_S} = \frac{l}{l_S}$$

$$\therefore \quad \frac{E}{1.08} = \frac{75}{54}$$

$$\therefore \quad E = 1.08 \times \frac{75}{54} = 1.5\text{ V}$$

(b) (i) A complete circuit is formed which contains the Daniell cell, galvanometer and 2 kΩ resistor only. (See Fig 29.4 with sliding contact moved to A.) The current is

$$I\left(= \frac{V}{R}\right) = \frac{1.08}{2000}\text{ A} = 0.54\text{ mA}$$

(ii) The situation is depicted in Fig 29.4.

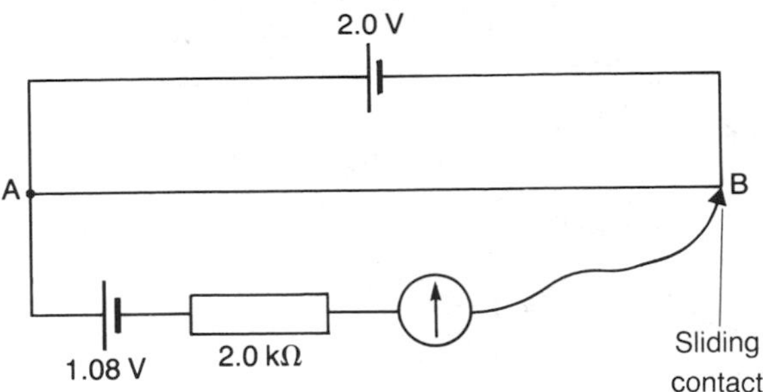

Fig 29.4 Circuit diagram for Example 2

The PD between A and B is 2.0 V, with A at higher potential. The current from A to B *via the galvanometer* is due to this 2.0 V minus the 1.08 V of the Daniell cell. Thus the current is

$$\frac{2.0 - 1.08}{2000} = 0.46 \times 10^{-3}\,\text{A} = 0.46\,\text{mA}.$$

Answer
(a) 1.5 V, (b) (i) 0.54 mA, (ii) 0.46 mA in opposite direction through galvanometer.

Comparison of two resistors by use of a potentiometer

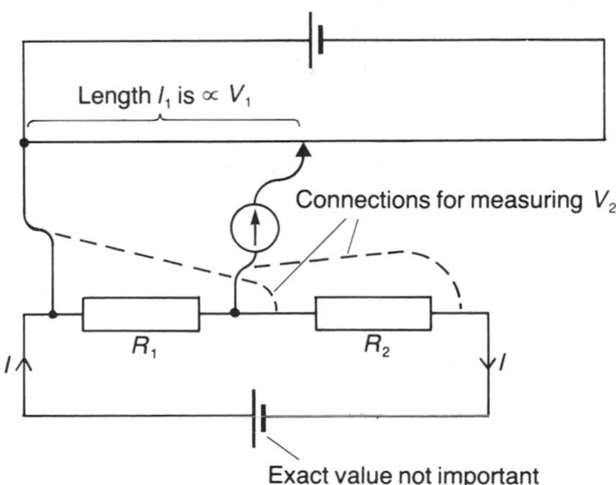

Fig 29.5 Comparing two resistors

As shown in Fig 29.5 two resistors R_1 and R_2 can be compared by passing the same current I through both and comparing the PDs, V_1 and V_2, across them by use of the potentiometer.

The balance lengths recorded are l_1 and l_2 corresponding to PDs V_1 and V_2.

$$\frac{l_1}{l_2} = \frac{V_1}{V_2} = \frac{IR_1}{IR_2} = \frac{R_1}{R_2}$$

Fitting a resistor in series with the driver cell

This has the effect of reducing the PD across the slide wire, so that the length of slide wire needed to balance a given PD is increased. It is essentially a resistance which drops the voltage across the slide wire.

Example 3
What value of resistance is needed in series with a driver cell of negligible internal resistance and approximately 2 V EMF to arrange that $\frac{3}{4}$ of the 4 Ω slide wire is required to balance a PD of 1 V?

Method
The potentiometer is shown in Fig 29.6.

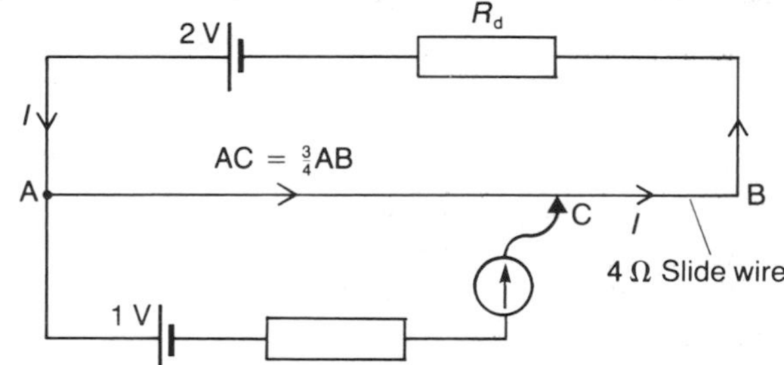

Fig 29.6 Potentiometer with a series resistance

At balance no current flows via the galvanometer part of the circuit, so that the same current I flows in all parts of the upper loop.

Since V_{AC} must equal 1 V and equals $\frac{3}{4}V_{AB}$,

$V_{AB} = 4/3$ V.

$$I = \frac{V_{AB}}{R_{AB}} = \frac{4/3}{4} = \frac{1}{3}\,\text{A}$$

But $\quad I = \dfrac{2}{R_d + 4} \qquad \therefore \quad \dfrac{1}{3} = \dfrac{2}{R_d + 4}$

$\therefore \quad R_d + 4 = 6 \qquad \therefore \quad R_d = 2\,\Omega$

Answer
2 Ω.

Measurement of high voltage using a potentiometer

If the high voltage V to be measured is connected to a voltage divider (see Fig 29.1a) a voltage $VR_1/(R_1 + R_2)$ is obtained and this can be made small enough to be easily measured by a simple potentiometer method.

Measurement of a small EMF, e.g. from a thermocouple

A simple potentiometer such as that shown in Fig 29.3 would give a negligible balance length (AC) when the PD to be measured is only the few millivolts or less that might be produced by a thermocouple. The answer to this problem is the use of a resistance in series with the driver cell. This should be clear from the following example.

Example 4

A potential difference of approximately 4 mV is required across a uniform potentiometer wire whose resistance is 3.00 Ω. A driver cell is to be used which has an EMF of approximately 2 V and negligible internal resistance. Calculate the value of a suitable resistance to be placed in series with the cell and wire.

Method

A suitable diagram is shown in Fig 29.7

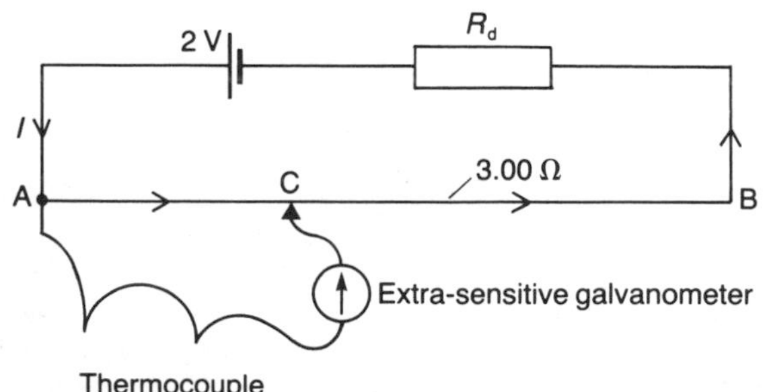

Fig 29.7 Use of potentiometer with a thermocouple

$$V_{AB} = 4 \times 10^{-3}\,\text{V}, \qquad R_{AB} = 3.00\,\Omega$$

$$I = \frac{V_{AB}}{R_{AB}} = \frac{4 \times 10^{-3}}{3.00} = 1.33 \times 10^{-3}\,\text{A}$$

But $$I = \frac{2}{R_d + 3.00}$$

$$\therefore \quad 1.33 \times 10^{-3} = \frac{2}{R_d + 3.00}$$

$$\therefore \quad R_d = 1497\,\Omega$$

Answer

1.5 kΩ.

Calibrating the slide wire for small EMF measurement

After the series resistance has been chosen to suit the maximum thermocouple voltage expected, it is fitted in place and the sliding contact is moved to obtain balance with the thermocouple EMF. The length AC (Fig 29.7) is recorded.

The value of the thermocouple EMF is equal to (AC/AB) × PD across the slide wire, but the PD across the slide wire is usually not accurately known (because the driver cell EMF and internal resistance are usually not accurately known). This problem is overcome by calibrating the slide wire accurately with the help of a standard cell such as a Weston cell. The circuit for this is seen in Fig 29.8.

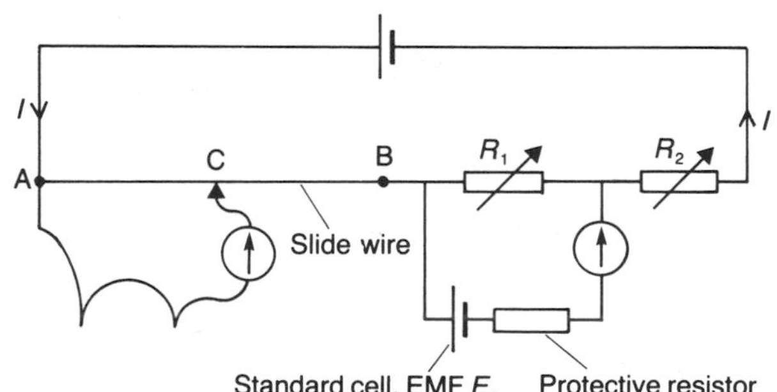

Fig 29.8 Calibration of potentiometer for small EMF measurement

Calibration would be simple if a millivolt source of constant and accurately known value could be used. Assuming, however, that the standard available is a cell of about 1 V EMF, we realise that no length of the slide wire will balance this. Now the series resistor, R_d in Fig 29.7, must not be altered in value, but it can be replaced by two variable, calibrated resistance boxes whose total value is made, and kept, equal to R_d. By suitable adjustment of R_1 (and of R_2 to keep $R_1 + R_2 = R_d$) a balance is obtained where $E_S = IR_1$. This allows an accurate value for I to be deduced. Knowing I and R_{AB}, the PD V_{AB} across the slide wire becomes known. Example 5 is concerned with this method. If balance against E_S is carried out before balancing the thermocouple, the method can be modified as follows.

R_1 is *calculated* so that exactly the required current I will be flowing when E_S is balanced and the balancing is achieved by small adjustments to R_2 only. When subsequently the thermocouple is being balanced, the value used for $R_1 + R_2$ must of course not be changed. Example 6 illustrates this technique.

A small variation of the methods so far described is the balancing of E_S against R_1 plus some length of the slide wire.

Example 5

A slide wire potentiometer has an accumulator cell of approximately 2 V EMF and negligible internal resistance connected in series with a 1.00 m slide wire of 2.00 Ω resistance and two resistances R_1 and R_2 which may be varied in 1.0 Ω steps.

(a) Calculate the approximate value for $R_1 + R_2$ if a PD of about 4 mV is required across the slide wire.

(b) A standard cell of 1.02 V EMF and a suitable galvanometer are fixed in parallel with R_1, and balance is obtained with R_1 set at exactly 500 Ω, $R_1 + R_2$ still being kept at the value chosen in (a). What is the exact value of the slide wire current?

(c) With the potentiometer set up as described above, a certain thermocouple was balanced by 20.0 cm of slide wire. What EMF was the thermocouple producing?

Method

(A diagram like Fig 29.8 could usefully be drawn and labelled with the given values.)

(a) 4 mV across the 2.00 Ω wire requires a current I given by

$$I = \frac{4 \times 10^{-3}}{2} = 2 \times 10^{-3}\,\text{A} = 2\,\text{mA}$$

But the required $R_1 + R_2$ is given by

$$I = \frac{2}{R_1 + R_2 + 2}$$

$$\therefore \quad 2 \times 10^{-3} = \frac{2}{R_1 + R_2 + 2}$$

$$\therefore \quad R_1 + R_2 = 1000 - 2 = 998\,\Omega$$

(b) E_S equals the PD across R_1 which equals $I \times R_1$:

$$1.02 = 500I$$

$$\therefore \quad I = \frac{1.02}{500} = 2.04 \times 10^{-3}\,\text{A}$$

(This is the exact value of I, not the 2 mA predicted.)

(c) The exact PD across the slide wire is $I \times 2.00\,\Omega$ or 4.08 mV. The thermocouple balances one-fifth of this (20.0 cm), so that its EMF is (4.08/5) mV or 0.816 mV.

Answer

(a) 998 Ω, (b) 2.04 mA, (c) 0.816 mV.

Example 6

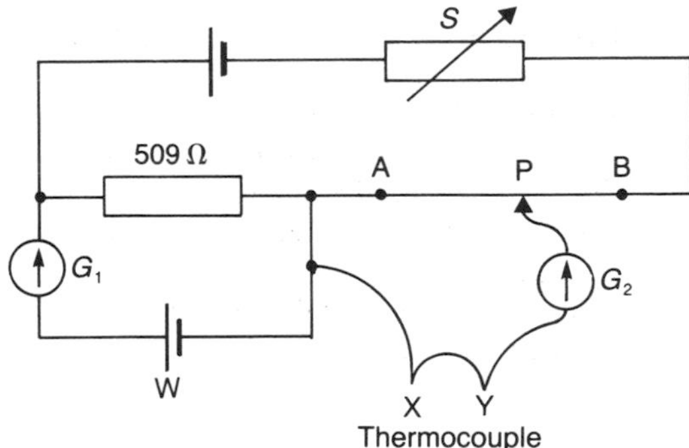

Fig 29.9 Circuit diagram for Example 6

The circuit shown is used to measure the EMF of a thermocouple whose junctions X and Y are maintained at temperatures 0 °C and 150 °C respectively. W is a standard cell having an EMF of 1.018 V. AB is a potentiometer wire having a resistance of 1.50 Ω m^{-1}. Explain how this circuit is used, and calculate the EMF of the thermocouple if the length AP of the potentiometer wire is 40 cm when the galvanometer G_2 shows no deflection. [SUJB 79, p]

Method

Calculation of the thermocouple EMF:

The current I through the 509 Ω gives a PD equal to 1.018 V

$$\therefore \quad 509I = 1.018$$

$$\therefore \quad I = \frac{1.018}{509} = 2.00 \times 10^{-3}\,\text{A}$$

The PD per metre of slide wire (i.e. per 1.50 Ω) is $I \times 1.50$, i.e. 3.00×10^{-3} V.

The PD across 40 cm, i.e. 0.40 m, is $3.00 \times 10^{-3} \times 0.40$ V. which equals 1.20×10^{-3} V or 1.20 mV.

Answer

1.2 mV.

Exercise 103

1 A 12 V battery of negligible internal resistance is connected to a 5.0 Ω and a 10 Ω resistor in series. What is the PD across the 5.0 Ω resistor (a) when measured by a high-resistance voltmeter, (b) when apparatus with a resistance of 20 Ω is connected in parallel with the 5.0 Ω resistor?

2 Using a simple slide wire potentiometer, a standard cell of 1.02 V EMF was balanced by 55.0 cm of the 100.0 cm slide wire and another cell was balanced by 60.0 cm.

(a) What was the EMF of this other cell?

(b) What current was flowing through the slide wire, given that the slide wire resistance was 4.00 Ω and the driver cell internal resistance is negligible.

Calculate the value of the series resistor that will double the balance lengths.

3

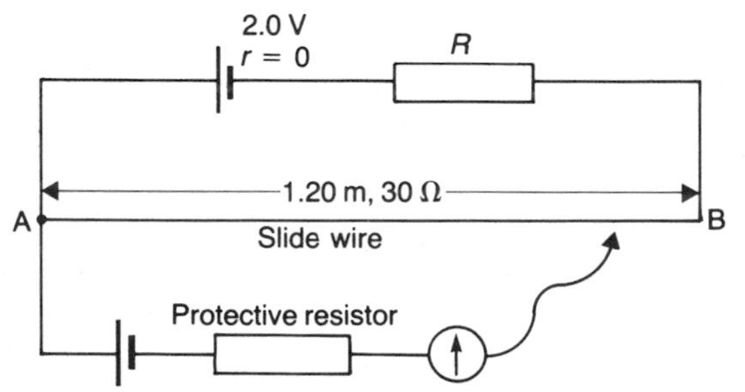

Fig 29.10 Circuit diagram for Question 3

The potentiometer in Fig 29.10 is to have, at balance, a potential drop of 10 mV along each 1.0 cm of slide wire. Explain carefully how this can be achieved, and find the value of R.

Find the power provided by the driver cell, and the power dissipation in the potentiometer wire.

4

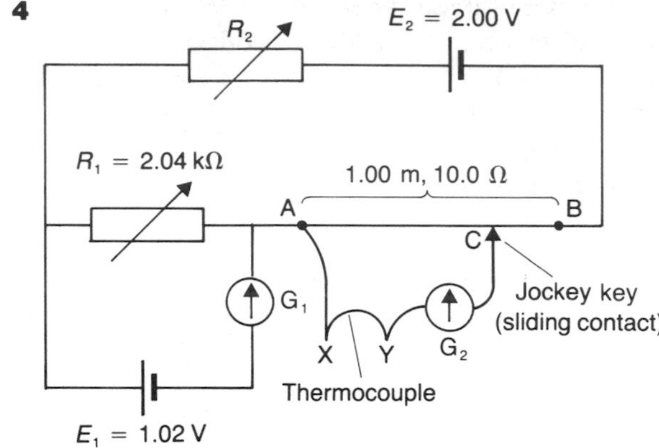

Fig 29.11 Circuit diagram for Question 4

In Fig 29.11 G_1 and G_2 are sensitive galvanometers. X and Y are the junctions of a thermocouple. With C not touching the wire, R_2 is adjusted until no current flows through G_1.

(a) What is the current through R_1?

The jockey key C is now touched on to the potentiometer wire and moved along it until it is 0.600 m from A, when G_2 registers no current.

(b) What is the EMF of the thermocouple?

The Wheatstone bridge circuit

Fig 29.12a shows the basic Wheatstone bridge circuit and 29.12b shows how two resistors may be replaced by a slide wire to produce the so-called metre bridge (traditionally using a 1-metre-long slide wire).

In the circuit of Fig 29.12a it is usual to keep R_1 and R_2 fixed in value (e.g. equal), while R_3 is varied until the galvanometer reads zero (i.e. the bridge is balanced). Then

$$\frac{R_1}{R_2} = \frac{R_3}{R_4} \qquad (29.3)$$

(a)

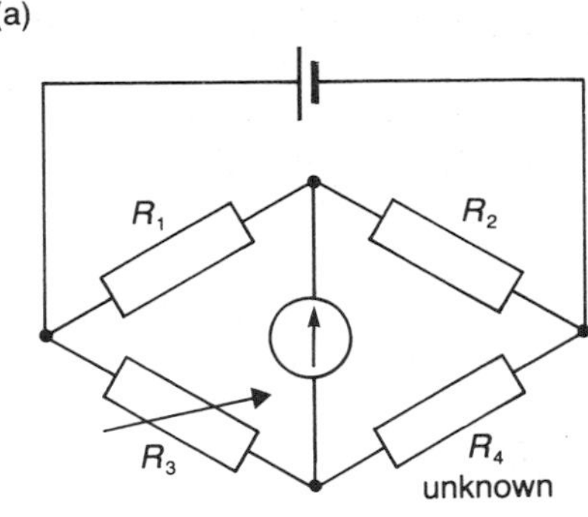

(b)

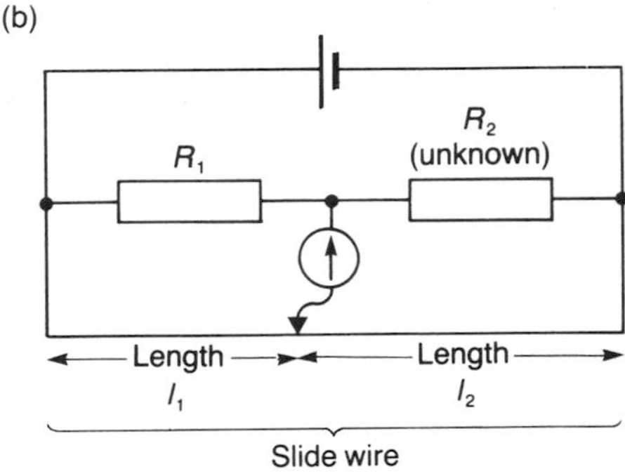

Fig 29.12 (a) Wheatstone bridge, (b) Metre bridge

When the *metre bridge* is balanced

$$\frac{R_1}{R_2} = \frac{R_3}{R_4} = \frac{l_1}{l_2}$$

(provided that the slide wire is uniform).

The purpose of these bridges is to evaluate the unknown resistance when the other resistances are known.

In the metre bridge, R_1 should be of similar value to the unknown resistance. In the basic Wheatstone bridge it is convenient to have $R_1 = R_2$ initially, but making $R_1 = 10R_2$ and $R_1 = 100R_2$ can give more accurate answers for R_4. (This is illustrated below.)

Example 7

A Wheatstone bridge comprises a driver cell, galvanometer, three resistance boxes adjustable in 1 Ω steps and a resistor whose resistance is to be determined (believed to be a few ohms). See Fig 29.12a.

(a) If $R_1 = R_2 = 10\,\Omega$, and the minimum deflection of the galvanometer is obtained with $R_3 = 4\,\Omega$, calculate a value for R_4.

(b) When $R_1 = 100\,\Omega$, $R_2 = 10\,\Omega$, minimum deflection required $R_3 = 39\,\Omega$. Calculate R_4.

(c) With $R_1 = 1000\,\Omega$, $R_2 = 10\,\Omega$, it was just possible to see that R_3 needed to be 392 Ω rather than 393 Ω. Calculate R_4.

Method

Using $R_1/R_2 = R_3/R_4$, we have

(a) $\frac{10}{10} = \frac{4}{R_4}$ $\quad \therefore R_4 = 4\,\Omega$

(b) $\frac{100}{10} = \frac{39}{R_4}$ $\quad \therefore R_4 = \frac{39}{10}$ $\quad \therefore R_4 = 3.9\,\Omega$

(c) $\frac{1000}{10} = \frac{392}{R_4}$ $\quad \therefore R_4 = \frac{392}{100}$ $\quad \therefore R_4 = 3.92\,\Omega$

Answer

(a) 4 Ω, (b) 3.9 Ω, (c) 3.92 Ω (3.9 Ω to the two significant figures justified.)

Wheatstone bridge when out of balance

To determine the galvanometer current when the balance condition has not been obtained can be difficult. However, if the galvanometer resistance is much greater than the values of R_1, R_2, R_3 and R_4 (the resistances in the 'arms' of the bridge) then the calculations are easy.

Example 8

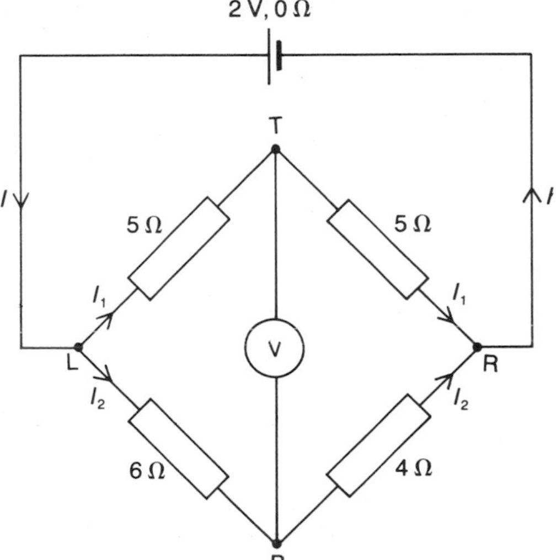

Fig 29.13 Circuit diagram for Example 8

In Fig 29.13, what is the out-of-balance PD recorded by the voltmeter assumed to have high resistance?

Method

The current I_1 flows between L and R through 10 Ω and the PD between L and R is 2 V.

$$I_1 = \frac{2}{10} = 0.2\text{ A}$$

The PD between L and T is $0.2 \times 5 = 1.0$ V (and similarly the PD between T and R is 1 V, as expected). For current I_2 flowing from L to R through 6 plus 4 ohms

$$I_2 = \frac{2}{10} = 0.2\text{ A}$$

The PD between L and B is $I_2 \times 6 = 1.2$ V (and the PD between B and R is 0.8 V).

B is 1.2 V different from L, while T is 1.0 V different from L.

PD between T and B is 0.2 V with T being + with respect to B.

Answer

0.2 V.

Exercise 104

1 A Wheatstone bridge has a resistance of 5.0 Ω in each of two adjacent arms, a resistance adjusted to 10 Ω in a third arm and an unknown resistance in the fourth arm. If the bridge is balanced with these values (a) what is the value of the unknown resistance and (b) what power is dissipated in each of the resistances if the supply voltage is 2.0 V and connects to the ends of the 5.0 Ω resistors?

2 A Wheatstone bridge has resistances of 5.0 Ω, 5.0 Ω, 2.0 Ω and 4.0 Ω in its four arms AB, BC, CD and DA respectively. The driver cell is producing 2.0 V. What current, approximately, flows through the galvanometer, whose resistance is 100 Ω when it is connected between B and D? The galvanometer resistance may be assumed high compared with the other resistances.

3 Fig 29.14 represents a simple metre-wire form of Wheatstone bridge. X is an unknown resistance, S a standard resistance and AB a metre-length of uniform-resistance wire along which the contact C can be moved.

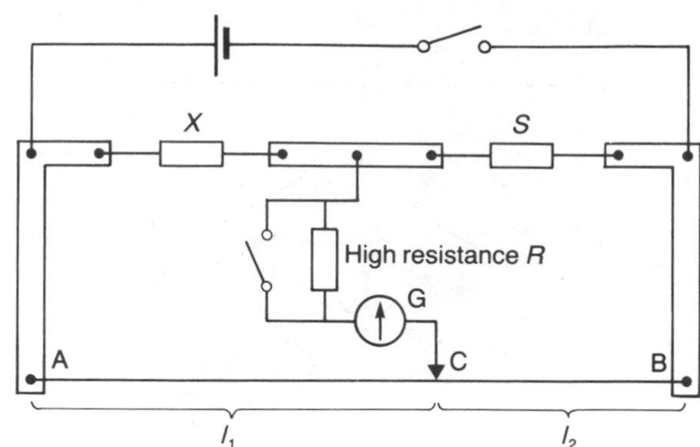

Fig 29.14 Circuit diagram for Question 3

(a) If $S = 5.00$ Ω and the bridge is found to balance with $l_1 = 60$ cm and $l_2 = 40$ cm, what is the value of X?
(b) If the PD applied across AB is 2.0 V, what is the PD across S at balance?
(c) Estimate the current passing through the galvanometer G when the sliding contact C is moved to end B of the wire, if the resistance of R plus galvanometer resistance is 2000 Ω.

4

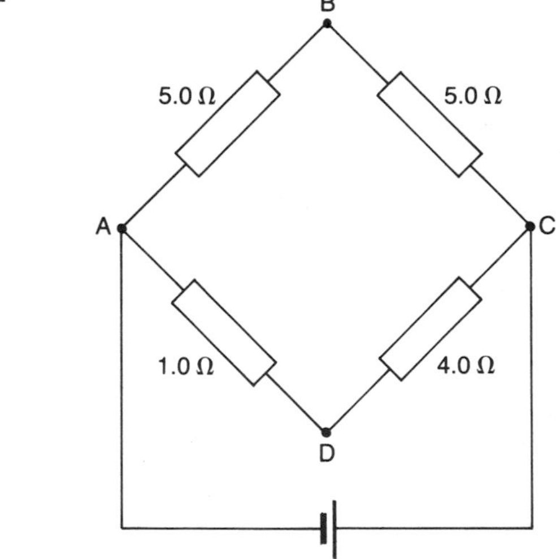

Fig 29.15 Circuit diagram for Question 4

In the circuit shown (Fig 29.15) A is at a potential of −2.0 V with respect to C. The potential of B with respect to D is

A +0.50 V **B** zero **C** +1.4 V
D +0.60 V **E** +2.6 V

Exercise 105: Questions of A-level standard

1

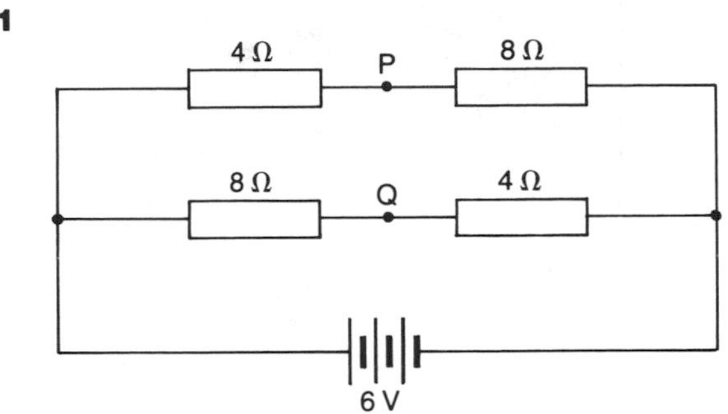

Fig 29.16 Circuit diagram for Question 1

If the 6-volt battery has negligible internal resistance, what is the potential difference between points P and Q in the circuit in Fig 29.16?

A 0 V **B** 1 V **C** 2 V
D 3 V **E** 4 V [L 78]

2

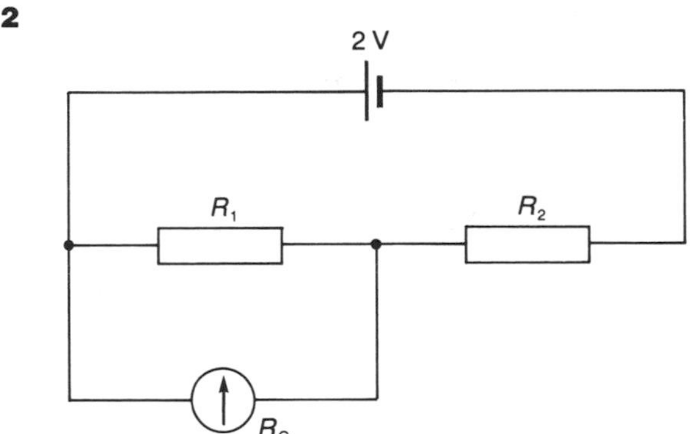

Fig 29.17 Circuit diagram for Question 2

In the circuit in Fig 29.17 the cell has zero resistance.

(a) What current flows through the galvanometer when $R_1 = 5\,\Omega$, $R_2 = 95\,\Omega$ and $R_G = 95\,\Omega$?

(b) A cell is introduced in series with the galvanometer such that the reading on the galvanometer becomes zero. What EMF has the cell, and which way round must it be connected?

(c) If the galvanometer is replaced by another having double the original resistance (i.e. new $R_G = 190\,\Omega$), what EMF cell is now needed to produce zero galvanometer deflection?

(d) Show on a diagram how a 1 m length of wire of total resistance 2 Ω could be incorporated in this circuit to produce a direct-reading potentiometer. What values must R_1 and R_2 have so that the total potential drop along the wire is 5 mV and the device is standardised by connecting a 1.018 V standard cell across R_2 and the wire? [WJEC 79]

3 A certain thermocouple thermometer is calibrated by placing its hot and cold junctions in steam and melting ice respectively and measuring an EMF of 5.6 mV with a potentiometer. Subsequently the thermocouple, of resistance 10 Ω, is used in series with a millivoltmeter of resistance 100 Ω. If the millivoltmeter reads 2.8 mV when the cold junction is in melting ice and the hot junction is in a liquid bath, what is the temperature of the bath on the centigrade scale of this thermometer? [C 77]

4

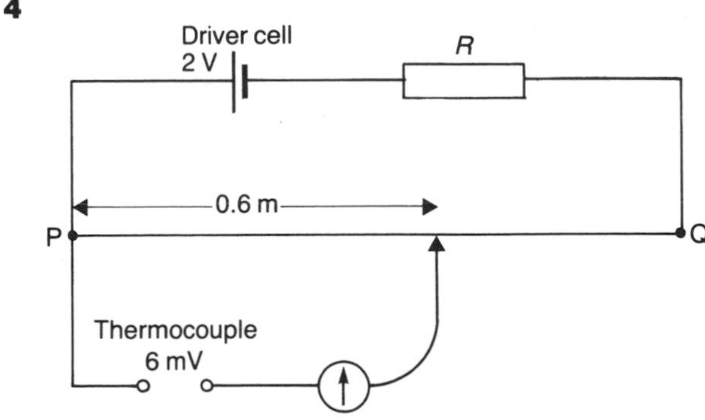

Fig 29.18 Circuit diagram for Question 4

Fig 29.18 shows a simple potentiometer circuit for measuring the EMF of a thermocouple. The metre wire PQ has a resistance of 5 Ω and the driver cell has an EMF of 2 V. If a balance point is obtained 0.6 m along PQ when measuring a thermocouple EMF of 6 mV, what is the value of the resistance R?

A 95 Ω **B** 195 Ω **C** 495 Ω
D 995 Ω **E** 1995 Ω [O 81]

5

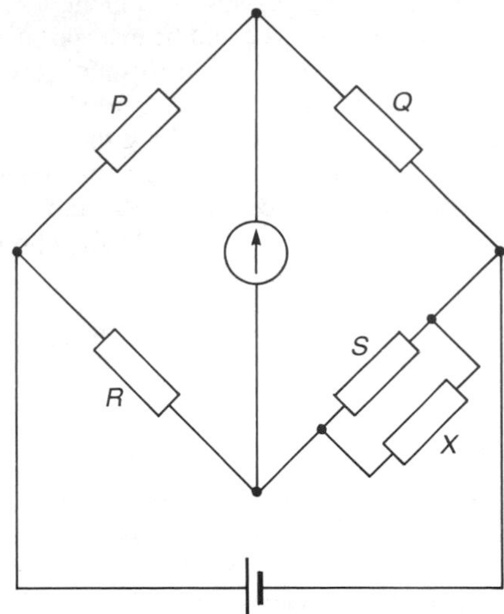

Fig 29.19 Circuit diagram for Question 5

The Wheatstone bridge circuit shown in Fig 29.19 is balanced when $P = 10\,\Omega$, $Q = 50\,\Omega$, $R = 30\,\Omega$ and $X = 1050\,\Omega$. The resistor S consists of a 3.5 m length of nichrome wire of cross-section area $1.5 \times 10^{-8}\,\text{m}^2$. Calculate (a) the resistance of S, (b) the resistivity of nichrome. [SUJB 81]

6 In the circuit in Fig 29.20 find (a) the current through each resistor, (b) the PD between A and B, (c) the power dissipated in the circuit.

If a milliammeter of resistance $75\,\Omega$ is connected between A and B, find an approximate value of the current that flows and state the direction of this current. Discuss whether the actual current is greater or less than the approximate value you have calculated. (Do not attempt to find the precise current.)

To what value must R_2 be altered for the bridge to balance?

The bridge is now in balance, but when a longitudinal stress is applied to R_2 the bridge goes out of balance. Explain this.

A $3\,\Omega$ resistor is connected in parallel with R_1 in the original circuit shown. To what value should R_4 be adjusted for balance to be obtained? [WJEC 82]

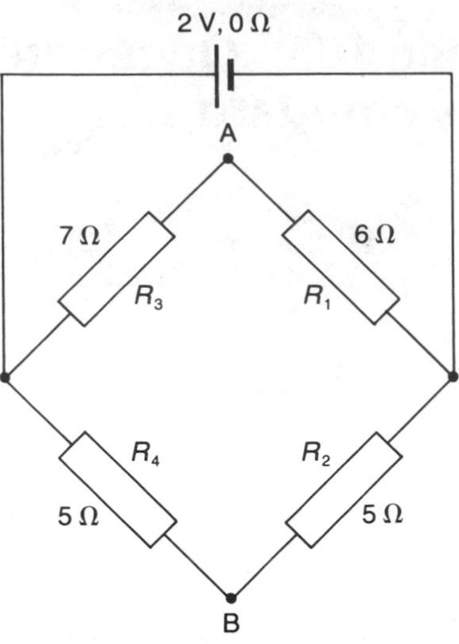

Fig 29.20 Circuit diagram for Question 6

30 ELECTROSTATICS

Electric charges

Charges have already been discussed in Chapter 27.

The SI unit for charge is the coulomb (C).

Force between charges

The force F between two small conducting spheres with charges Q_1 and Q_2 is given by

$$F = \frac{Q_1 Q_2}{4\pi\varepsilon R^2} \tag{30.1}$$

where R is the distance between the centres of the spheres and ε is the permittivity of the medium in which the spheres lie. ε for vacuum is denoted by ε_0 and ε for air is so close to ε_0 that we take it as equal to ε_0. The SI unit for ε is farad per metre ($F\,m^{-1}$) (see p.207).

The above formula applies also to the forces between any charged objects provided that their sizes are small compared to the separation R, i.e. they are 'point charges'. The fact that F is proportional to $1/R^2$ is called the inverse square law of electrostatics.

Example 1

Calculate the force between two small metal spheres with charges $+1.0 \times 10^{-9}$ C and $+9.0 \times 10^{-9}$ C whose centres are 30 cm apart in air, for which the permittivity is $8.9 \times 10^{-12}\ F\,m^{-1}$. Is the force attractive or repulsive?

Method

The force is

$$F = \frac{Q_1 Q_2}{4\pi\varepsilon R^2} = \frac{1.0 \times 10^{-9} \times 9.0 \times 10^{-9}}{4\pi \times 8.9 \times 10^{-12} \times (0.3)^2}$$

Note the conversion from centimetres to SI units, i.e. metres

$$\therefore \quad F = \frac{9.0 \times 10^{-18}}{4\pi \times 8.9 \times 10^{-12} \times 9 \times 10^{-2}}$$

$$= \frac{1}{4\pi \times 8.9} \times 10^{-4}$$

$$= 8.94 \times 10^{-7}\ N$$

$$= 0.89\ \mu N$$

Note too that it may be found helpful to collect together the tens to various powers, as shown in the equation above.

The force is repulsive because both charges are positive.

Answer

0.89 μN. The force is repulsive.

Electric intensity

In the vicinity of any charge Q there is a region within which other charges may be attracted or repelled by it. This region is called the 'field' of the charge Q. We can describe the field strength at any point in an electric field by the value of F/q, where q is the size of a small charge placed at the point concerned and F is the force it experiences due to the presence of Q. This ratio is called the electric intensity E of the field:

$$E = \frac{F}{q} \tag{30.2}$$

The unit for E could be $N\,C^{-1}$ but volt per metre (see p.201) is preferred.

Intensity *E* due to an isolated charged conducting sphere

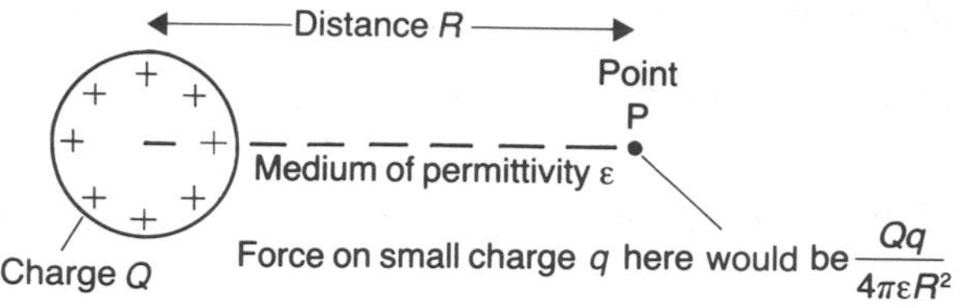

Fig 30.1 Intensity due to a charged sphere

In Fig 30.1

$$E = \frac{F}{q} = \frac{Q\cancel{q}}{4\pi\varepsilon R^2 \cancel{q}}$$

$$\therefore \quad E = \frac{Q}{4\pi\varepsilon R^2} \qquad (30.3)$$

The same formula applies if Q is a point charge.

Electric lines of force

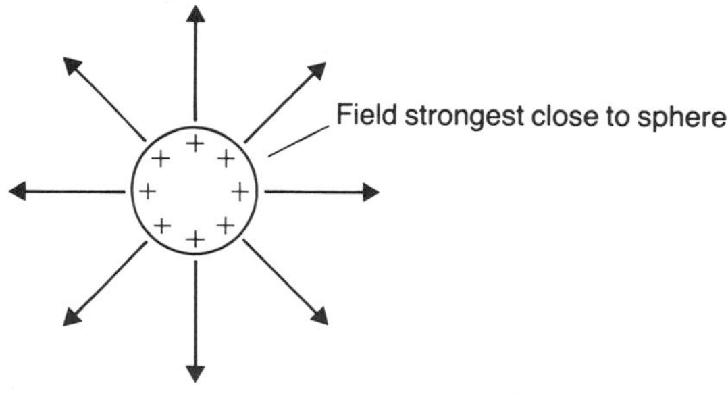

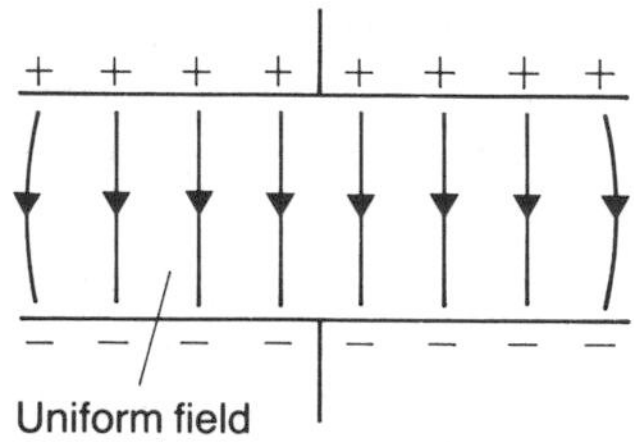

Fig 30.2 Electric lines of force

Intensity has direction. The direction is that of the force experienced by a small positive charge. Lines of force are lines which show the directions of E in an electric field. Two examples are shown in Fig 30.2.

Example 2

Point charges are located in air at points A and B as shown in Fig 30.3. Calculate the magnitude of the intensity at P and the direction of the intensity. (Take $1/4\pi\varepsilon_0$ as 9.0×10^9 m F^{-1}).

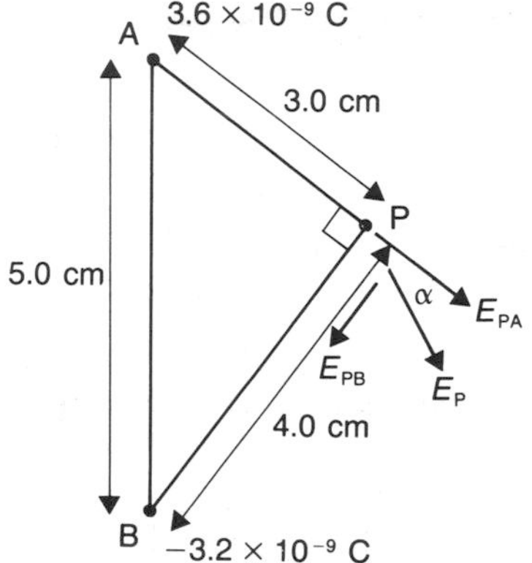

Fig 30.3 Diagram for Example 2

Method

The intensity E_{PA} at P due to the charge at A is given by

$$E_{PA} = \frac{Q}{4\pi\varepsilon_0 R^2} = \frac{9.0 \times 10^9 \times 3.6 \times 10^{-9}}{0.03^2}$$

This gives

$$E_{PA} = 36\,000 \text{ V m}^{-1}$$

The intensity E_{PB} at P due to the charge at B works out by the same method to be 18 000 V m^{-1}.

The directions of E_{PA} and E_{PB} are shown by the arrows in the diagram and the combined effect (intensity E_P) at P is found by vector addition (parallelogram rule, see page 31). Since E_{PA} and E_{PB} are perpendicular this addition can be done by use of Pythagoras' equation.

$$E_P{}^2 = 36\,000^2 + 18\,000^2 = 1620 \times 10^6$$

whence

$$E_P = 40.2 \times 10^3 \text{ V m}^{-1}$$

To find the direction of E we have $\tan\alpha = \dfrac{E_{PB}}{E_{PA}} = 0.5$.

This gives $\alpha = 26.6°$.

Answer

40 kV m^{-1}, 27° to direction AP, 63° to PB.

A relationship between intensity and potential

Consider first a small charge $+q$ being moved from close to the negative plate in Fig 30.2b up to the positive plate through distance d. Let the PD between the plates be V and the intensity E. The work done is $W = Fd$ (see p.44) and equals Eqd. Also, by definition of PD (see p.176) $W = Vq$ (Equation 27.3). Hence $Eqd = Vq$ or $E = V/d$.

$$\text{Intensity } E = \frac{V}{d} \qquad (30.4)$$

This is an important result. It also justifies our measuring E in volt per metre. Example 1 in Chapter 39 illustrates the use of this formula.

Potential at a distance *R* from a sphere or point charge

It is common, in GCE work particularly, to take as zero for potential measurements the potential at a large distance away from any charge, i.e. at infinity.

The potential difference between infinity (e.g. at far right of Fig 30.1) and position P can be shown to equal $Q/4\pi\varepsilon R$, i.e.

$$\text{Potential at P is } V = \frac{Q}{4\pi\varepsilon R} \qquad (30.5)$$

Note the R (not R^2).

Example 3

Point charges of -2.0×10^{-10} C and -3.0×10^{-10} C are located in air at A and B which are 4.0 cm apart. Calculate the electric intensity and potential midway between A and B. ($1/4\pi\varepsilon_0$ may be taken as 9.0×10^{9} m F^{-1})

Method

We are interested in a point which is 2.0 cm or 2.0×10^{-2} m from A and from B.

The intensity there caused by the -2.0×10^{-10} C is given by $E = Q/4\pi\varepsilon_0 R^2$ and so equals

$$\frac{2.0 \times 10^{-10} \times 9.0 \times 10^{9}}{(2.0 \times 10^{-2})^2} \quad \text{or} \quad 4.5 \times 10^{3}\ \text{V m}^{-1}$$

The direction of this intensity, because of the negative charge at A, is from B to A.

The intensity at the midpoint due to the -3.0×10^{-10} C charge at B is similarly

$$\frac{3.0 \times 10^{-10} \times 9.0 \times 10^{9}}{(2.0 \times 10^{-2})^2} \quad \text{or} \quad 6.75 \times 10^{3}\ \text{V m}^{-1}$$

The direction of this intensity, because the charge at B is negative, is towards B.

The total intensity at the midpoint caused by the two charges is obtained by adding the two intensities vectorially, i.e. with consideration of their directions.

Total intensity is, in the direction towards B, $6.75 \times 10^{3} - 4.5 \times 10^{3}$ V m^{-1} or 2.25×10^{3} V m^{-1}. To two significant figures we have 2.2 kV m^{-1}.

The potential at the midpoint due to the charge -2.0×10^{-10} C at A is given by $V = Q/4\pi\varepsilon_0 R$ and so equals

$$\frac{-2.0 \times 10^{-10} \times 9.0 \times 10^{9}}{2.0 \times 10^{-2}} \quad \text{or} \quad -90\ \text{V}$$

The potential due to the -3.0×10^{-10} C at B is

$$\frac{-3.0 \times 10^{-10} \times 9.0 \times 10^{9}}{2.0 \times 10^{-2}} \quad \text{or} \quad -135\ \text{V}$$

Since potential is a scalar quantity (no direction) we add the two contributions to the potential algebraically to get

$$-90 + -135 \quad \text{or} \quad -225\ \text{V}$$

To two significant figures this is -0.22 kV.

Answer

2.2 kV m^{-1}, -0.22 kV.

The Van de Graaff electrostatic generator

In this machine (Fig 30.4), charge is delivered on a moving belt to the inside of a spherical hollow metal dome. The charge is continuously given to the lower end of the belt, being either created as frictional charges or supplied from a high-voltage generator apparatus. Work is done in adding the charge (e.g. positive charge) to the charge already accumulated in the dome. If the PD between the dome and the distant end of the belt is V, then the work done per second equals VI, where I is the charge delivered per second. The charge passes

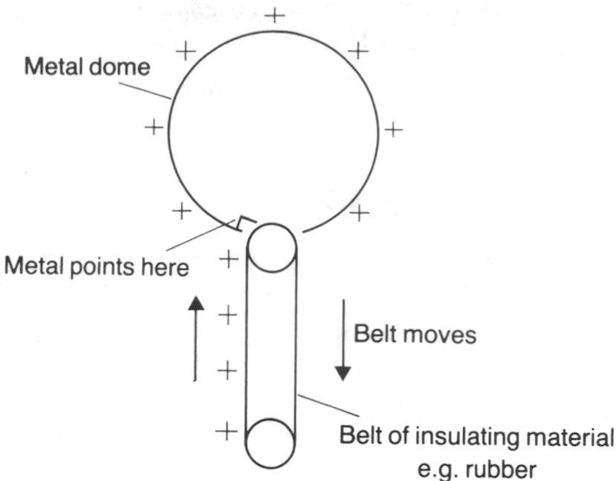

Fig 30.4 The principle of the Van de Graaff generator

from the belt to metal points within the dome (the points creating a high intensity that causes conduction in the air at the points).

This machine is discussed again in Chapter 31.

Example 4

A certain Van de Graaff generator is producing a PD of 0.5 MV. If the belt is 0.1 m wide, is moving at $5\,\mathrm{m\,s^{-1}}$ and carries a charge density of $2 \times 10^{-5}\,\mathrm{C\,m^{-2}}$, what is (a) the current being delivered to the dome and (b) the minimum power of the machine's motor. (c) If the dome has a diameter of 0.5 m, what is the electric intensity in the air adjacent to the dome's outer surface?

Method

(a) $$\text{Current } I = \begin{pmatrix}\text{Area of belt}\\ \text{entering dome}\\ \text{per second}\end{pmatrix} \times 2 \times 10^{-5}$$

$$= 5 \times 0.1 \times 2 \times 10^{-5}\,\mathrm{A}$$

$$= 10^{-5}\,\mathrm{A} \quad \text{or} \quad 10\,\mu\mathrm{A}$$

(b) $$\text{Power} = \text{Current} \times \text{PD} = 10^{-5} \times 0.5 \times 10^{6} = 5\,\mathrm{W}$$

(c) $V = Q/4\pi\varepsilon R$ but $E = Q/4\pi\varepsilon R^2$

$$\therefore \quad E = \frac{V}{R} \quad \text{and} \quad R = 0.25\,\mathrm{m}$$

$$\therefore \quad E = \frac{0.5 \times 10^{6}}{0.25}$$

$$= 2 \times 10^{6}\,\mathrm{V\,m^{-1}} \quad \text{or} \quad 2\,\mathrm{MV\,m^{-1}}$$

Answer

(a) $10\,\mu\mathrm{A}$, (b) 5 W, (c) $2\,\mathrm{MV\,m^{-1}}$.

Exercise 106

(Take ε for air to be $8.9 \times 10^{-12}\,\mathrm{F\,m^{-1}}$ unless otherwise stated.)

1 Calculate (a) the force between two charges of +1.4 nC and +1.6 nC on point conductors 40 cm apart in air. (b) What size of charge on a third point conductor placed midway between the first two conductors would result in doubling of the magnitude of the force on the 1.4 nC charge?

2 (a) Calculate (i) the electric intensity and (ii) the potential at a point midway between two point charges of $+10^{-9}$ C and -10^{-9} C which are 20 cm apart in air.
(b) To produce an equally large electric intensity midway between two large-area, parallel plates 2.0 cm apart in air, what PD would be needed between the plates?

3 When a charge of $50\,\mu\mathrm{C}$ is moved between two points P and Q in a uniform electric field, $100\,\mu\mathrm{J}$ of work is done. What is the potential difference between P and Q?

4 Calculate the potential at the surface of an isolated metal sphere carrying a negative charge of 2.0×10^{-8} C and surrounded by air, if the sphere's radius is 2.0 cm.

How much work would be done in moving a positive charge of 1.6×10^{-6} C from the sphere's surface to a point 3.0 cm further from the centre?

5 Calculate the electric potential and electric field strength (or intensity) at C in Fig 30.5.

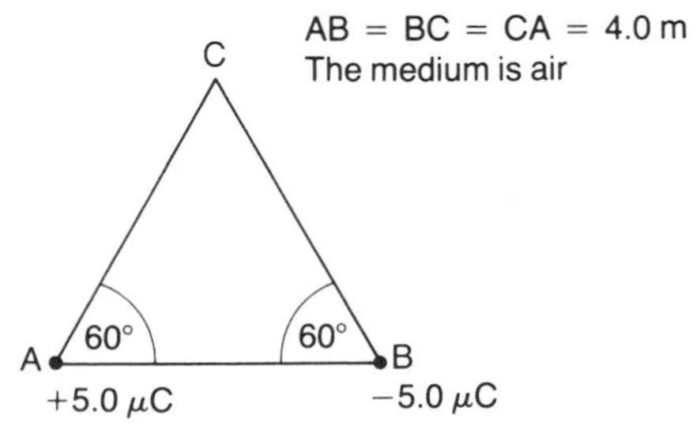

Fig 30.5 Diagram for Question 5

6 A Van de Graaff generator has a belt 10 cm wide which moves towards the metal dome of the generator through a PD of 0.4 MV and at a speed of $10\,\mathrm{m\,s^{-1}}$. (a) If the charge density on the belt is

$2.0 \times 10^{-5}\,\text{C m}^{-2}$, what is the minimum power output required of the motor that drives the belt? (b) If the potential of the dome fails to increase in spite of the arrival of this charge, what is the resistance through which discharge from the dome to earth is occurring? State any assumptions made.

Exercise 107: Questions of A-level standard

1 Two point charges of magnitude $-Q$ and $-2Q$ repel each other with a force of magnitude F. If an extra $-Q$ coulomb of charge is added to each of the point charges, and also their separation is doubled, what is the new force between them?

A $0.75F$ **B** $1.2F$ **C** $1.5F$
D $3F$ **E** $6F$

2 A charge of $2.0\,\mu\text{C}$ is moved through a uniform electric field, of strength $10\,\text{V m}^{-1}$, such that it traverses a path 3.0 m parallel to and then 4.0 m perpendicular to the field lines. The change in μJ in potential energy of the charge between the end points of the path is

A 30 **B** 40 **C** 60
D 80 **E** 100 [AEB 81]

3

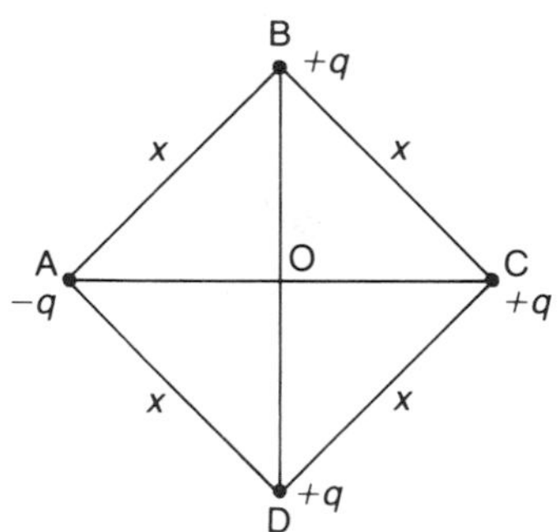

Fig 30.6 Circuit diagram for Question 3

Equal charges of $+q$ coulombs are placed at the corners B, C, D of a square ABCD, of side x, *in vacuo* (Fig 30.6). A charge of $-q$ coulombs is placed at the corner A. The field intensity at the centre O is

A $q/\pi\varepsilon_0 x^2$ towards C
B $q/\pi\varepsilon_0 x^2$ towards A
C $2q/\pi\varepsilon_0 x^2$ towards C
D $2q/\pi\varepsilon_0 x^2$ towards A
E $\sqrt{(2)}q/4\pi\varepsilon_0 x^2$ towards A [AEB 82]

4 Two identical positive point charges A and B, of mass $1.11 \times 10^{-27}\,\text{kg}$ and charge $2.22 \times 10^{-19}\,\text{C}$, are held at a distance $10^{-10}\,\text{m}$ apart *in vacuo*. What is their potential energy?
(a) If the charges are released, calculate their velocities when they are a very large distance apart.
(b) A third point charge is placed on the line AB so that, when A and B are released, none of the three charges moves.
(i) Determine where the third charge is placed, and its magnitude.
(ii) Calculate the total potential energy of the three charges.
(iii) If the charge A is slightly displaced along the line AB towards B and released, consider qualitatively what the motion of the three charges will be.
($4\pi\varepsilon_0$ is numerically equal to 1.11×10^{-10})
[WJEC 77, p]

5 Given that air becomes ionised if the electrostatic field strength exceeds $3 \times 10^6\,\text{V m}^{-1}$, calculate the maximum potential to which a small metal sphere of diameter 0.1 mm can be charged. [SUJB 81, p]

6 The diagram given with this question is shown in Fig 30.7.

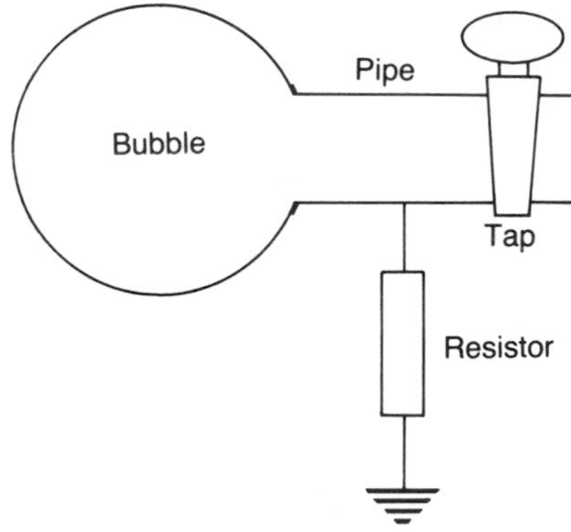

Fig 30.7 Diagram for Question 6

A charged spherical soap bubble on the end of a small metal pipe is losing gas through the tap and losing charge through the resistor. The tap is continuously adjusted so that the current in the resistor is kept constant. Show that, under these conditions, the radius of the bubble will decrease at a constant rate. If the resistor has the value $10^{12}\,\Omega$, calculate how long a bubble of 5 cm radius would take to collapse, assuming that it did not burst.
($\varepsilon_0 = 8.85 \times 10^{-12}\,\text{F m}^{-1}$.) [C 77, p]

7 In an electrostatic machine of the Van de Graaff type a belt of width w carrying electric charge of σ coulomb per metre^2 of belt moves up at a speed s and reaches the inside of a metal dome where it gives up all its charge. The voltage through which the charge is carried to the dome is V. What are the values of the current I delivered to the dome and the work done per second (P) in delivering the charge?

A $I = \sigma s$, $P = \frac{1}{2}\sigma V^2$
B $I = s\sigma/w$, $P = sw\sigma/V$
C $I = \sigma s$, $P = \sigma s V$
D $I = sw\sigma$, $P = sw\sigma V$
E $I = sw\sigma$, $P = \frac{1}{2}\sigma V^2$

31 CAPACITORS

Capacitance

This is the ability of a capacitor to store charge. It is defined as the final charge stored per volt used, i.e. final charge divided by the supply PD, or charge stored divided by the PD between the plates.

$$\text{Capacitance } C = \frac{Q}{V} \qquad (31.1)$$

The SI unit for capacitance is called the farad (F).

Capacitances are mostly met in microfarad (μF) and smaller sizes.

(a) Circuit symbol for capacitor

(b) Capacitor charging

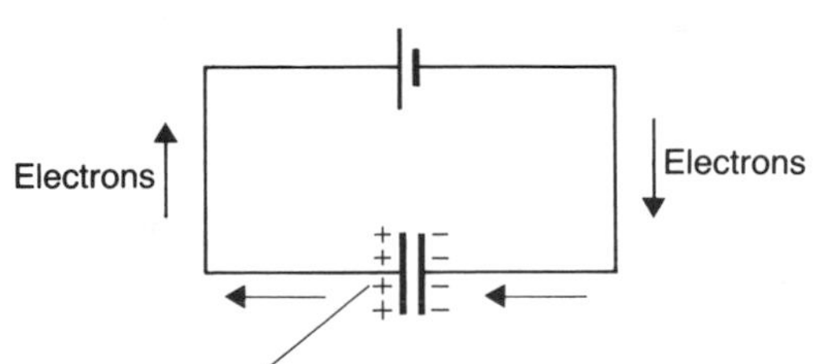

(c) Capacitor discharging

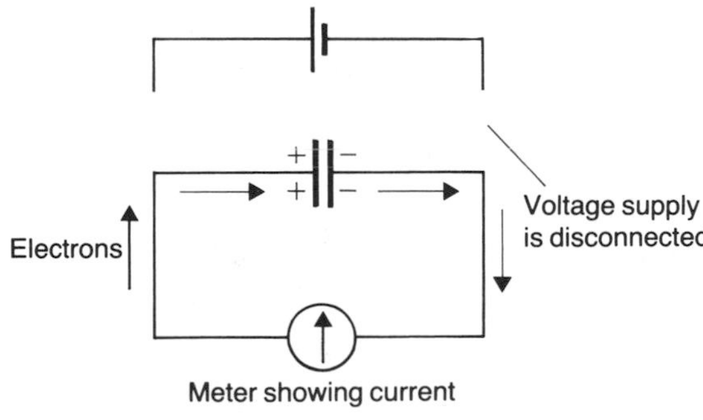

Fig 31.1 Capacitors

Energy stored in a charged capacitor

This equals $\frac{1}{2}QV$ because, during charging, Q coulombs of electrons have in effect been taken from one conductor of the capacitor to the other through a PD which was initially zero, is finally V, and has an average value of $\frac{1}{2}V$. So the work done (or energy stored) is $Q \times \frac{1}{2}V$. Note that because $Q = CV$ we can also write $\frac{1}{2}QV$ as $\frac{1}{2}CV^2$ or $\frac{1}{2}Q^2/C$.

$$\text{Energy stored} = \tfrac{1}{2}QV \qquad (31.2)$$

Example 1
A capacitor is charged by a 20 V DC supply and when it is discharged through a charge meter it is found to have carried a charge of 5.0 μC. What is its capacitance, and how much energy was stored in it?

Method

$$C = \frac{Q}{V}$$

$$\therefore \quad C = \frac{5.0 \times 10^{-6}}{20} = 0.25 \times 10^{-6}\,\text{F}$$

$$= 0.25\,\mu\text{F}$$

$$\text{Energy stored} = \tfrac{1}{2}QV$$
$$= \tfrac{1}{2} \times 5.0 \times 10^{-6} \times 20$$
$$= 5.0 \times 10^{-5}\,\text{J}$$

Or, using the formula $\frac{1}{2}CV^2$, the energy is

$$\tfrac{1}{2} \times 0.25 \times 10^{-6} \times 20^2$$

This equals 50×10^{-6} or 5.0×10^{-5} J.

Answer
0.25 μF, 5.0×10^{-5} J

Experimental measurements of capacitance

Electric charge meters are now available so that it is convenient to charge a capacitor using a known PD V and then discharge it through the meter to measure the charge Q. Then C can be calculated from $C = Q/V$.

In the past a ballistic galvanometer would be used for charge measurements. Following the quick discharge of the capacitor through the ballistic galvanometer the first deflection θ of the meter is proportional to the charge Q. However, such a meter typically has a high, but somewhat variable, sensitivity, so that the meter is not permanently calibrated. The usual practice therefore is to obtain another reading θ_S using a capacitor of known value C_S, i.e. a standard capacitor, and doing this with the same PD V. Thus

$$\frac{Q}{Q_S} = \frac{CV}{C_S V} \text{ and equals } \frac{\theta}{\theta_s}$$

$$\therefore \quad C = \frac{\theta}{\theta_S}C_S \qquad (31.3)$$

Quite different is the repeated discharge method illustrated in Fig 31.2.

The switch, usually a reed switch operated by an alternating current of frequency f, causes the capacitor to be charged to a PD V, and then the capacitor is connected to the current meter through which it discharges its charge Q. This cycle is repeated f times per second so that the charge per second (the current I) through the meter is fQ or fCV.

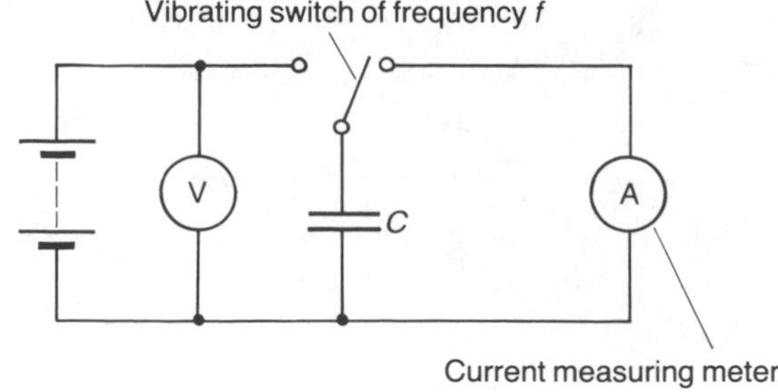

Fig 31.2 Measurement of capacitance by the repeated discharge method

$$C = \frac{I}{fV} \qquad (31.4)$$

Example 2

A capacitor of capacitance C_1 is connected to a 2 V supply and is discharged through a ballistic galvanometer. An initial deflection of 10 divisions is observed. A second capacitor of capacitance C_2 is connected in parallel with C_1 and the experiment is repeated. This time the initial deflection is 15 divisions. Calculate the ratio C_1/C_2.

Method
The first charge measured Q_1 is 10 units compared with 15 units for the second charge Q_2, the actual size of the unit being unimportant.

$$\frac{Q_1}{Q_2} = \frac{10}{15} = \frac{2}{3}$$

But $Q_1 = C_1V$ and $Q_2 = C_2V + C_1V$

$$\therefore \quad \frac{Q_1}{Q_2} = \frac{C_1V}{C_2V + C_1V} = \frac{C_1}{C_1 + C_2}$$

$$\therefore \quad \frac{C_1}{C_1 + C_2} = \frac{2}{3}$$

$$\therefore \quad 3C_1 = 2C_1 + 2C_2$$

$$\therefore \quad C_1 = 2C_2$$

$$\frac{C_1}{C_2} = 2$$

Answer
2.

Example 3

A capacitor repeatedly charged to 15 V and discharged through a milliammeter by use of a reed switch working at 120 cycles per second causes a meter reading of 3.6 mA. Calculate the capacitance of the capacitor.

Method

$$I = fCV$$

$$\therefore\ 3.6 \times 10^{-3} = 120 \times C \times 15$$

$$\therefore\quad C = \frac{3.6 \times 10^{-3}}{120 \times 15} = \frac{3.6}{1800} \times 10^{-3}$$

$$= 2.0 \times 10^{-6}\,\text{F} \quad \text{or} \quad 2.0\,\mu\text{F}$$

Answer

2.0 μF.

Formula for the capacitance of a parallel-plate capacitor

When the two conductors of a capacitor are parallel as in a 'parallel-plate' capacitor or a waxed-paper capacitor the capacitance C is given by the formula

$$C = \frac{\varepsilon A}{d} \qquad (31.5)$$

ε is the permittivity of the medium, called the dielectric, separating the conductors ('plates') of the capacitor, A is the area of the plates and d is their separation, i.e. the dielectric thickness.

The SI unit for ε

The SI unit for ε is farad per metre (F m^{-1}). This is seen to be appropriate from the formula $\varepsilon = Cd/A$ (equation 31.5).

Dielectric constant

ε denotes the permittivity of a medium, ε_0 is the permittivity of air or vacuum. The relative permittivity of a medium is the ratio of its permittivity to ε_0, i.e. $\varepsilon/\varepsilon_0$. When the medium is being used as a dielectric, its relative permittivity is often described as the dielectric constant.

Example 4

A 0.10 μF capacitor is to be constructed with metal foil and waxed paper (dielectric constant 2.0). The width of the foil is to be 4.0 cm, and the length no more than 5.0 m. What is the maximum thickness of the waxed paper? ($\varepsilon_0 = 8.9 \times 10^{-12}\,\text{F m}^{-1}$.)

Method

$$C = \frac{\varepsilon A}{d}$$

$$\therefore\quad 0.1 \times 10^{-6} = \frac{2.0 \times 8.9 \times 10^{-12} \times 0.04 \times 5.0}{d}$$

$$\therefore\quad d = 2.0 \times 8.9 \times 0.04 \times 5.0 \times 10^{-5}$$

which gives 3.6×10^{-5} m or 0.036 mm.

Answer

0.036 mm.

Exercise 108

1 Calculate the capacitance of a pair of semicircular brass discs, of 3.5 cm radius, with their planes parallel and 0.50 mm apart in air (a) with the maximum overlap of the plates, (b) when the area of overlap is halved. ($\varepsilon_{air} = 8.9 \times 10^{-12}\,\text{F m}^{-1}$.)

2 A 2.0 μF capacitor is charged using 0.40 V from a cell and voltage divider. It is then allowed to discharge quickly through a ballistic galvanometer. The meter gives a maximum deflection of 6.4 cm. What is the sensitivity of the meter in mm C^{-1}?

A second capacitor charged to the same PD gives a first deflection of 9.5 cm. What is its capacitance?

A third capacitor is charged to 40 V and gives a first deflection of 3.3 cm. What is its capacitance?

3 A simple capacitor is constructed from a pair of metal plates separated by insulating spacers so as to leave a 1.5 mm air space between them. The plates, each 20 cm square, are placed to achieve maximum capacitance. The capacitance is then measured by repeatedly charging it to 100 V and discharging it through a calibrated microammeter 200 times per second. The current recorded is 4.8 μA. Obtain a value for the capacitance of this capacitor and for the permittivity of air.

4 A 5.0 μF capacitor is alternately connected to a 20 V DC supply and then to a milliammeter by a vibrating switch working at a frequency of 80 Hz. What reading is expected on the meter?

Capacitors in series and parallel

Two capacitors C_1 and C_2 can be joined together either in series or parallel, as shown in Figs 31.3a and b, where the capacitance, charge and PD are labelled for each capacitor and for the combined capacitance C.

(a) In parallel

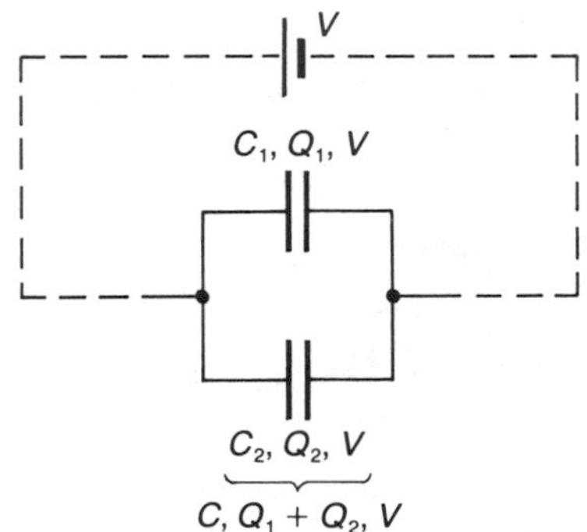

(b) In series

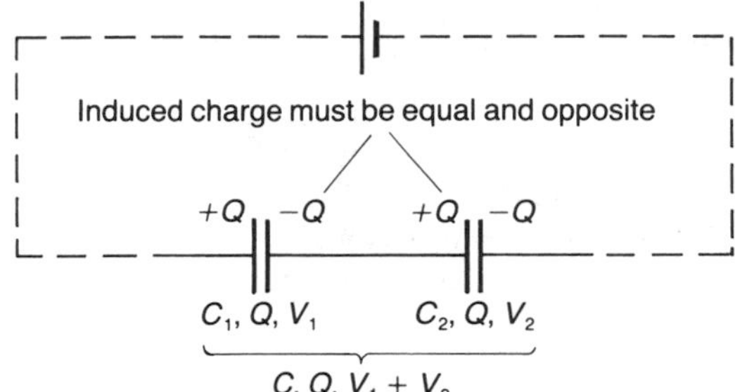

Fig 31.3 Combinations of capacitors

For C_1 and C_2 in parallel

$$C = C_1 + C_2 \tag{31.6}$$

$$\left(\text{because } C = \frac{Q_1 + Q_2}{V} = C_1 + C_2\right).$$

For C_1 and C_2 in series

$$\frac{1}{C} = \frac{1}{C_1} + \frac{1}{C_2}$$

$$\text{or}\quad C = \frac{C_1 C_2}{C_1 + C_2} \tag{31.7}$$

$$\left(\text{because } \frac{1}{C} = \frac{V_1 + V_2}{Q} = \frac{V_1}{Q} + \frac{V_2}{Q} = \frac{1}{C_1} + \frac{1}{C_2}\right).$$

Example 5

A 2.0 μF capacitor is charged to 12 V. The voltage supply is removed and then a 4.0 μF capacitor is fitted in parallel with the 2.0 μF one. Calculate the charge stored in the 2.0 μF capacitor (a) initially, (b) finally.

Method

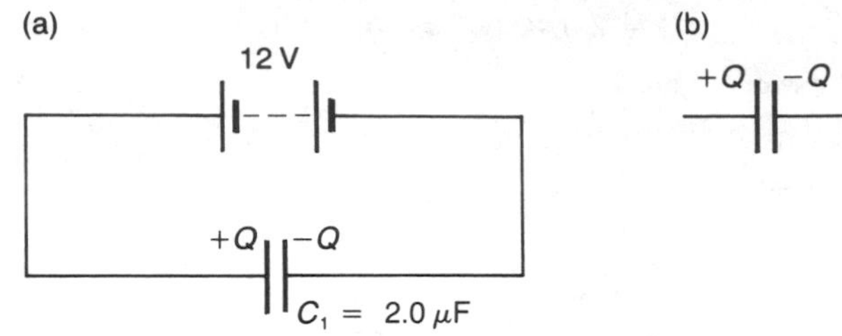

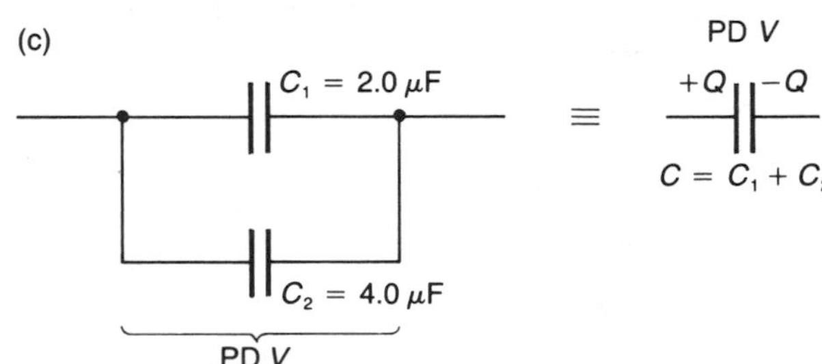

Fig 31.4 Diagrams for Example 5

(a) In Fig 31.4a Q is given by

$$Q = CV = 2.0 \times 10^{-6} \times 12$$
$$= 24 \times 10^{-6}\,\text{C}$$

In Fig 31.4b the battery has been removed and we have $+Q$ on the left and $-Q$ on the right.

(b) In Fig 31.4c the 4.0 μF has been connected in parallel. The total capacitance is, from Equation 31.6,

$$C = C_1 + C_2 = 2.0 + 4.0 = 6.0\,\mu\text{F}$$

Now we must realise that the charge on the final 6.0 μF combined capacitor is still $+Q$ on the left and $-Q$ on the right, i.e. 24 μC. The charge on the left in Fig 31.4b is now shared by C_1 and C_2, but it cannot escape from the left or be added to. Thus the PD V across the combined capacitor is

$$V = \frac{Q}{C} = \frac{24 \times 10^{-6}}{6.0 \times 10^{-6}} = 4.0\,\text{V}$$

This is the PD across C_1 and across C_2 in Fig 31.4c, so that the new charge on C_1 is given by

$$\text{Charge} = C_1 \times \text{PD}$$

i.e. $2.0 \times 10^{-6} \times 4.0$ or 8.0×10^{-6} C.

(The 8.0 μC on the 2.0 μF, and similarly the 16 μC on the 4.0 μF illustrate that, in a parallel combination, the charge is shared in proportion to the capacitances.)

Answer

(a) 24 μC, (b) 8.0 μC.

Example 6

(a) Calculate the charge stored in a 3.0 μF capacitor and a 6.0 μF capacitor joined in series and then connected across the terminals of an 18 V battery.

(b) What is the PD across each of these capacitors?

Method

(a) A diagram should be sketched (see Fig 31.3b).

The combined capacitance C is given by Equation 31.7:

$$\frac{1}{C} = \frac{1}{C_1} + \frac{1}{C_2}$$

$$\therefore \quad \text{in } \mu\text{F}, \quad \frac{1}{C} = \frac{1}{3.0} + \frac{1}{6.0} = \frac{1}{2}$$

So that $C = 2.0\,\mu$F.

Therefore the charge stored Q is

$$Q = CV = 2.0 \times 10^{-6} \times 18$$
$$= 36 \times 10^{-6}\,\text{C}$$

and, for capacitors in series, this is the same for both capacitors.

(b) The PD across the 3.0 μF is given by charge divided by capacitance and equals $36 \times 10^{-6}/3.0 \times 10^{-6}$ or 12 V. For the 6.0 μF, we have PD = $36 \times 10^{-6}/6.0 \times 10^{-6}$ or 6.0 V. (We note that the total PD, 18 V here, is shared by capacitors in series in *inverse proportion* to the capacitances.)

Answer

(a) 36 μC, (b) 12.0 V and 6.0 V.

Example 7

(a) A 5.0 μF capacitor is charged to 4.0 V and is removed from the voltage supply. How much energy is stored?

(b) If the 5.0 μF capacitor is connected in parallel with a 3.0 μF capacitor, what is the new energy stored in the capacitor combination, and how much energy was converted to heat by the movement of charge through the wires between the two capacitors?

Method

(a) From Equation 31.2

Energy = $\frac{1}{2}C_1V^2$

$$= \tfrac{1}{2} \times 5.0 \times 10^{-6} \times 4.0^2 = 40 \times 10^{-6}\,\text{J}.$$

(b) The new capacitance ($C = C_1 + C_2$) is 5.0 + 3.0 μF or 8.0 μF. We do not know the new PD, but we know that the charge is the same as in (a). This charge is given by $Q = C_1V$, so that it equals $5.0 \times 10^{-6} \times 4.0$ C or 20 μC.

The energy now in the 8 μF is given by $\frac{1}{2}Q^2/C$ as

$$\tfrac{1}{2} \times \frac{(20 \times 10^{-6})^2}{8 \times 10^{-6}} = \frac{200}{8} \times 10^{-6}\,\text{J} \quad \text{or} \quad 25\,\mu\text{J}.$$

The electrical potential energy has fallen from 40 μJ to 25 μJ, i.e. 15 μJ has become heat in the connecting wires.

Answer

(a) 40 μJ, (b) 25 μJ, 15 μJ.

Capacitor with two-layer dielectric

(a) The capacitor

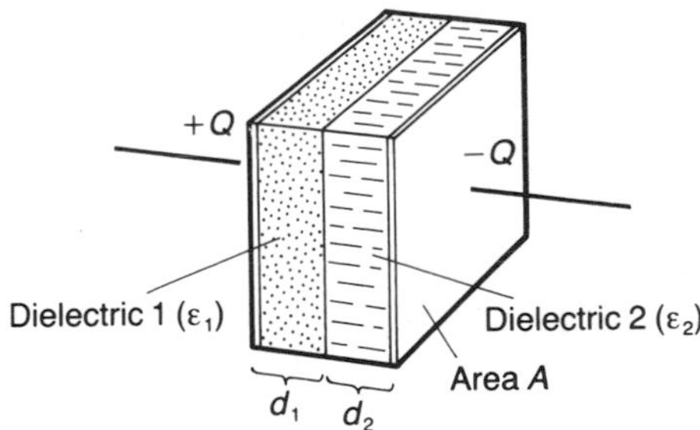

(b) Regarded as two separate capacitors

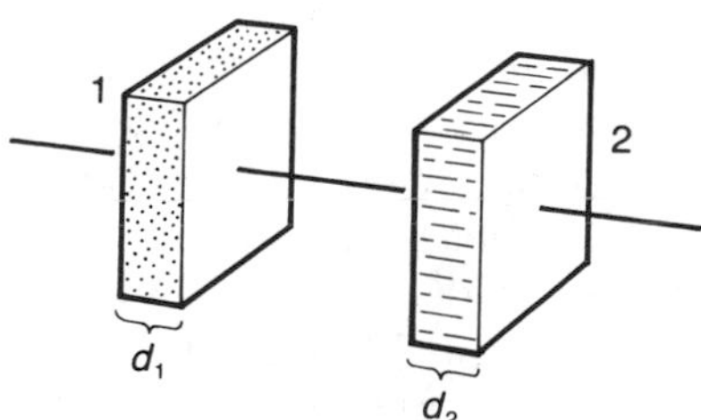

Fig 31.5 Two layers of dielectric

Fig 31.5a shows an example. How is the capacitance related to the permittivities ε_1 and ε_2 and to the thicknesses d_1 and d_2?

The simplest approach to this question is to regard the capacitor as two separate capacitors C_1 and C_2 in series, as shown in Fig 31.5b. Thus

$$\frac{1}{C} = \frac{1}{C_1} + \frac{1}{C_2} = \frac{d_1}{\varepsilon_1 A} + \frac{d_2}{\varepsilon_2 A}$$

Example 8

A parallel-plate capacitor of area 0.030 m² has a 2.0 mm gap between its two plates. The gap is filled with air. If a sheet of insulating material 1.5 mm thick and of relative permittivity 6.0 is placed between the plates, what is its capacitance? ($\varepsilon_0 = 8.9 \times 10^{-12}\,\mathrm{F\,m^{-1}}$.)

Method

Regarding the new capacitor as two capacitors C_1 and C_2 in series, we have, using $C = \varepsilon A/d$,

$$C_1 = \frac{6.0 \times \varepsilon_0 \times A}{1.5 \times 10^{-3}} \quad \text{and} \quad C_2 = \frac{\varepsilon_0 A}{0.5 \times 10^{-3}}$$

so that the required capacitance C is given by

$$\frac{1}{C} = \frac{1.5 \times 10^{-3}}{6.0\varepsilon_0 A} + \frac{0.5 \times 10^{-3}}{\varepsilon_0 A}$$

$$= \frac{10^{-3}}{\varepsilon_0 A}\left(\frac{1}{4} + \frac{1}{2}\right)$$

$$= \frac{0.75 \times 10^{-3}}{\varepsilon_0 A}$$

$$\therefore \quad C = \frac{8.9 \times 10^{-12} \times 0.03}{0.75 \times 10^{-3}} = 35.6 \times 10^{-11}\,\mathrm{F}$$

Answer

36×10^{-11} F or 0.36 nF.

Time constant

The time interval within which a capacitor becomes charged and finally acquires a PD equal to the supply PD depends upon the resistance R through which the charging current has to flow and also upon the capacitance C to be filled. The product RC is the deciding factor, and its value equals the time in seconds that it takes the charge stored Q and the capacitor PD V to rise to within about one-third of their final values (1/e to be precise; e = 2.718, see Chapter 2). This product RC is called the 'time constant' of the circuit. This is shown in Fig 31.6a. Similarly RC is the time required for a capacitor C discharging in a circuit of resistance R to reach a charge and PD that are 1/e of the initial values, as shown in Fig 31.6b.

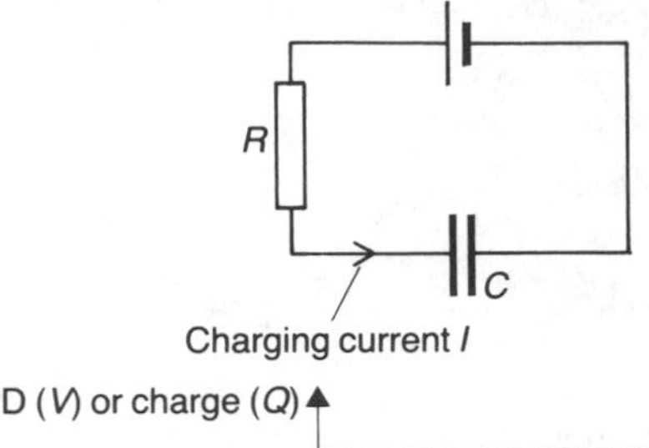

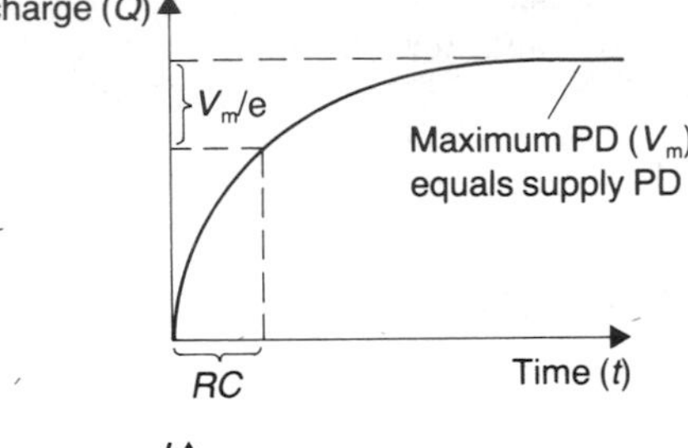

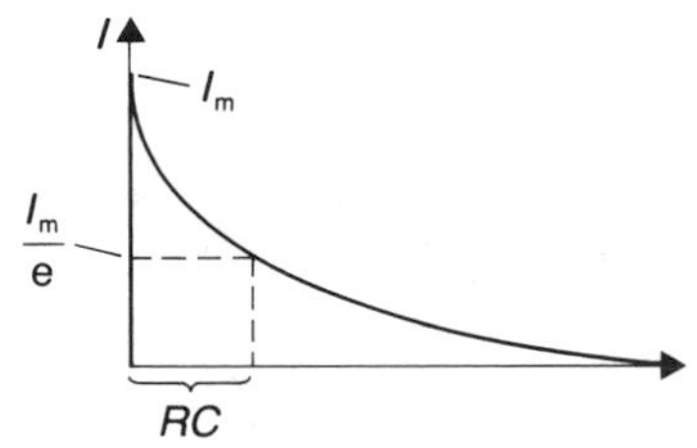

(b) Discharging

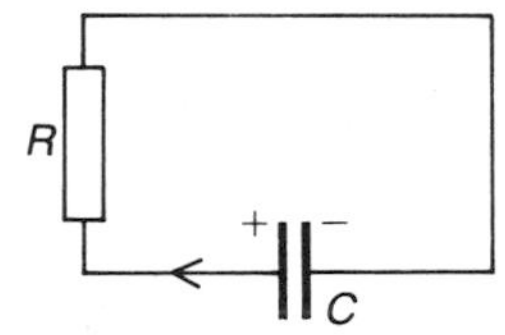

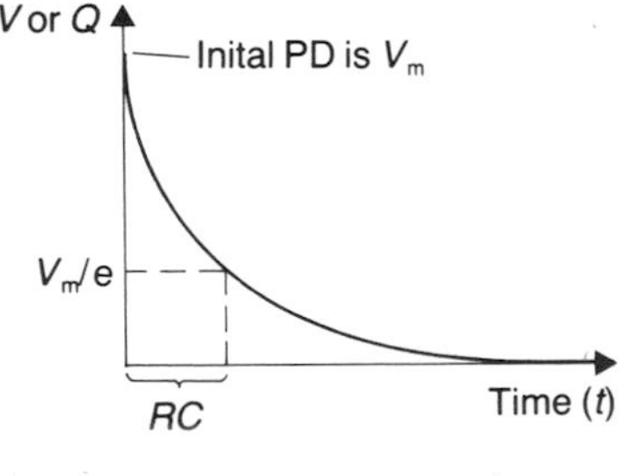

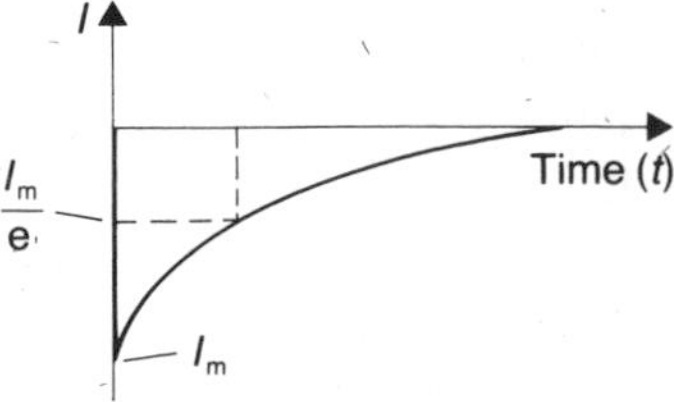

Fig 31.6 Time constant

Example 9

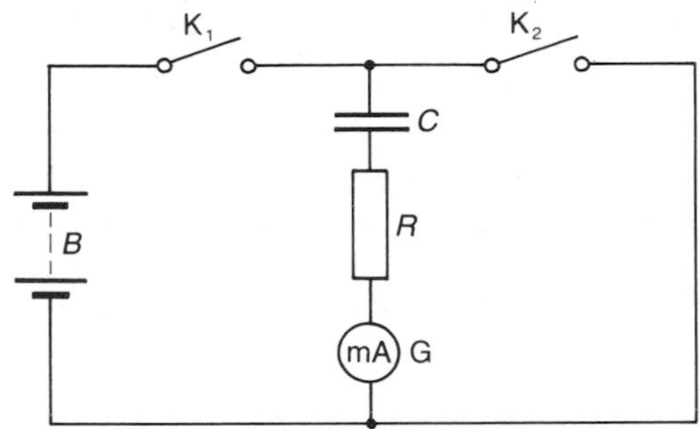

Fig 31.7 Circuit diagram for Example 9

In the circuit in Fig 31.7 B is a battery of EMF 10 V, K_1 and K_2 are switches, C is a capacitor of capacitance 1000 μF, R is a non-inductive resistor of resistance 9.9 kΩ and G is a centre-zero milliammeter of resistance 100 Ω. Initially the capacitor is discharged and both keys are open. What will be the reading of the milliammeter (a) immediately after closing K_1; (b) 10 s after closing K_1; (c) several minutes after closing K_1?

If after several minutes K_1 is opened, what will be the reading on the galvanometer (d) immediately after closing K_2; (e) 10 s after closing K_2; (f) several minutes after closing K_2? [AEB 80]

Method

(a) At the first instant of charging, the PD across C is zero, so that the current I is decided only by the supply PD and, of course, the circuit resistance R.

$$I = \frac{10}{9.9\,\text{k}\Omega + 100\,\Omega} = \frac{10}{10 \times 10^3} = 10^{-3}\,\text{A} \quad \text{or} \quad 1\,\text{mA}$$

(b) The time constant $= \text{RC}$

$$= 1000\,\mu\text{F} \times (9.9\,\text{k}\Omega + 100\,\Omega) = 10^{-3} \times 10 \times 10^3 = 10\,\text{s}$$

The time elapsed is exactly equal to the time constant so that the capacitor PD V is $V_m - V_m/\text{e}$, i.e. $10 - 10/2.718$, which equals $10 - 3.7$ or 6.3 V. Consequently $I = (10 - 6.3)/R = 3.7/(10 \times 10^3)$, which is 0.37 mA.

(c) After several minutes, i.e. many times CR, I will be effectively zero because battery EMF and capacitor PD will then be equal and opposite.

(d) During the first instant of discharge the capacitor PD is 10 V, and this causes the current to be $10/R$, i.e. $10/(10 \times 10^3)$ or 10^{-3} A.

(e) After a time equal to CR the capacitor PD will have fallen to V_m/e, i.e. to 10/2.718 or 3.7 V, which gives a discharge current of $3.7/(10 \times 10^3)$, i.e. 0.37 mA.

(f) For $t \gg CR$, the discharge current is effectively zero because the capacitor PD is then zero.

Answer

(a) 1.0 mA, (b) 0.37 mA, (c) zero, (d) 1.0 mA, (e) 0.37 mA, (f) zero.

Capacitance of an isolated spherical conductor

Isolated means 'far away from other objects'. However, any charge Q on a conductor must induce a charge $-Q$ in its surroundings, so we realise that this conductor is one 'plate' of a capacitor and its distant surroundings act as the other conductor.

The potential V of the spherical surface, radius R, is $Q/4\pi\varepsilon_0 R$ relative to the distant (zero potential) surroundings (see Chapter 30).

$$\therefore \quad C\left(= \frac{Q}{V}\right) = 4\pi\varepsilon_0 R \qquad (31.8)$$

Example 10

How much charge must be given to the spherical head of a Van de Graaff generator of 1.4 m radius and surrounded by air, to give it a potential of 1.0 MV (relative to distant surroundings). Take the permittivity of air to be $9.0 \times 10^{-12}\,\text{F m}^{-1}$.

Method

$$C = 4\pi\varepsilon R = 4\pi \times 9.0 \times 10^{-12} \times 1.4 = 1.6 \times 10^{-10}\,\text{F}$$

$$Q = CV = 1.6 \times 10^{-10} \times 10^6 = 1.6 \times 10^{-4} \quad \text{or} \quad 0.16\,\text{mC.}$$

Answer

0.16 mC.

Exercise 109

1 In the circuit shown in Fig 31.8, 10 μF and 20 μF capacitors are connected in series with a 30 V DC supply. What is the charge on each capacitor?

A 0.30 mC **B** 0.20 mC **C** 1.0 μC
D 0.90 mC **E** 4.5 μC

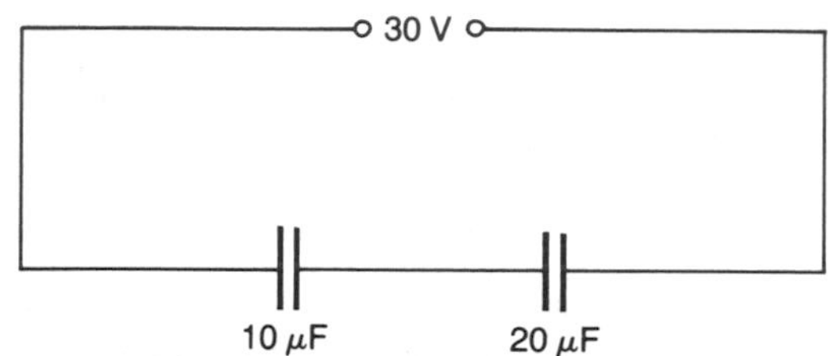

Fig 31.8 Circuit diagram for Question 1

2 A 2.0 μF capacitor is charged by connecting it across the terminals of a cell whose EMF is 1.5 V
(a) What is the charge Q stored in this capacitor, the energy E stored in it and the PD V across it?
(b) If the cell remains connected, and a second 2.0 μF capacitor is connected in parallel with the first one, what are Q, E and V for the second capacitor?
(c) The cell is removed without discharging the capacitors, and a third 2.0 μF capacitor is fitted in parallel with the other two. What are Q, E and V for this third capacitor?

3 A capacitor A of capacitance 4.0 μF is charged to a potential difference of 20 V. An uncharged capacitor B of capacitance 2.0 μF is then connected in parallel with A. What is (a) the energy initially stored in A, (b) the potential difference across A after B has been connected, (c) the energy finally stored in A and B?

4 A simple parallel-plate capacitor with a 2 mm-thick air dielectric has a capacitance of 5×10^{-10} F. A uniform sheet of material whose dielectric constant is 2 and thickness is 1 mm is now inserted between the plates throughout the capacitor area, the plates remaining 2 mm apart. What will the new capacitance be?

5 A 30 μF capacitor is initially charged with 15 mC. It is then discharged through a 200 Ω resistor. What is the maximum current during the discharge?

A 2.5 A **B** 10 μA **C** 2.5 mA
D 2.2×10^{-12} A **E** 2.5×10^{-12} A

6 A 2.0 μF capacitor initially charged to 20 V is discharged through a 500 kΩ resistance. Calculate the rate of fall of the capacitor PD (a) at the first instant of discharge, (b) after 1 second. (e = 2.718.)

7 The high-voltage conductor of a Van de Graaff generator consists of an insulated spherical metal shell of radius 0.50 m. The shell carries a charge of 60 μC. Calculate (a) the potential of the shell, (b) the field strength (or intensity) at the surface. (The permittivity of vacuum is 8.85×10^{-12} F m^{-1}.)

Exercise 110: Questions of A-level standard

1 A capacitor charged from a 50 V DC supply is discharged across a charge-measuring instrument and found to have carried a charge of 10 μC. What is the capacitance of the capacitor, and how much energy was stored in it? [L 79]

2 If the energy of a 100 μF capacitor, when charged to a voltage of 6 kV, could all be used to lift a 50 kg mass, the greatest vertical height through which it could be raised would be. (Take g to be 10 m s^{-2}.)

A 0.6 mm **B** 1.2 mm **C** 3.6 m
D 12 m **E** 600 m [O 80]

3 A capacitor is alternately charged to 20 V and then fully discharged through a DC milliammeter by use of a vibrating switch working at a frequency of 80 Hz. The meter gives a reading of 3.2 mA. Calculate the capacitance of the capacitor.

4 A 4.0 μF capacitor is charged to 12 V. The voltage supply is removed without discharging the capacitor. A second capacitor is now connected across the 4.0 μF capacitor, with the result that the potential difference across the capacitors falls to 8.0 V. What is the capacitance of the second capacitor?

5 Two capacitors, of capacitance 2.0 μF and 4.0 μF respectively, are each given a charge of 120 μC. The positive plates are now connected together, as are the negative plates. Calculate the new potential difference between the plates of the capacitors. [AEB 79, p]

6 The diagram for this question is shown in Fig 31.9

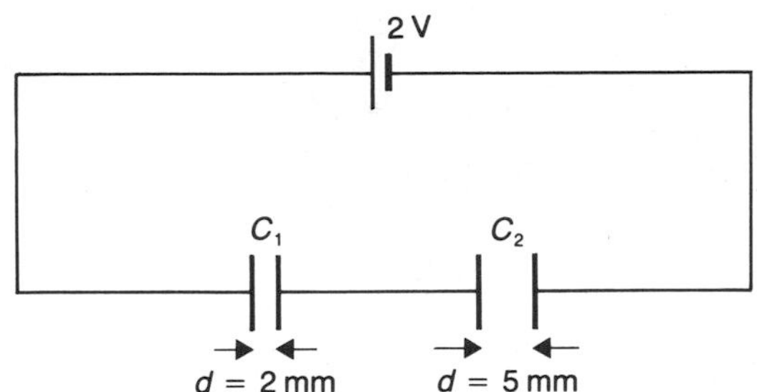

Fig 31.9 Diagram for Question 6

In the circuit, the parallel-plate capacitors are identical except that the distances apart *d* of the plates are as shown. Find the potential difference across each capacitor and the electric field intensity between *each* pair of plates. [WJEC 79]

7 The plates of a parallel-plate capacitor each have an area of 25 cm^2 and are separated by an air gap of 5 mm. The electric field intensity between the plates is $7 \times 10^4\, V\, m^{-1}$. If one plate is at zero potential relative to earth, find the potential of the other plate, and indicate on a sketch some equipotential surfaces in the gap. What is the potential half-way between the plates? Ignore end effects.

What is the capacitance of the capacitor, and how much electrical energy is stored in it?

A slab of dielectric of relative permittivity 15 is introduced into the isolated capacitor so as to exactly fill the gap. What are the new values of (a) the potential difference, (b) the capacitance, (c) the energy?
($\varepsilon_0 = 8.85 \times 10^{-12}\, F\, m^{-1}$.) [WJEC 80, p]

8 A 25 μF capacitor is charged by passing a constant current of 5 μA through it. How long does it take for the potential difference across the capacitor to rise from zero to 20 V? How would you produce the constant 5 μA current?

A capacitor A consists of two parallel metal plates, each of area $2 \times 10^{-3}\, m^2$, separated by a distance of 3 mm in air. Another capacitor B consists of two metal plates, also of area $2 \times 10^{-3}\, m^2$, separated by a distance of 5 mm, the space between them being filled with a material of relative permittivity (dielectric constant) 5. The capacitors are connected in series across a source of potential difference 88 V. What is (a) the capacitance of the system, (b) the electric field strength (potential gradient) between the plates of the capacitor A?
($\varepsilon_0 = 9 \times 10^{-12}\, F\, m^{-1}$.) [SUJB 82, p]

9 A large Van de Graaff generator has a top terminal in the form of a sphere of diameter 4 m. When the terminal is at the operating potential of 5×10^6 V, what is (a) the stored charge, (b) the stored energy, (c) the electric field (potential gradient) at the surface of the sphere?
($\varepsilon_0 = 8.85 \times 10^{-12}\, F\, m^{-1}$.) [SUJB 80, p]

32 MAGNETIC FORCES

Force between parallel wires carrying currents

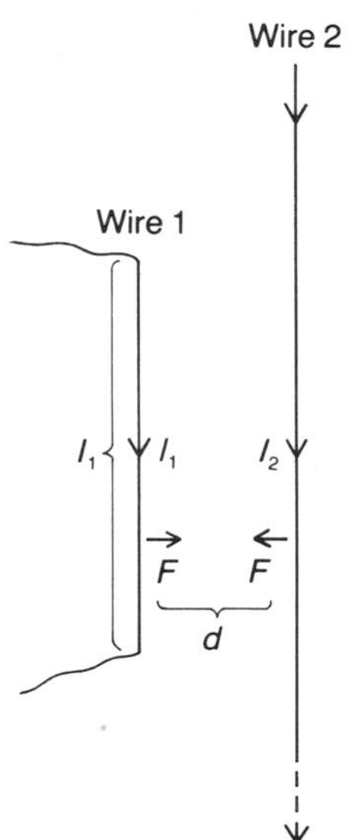

Fig 32.1 Force between parallel wires

As shown in Fig 32.1, a straight wire 1 lying parallel to a long straight wire 2 experiences a force per unit length F/l_1 given by

$$\frac{F}{l_1} = \frac{\mu I_1 I_2}{2\pi d} \qquad (32.1)$$

where μ is a constant for the medium in which the conductors lie. It is called the permeability of the medium.

For air or vacuum the permeability is denoted by μ_0 and has a value of $4\pi \times 10^{-7}$ H m^{-1} (see p.230).

If both wires are very long, i.e. extend well away from the place of interest, then the force per metre is the same for both wires.

Note that if I_1 and I_2 are in the same direction, then the force is an attraction, i.e. like currents attract and opposite currents repel (quite opposite to the rule for poles).

Example 1

Currents of 5 A and 15 A respectively flow down two long, straight vertical wires which are 10 cm apart. Draw a diagram to show the form of the magnetic field in a plane perpendicular to these wires. (Assume there is no other magnetic field present.) Is there any point where the magnetic field is zero, and, if so, where is it situated?

What is the magnitude and direction of the force acting on unit length of each wire due to the currents?
[SUJB 81]

Method

For a straight wire on its own the lines of force due to its current are circles whose centres coincide with the wire and they lie perpendicular to the wire. The directions of the resultant forces when the two wires are present are shown by the lines of force in Fig 32.2. See also p.228.

A point where the field strength is zero (a weak magnet pole put there would experience zero force) occurs between the two wires as shown in the diagram. Its

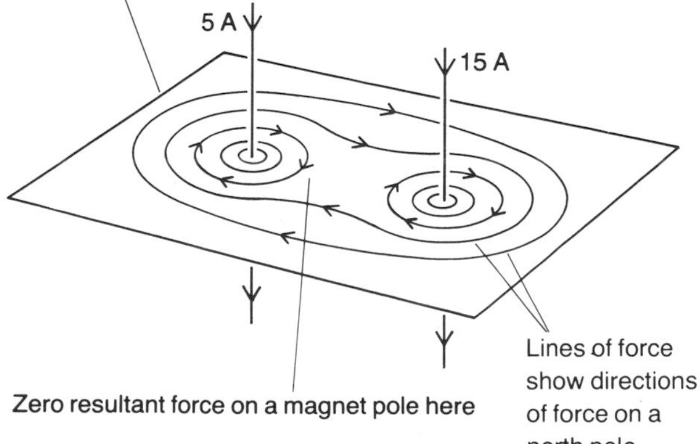

Fig 32.2 Diagram for Example 1

position is 2.5 cm from the 5 A wire and 7.5 cm from the 15 A wire, a 1 : 3 ratio because this is the ratio of the currents — see Chapter 34, Equation 34.1.

The size of the force per metre on each wire is given by Equation 32.1 as

$$\frac{F}{l_1} = \frac{\mu I_1 I_2}{2\pi d} = \frac{4\pi \times 10^{-7} \times 5 \times 15}{2\pi \times 0.1}$$
$$= 1.5 \times 10^{-4}\,\mathrm{N\,m^{-1}}$$

The *forces are inwards*, from one wire towards the other, because *currents in the same direction attract.*

Answer
$1.5 \times 10^{-4}\,\mathrm{N\,m^{-1}}$.

Force on a straight wire related to field applied to it

The equation $F/l = \mu I_1 I_2/2\pi d$ can be written as $F = BI_1 l$, where B is a characteristic of the place where wire 1 lies. B is decided by the current I_2 and the distance away d of the other conductor 2, and by the permeability of the air between. B is a measure of the magnetic field strength affecting wire 1 caused by wire 2. Thus

$$F = BIl \qquad (32.2)$$

where F is the force on a straight wire carrying a current I and having a length l when it lies in a magnetic field whose strength is described by B. We call B the magnetic flux density of the field. It is measured in tesla (T). The direction of B is the direction of the lines of force caused here by wire 2, but note that this applied field could be provided by a permanent magnet instead. In this equation, B must be perpendicular to I.

The direction of F is related to the direction of B by the 'left-hand rule'. According to this the direction of the magnetic **F**ield (flux density B) is indicated by the **F**irst finger, **C**urrent by the se**C**ond finger and the force F, or the **M**otion produced by it, by the thu**M**b when these fingers of the left hand are held mutually at right angles.

If the flux density B is not perpendicular to I but is at an angle θ to it, then our formula for F needs the component perpendicular to I, which is $B \sin\theta$ (Fig 32.3). The formula is then $F = BIl \sin\theta$.

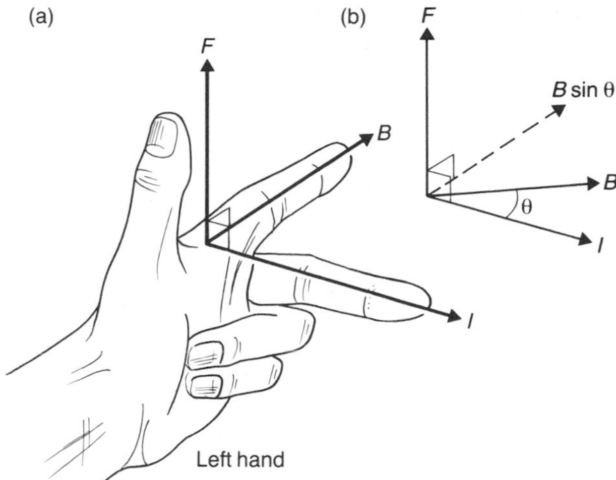

Fig 32.3 The left-hand rule

Example 2

Fig 32.4 shows a simple current balance which is kept horizontal by having the 5.0 g mass at a distance l_2 of 48 mm from the pivots. If the current is 4.8 A, the length l_1 is 100 mm and the length of the horizontal wire in the magnetic field is 40 mm, calculate the average flux density between the magnet's poles. Acceleration due to gravity may be taken as $10\,\mathrm{m\,s^{-2}}$.

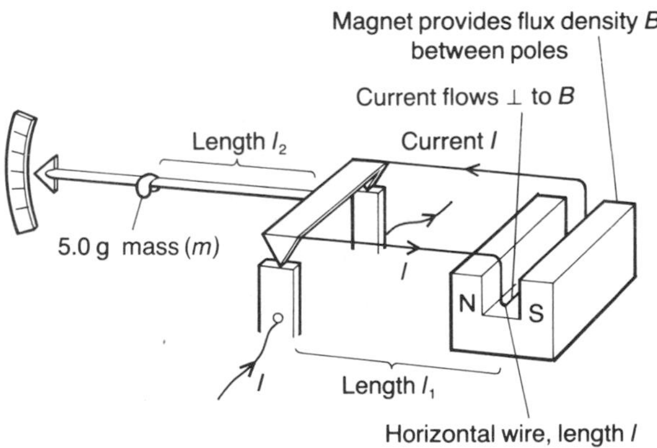

Fig 32.4 Diagram for Example 2

Method
$F = BIl$ and, because we have no rotation, $F \times l_1 = mgl_2$

$$\therefore \quad BIll_1 = mgl_2$$

$$B \times 4.8 \times 40 \times 10^{-3} \times 0.1 = 5.0 \times 10^{-3} \times 10 \times 48 \times 10^{-3}$$

which gives $B = 0.125\,\mathrm{T}$

Answer
0.12 T.

Example 3

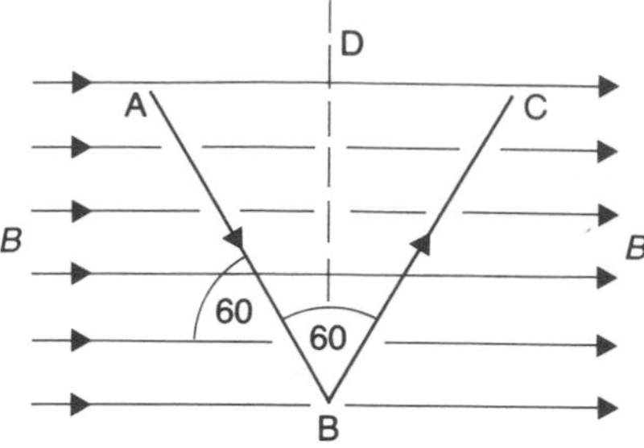

Fig 32.5 Diagram for Example 3

Fig 32.5 shows two straight conductors AB and BC, joined at B, carrying a current of 2.0 A and subjected to a uniform magnetic field of flux density 0.01 T whose direction lies in the plane ABC at 60° to AB. Both AB and BC are 5.0 cm long. The angle ABC is 60°. Calculate the forces on AB and BC. What movement do the two forces together try to produce?

Method

The component of B perpendicular to AB (namely $B\cos 30$ or $B\sin 60$)

$$= 0.01 \times \cos 30 = 0.01 \times 0.866 = 0.0087\,\text{T}$$

The force on AB is

$$F = 0.0087 \times 2.0 \times 5 \times 10^{-2} = 0.000\,87\,\text{N}$$

The force on BC is the same but, while the force on AB is upwards out of the diagram, the left-hand rule gives the force on BC to be downwards. These forces therefore produce a couple about the line BD shown in the diagram, and rotation about this line is expected.

Answer

8.7×10^{-4} N. Rotation about BD.

Force on a charged particle moving through a magnetic field

The formula for this is

$$F = Bqv \qquad (32.3)$$

Here q is the charge carried by the particle, v is the particle's velocity and B is the magnetic flux density perpendicular to v.

We are assuming B to be perpendicular to v. Otherwise B must be replaced in the formula by its component perpendicular to v, since only this component is effective. The direction of F is of course given by the left-hand rule (and note that this rule considers current direction, which for the movement of negative particles such as electrons will be opposite to that of the particle velocity).

The formula $F = Bqv$ is used later in this chapter and also in Chapter 39.

Couple on a coil

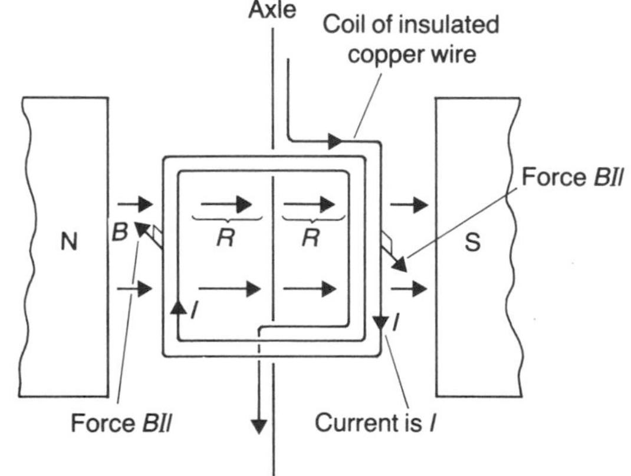

Fig 32.6 Couple on a coil

In Fig 32.6 a force BIl acts on each vertical wire. If there are n turns of wire on the coil, the total force is $BIln$ on each side of the coil.

The couple is $C = 2BIlnR$. However, the coil area A equals $2Rl$, so that

$$C = BAIn \qquad (32.4)$$

If the magnetic field is radial (see Fig 32.7a), then B is always parallel to the plane of the coil even when the coil is allowed to rotate, and $C = BAIn$ still. If instead the field is uniform (see Fig 32.7b and c), then the component of B which is effective is $B\cos\theta$ (see diagram), and the couple is $C = BAIn\cos\theta$.

(a) Radial field (seen from above)

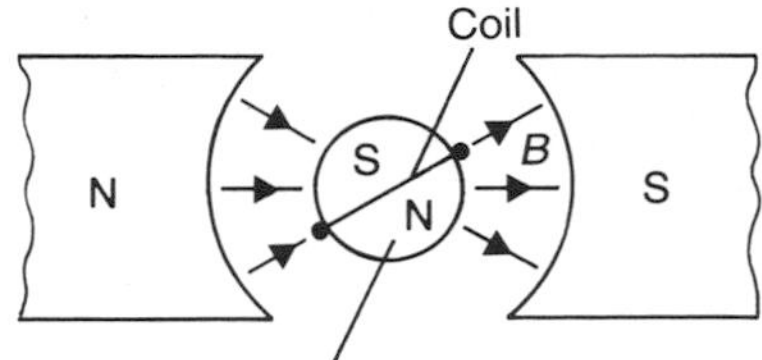

(b) Uniform field

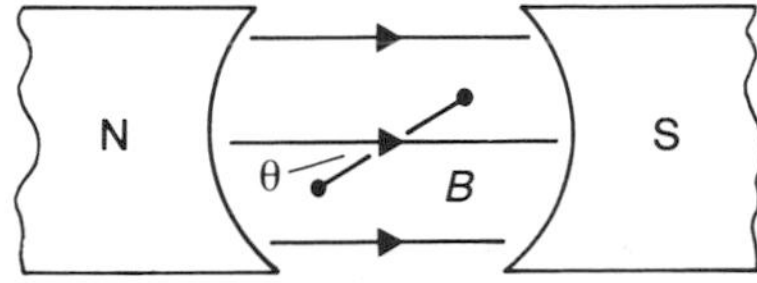

(c) Uniform field with large soft iron cylinder

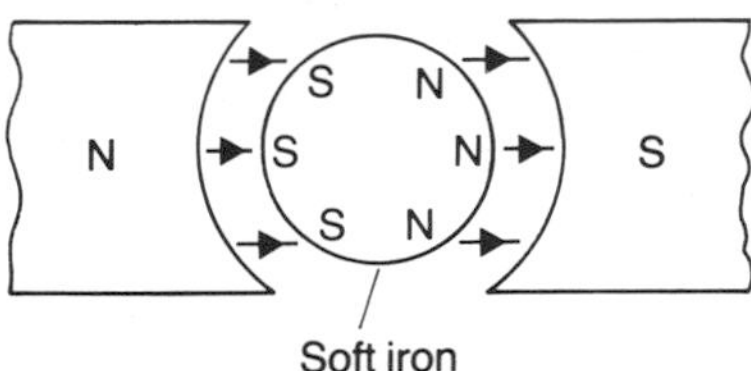

Fig 32.7 Radial and uniform magnetic fields

In a moving-coil meter the current to be measured flows through the coil, the field is radial, and the couple $BAIn$ turns the coil. The turning tightens a spring which therefore produces an opposing couple of k newton metre per unit angle of rotation.

The coil comes to rest when $BAIn = k\theta$, where the angle of rotation of the coil θ is in degrees or radians.

$$BAIn = k\theta \qquad (32.5)$$

In more sensitive meters the coil is supported on a fine wire instead of an axle. When the coil rotates, the twist (torsion) in the wire causes the opposing couple $k\theta$, and a spring is not needed.

Example 4

A coil of area 50 cm^2 is suspended with its plane vertical in a horizontal magnetic field of flux density 8.0×10^{-3} T. The coil has 100 turns, and its plane makes an angle of 30° with the magnetic field. What is the couple acting on the coil when the current through it is 80 mA? [AEB 82, p]

Method

A diagram could be useful here.

Note that the field is uniform.

The couple is given by Equation 32.4 $C = BAIn$.

B here is the component of the 8.0×10^{-3} T in the plane of the coil, i.e. $8.0 \times 10^{-3} \times \cos 30$ which is $8.0 \times 10^{-3} \times 0.866$ or 6.93×10^{-3} T. The current is 80×10^{-3} A and $n = 100$. Area $A = 50 \times 10^{-4}$ m^2.

$$\therefore \quad C = 6.93 \times 10^{-3} \times 50 \times 10^{-4} \times 80 \times 10^{-3} \times 100$$

$$= 27\,720 \times 10^{-8}\,\text{N m} \quad \text{or} \quad 2.8 \times 10^{-4}\,\text{N m}$$

Answer

2.8×10^{-4} N m.

Example 5

A moving-coil galvanometer has a coil of 80 turns, each of area 50 mm^2, suspended in a radial magnetic field of 0.3 T by a wire of torsional constant 6×10^{-9} N m rad^{-1}. The resistance of the coil is 20 Ω. Calculate the angular deflection of the coil produced by (a) a current of 1 μA, (b) an applied potential difference of 1 μV.

The galvanometer described above is modified by rewinding the coil using 160 turns of thinner wire having half the cross-section area of that previously employed. The turns on the new coil have each the same area as those on the old one. What effect have these modifications on the angular deflections by (a) a current of 1 μA, (b) a potential difference of 1 μV? [SUJB 80, p]

Method

(a) Equation 32.5 can be used.

$$BAIn = k\theta$$

where $n = 80$, $A = 50 \times 10^{-6}$, $B = 0.3$, $k = 6 \times 10^{-9}$, $I = 10^{-6}$.

$$\theta = \frac{BAIn}{k}$$

$$= \frac{0.3 \times 50 \times 10^{-6} \times 10^{-6} \times 80}{6 \times 10^{-9}}$$

$$= 0.2\,\text{rad}$$

This answer for θ is in radians because k was in N m per *radian*.

(b) An applied PD V will give a current I given by $I = V/R$, and here the resistance R is 20 Ω. Now $V = 10^{-6}$ volt so

$$I = \frac{10^{-6}}{20} = 0.05 \times 10^{-6}\,\text{A}$$

We know from part (a) that 1 μA gives $\theta = 0.2$ so that 0.05 μA gives 0.2×0.05 or 0.01 rad.

The effect of the modifications upon the current sensitivity is that n is doubled in the formula $\theta = BAIn/k$, so that θ is doubled. Halving the wire's area increases its resistance in accordance with the equation $R = \rho l/A$ (Equation 27.8), and l is doubled too. Therefore 1 μV produces only a quarter of the previous current, but the current sensitivity has been doubled. So the deflection is halved.

Answer
(a) 0.2 rad, (b) 0.01 rad and, after modifications:
(a) 0.4 rad, (b) 0.005 rad.

The earth's magnetism

In the United Kingdom the direction of the magnetic flux density due to the earth's magnetism makes an angle θ (called the 'angle of dip') of about 70° to the horizontal. The horizontal component of this flux density B is $B_0 = B\cos\theta$, and the vertical component is $B_V = B\sin\theta$.

Note also that

$$B_V \;(= B\sin\theta) = \frac{B_0 \sin\theta}{\cos\theta} = B_0 \tan\theta,$$

where θ is the angle of dip as shown in Fig 32.8.

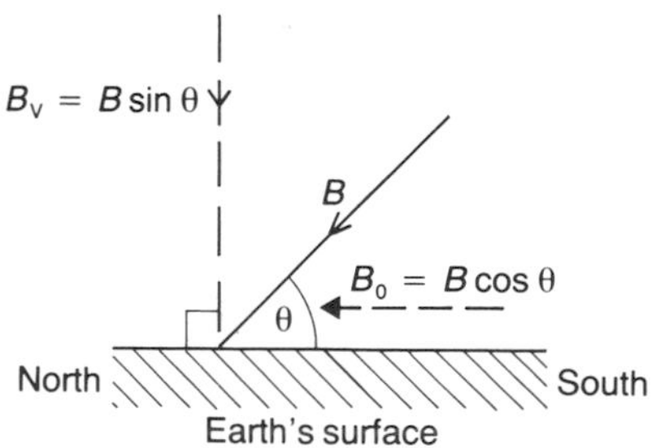

Fig 32.8 The angle of dip

Example 6

Calculate the size and direction of the force per metre length on a straight, horizontal wire lying with 2.0 A flowing through it in direction north to south. (Earth's horizontal field component = 1.6×10^{-5} T. Angle of dip = 70°.)

Method

$$B_v = B_0 \tan\theta = 1.6 \times 10^{-5} \times \tan 70$$
$$= 4.4 \times 10^{-5}\,\text{T}$$
$$F = B_v Il = 4.4 \times 10^{-5} \times 2.0 \times 1$$
$$= 8.8 \times 10^{-5}\,\text{N}$$

By the left-hand rule, with B_v downwards, F is eastwards.

Answer
88 μN, eastwards.

Neutral points

In the vicinity of a magnet or current-carrying conductor there may be places where the flux density due to the magnet is exactly cancelled by an equal and oppositely directed flux density caused by the field of the earth or of some other magnet. Such points are known as neutral points. (See Chapter 34 Example 1.)

Exercise 111

(Where necessary take μ_0 to be $4\pi \times 10^{-7}\,\text{H m}^{-1}$.)

1 Two very long parallel wires 0.4 m apart in air each carry a current of 5.0 A. What is the force, in newtons, on each metre length of wire?

2 A horizontal wire of length 4.0 cm is moving vertically downwards, with a current of 1.0 A flowing through it. If the plane in which the wire moves is perpendicular to a magnetic flux density of 0.1 T, calculate the force on the wire due to the current.

3 A 5.0 cm long, horizontal straight wire is fitted to the end of a pivoted frame so as to form a simple current balance which weighs 25 g. The distance from the wire to the pivot is 15 cm. The wire has a horizontal magnetic field of flux density 0.2 T perpendicular to it and it carries a current of 5.0 A.
(a) Calculate the force on the wire caused by this current flow.
(b) It is found that the 5.0 A current keeps the frame balanced horizontally without any extra weights being added. (i) How far from the pivot

is the centre of gravity? (ii) If the current direction is reversed, where would a 15 g weight have to be hung to restore balance? (g may be taken as $10\,\mathrm{m\,s^{-2}}$.)

4 A moving-coil meter has a 50-turn coil measuring 1.0 cm by 2.0 cm which is supported on a torsion wire whose torsional constant is $3.0 \times 10^{-8}\,\mathrm{N\,m\,rad^{-1}}$. The radial magnetic field has a flux density of 0.15 T. What current is required to give a deflection of 0.5 rad?

5 A certain electric motor rotates in a uniform magnetic field of 0.1 T between the poles of a magnet. When it first starts a current of 4.0 A flows. If the coil is in the form of a flat, rectangular winding of 10 turns and $6.0\,\mathrm{cm^2}$ area, calculate (a) the maximum torque on the coil and (b) the maximum angular acceleration of the coil if there are no retarding forces, and given that the moment of inertia of the coil about the axle is $8.0 \times 10^{-5}\,\mathrm{kg\,m^2}$.

Exercise 112: Questions of A-level standard

1 A current of 3 A flows down each of two long vertical wires which are mounted side by side 5 cm apart. Show on a diagram the magnetic field pattern in a horizontal plane, indicating clearly the direction of the magnetic field at any point. What is the magnitude and direction of the force on a 25 cm length of wire? (Take μ_0 to be $4\pi \times 10^{-7}\,\mathrm{H\,m^{-1}}$.) [SUJB 79]

2 Two long parallel vertical wires 0.3 m apart are placed east–west of one another. The current in the westerly wire is 20 A and the other is 30 A. Both currents flow upwards. If the horizontal component of the earth's magnetic flux density is $2.0 \times 10^{-5}\,\mathrm{T}$, calculate the resultant force on 1 metre of each wire. ($\mu_0 = 4\pi \times 10^{-7}\,\mathrm{H\,m^{-1}}$.) [C 78, p]

3 The diagram for this question is given in Fig 32.9. The diagram represents a cylindrical aluminium bar AB resting on two horizontal aluminium rails which can be connected to a battery to drive a current through AB. A magnetic field, of flux density 0.10 T, acts perpendicularly to the paper and into it. In which direction will AB move if the current flows from A to B?

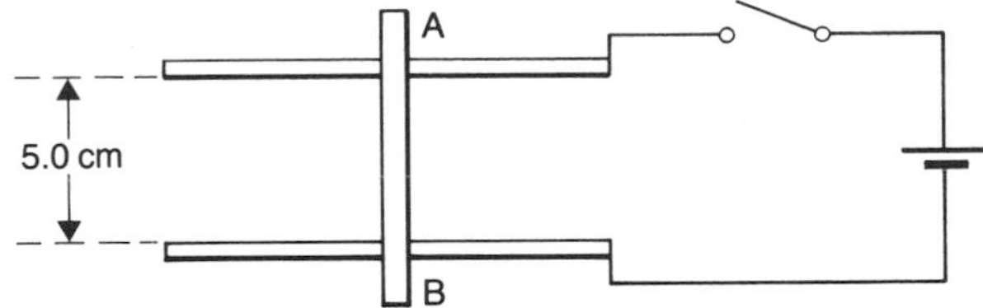

Fig 32.9 Diagram for Question 3

Calculate the angle to the horizontal to which the rails must be tilted to keep AB stationary if its mass is 5.0 g, the current in it is 4.0 A and the direction of the field remains unchanged. (Acceleration of free fall $g = 10\,\mathrm{m\,s^{-2}}$.) [L 77]

4 Two moving-coil meters X and Y are of identical construction except that the coil of X has 100 turns and its resistance is 10 Ω, and the coil of Y has 200 turns and its resistance is 60 Ω. What is the ratio of the deflections $\theta_X : \theta_Y$ when each is connected in turn across a DC source of internal resistance 40 Ω?

	$\theta_X : \theta_Y$
A	1 : 3
B	1 : 1
C	2 : 1
D	3 : 1
E	12 : 1

[O 81]

5 A moving-coil galvanometer has a coil of 100 turns, each of area $5\,\mathrm{cm^2}$, mounted in a radial B field of $20 \times 10^{-3}\,\mathrm{T}$. It is desired to use this as a milliammeter giving a full scale deflection of $\frac{1}{3}\pi$ rad for a current of 10 mA. Calculate (in appropriate units) the necessary restoring couple per unit angle of twist of the suspension.

If the coil resistance is 30 Ω, how can the above milliammeter be adapted to produce (a) an ammeter of FSD 5 A and (b) a voltmeter of FSD 150 V? [WJEC 81, p]

33 ELECTROMAGNETIC INDUCTION

Straight conductors

Induced EMF in a straight wire

Consider a straight wire of length l moving with a velocity v perpendicular to its length in a magnetic field of flux density B which is perpendicular to both the wire's length and the velocity (Fig 33.1a).

The metal wire contains free electrons, so that movement of the wire means movement of these electrons at a velocity v. We can use the formula $F = Bev$ (Chapter 32) and deduce that each free electron is moved by this force along the wire until a PD is established between the two ends of the wire sufficient to stop any further movement of the electrons. This PD is produced almost instantly and is given by

$$E = Blv \qquad (33.1)$$

This is the induced EMF. If the ends of the wire were joined by conductors of resistance R to form a complete circuit of resistance $R + r$, where r is the resistance of the straight wire itself (Fig 33.1b) then the current I resulting from the electromagnetic induction is

$$I = \frac{Blv}{R + r}$$

and the terminal potential difference between the wire's ends is

$$V = \frac{BlvR}{R + r}$$

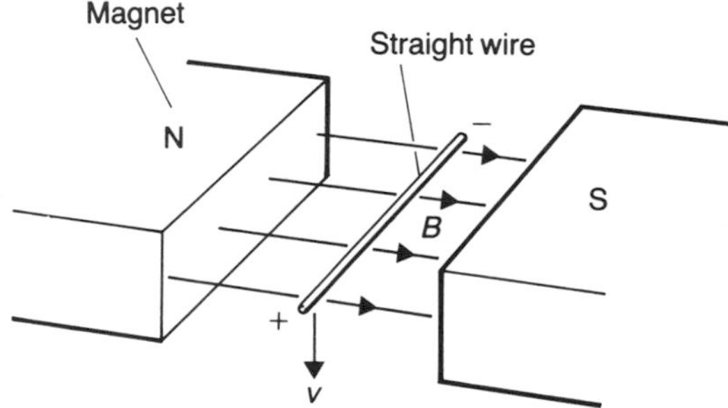

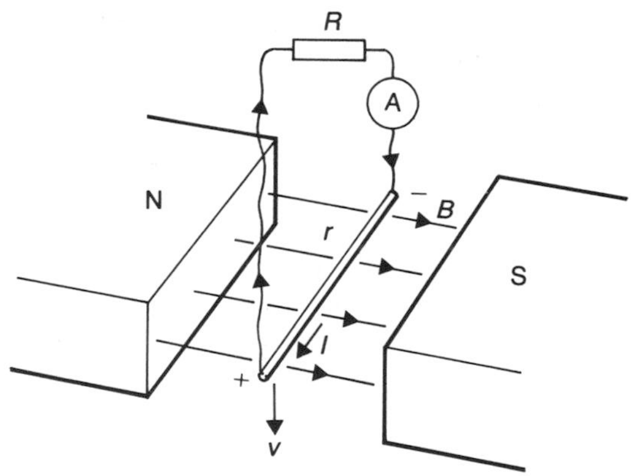

Fig 33.1 Electromagnetic induction in a straight wire

The direction in which the induced current flows (Fig 33.1b) can be deduced by use of Lenz's law together with the left-hand rule (Chapter 32) or, alternatively, the right-hand rule may be used with the first finger for the B direction, thumb for movement direction and the second finger for the induced current.

Magnetic flux Φ

If an area A lies perpendicular to a magnetic flux density B then the product BA is called the magnetic flux and is usually denoted by the symbol Φ. The direction of the flux is the same as the direction of the magnetic field.

The unit for Φ is the weber, and 1 tesla = 1 weber per metre^2.

$$\Phi = BA \qquad (33.2)$$

Induced EMF in a straight wire in terms of magnetic flux

In the formula $E = Blv$ the product lv is the area cut through per second by the wire moving perpendicular to B, so that $B \times lv$ is $B \times$ the area per second. Therefore E = flux cut per second by the moving wire. Thus $E = \Phi/t$ where Φ is the flux cut in time t, and we are assuming that E is constant. If E is not constant then its value at any instant is $d\Phi/dt$:

$$E = \frac{d\Phi}{dt} \qquad (33.3)$$

B not perpendicular to the area

If the direction of B is inclined to the area at an angle θ, then the effective value of B is $B\cos(90 - \theta)$ or $B\sin\theta$. (The same result is obtained if we say that there is an area $A\sin\theta$ perpendicular to B. Either way $E = Blv\sin\theta$.)

Example 1

An aeroplane is travelling at $100\,\text{m}\,\text{s}^{-1}$ in a direction which is horizontal and northwards. Calculate the EMF induced between the tips of its wings, which have a span of 20 m. Take the earth's magnetic flux density to be 5.0×10^{-5} T and the angle of dip 71° at the place concerned.

Method

From Fig 33.2 the component of the flux density perpendicular to the aeroplane wing's movement is $5.0 \times 10^{-5}\cos 19°$. Using the formula $E = Blv$:

$$E = (5.0 \times 10^{-5}\cos 19) \times 20 \times 100$$
$$= 10^{-1} \times \cos 19 = 10^{-1} \times 0.946 = 0.0946\text{ V}$$

or 95 mV to two significant figures.

Answer

95 mV.

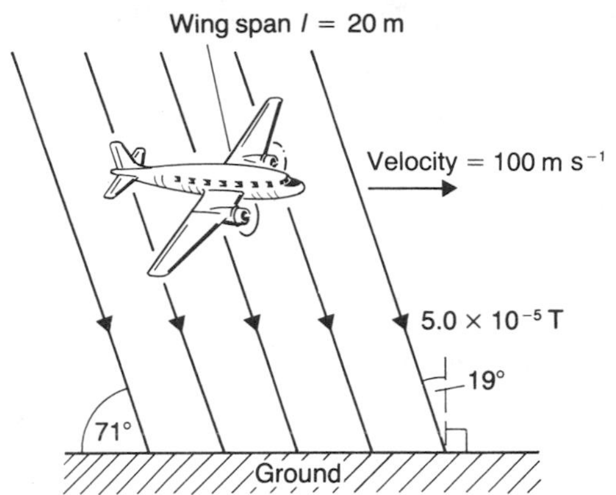

Fig 33.2 Suitable diagram for Example 1

Example 2

A circular metal disc of area $3.0 \times 10^{-3}\,\text{m}^2$ is rotated at 50 rev/s about an axle through its centre and perpendicular to its plane. The disc is in a uniform magnetic field of flux density 5.0×10^{-3} T in the direction of the axle. Between which points on the disc is the maximum EMF induced? What is the value of this EMF? [L 81, p]

Method

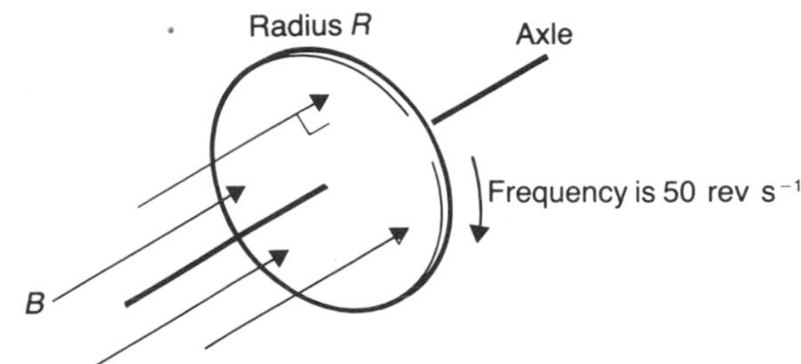

Fig 33.3 Diagram for Example 2

The maximum EMF is obtained between the rim and the axle. A suitable diagram is shown in Fig 33.3. We may look at this problem in two ways.

(1) Consider a radial strip (like a spoke of a wheel). As it rotates, the induced EMF in it is $E = Blv$ where l is its length (= radius of the disc) and v is the *average* velocity of rotation. Now the velocity of its outer tip is $2\pi R \times$ frequency $= 2\pi R \times 50$.

$$\therefore \quad \text{Average velocity} = \frac{2\pi R \times 50 - 0}{2}$$

or $50\pi R$.

$$\therefore \quad E = B \times R \times 50\pi R = B \times \pi R^2 \times 50$$
$$= 5.0 \times 10^{-3} \times 3.0 \times 10^{-3} \times 50$$
$$= 750 \times 10^{-6} \text{ volt} \quad \text{or} \quad 0.75 \text{ mV}$$

(2) This time we think of the formula $E = d\Phi/dt$. Again we consider a radial strip or spoke. It cuts through an area πR^2 in each revolution. Therefore the flux cut per second is $\pi R^2 B \times$ frequency. $E = 3.0 \times 10^{-3} \times 5.0 \times 10^{-3} \times 50$, agreeing with (1) and giving 0.75 mV.

Answer
0.75 mV.

Exercise 113

1 Calculate the induced EMF in a straight wire when it is moving at 5.0 m s^{-1} perpendicular to its length in a magnetic field of flux density 0.10 T if the field direction is (a) perpendicular to the plane of movement, (b) parallel to it, (c) at 60° to it. The wire length is 1.0 cm.

2 (a) Calculate the EMF induced between the axle and the rim of a spoked metal wheel if the wheel radius is 20 cm and the uniform field in which it lies is 0.020 T perpendicular to the plane of the wheel, the speed of rotation being 10 revolutions per second.

(b) What is the expected current size through a 10 Ω resistor connected between the axle and the rim if the wheel's resistance is negligible?

3 Calculate the flux cut through in 1.0 ms by a straight wire 3.0 cm long moving at 2.0 m s^{-1} perpendicular to its length and to a magnetic field of flux density 10 mT.

Coils

Induced EMF in a coil

If a flat coil lies with its plane, of area A, perpendicular to a magnetic field whose flux density is B, then the flux Φ 'passing through'* the coil is $B \times A$.

'Passing through' as if flux were a flow of something through the coil, along the lines of force of the magnetic field.

$$\text{Flux } \Phi = B \times A \tag{33.4}$$

Φ can be changed — e.g. by changing B or by rotating the coil so that less flux passes through it. Now the change of this flux Φ through the coil is also the flux cut through by the wires of the coil. So we can use the formula (Equation 33.3) obtained earlier for the induced EMF, namely $E = d\Phi/dt$.

However, for the coil it is appropriate to describe $d\Phi/dt$ as the *rate of change of flux through the coil.*

For a coil of n turns the induced EMF is n times greater.

$$E = n\frac{d\Phi}{dt} \tag{33.5}$$

where Φ is the flux through the coil.

Alternatively we write

$$E = \frac{d\Phi}{dt} \tag{33.6}$$

even though the coil has n turns and Φ now represents the 'effective flux' through the coil, called the flux linkage, this quantity being the product of flux through coil × number of turns.

$$\text{Flux linkage} = n \times \text{Flux}$$

The SI unit for flux linkage is also the weber (Wb).

Note that, since both flux and flux linkage are usually denoted by the same symbol Φ and have the same unit, it will sometimes be necessary to distinguish between them, e.g. by writing 'flux Φ' or 'flux linkage Φ'.

Frequently Equation 33.6 is written as

$$E = -\frac{d\Phi}{dt}$$

so that, using a suitable sign convention for $d\Phi/dt$, the polarity of E is obtained. This is not essential for A-level calculations.

For a coil of n turns and area A, perpendicular to a uniform flux density B the flux linkage is

$$\Phi = BAn \tag{33.7}$$

A typical example of induced EMF in such a coil is the steady reduction to zero in time t of the flux density B. The flux linkage change is $BAn - 0$ so that $E = BAn/t$.

Self-induction

If the current I in a coil changes, then the magnetic flux density B within the coil changes (as well as the field around the coil of course), and this causes electromagnetic induction in the coil. The induced EMF is

$$E = L\frac{dI}{dt} \tag{33.8}$$

where L is called the self-inductance of the coil. L is decided by the coil's geometry and number of turns and also by the presence of magnetic material within or around the coil. The SI unit for self-inductance is the henry (H).

Mutual induction

When two coils are close so that a change of current I_1 in one of the coils causes a change in the flux density inside the second coil, an EMF is induced in the second coil, given by

$$E_2 = M\frac{dI_1}{dt} \tag{33.9a}$$

where M is called the mutual inductance of the pair of coils. Similarly a change of current in coil 2 will cause an induced EMF in coil 1 given by

$$E_1 = M\frac{dI_2}{dt} \tag{33.9b}$$

where M is the same mutual inductance value.

The unit for M is also the henry (H).

Induced charge

When a coil which is part of a complete circuit is subjected to a change of flux linkage, the charge Q that flows is independent of the *rate* of change of flux linkage, and is given by

$$Q = \frac{\text{Total flux linkage change}}{\text{Total circuit resistance}} \tag{33.10}$$

Example 3

The flux threading a coil of 3 turns drops from 5 Wb to zero in 8 s. Which of the following gives the magnitude of the EMF induced in the coil?

A $5 \times 8 \times 3$ V **B** $\frac{5 \times 8}{3}$ V **C** $\frac{5 \times 3}{8}$ V

D $\frac{3 \times 8}{5}$ V **E** $\frac{5}{8 \times 5}$ V [L 77]

Method

A suitable diagram for this example is shown in Fig 33.4

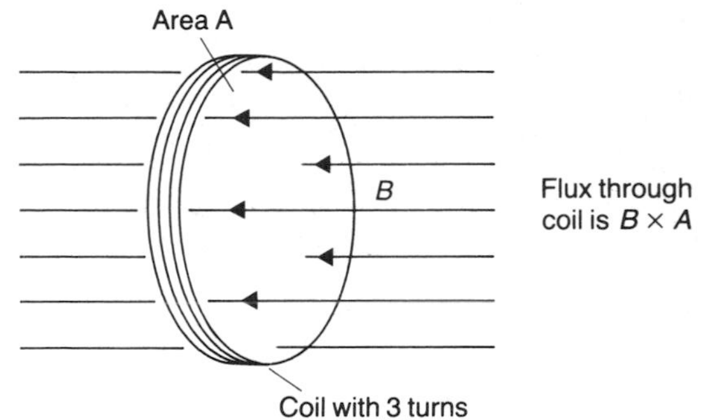

Fig 33.4 Diagram for Example 3

The flux $\Phi = B \times A$ (Equation 33.4)

$= 5$ Wb initially.

Initial flux linkage is $\Phi = BAn$ (Equation 33.7) $=$ 5×3 Wb. Final flux linkage is zero.

$$\text{Induced EMF} = \frac{\text{Change of flux linkage}}{\text{Time taken}} \quad \text{(Equation 33.6)}$$

$$= \frac{BAn}{t} = \frac{5 \times 3}{8} \quad \text{i.e. answer } \mathbf{C}$$

Answer

C.

Example 4

If the current in a coil falls at 2.0 A s^{-1} and consequently induces an EMF of 4.0 mV in a second coil close to it, what induced voltage would occur in the first coil due to a 4.0 A s^{-1} fall of current in the second coil?

Method

$$E_2 = M\frac{dI_1}{dt} \quad \text{and} \quad E_1 = M\frac{dI_2}{dt}$$

$$\therefore \quad 4.0 \times 10^{-3} = M \times 2.0$$

$$\therefore \quad M = 2.0 \times 10^{-3}\,\text{H}$$

$$E_1\left(= M\frac{dI_2}{dt}\right) = 2.0 \times 10^{-3} \times 4.0 = 8.0 \times 10^{-3},$$

i.e. 8.0 mV.

Answer

8.0 mV.

Example 5

A ballistic galvanometer was connected in series with a flat circular coil (having 100 turns of radius 1.5 cm) and a resistor, such that the resistance of the circuit was 550 Ω. The coil was held in the uniform field of a magnet with its plane perpendicular to the field direction and then the coil was quickly removed. The resulting first deflection of the galvanometer was 4.5 cm. The same deflection was obtained by discharging through the meter a 1.0 μF capacitor that had been charged to 9.0 V. Obtain a value for the flux density of the magnet's field.

Method

Suitable diagrams are shown in Fig 33.5

(a) Using the search coil

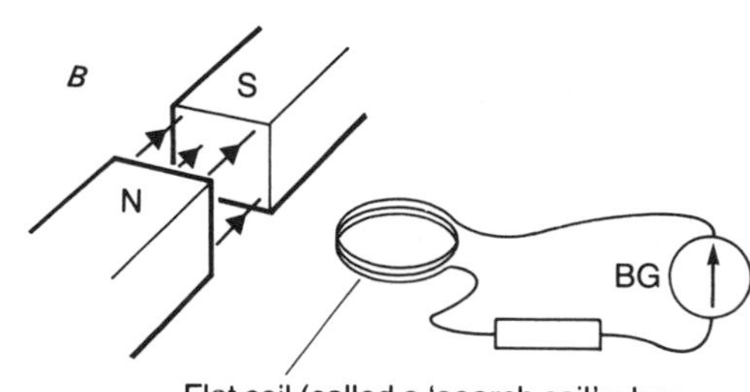

(b) Using the capacitor

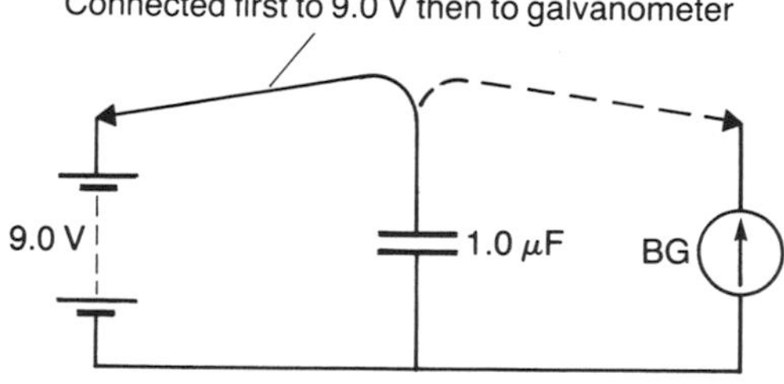

Fig 33.5 Diagram for Example 5

The number of turns (n) is 100.

The area A of each turn is $\pi \times (1.5 \times 10^{-2})^2$ or $2.25\pi \times 10^{-4}$.

The charge stored in the capacitor (see Chapter 31) is

$$Q = CV = 1.0 \times 10^{-6} \times 9.0 = 9.0 \times 10^{-6}\,\text{C}$$

For the induced charge in the search coil circuit we use Equation 33.10.

$$Q = \frac{\text{Total flux linkage change}}{\text{Total resistance}}$$

which gives

$$Q = \frac{BAn}{550}$$

$$= \frac{B \times \pi \times 2.25 \times 10^{-4} \times 100}{550}$$

$$= B \times 1.28 \times 10^{-4}\,\text{C}$$

But this must equal 9.0×10^{-6} C because it also gives a 4.5 cm reading.

$$\therefore \quad 9.0 \times 10^{-6} = 1.28 \times 10^{-4} \times B$$

$$\therefore \quad B = \frac{9.0}{1.28} \times 10^{-2} = 7.0 \times 10^{-2}\,\text{T}$$

Answer

7.0×10^{-2} T.

Exercise 114

1 A flat coil having an area of 8.0 cm^2 and 50 turns lies perpendicular to a magnetic field of 0.20 T. If the flux density is steadily reduced to zero, taking 0.50 second, what is (a) the initial flux through the coil, (b) the initial flux linkage, (c) the induced EMF?

2 Calculate the self-inductance of a coil that experiences an induced EMF of 20 mV when the current through it changes at a rate of 2.0 A s^{-1}.

3 The current in one coil of a pair changes at 0.10 A s^{-1}. Calculate the induced EMF in the other coil if this second coil is open circuit and the mutual inductance is 100 mH.

Rotating coils

Induced charge in a rotated coil

The flux linkage for a coil with its axis inclined at angle θ to a flux density B (Fig 33.6) is $B \cos\theta \times An$ where A is the coil area and n its number of turns.

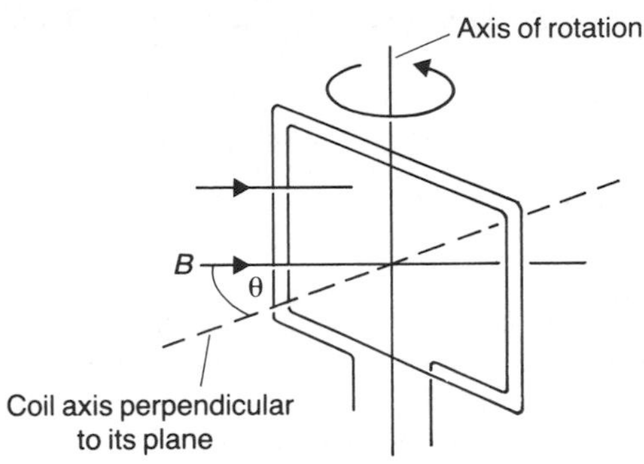

Fig 33.6 Induced charge in a rotated coil

The induced charge is given by flux linkage change divided by the resistance of the circuit (Equation 33.10).

Note that when the plane of the coil becomes perpendicular to B, the induced voltage and current are zero and then reverse. Consequently a rotation made up of equal movements towards and away from this position will produce no net induced charge. For example, a rotation starting from this position and moving through 180° gives an induced charge of $2 \times BAn$; but starting from a position where the plane of the coil is parallel to B and turning through 180° gives a net induced charge of zero because the current reverses half-way through the rotation.

Induced EMF in a rotating coil in a uniform field

When a coil rotates as in Fig 33.7a the formula for the induced EMF can be obtained by use of the equation $E = \mathrm{d}\Phi/\mathrm{d}t$ and is

$$E = 2\pi fBAn \sin 2\pi ft \qquad (33.11)$$

in which f is the number of revolutions per second (i.e. the frequency of rotation), A is the coil area, n the number of turns of wire on the coil and t is the time. The magnetic flux density B is assumed to be uniform (the same everywhere) and perpendicular to the axis of rotation.

As t increases, $\sin 2\pi ft$ will reach a maximum value of unity, so that the maximum, or peak, value of E is $2\pi fBAn$ and we can write $E = E_p \sin \omega t$ where

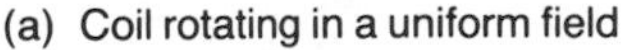

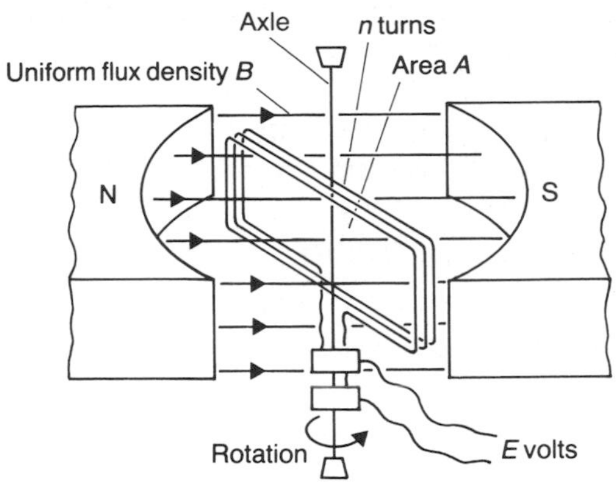

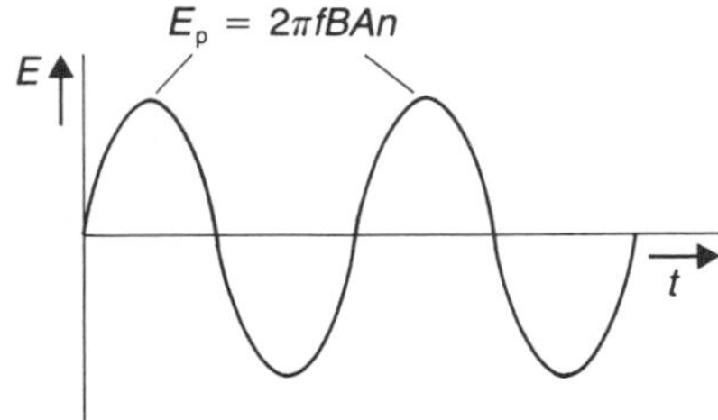

Fig 33.7 Induced EMF in a rotating coil

E_p is the peak value and ω is the angular frequency ($2\pi f$) of the EMF or the angular velocity of the rotating coil (Fig 33.7b). In these equations t is zero when the plane of the coil is perpendicular to B, and ωt is the angle between the coil axis (not rotation axis) and B.

In a radial field E would equal E_p for a half revolution, then suddenly reverse in polarity when B reverses direction, but again will equal E_p in size.

Example 6

A coil of 200 turns and 12 cm^2 area is rotating at 20 revolutions per second in a uniform magnetic field of flux density 0.020 T. Calculate (a) the induced EMF when the coil's plane is momentarily (i) parallel to B, (ii) at 20° to B; (b) the charge that flows during the coil's movement through the 20° from position (i) to position (ii), given that the coil is part of a circuit of 120 Ω resistance.

What would the induced EMF be if a radial field of 0.020 T were used instead?

Method

(a) (i) The induced EMF is $E = 2\pi fBAn \sin 2\pi ft$ (Equation 33.11).

We have $f = 20\,s^{-1}$, $B = 0.020\,T$, $A = 12 \times 10^{-4}\,m^2$, $n = 200$ and $\sin 2\pi ft = 1$ when the coil's plane is parallel to B.

$$E = 2\pi \times 20 \times 0.020 \times 12 \times 10^{-4} \times 200 \times 1$$

$$= 0.603\,V \text{ or } 0.60\,V$$

(ii) The angle $2\pi ft$ equals 90° when the coil's plane is parallel to B, and movement through 20° from that position means that the angle $2\pi ft = 70°$ (or 110°).

$$E = 2\pi \times 20 \times 0.02 \times 12 \times 10^{-4} \times 200 \times \sin 70°$$

$$= 0.603\,V \times 0.94$$

$$= 0.57\,V$$

(b) The flux linkage change is from zero, when the coil's plane is parallel to B, to $B \cos 70° \times An$, when the plane is at 20° to B. Using Equation 33.10 we get

$$Q = \frac{0.020 \times 12 \times 10^{-4} \times 200 \times \cos 70°}{120}$$

$$= 4.0 \times 10^{-5} \times \cos 70$$

$$= 4.0 \times 10^{-5} \times 0.342$$

$$= 1.368 \times 10^{-5} \quad \text{or} \quad 14\,\mu C$$

For the radial field the EMF would be 0.60 V because the coil's plane is in line with B at all times. ($\sin 2\pi ft = 1$.)

Answer

(a) (i) 0.60 V, (ii) 0.57 V, (b) 14 μC.
0.60 V for radial field.

Exercise 115

1 A flat rectangular coil measuring 2.0 cm by 3.0 cm and having 60 turns is connected to a ballistic galvanometer or other charge meter. If the total circuit resistance is 100 ohm, and the coil is turned through 90°, calculate the charge that should be indicated by the meter, given that the coil lies in a field of flux density 0.30 T, with the coil plane (a) initially perpendicular to B, (b) initially parallel to B.

What would the answers be if the rotation were continued to 180°?

2 A flat coil of area 4.5 cm² having 200 turns of resistance 20 Ω lies with its area (a) perpendicular to the field, (b) at 30° to the field. If the field is reduced from 1.0 T to 0.60 T, what is the induced charge in each case if the external circuit resistance is negligible?

3 A coil is rotating in a uniform field of 0.01 T perpendicular to the axis of rotation (as in Fig 33.7). The coil area is 2.0 cm², the number of turns is 50 and the steady speed of rotation is 20 revolutions per second. Calculate (a) the maximum induced EMF, (b) the induced EMF at the instant when the plane of the coil lies at 20° to the field direction.

Exercise 116: Questions of A-level standard

1 A metal-framed aeroplane with a wing span of 30 m is flying horizontally, due west, over the British Isles. Given that its speed is 720 km h^{-1}, calculate the EMF induced between its wing tips on account of the earth's magnetic field. At the latitude concerned, this field has a horizontal component of 1.6×10^{-5} T and the angle of dip may be taken as 70°.

2 A circular disc of radius 0.010 m is rotated at 100 revolutions per minute about an axis through its centre and normal to its plane. A uniform magnetic field of flux density 5.0 T (Wb m^{-2}) exists normal to the plane of the disc. Calculate the value of the potential difference developed between the centre and the rim of the disc. [AEB 79, p]

3 A closed wire loop in the form of a square of side 4.0 cm is mounted with its plane horizontal. The loop has a resistance of 2.0×10^{-3} Ω, and negligible self-inductance. The loop is situated in a magnetic field of strength 0.70 T directed vertically downwards. When the field is switched off, it decreases to zero at a uniform rate in 0.80 s. What is (a) the current induced in the loop, (b) the energy dissipated in the loop during the change in the magnetic field? Show on a diagram, justifying your statement, the direction of the induced current [SUJB 80]

4 What is meant by the statement that a solenoid has an inductance of 2.0 H? A 2.0 H solenoid is connected, in series with a resistor, so that the total resistance is 0.50 Ω, to a 2.0 V DC supply. Sketch the graph of current against time when the current is switched on.

What is (a) the final current, (b) the initial rate of change of current with time and (c) the rate of change of current with time when the current is 2.0 A?

Explain why an EMF greatly in excess of 2.0 V will be produced when the current is switched off.

[L 77, p]

5 Two coils have a mutual inductance of 1.5 H. What change of current in one coil will induce an EMF of 3.0 V in the other coil?

6 A flat search coil containing 50 turns each of area $2.0 \times 10^{-4}\,\mathrm{m}^2$ is connected to a galvanometer; the total resistance of the circuit is 100 Ω. The coil is placed so that its plane is normal to a magnetic field of flux density 0.25 T.

(a) What is the change in magnetic flux linking the circuit when the coil is moved to a region of negligible magnetic field?

(b) What charge passes through the galvanometer?

[C 78]

34 MAGNETIC FIELD CALCULATIONS

Field due to current in a long straight wire

As shown in Chapter 32 (p.215) the field strength,* called the magnetic flux density, is given by

$$B = \frac{\mu I}{2\pi d} \tag{34.1}$$

where I is the current through the straight wire and B is the resulting flux density at a point distance d from the wire. μ equals permeability of the medium. The lines of force of this field are circles centred upon the wire (as stated in Chapter 32), and this is shown in Fig 34.1. The directions of the lines of force are given by the 'corkscrew rule' according to which these directions are clockwise when one looks along the wire in the direction of the current.

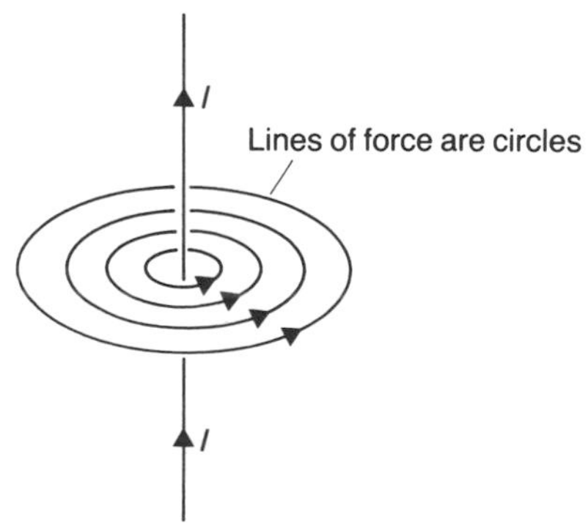

Fig 34.1 Field around a long straight wire

**Magnetic intensity* H *is a different quantity that is also used to describe field strength.*

Example 1

At what distance from a long vertical wire carrying a current of 8.0 A can a neutral point be found if the earth's magnetic field has a horizontal component B of 1.6×10^{-5} T and the permeability of air is $4\pi \times 10^{-7}$ H m^{-1}? (The neutral point is to be detected by a small plotting compass whose needle can turn only in a horizontal plane.)

Method

The required distance d is such that the flux density $B = \mu I/2\pi d$ is equalled by the earth's horizontal flux density B_0. Therefore

$$\frac{4\pi \times 10^{-7} \times 8.0}{2\pi d} = 1.6 \times 10^{-5}$$

$$\therefore \quad \frac{16 \times 10^{-7}}{d} = 1.6 \times 10^{-5}$$

$$\therefore \quad d = 10^{-1}\,\text{m} \quad \text{or} \quad 10\,\text{cm}$$

Answer

10 cm.

Flux density at the centre of a flat circular coil

If the coil radius is R and the current flowing through the coil is I then the flux density at the centre of the coil is given by

$$B = \frac{\mu I n}{2R} \tag{34.2}$$

where n is the number of turns on the coil and μ is the permeability of the surrounding medium. The direction of B is shown in Fig 34.2.

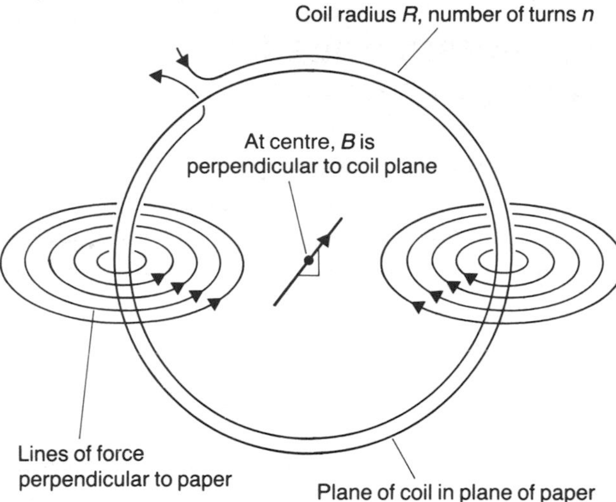

Fig 34.2 Field due to a flat circular coil

Example 2

A circular coil of 4 turns and 11 cm diameter has its plane vertical, and its axis lies in an east to west direction. Calculate the magnitude and direction of the resultant magnetic flux density at the centre of the coil if the earth's horizontal flux density (B_0) is 1.6×10^{-5} T and the coil carries a current of 0.35 A. ($\mu_0 = 4\pi \times 10^{-7}$ H m^{-1}.)

Method

Due to the current in the coil we have flux density $B = \mu In/2R$. I is 0.35 A, $n = 4$, R is 5.5 cm and μ for air (denoted by μ_0) is $4\pi \times 10^{-7}$ H m^{-1}.

$$\therefore \quad B = \frac{4\pi \times 10^{-7} \times 0.35 \times 4}{2 \times 5.5 \times 10^{-2}}$$

$$= 1.6 \times 10^{-5}\ \text{T}$$

The direction of this flux density is along the coil axis, i.e. due east or due west. Let us assume it is due west.

This value of 1.6×10^{-5} T just happens to equal B_0.

The direction of B_0 is of course due north.

The resultant of B and B_0, which are two vectors at right angles, is obtained, as explained in Chapter 6, by use of the cosine rule, which becomes Pythagoras' rule when the vectors are perpendicular.

$$\therefore \quad \text{Resultant}^2 = B^2 + B_0^2$$

$$= 2(1.6 \times 10^{-5})^2$$

$$= 5.12 \times 10^{-10}$$

$$\therefore \quad \text{Resultant} = 2.26 \times 10^{-5}\ \text{T}$$

Since B and B_0 are equal, the direction of the resultant is clearly at 45° to B and to B_0, i.e. it is exactly north-west. If we had assumed B to be in the opposite direction, then the resultant would have been north-east.

Answer

2.3×10^{-5} T north-west.

Magnetic field at a point within a toroid or well inside a solenoid

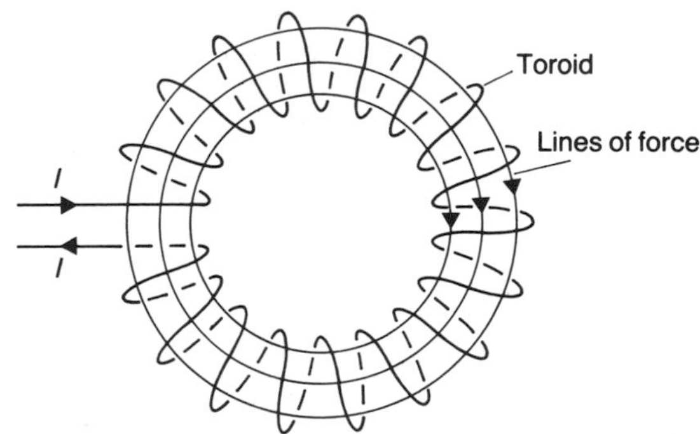

Fig 34.3 Field within a toroid

Within a toroid (an endless coil, see Fig 34.3) the magnetic flux density is given by

$$B = \mu I \times \text{number of turns per metre}$$

or
$$B = \mu I \times \frac{n}{l} \tag{34.3}$$

A solenoid is a long coil, i.e. its length is considerably greater than its diameter, as shown in Fig 34.4.

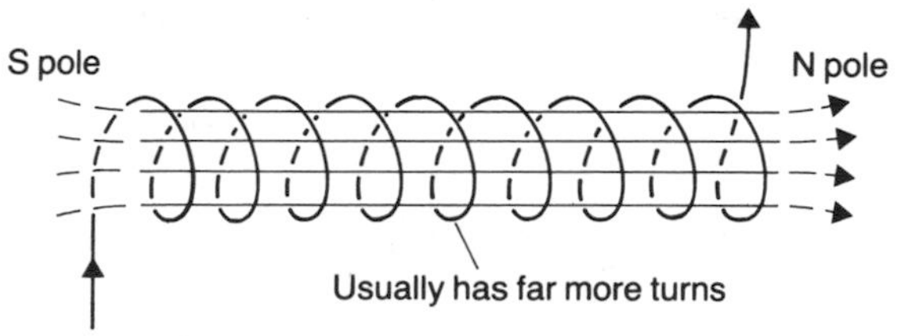

Fig 34.4 A solenoid

A solenoid can be thought of as part of a large toroid, and the turns of the remainder of the toroid are too far from the middle of the solenoid to affect the flux density there. Hence the same formula (34.3) applies to a solenoid.

Example 3

(a) Calculate the flux density in the middle of a solenoid having 10 turns per centimetre and carrying a current of 0.50 A. The medium within the solenoid is air, for which the permeability is $4\pi \times 10^{-7}$ H m^{-1}.

(b) If a coil of 100 turns and 1.0 cm radius is wound around the middle of the solenoid, what charge will be induced in this coil when the solenoid current is stopped, given that the coil is part of a complete circuit having a resistance of 10 Ω?

Method

(a) The flux density is given by Equation 34.3:

$$B = \frac{\mu In}{l}$$

$\mu = 4\pi \times 10^{-7}$, $I = 0.50$, $n/l = 10$ per cm, i.e. 1000 m^{-1}.

$$\therefore \quad B = 4\pi \times 10^{-7} \times 0.5 \times 1000$$
$$= 2\pi \times 10^{-4}$$
$$= 6.3 \times 10^{-4}\ \text{T}$$

(b) The induced charge is calculated from Equation 33.10

$$Q = \frac{\text{Total flux linkage change}}{\text{Total circuit resistance}}$$

The flux linkage for the 100 turn coil is BAn where $n = 100$. The coil's radius is 1.0 cm, so that its area A is $\pi \times 10^{-4}$ m^2. B is the flux density caused by the solenoid, $2\pi \times 10^{-4}$ T. The resistance concerned is 10 Ω.

$$\therefore \quad Q = \frac{2\pi \times 10^{-4} \times \pi \times 10^{-4} \times 100}{10}$$
$$= 1.97 \times 10^{-6}\ \text{C} \quad \text{or} \quad 2.0\ \mu\text{C}$$

Answer

(a) 6.3×10^{-4} T, (b) 2.0 μC.

Self-inductance L of a solenoid

As explained in Chapter 33 the self-inductance (often called just 'inductance') of a coil is defined by $E = L\,\mathrm{d}I/\mathrm{d}t$. It is useful to obtain a formula by which L can be calculated from its construction details.

From Equations 33.6 and 33.7 we have

$$E = \frac{\mathrm{d}\phi}{\mathrm{d}t} = \frac{\mathrm{d}(BAn)}{\mathrm{d}t} = An\left(\frac{\mathrm{d}B}{\mathrm{d}t}\right)$$

where A and n are the area and number of turns of the solenoid. B is the flux density caused by the solenoid's own current I and is given by Equation 34.3 as $\mu In/l$, so that

$$E = An\frac{\mu n}{l}\frac{\mathrm{d}I}{\mathrm{d}t}.$$

This means that

$$L = \frac{\mu A n^2}{l} \tag{34.4}$$

This equation agrees with our measuring μ in H m^{-1}.

Energy stored in an inductor

An inductor is a conductor constructed for the purpose of providing self-inductance. Its circuit symbol is shown in Fig 34.5a.

When a current is being established in a conductor, the current is opposed by self-induced EMF so that work is done. To establish a current I the work done, and therefore energy stored (in the magnetic field of the conductor), is given by

$$\text{Energy stored} = \tfrac{1}{2}LI^2 \tag{34.5}$$

Example 4

(a) Calculate the inductance of a toroid, having 200 turns and length 20 cm, that is uniformly wound over an iron alloy core of 1.0 cm^2 cross-section area and permeability 1.0×10^{-2} H m^{-1}.

(b) What would the self-induced EMF be when the current in the toroid is changing at 1.0 A s^{-1}?

(c) How much energy is stored by the inductor when it carries a current of 100 mA?

Method

(a) The inductance L is given by Equation 34.4 as

$$L = \frac{\mu A n^2}{l}$$

$\mu = 1.0 \times 10^{-2}$, $A = 10^{-4}$, $n = 200$ and $l = 20 \times 10^{-2}$.

$$\therefore \quad L = \frac{1.0 \times 10^{-2} \times 10^{-4} \times 200^2}{20 \times 10^{-2}}$$

$$= \frac{4.0 \times 10^{-2}}{2.0 \times 10^{-1}}$$

$$= 0.20\,\text{H}$$

(b) The induced EMF is given by Equation 33.8 as

$$E = L\frac{dI}{dt}$$

$L = 0.20\,\text{H}$ and $dI/dt = 1.0\,\text{A s}^{-1}$.

$$\therefore \quad E = 0.20 \times 1.0 = 0.20\,\text{V}$$

(c) The energy stored is given by Equation 34.5 as $\frac{1}{2}LI^2$. Using $L = 0.20\,\text{H}$ and $I = 0.1\,\text{A}$ we get

$$\text{Energy stored} = \tfrac{1}{2}LI^2$$

$$= \tfrac{1}{2} \times 0.20 \times 0.1^2$$

$$= 0.0010\,\text{J}$$

Answer

(a) 0.20 H, (b) 0.20 V, (c) 0.0010 J.

Time constant for an inductive circuit

Any growth or decay of the current through an inductor causes an induced EMF that opposes the change. Consequently the change is slowed down by the presence of inductance. The application of a fixed voltage E sufficient to produce a steady current I_{max} will result in a current that rises to the final value I_{max}, as shown in Fig 34.5a and b.

Decay of the current I after the supply voltage is suddenly reduced to zero is illustrated in Fig 34.5c.

The time constant here is L/R (see also p.210).

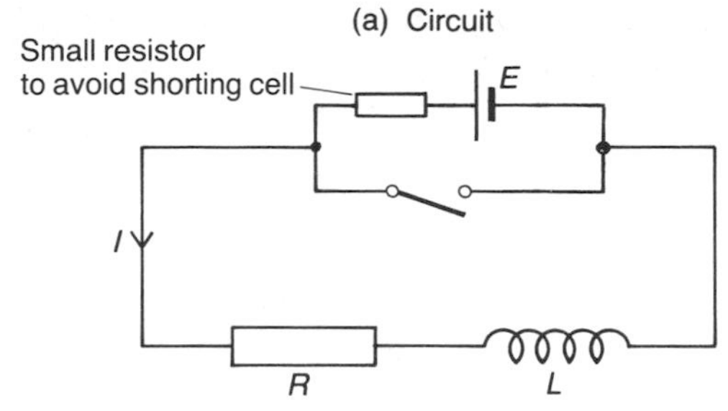

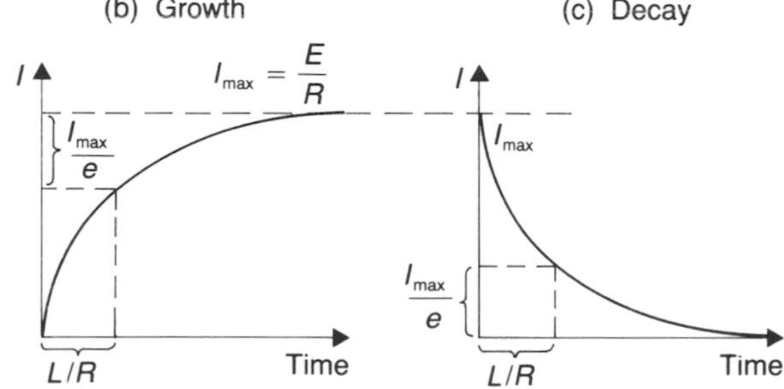

Fig 34.5 Growth and decay of current in an inductive circuit

Exercise 117

(The permeability of air may be taken as $4\pi \times 10^{-7}\,\text{H m}^{-1}$.)

1 A vertical wire carries a downward current of 5.0 A, and 12 cm east of this there is another vertical wire carrying an equal downward current. The earth's horizontal component is 1.6×10^{-5} T. What is the flux density at a distance 2.0 cm from the first wire and 10.0 cm from the other?

2 A flat, circular coil has a radius of 1.0 cm and is made up of 20 turns. A second coil of 30 turns and 5.0 cm radius encircles the first coil, both coils lying in the same plane and in air. (a) What EMF is induced in the first coil when the current in the second coil changes at a steady rate of 5.0 A in 5.0 s? (b) What is the value of the mutual inductance for these coils?

3 A solenoid having 200 turns per metre and carrying a current of 0.050 A lies with its axis east–west. Well inside the solenoid is a small compass whose needle points 37° west of north. Calculate the earth's horizontal magnetic field component B_0.

4 Calculate the inductance of a solenoid having 300 turns over a 15.0 cm length that has a cross-section area of 4.0 cm^2 and is air-cored.

How much energy is stored in its magnetic field when it is carrying a steady 200 mA current?

Exercise 118: Questions of A-level standard

1 (a) Draw a diagram showing the magnetic field pattern due to a long, straight wire carrying a current. Mark clearly the relative directions of the current I (conventional) and the magnetic flux density B.

(b) At a distance r from the wire, the magnitude of B is given by

$$B = \frac{\mu_0 I}{2\pi r}$$

where μ_0 is the permeability of free space. Calculate the value of B at a distance 50 cm from a long, straight cable carrying a direct current of 1000 A. ($\mu_0 = 4\pi \times 10^{-7}\,\mathrm{H\,m^{-1}}$.)

(c) Ions in the atmosphere will be affected by the magnetic field produced by such a cable. Calculate the force acting on an ion travelling directly towards the cable at a speed of $10^4\,\mathrm{m\,s^{-1}}$ when it is 1.0 m away from the cable. Assume that the ion carries a charge of $+1.6 \times 10^{-19}$ C.

[L 81, p]

2 Fig 34.6 represents a plane, circular coil of radius 0.25 m, with 2 turns and carrying a current of 10 A, set with its plane vertical and at right angles to the magnetic meridian. It forms the fixed part of a simple current balance. The only moving part of the balance which is of interest is the horizontal cross-piece 40 mm long, as shown; this is in the plane of the circular coil and passes through its centre; the same current, 10 A, flows along it. Taking the horizontal component of the earth's magnetic flux density to be 2×10^{-5} T, and assuming that the current through the 40 mm wire is always in the direction shown, find the magnitude and direction of the force on the cross-piece when the current through the coil flows (a) clockwise and (b) anti-clockwise in the coil (as seen from the left of the figure). (The value of μ_0 is $4\pi \times 10^{-7}\,\mathrm{H\,m^{-1}}$.)

[O 80, p]

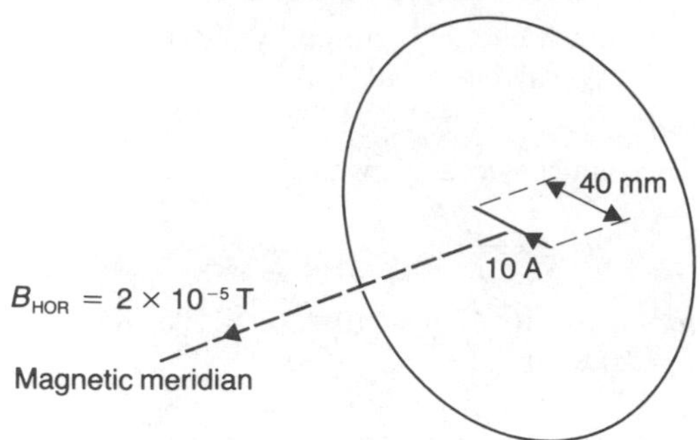

Fig 34.6 Diagram for Question 2

3 A circular coil A, 40 cm diameter, consisting of 150 turns of wire is mounted with its plane vertical. What current in the wire will produce a magnetic field of 2.5×10^{-3} T at its centre? (The earth's magnetic field may be neglected.)

A second small coil B, consisting of 100 turns each of area $7.5 \times 10^{-5}\,\mathrm{m^2}$, is mounted so that the centres of A and B coincide, and their planes are mutually perpendicular. Calculate the couple acting on B when a current of 0.2 A is flowing in B, and the current previously calculated is flowing in A. ($\mu_0 = 4\pi \times 10^{-7}\,\mathrm{H\,m^{-1}}$.) [SUJB 82, p]

4 A long air-cored solenoid has 1000 turns of wire per metre and a cross-section area of 8.0 cm^2. A secondary coil of 200 turns is wound around its centre and connected to a ballistic galvanometer, the total resistance of coil and galvanometer being 60 Ω. The sensitivity of the galvanometer is 2.0 divisions per microcoulomb. If a current of 4.0 A in the primary solenoid were switched off, what would be the deflection of the galvanometer?
(Permeability of free space = $4\pi \times 10^{-7}\,\mathrm{H\,m^{-1}}$)

[L 77, p]

35 ALTERNATING CURRENTS

Variation of voltage with time

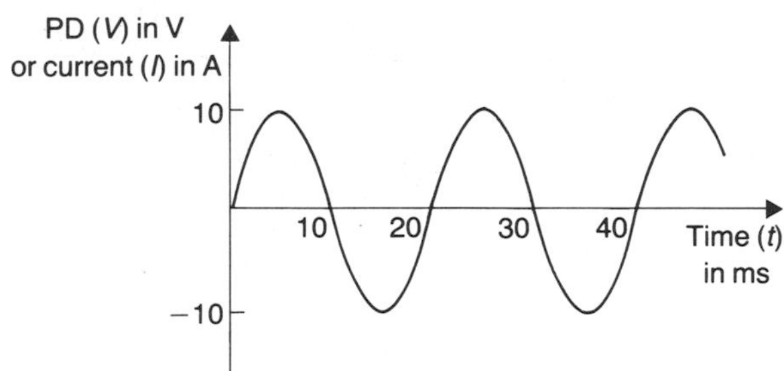

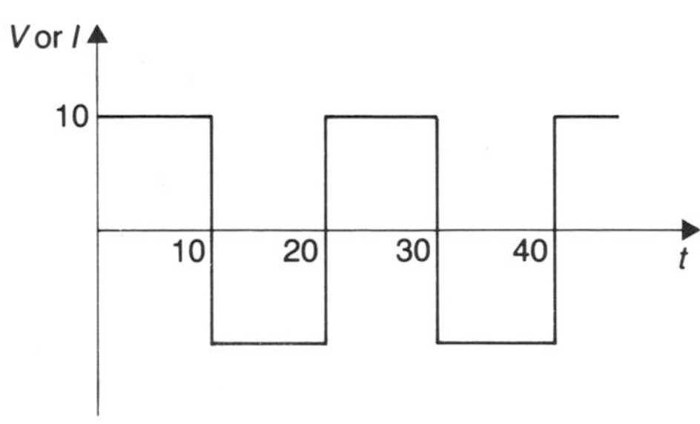

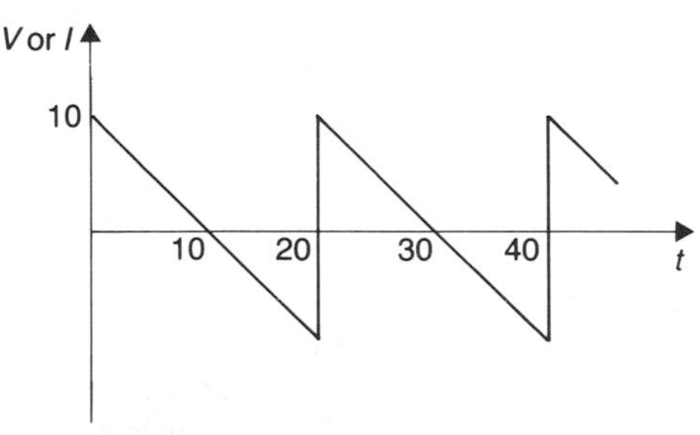

Fig 35.1 Variation of voltage and current with time

Fig 35.1 shows some examples of alternating voltages and currents.

The sinusoidal variation with time is very common, and mains AC supply is sinusoidal in waveform. Unless otherwise stated 'alternating current' or 'alternating voltage' means sinusoidal current or PD. As shown in Chapter 33 a uniform-field generator produces a sine-wave voltage.

The variation of voltage with time is described by the formula

$$V = V_p \sin 2\pi ft \qquad (35.1)$$

where f is the number of cycles (i.e. repeats) per second and is the frequency, t is the time measured from an instant when $V = 0$, and V_p is the maximum or peak value of the voltage. Note that $2\pi f$ may be written as ω, known as the angular frequency. If the voltage is produced by a rotating-coil generator, then ω may be identified with the angular frequency of the coil's rotation and $\omega = \theta/t$, where θ is the angle through which the coil rotates. The unit for ω is rad s^{-1}.

$$V = V_p \sin \omega t \quad \text{or} \quad V_p \sin \theta \qquad (35.2)$$

However, regardless of the cause of the voltage, the value of θ ($2\pi ft$) is important for describing the stage reached by the voltage variation and is called the phase angle, as explained in Chapter 14.

Fig 35.2 shows how the variation of voltage is described by a rotating phasor (see also Fig 14.2, describing simple harmonic motion).

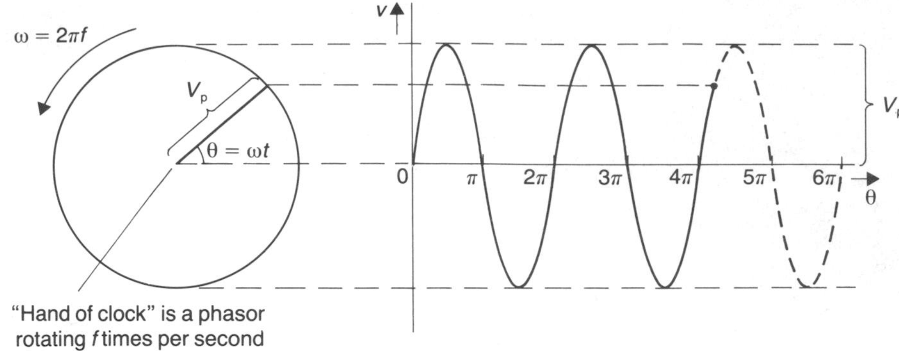

Fig 35.2 Use of a phasor for voltage or current

Size of current in a purely resistive circuit

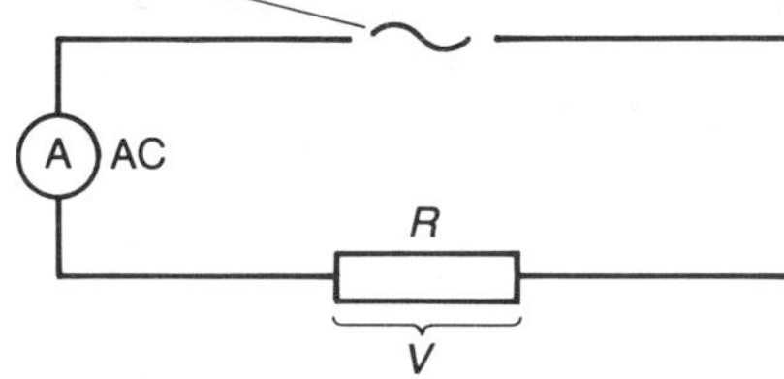

Fig 35.3 A purely resistive circuit

If the circuit concerned contains no significant capacitance or inductance, only resistance R, then at all times $I = V/R$ so that

$$I = \frac{V_p \sin 2\pi ft}{R}$$

or $$I = I_p \sin 2\pi ft \qquad (35.3)$$

where $I_p = V_p/R$.

The current rises and falls in step with the voltage, i.e. I and V are in phase. $I = I_p$ when $V = V_p$.

Average and RMS values

The effects produced by alternating currents will often depend on some kind of average of the current. A simple average over a half-cycle is known as the average (or mean) value of the current or voltage. However, the heating by a current is decided by the mean value of I^2R or V^2/R, and mean I^2 ('mean square current') or mean V^2 is important. The square root of mean I^2 or V^2 is the root mean square (RMS) value. The sizes quoted for alternating voltages and currents, unless otherwise stated, are always RMS values.

For a sine wave variation the mean value equals $(2/\pi) \times$ peak value and the RMS value equals $(1/\sqrt{2}) \times$ peak value.

i.e. $$I_{RMS} = \frac{1}{\sqrt{2}} I_p \quad \text{and} \quad \overline{I} = \frac{2}{\pi} I_p \qquad (35.4)$$

In Fig 35.1 for example the peak voltage is 10 V and for the sine wave $V_{RMS} = 7.1$ V, $\overline{V} = 6.3$ V.

For the square wave $V_p = V_{RMS} = 10$ V. For the triangular wave $V_p = 10$ V, $V_{RMS} = 10/\sqrt{3}$ V.

Example 1

A sinusoidal alternating voltage displayed on a cathode ray oscilloscope is seen to have a peak value of 75 V. What reading should be obtained with a voltmeter indicating RMS voltage?

Method

$V_p = 75$ V but $V_{RMS} = V_p/\sqrt{2}$.
Therefore $V_{RMS} = 75/1.414 = 53$ V.

Answer

53 V.

Example 2

Calculate the value of a sinusoidal voltage having a peak value of 30 V at a time of one-tenth of a cycle after a peak has been reached. What current will be present at this instant if the total resistance of the circuit is 9.0 Ω?

Method
$V = V_p \sin \theta$ (Equation 35.2).

One-tenth of a cycle is 360/10 degrees, i.e. 36°. Therefore we need V when θ is 36° greater than 90°, i.e. $\theta = 126°$. However, we should realise that V will have the same value at 36° less than 90°, namely $\theta = 54°$.

$$V = 30 \times \sin 54 \text{ or } 30 \times \sin 126$$

Hence $\quad V = 30 \times 0.81 = 24.3\ \text{V}$

$$\text{Current} = \frac{V}{R} = \frac{24.3}{9.0} = 2.7\ \text{A}$$

Answer
24 V, 2.7 A.

Impedance

This is the opposition of a circuit to the flow of alternating current. It is denoted by Z and is defined by

$$Z = \frac{V_{\text{RMS}}}{I_{\text{RMS}}} \qquad (35.5)$$

where V_{RMS} is the RMS supply voltage and I_{RMS} the resulting current. Clearly we could use peak values or mean values in place of RMS in the above equation. Z is decided not only by the resistance R of the circuit but, as we shall soon see, by the presence of inductance or capacitance in the circuit also. In a purely resistive circuit Z equals R because $V_{\text{RMS}}/I_{\text{RMS}} = R$.

Inductive reactance

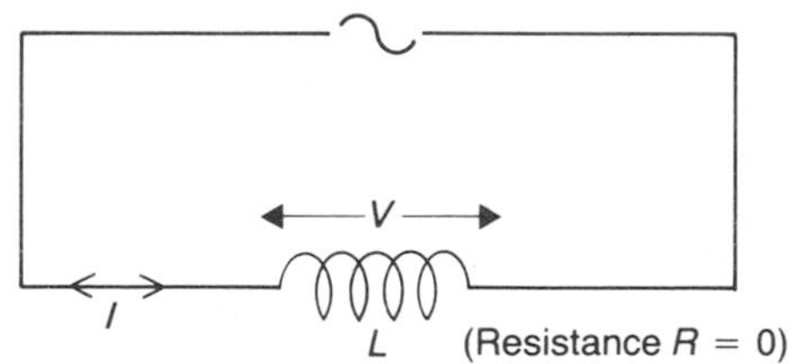

Fig 35.4 A purely inductive circuit

Suppose that an alternating voltage is applied to a copper coil of appreciable inductance L (see Chapter 33) but negligible resistance (Fig 35.4). The continual changes of current I cause induced voltages that oppose every rise and fall of current. Consequently there is opposition to the flow of the alternating current. This opposition due to inductance is called inductive reactance X_L. It is, of course measured in ohms and its magnitude is given by

$$X_L = \omega L \qquad (35.6)$$

where ω is the angular frequency of the alternating current.

Capacitive reactance

When an alternating voltage is applied to a capacitor C, it repeatedly charges, discharges and recharges the capacitor with opposite polarity for each successive charging. Thus alternating current is flowing in the circuit. The extent of each charging of the capacitor, and hence the size of the current obtained, is limited by the PD that builds up across the capacitor. The current is greater if C is large and the process is rapid (i.e. the frequency is high). The opposition to alternating-current flow due to the presence of a capacitance is called capacitive reactance (X_C) and its size is given by

$$X_C = \frac{1}{\omega C} \qquad (35.7)$$

So we see that the impedance Z is equal to R or ωL or $1/\omega C$ if the circuit contains only resistance, only inductance or only capacitance respectively.

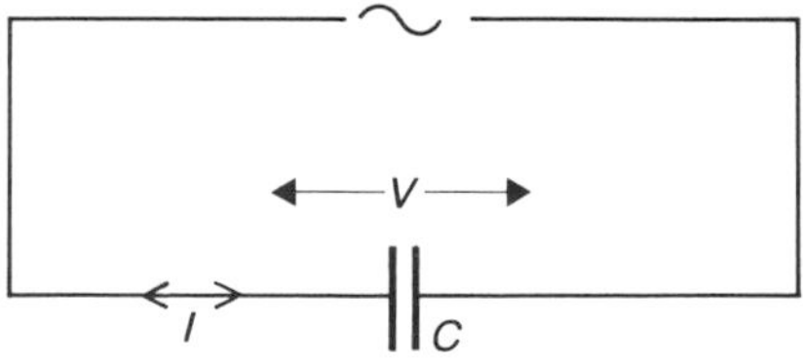

Fig 35.5 AC circuit containing capacitance but negligible inductance or resistance

Example 3
A sinusoidal alternating voltage of 6.0 V RMS and frequency 1000 Hz is applied to a coil of 0.5 H inductance and negligible resistance. What is the expected value for the RMS current?

Method

$$\frac{V_{RMS}}{I_{RMS}} = Z = X_L = \omega L = 2\pi f L$$

$$\therefore \quad I_{RMS} = \frac{V_{RMS}}{2\pi f L} = \frac{6.0}{2\pi \times 1000 \times 0.5}$$

$$= 0.0019 = 1.9 \times 10^{-3}\,\text{A}$$

Answer

1.9 mA.

Example 4

A 25 V peak, 50 Hz sinusoidal voltage is applied to a capacitor. If the peak current is 15.7 mA, what is the value of the capacitance?

Method

$$\frac{V_p}{I_p} = \frac{V_{RMS}}{I_{RMS}} = Z = X_C = \frac{1}{\omega C} = \frac{1}{2\pi f C}$$

$$\therefore \quad \frac{25}{15.7 \times 10^{-3}} = \frac{1}{2\pi \times 50 \times C}$$

$$C = \frac{15.7 \times 10^{-3}}{25 \times 2\pi \times 50}$$

$$= 2.0 \times 10^{-6}\,\text{F} \quad \text{or} \quad 2.0\,\mu\text{F}$$

Answer

2.0 μF.

Series *LCR* circuits

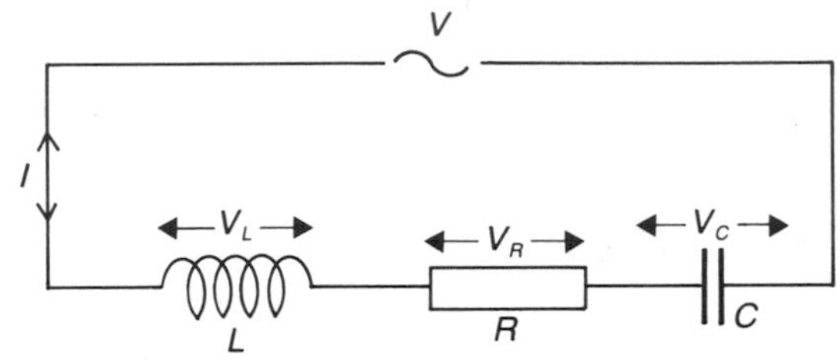

Fig 35.6 The series *LCR* circuit

(a) A 'clock with three hands' for V_{Lp}, V_{Rp} and V_{Cp}

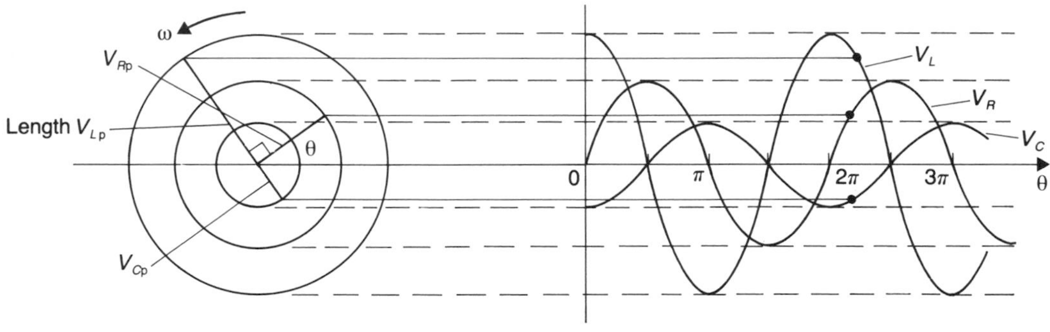

(b) The total of the three voltages

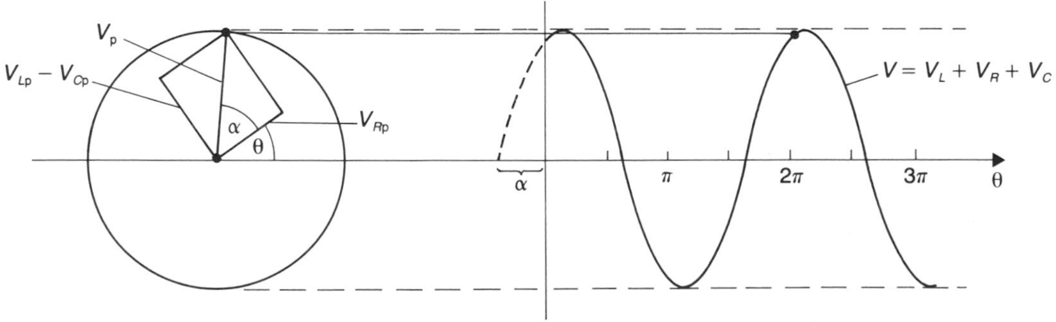

Fig 35.7 Use of rotating phasors with an *LCR* circuit (see Fig 35.6)

An AC circuit may contain a combination of resistances, inductances and capacitances. Assume that they are all in series. We can combine the resistances ($R = R_1 + R_2$, etc.) and the capacitors ($1/C = 1/C_1 + 1/C_2$, etc.) and the inductances ($L = L_1 + L_2$, etc.). However, the values of R, ωL and $1/\omega C$ cannot simply be added to find the impedance Z of the circuit.

In fact Z is less than what would be obtained by simple addition of these ohms, because the voltages V_L and V_C across L and C, which are responsible for the reactances ωL and $1/\omega C$ respectively, are not in phase, and neither is V_L nor V_C in phase with the PD across (and current through) R. In fact V_L reaches its peak a quarter-cycle before the current I reaches its peak, the current changing most rapidly at this time. V_C reaches its peak a quarter-cycle after I (when I is zero and C is about to start discharging). These facts can be illustrated using the rotating phasor method, as in Fig 35.7a (opposite). The word 'CIVIL' may be used as a reminder that, for C, I precedes V but V precedes I for L.

In Fig 35.7 V_{Lp} is shown greater than V_{Rp}, and V_{Cp} is smallest. As a result the total voltage V leads V_R, and so leads the current I (by the phase angle α). If the capacitive reactance played a larger part in the circuit, α would be negative, i.e. the current would reach its peak *before* the total voltage (or supply voltage). α should be remembered as the lag of current behind the supply PD.

From Fig 35.7b it can be seen that, to satisfy $V = V_L + V_R + V_C$, the phasor size must be given by $V_p^2 = V_{Rp}^2 + (V_{Lp} - V_{Cp})^2$. It follows, since I_p is the same throughout, that

$$Z^2 = R^2 + \left(\omega L - \frac{1}{\omega C}\right)^2 \qquad (35.8)$$

So we see that the ohms combine like vectors by the parallelogram rule (Chapter 6), but a little simpler because we have a rectangle. Also

$$\tan\alpha = \frac{V_{Lp} - V_{Cp}}{V_{Rp}}$$

or $$\tan\alpha = \frac{\omega L - 1/\omega C}{R} \qquad (35.9)$$

These results are summarised in Fig 35.8.

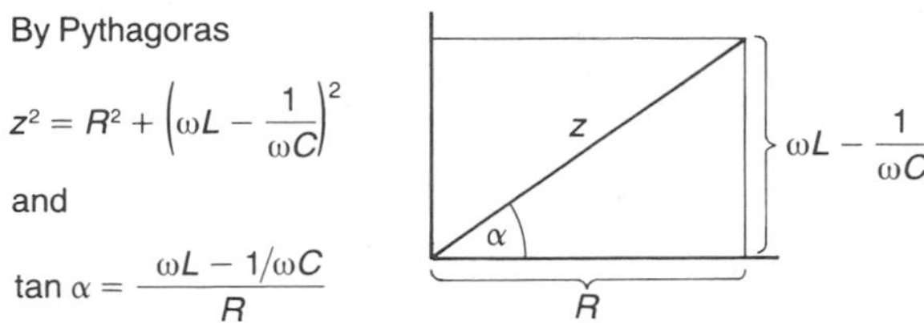

Fig 35.8 Combining R, ωL and $1/\omega C$ in series

It is useful to note that

$$\omega L = \frac{V_{LRMS}}{I_{RMS}}, \qquad \frac{1}{\omega C} = \frac{V_{CRMS}}{I_{RMS}},$$

$$R = \frac{V_{RRMS}}{I_{RMS}}, \qquad Z = \frac{V_{RMS}}{I_{RMS}} \qquad (35.10)$$

Example 5

Calculate the current expected when a 0.30 H coil having 55 Ω resistance is connected to a 22 V RMS, 70 Hz voltage supply.

Method

$1/\omega C = 0$ here because, where a capacitor might have been, we have a low resistance connecting wire instead. Equation 35.8 becomes

$$Z^2 = R^2 + (\omega L)^2,$$

and $Z = V_{RMS}/I_{RMS}$, where V_{RMS} and I_{RMS} are the voltage supply and current.

Also $\omega = 2\pi f$

$L = 0.3$, $\omega = 2\pi f = 2\pi \times 70$, $R = 55$, $V_{RMS} = 22$.

$$\therefore \quad Z^2 = 55^2 + (2 \times \pi \times 70 \times 0.3)^2 = 20\,449$$

$$\therefore \quad Z = 143\ \Omega$$

I_{RMS} $(= V_{RMS}/Z) = 22/143 = 2/13 = 0.154$ A RMS

Answer

0.15 A RMS.

Example 6

A 16 μF capacitor and an inductive coil of 300 Ω resistance are connected in series across a 20 V, 50 Hz AC supply. The current obtained is 40 mA RMS. What is the inductance of the coil?

Method

$$Z^2 = R^2 + \left(\omega L - \frac{1}{\omega C}\right)^2$$

$$\omega = 2\pi f \quad \text{and} \quad Z = \frac{V_{RMS}}{I_{RMS}}$$

$V_{RMS} = 20$, $I_{RMS} = 40 \times 10^{-3}$, $f = 50$, $R = 300$, $C = 16 \times 10^{-6}$.

$$Z = \frac{20}{40 \times 10^{-3}} = 500\,\Omega, \qquad \omega = 2\pi \times 50 = 314.2$$

$$\therefore \quad Z^2 = 500^2 = 300^2 + \left(314.2L - \frac{10^6}{314.2 \times 16}\right)^2$$

$$(314.2L - 199)^2 = 500^2 - 300^2 = 160\,000$$

$$314.2L - 199 = 400 \quad \text{and} \quad L = \frac{599}{314} = 1.9\text{ H}$$

Answer

1.9 H.

Example 7

Calculate the time interval by which the current lags on the 50 Hz supply voltage for a circuit in which a 10 H, 1000 Ω coil only is connected to the supply. This supply has negligible internal resistance and reactance.

Method

$$\tan\alpha = \frac{\omega L - 1/\omega C}{R} \quad \text{(Equation 35.9)}$$

and $\omega = 2\pi f$, $f = 50$, $L = 10$, $R = 1000$, $1/\omega C = 0$.

$$\tan\alpha = \frac{2\pi \times 50 \times 10}{1000} = 3.14$$

$$\therefore \quad \alpha = 72.3°$$

But 360° is a whole cycle, i.e. one-fiftieth of a second.

Therefore the lag is

$$\frac{1}{50} \times \frac{72.3}{360} = 4.0 \times 10^{-3}\text{ s}$$

Answer

4.0 ms.

Example 8

In a series LCR circuit $L = 2.0$ H, $C = 100\,\mu$F, $R = 75\,\Omega$. At what angular frequencies of the applied alternating supply voltage will the circuit's impedance be 125 Ω?

Method

$$Z^2 = R^2 + \left(\omega L - \frac{1}{\omega C}\right)^2$$

$Z = 125$, $R = 75$, $L = 2.0$, $C = 100 \times 10^{-6} = 10^{-4}$.

$$\left(\omega L - \frac{1}{\omega C}\right)^2 = 125^2 - 75^2 = 10\,000$$

which gives

$$\left(\omega L - \frac{1}{\omega C}\right) = \pm 100$$

$$\therefore \quad 2.0\omega - \frac{10^4}{\omega} = \pm 100$$

$$\therefore \quad 2\omega^2 - 10^4 \pm 100\omega = 0$$

If we use the + sign, $(2\omega - 100)(\omega + 100) = 0$ so that $\omega = 50$ or -100.

If we use the − sign, $(2\omega + 100)(\omega - 100) = 0$ so that $\omega = -50$ or 100.

The two possible answers are $\omega = 50$ or 100.

Answer

50 and 100 rad s^{-1}.

Heating by an alternating current

In a resistance R the heat produced per second (i.e. the electrical energy per second or power converted into internal energy within the resistance) is the mean value of I^2R, i.e. $I^2_{RMS}R$.

In a pure inductance or capacitance there is no production of heat. So the power dissipated in an LCR circuit is

$$P = I^2_{RMS}R \quad \text{or} \quad I^2_{RMS}Z\cos\alpha$$

$$\text{or} \quad V_{RMS}I_{RMS}\cos\alpha \qquad (35.11)$$

($R = Z\cos\alpha$, as shown in Fig 35.8.)

The product $V_{RMS}I_{RMS}$ is often called the 'apparent power' and $\cos\alpha$, called the power factor, tells us the ratio of true to apparent power.

Example 9

Calculate the true power and the apparent power in Example 6.

Method

The true power is $I^2_{RMS}R$. Using $I_{RMS} = 40 \times 10^{-3}$ A and $R = 300\ \Omega$ we get $(40 \times 10^{-3})^2 \times 300$ which equals 0.48 W.

The apparent power is $V_{RMS} \times I_{RMS}$. Using $V_{RMS} = 20$ V and $I_{RMS} = 40 \times 10^{-3}$ A we get $20 \times 40 \times 10^{-3}$, which equals 0.80 W.

Answer

0.48 W, 0.80 W.

Inductors and chokes

A coil constructed for the purpose of producing inductance is called an inductor. Its circuit symbol is shown in Fig 35.4.

An inductor used for the purpose of limiting the size of an alternating current by virtue of its inductive reactance is called a choke.

Exercise 119

1 A sinusoidal alternating voltage supply has an RMS value of 2.0 V. Calculate (a) the peak voltage, (b) the expected peak current if the circuit's resistance is 20 Ω.

2 What is the shortest time it takes for a 100 Hz alternating current to change from zero to (a) its peak value, (b) half of its peak value?

3 A sinusoidal voltage supply having an angular frequency ω of 200 rad s^{-1} and a peak voltage of 100 V is connected to an inductor of 0.50 H and negligible resistance. Calculate (a) the inductive reactance of the inductor, (b) the peak current, (c) the PD at a time of one-sixth of a cycle after the PD was zero, (d) the current at this time.

4 A coil having inductance 0.040 H and resistance X is connected in series to a 25 Ω resistor and a sinusoidal voltage supply with a frequency of 50 Hz. If the RMS PD across the coil equals that across the resistor, calculate (a) the impedance of the coil, (b) the value of X.

5 A 6.0 V RMS alternating voltage supply with a frequency of 700 Hz and negligible impedance is connected to a 35 Ω resistor and a 3.5 μF capacitor in series. Calculate (a) the impedance of the circuit, (b) the peak PD across the resistor.

6 A 25 W, 100 V heater is to be run from a 250 V 50 Hz sinusoidal AC supply. Calculate the inductance of a suitable choke to be included in the circuit. Assume its resistance to be negligible.

Resonance in an *LCR* series circuit

In the formula $Z^2 = R^2 + (\omega L - 1/\omega C)^2$ it can be seen that $Z = R$ if $\omega L = 1/\omega C$, but under all other circumstances Z is greater. Thus for a given PD applied to an LCR series circuit the current is exceptionally high when $\omega L = 1/\omega C$. This condition usually arises as a result of the supply's frequency being varied until $\omega^2 = 1/LC$ (or $\omega = 1/\sqrt{LC}$) or

$$f = \frac{1}{2\pi\sqrt{(LC)}} \qquad (35.12)$$

This phenomenon is called *resonance*. It is the result of the applied frequency matching the circuit's own (or natural) frequency of $\frac{1}{2\pi}\sqrt{(LC)}$. The high current occurs because V_L becomes equal to V_C, so that the would-be opposition to current flow due to L is cancelled by that due to C.

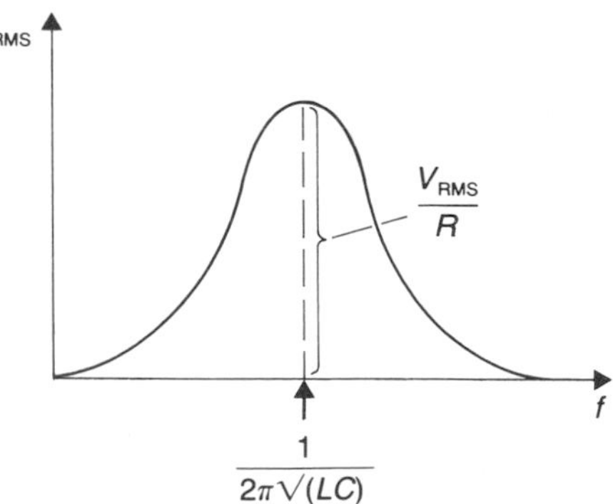

Fig 35.9 Current in an *LCR* series circuit, showing resonance

At resonance

$$\tan\alpha \left(= \frac{\omega L - 1/\omega C}{R}\right) = 0$$

i.e. α is zero and I is in phase with the supply PD V. Resonance is illustrated in Fig 35.9.

Example 10

(a) Calculate the resonant frequency for a series LCR circuit in which $L = 0.01$ H, $C = 1.0\,\mu$F and $R = 20\,\Omega$.

(b) If the voltage supply is 12 V RMS, what current flows at resonance?

(c) What is the RMS PD across L and across C at resonance?

Method

(a) We use Equation 35.12:
$f = 1/2\pi\sqrt{(LC)}$, $L = 0.01$, $C = 1.0 \times 10^{-6}$.

$$f = \frac{1}{2\pi\sqrt{(0.01 \times 1.0 \times 10^{-6})}}$$

$$= \frac{1}{2\pi\sqrt{10^{-8}}} = \frac{10^4}{2\pi} = 1591 \text{ Hz}$$

(b) $$I_{RMS} = \frac{V_{RMS}}{Z} = \frac{V_{RMS}}{R} = \frac{12}{20} = 0.60 \text{ A RMS}$$

(c) $$\frac{V_{LRMS}}{I_{RMS}} = \omega L$$

$$\therefore \quad V_{LRMS} = 2\pi f \times 0.01 \times I_{RMS}$$

$$= 2\pi \times 1591 \times 0.01 \times 0.6$$

$$= 60 \text{ V RMS}$$

$$\frac{V_{CRMS}}{I_{RMS}} = \frac{1}{\omega C} \quad \text{gives 60 V also for } V_{CRMS}$$

(not surprisingly since, at resonance, V_C is equal and opposite in polarity to V_L at all times).

Answer

1.6 kHz, 0.60 A RMS, 60 V, 60 V.

Exercise 120

1 Calculate the resonant frequency for a 0.20 H inductor in series with a 2.0 μF capacitor.

2 What size of capacitor is needed in series with a 2.0 H, 100 Ω coil in order to get the current in phase with the 50 Hz, 240 V supply voltage?

What size of current will then flow?

3 A 100 Ω resistor, a 1.0 μF capacitor and a 0.20 H inductor of negligible resistance are connected in series with a supply of sinusoidal alternating EMF of 20 V RMS whose frequency f can be varied (see Fig 35.10). Calculate (a) the resonant frequency, (b) the value of the maximum RMS current, (c) the RMS voltage across each of the components at this frequency.

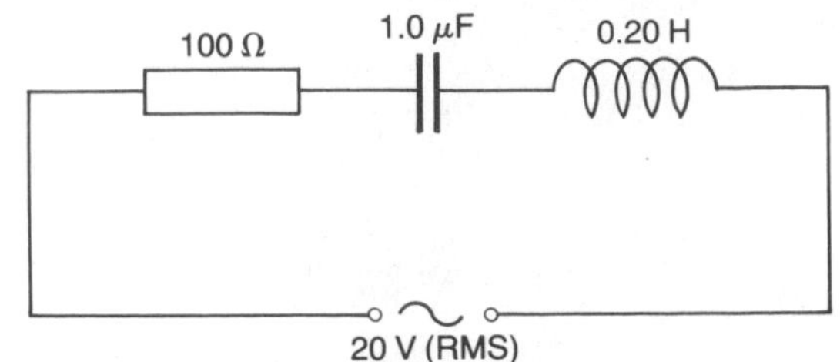

Fig 35.10 Circuit for Question 3

The transformer

A transformer consists of a primary coil to which an alternating voltage V_1 is applied, and a secondary coil from which the required alternating voltage V_2 is obtained as the result of mutual induction between the two coils. The coils are often wound on a core of magnetic material such as iron.

In an ideal transformer the coil resistances are negligible and eddy-current heating in the core is negligible (so that there is no energy wastage as heat). Also the flux Φ passes through ('links with') all the turns of the coils. For this ideal transformer V_2 and V_1 (RMS values) are related by

$$\frac{V_2}{V_1} = \left(\frac{n_2 \mathrm{d}\Phi/\mathrm{d}t}{n_1 \mathrm{d}\Phi/\mathrm{d}t}\right) = \frac{n_2}{n_1} = \text{the turns ratio } n \qquad (35.13)$$

where n_2 and n_1 are the number of turns on the secondary coil and the primary coil respectively. A step-up transformer produces a secondary voltage greater than the primary voltage: the opposite is step-down.

Assuming a small resistive load on the secondary (Fig 35.11), the currents I_1 and I_2 are approximately in phase with their respective voltages so that the

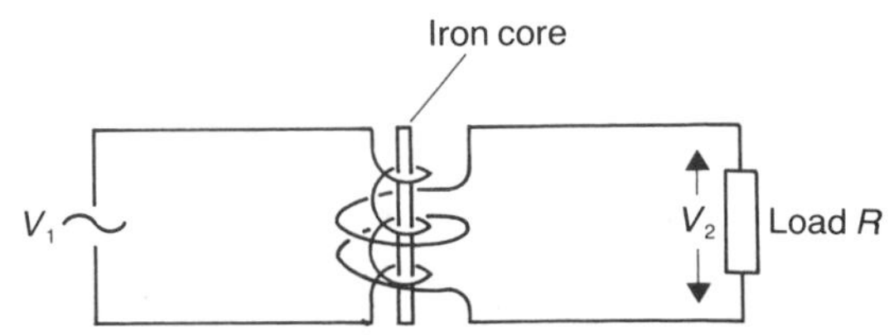

Fig 35.11 Principle of the transformer

power dissipated in the primary is V_1I_1 and the power dissipated in the load is V_2I_2. Since we are assuming no losses these powers can be equated, giving $V_1I_1 = V_2I_2$, from which

$$\frac{I_2}{I_1} = \frac{V_1}{V_2} = \frac{n_1}{n_2} = \frac{1}{n} \tag{35.14}$$

If the load on the secondary is resistance R, then $V_2/I_2 = R$, and so

$$\frac{V_1}{I_1}\left(= \frac{V_2/n}{nI_2}\right) = \frac{R}{n^2} \tag{35.15}$$

Transformers used for high-voltage power distribution

The product of RMS voltage and RMS current is the power. (We will assume no phase difference.) Consequently the same power is obtainable with high voltage and low current or with lower voltage and higher current. A generator in a power station produces its power for the consumer by means of a large current at low voltage, but for transmission of this power it is more convenient and economical to use small currents. Transformers are employed to step up the voltage and subsequently to step down the voltage again before it reaches the consumer.

Example 11

(a) What turns ratio would be needed for an ideal transformer to provide 12 V RMS when connected to 240 V RMS mains supply?

(b) If the transformer were loaded with a non-inductive 12 V, 60 W heater, what current would flow in the mains supply lead?

(c) If the transformer in practice gives 11.8 V RMS and 4.5 A RMS when the primary current is 0.26 A RMS, what is the efficiency of energy conversion by the transformer?

Method

(a) According to Equation 35.13,

$$\frac{V_2}{V_1} = \frac{n_2}{n_1} = n$$

$V_2 = 12,\ V_1 = 240.$

$$\therefore\quad n = \frac{12}{240} = \frac{1}{20}$$

(The primary has the greater number of turns.)

(b) From Equation 35.14

$$\frac{I_2}{I_1} = \frac{1}{n}$$

I_2 is known from the fact that, for a power of 60 W using 12 V, the current must be 60 W divided by 12 V, i.e. 5.0 A. $n = \frac{1}{20}$

$$\frac{I_2}{I_1} = \frac{5.0}{I_1}$$

but $$\frac{I_2}{I_1} = \frac{1}{n} = \frac{1}{1/20} = 20$$

$$\therefore\quad \frac{5.0}{I_1} = 20 \text{ or } I_1 = 0.25\text{ A}$$

(c) The input power is $V_1I_1 = 240 \times 0.26 = 62.4$ W.

The output power is $V_2I_2 = 11.8 \times 4.5 = 53.1$ W.

$$\text{Efficiency} = \frac{\text{Useful power output}}{\text{Power input}}$$

$$= \frac{53.1}{62.4} = 0.85 \quad\text{or}\quad 85\%$$

Answer

(a) $\frac{1}{20}$, (b) 0.25 A, (c) 85%.

Exercise 121

1 A transformer with a 100-turn primary winding and a 500-turn secondary winding is connected to a 2.0 V RMS supply. Calculate values for the output voltage from the secondary, and the maximum secondary current if the primary winding is to be limited to 0.10 A. State the assumptions made.

2 10 kW of electrical power at 100 V RMS is to be delivered by use of a 5.0 Ω cable. What is the rate of heat production in the cable?

If, instead, the power were transmitted, again through 5.0 Ω, but at 30 kV stepped down to 100 V at the user-end of the cable, what would be the new heat dissipation in the cable? Assume the transformer to be ideal, i.e. no energy loss occurs.

Exercise 122: Questions of A-level standard

1 A direct current I passing through a resistor produces a certain heating effect. The resistor value has to be halved to obtain the same heating effect when an alternating current is passed through it. The peak value of this alternating current is

A $\frac{I}{2}$ **B** $\frac{I}{\sqrt{2}}$ **C** I
D $\sqrt{(2)}I$ **E** $2I$ [AEB 81]

2 The RMS (root mean square) value of the sinusoidal voltage described by Fig 35.12a when divided by the RMS value of the square wave voltage described by Fig 35.12b equals

A $\frac{1}{2\sqrt{2}}$ **B** 1 **C** $\frac{1}{\sqrt{2}}$
D $\sqrt{2}$ **E** $2\sqrt{2}$

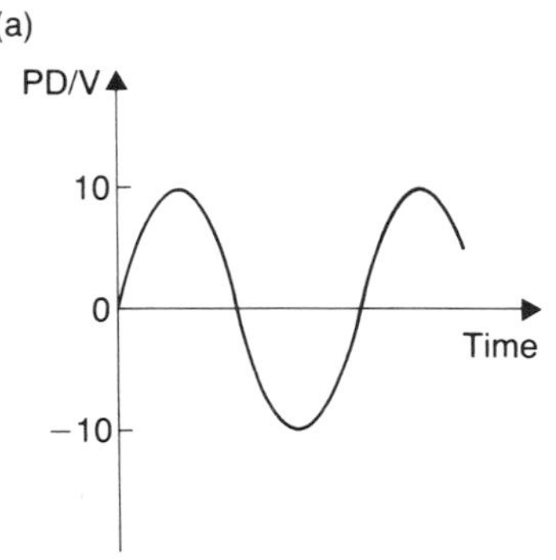

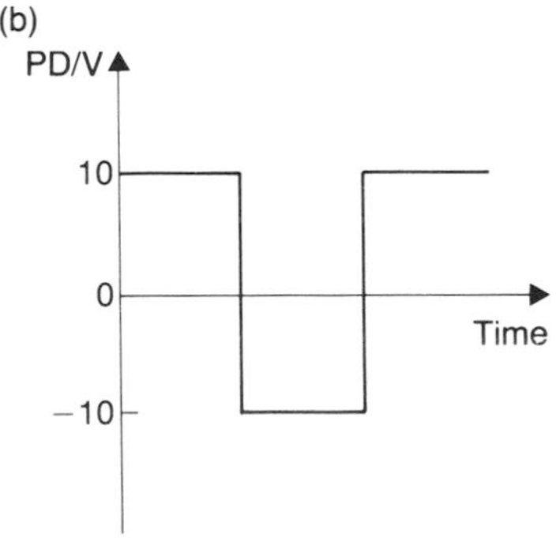

Fig 35.12 Graphs for Question 2

3 A coil having inductance and resistance is connected to an oscillator giving a fixed sinusoidal output voltage of 5.00 V RMS. With the oscillator set at a frequency of 50 Hz the RMS current in the coil is 1.00 A, and at a frequency of 100 Hz the RMS current is 0.625 A.
(a) Explain why the current through the coil changes when the frequency of the supply is changed.
(b) Determine the inductance of the coil.
(c) Calculate the ratio of the powers dissipated in the coil in the two cases. [JMB 80]

4 An alternating current having an angular frequency of 1.0×10^4 rad s^{-1} flows through a 10 kΩ resistor and a 0.10 μF capacitor in series. If the RMS potential difference across the resistor is 20 V, what is the RMS potential difference across the capacitor?

5 A series circuit consists of a 100 Ω resistor, a 20 μF capacitor and a 0.2 H inductor driven by an alternating source of frequency f. The potential difference across the inductor, measured with a high-impedance voltmeter, is 50 V RMS, and that across the capacitor is 200 V RMS. Find f, and the RMS current in the circuit. [WJEC 81, p]

6 A heater is rated 1.00 kW, 200 V. It is required to use it on a 250 V, 50 Hz supply. This may be done by placing either a resistor or an inductor in series with the heater. Calculate (a) the resistance of the heater, (b) the inductance of the inductor if its resistance is 5.0 Ω. State an advantage of using an inductor rather than a resistor. [AEB 82]

7 When a 2 V accumulator is connected in series with two basic circuit elements, the current in the circuit is 200 mA. If a 2 V, 50 Hz AC source replaces the accumulator, the current is 100 mA. Identify the circuit elements and determine their magnitudes. A capacitor is added in series with the AC source and the existing two elements, and the current is found to remain at 100 mA. Explain this, and find the value of this capacitor. Has the current changed in any way? [WJEC 77]

8 A small transformer is used to operate a 12 V lamp from the 240 V, 50 Hz mains. The voltage across the lamp is 12.0 V and the current through it is 2.0 A (both RMS). Assuming no transformer losses, the current in the primary is

A 1.00 A **B** 0.12 A **C** 0.10 A
D 0.08 A **E** 0.01 A [L 78]

9 A factory requires power of 144 kW at 400 V. It is supplied by a power station through cables having a total resistance of 3 Ω. If the power station were connected directly to the factory,

(a) show that the current through the cables would be 360 A;
(b) calculate the power loss in the cables;
(c) calculate the generating voltage which would be required at the power station;
(d) calculate the overall efficiency.

If the output from the power station provided a 10 000 V input to a transformer at the factory, the transformer having an efficiency of 96%,

(e) show that the current through the cables would be 15 A;
(f) calculate the power loss in the cables;
(g) calculate the overall efficiency. [AEB 79]

36 SEMICONDUCTORS, TRANSISTORS AND CONDUCTION IN SOLUTIONS

Charge carriers

An electric current is a flow of charged particles, i.e. of charge carriers, which in the case of solids are usually free electrons. In metals there are plenty of free electrons, and so metals are good conductors of electricity. In semiconductors the carriers may be free electrons or they may be holes or both.

In conducting non-metallic liquids the carriers are usually ions. Gases are discussed in Chapter 39.

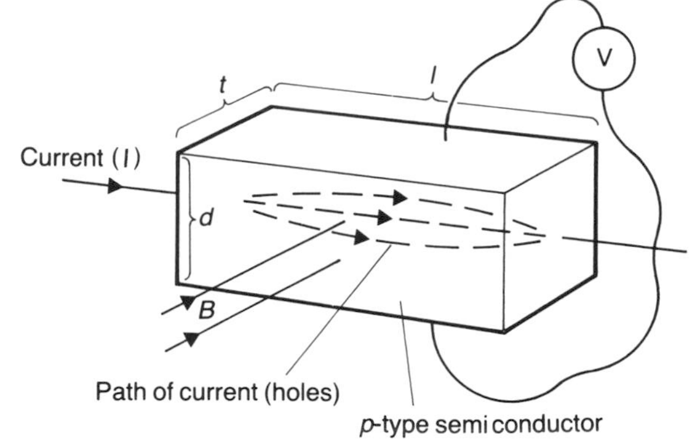

Fig 36.1 The Hall effect

The Hall effect

Fig 36.1 shows a current flowing through a rectangular block or slice of semiconductor with a flux density B perpendicular to the area ld. The holes, moving with velocity v, each experience an upwards force Bqv (left-hand rule), and so initially they are deflected upwards to make the top face positive and leave the lower face negative. This continues only until a PD V (the Hall voltage) is reached sufficient to cause an electric field intensity E $(= V/d)$ that exactly balances the magnetic force. Thus $Eq = Bqv$ or $Vq/d = Bqv$ and, using $I = nqvA$ (Equation 27.2) with $A = td$ we get $I = nqvtd$. Therefore the Hall PD is

$$V = \frac{BI}{nqt} \tag{36.1}$$

The Hall effect is particularly noticeable with semiconductors because they have small n values, giving large I/n and hence relatively large V. The polarity of the Hall PD is reversed if an n-type semiconductor is used, and if B or I are reversed.

Example 1

Calculate the Hall voltage when a magnetic flux density of 0.50 T is used, and a current of 48 mA, (a) for a copper sheet sample 0.50 mm thick with 10^{29} free electrons per m^3, (b) for a slice of p-type semiconductor also 0.50 mm thick but with 1.0×10^{25} holes per m^3. The magnetic field is parallel to the thickness in both cases. (Electronic charge $= 1.6 \times 10^{-19}$ C.)

Method

(a) Using Equation 36.1 we have $V = BI/nqt$ and $B = 0.50$, $I = 48 \times 10^{-3}$, $n = 1.0 \times 10^{29}$, q is the electronic charge 1.6×10^{-19}, $t = 0.50 \times 10^{-3}$.

$$\therefore \quad V = \frac{0.50 \times 48 \times 10^{-3}}{10^{29} \times 1.6 \times 10^{-19} \times 0.50 \times 10^{-3}}$$

$$= 30 \times 10^{-10}\,\text{V} \quad \text{or} \quad 3.0\,\text{nV}$$

(b) For the semiconductor $n = 1.0 \times 10^{25}$, i.e. 10^4 times smaller, so that V is 10^4 times bigger, i.e. $V = 30 \times 10^{-6}$ V or 30 μV.

Answer

(a) 3.0 nV, (b) 30 μV.

Semiconductor diodes

A simple semiconductor diode allows current to flow only in one direction through it, i.e. it can function as a rectifier. (It allows only direct current when an alternating voltage is used.)

The diode comprises a junction between n-type semiconductor and p-type semiconductor. The semiconductor is usually silicon.

(a) The diode comprises p and n regions

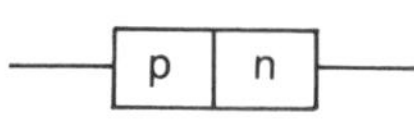

(b) Symbol for the diode

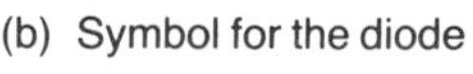

(c) Characteristic curve for germanium diode

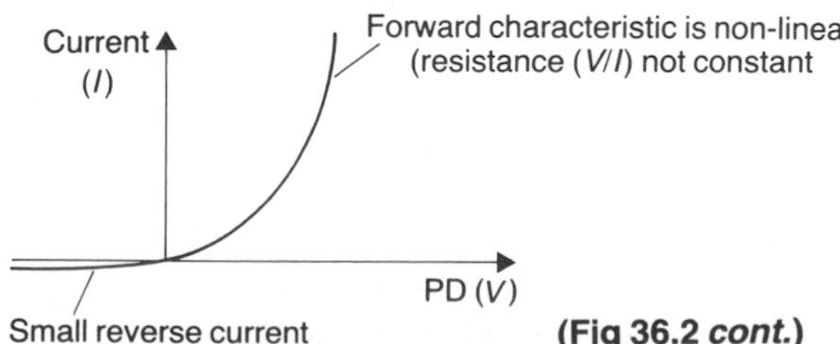

(Fig 36.2 *cont.*)

(d) Characteristic curve for silicon diode

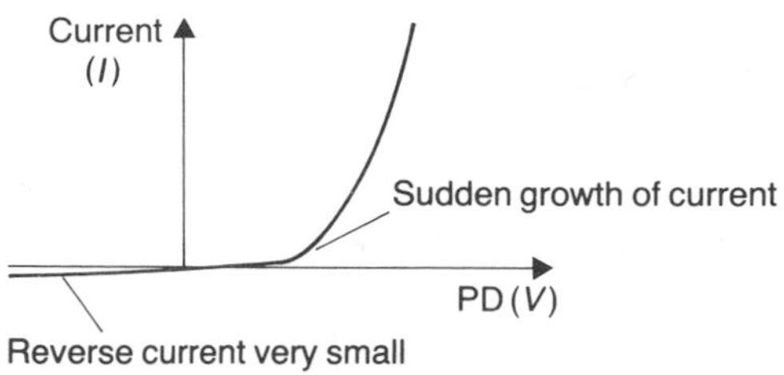

Fig 36.2 Rectification by a pn diode

Diodes used for rectification

Fig 36.3a shows a simple circuit for rectification.

(a) Simple circuit

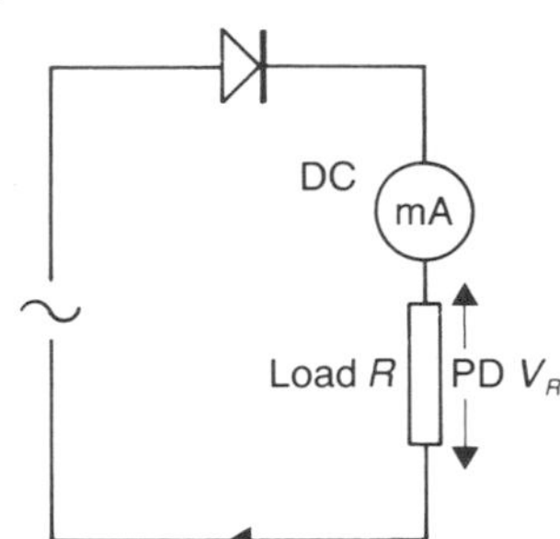

(b) Current obtained

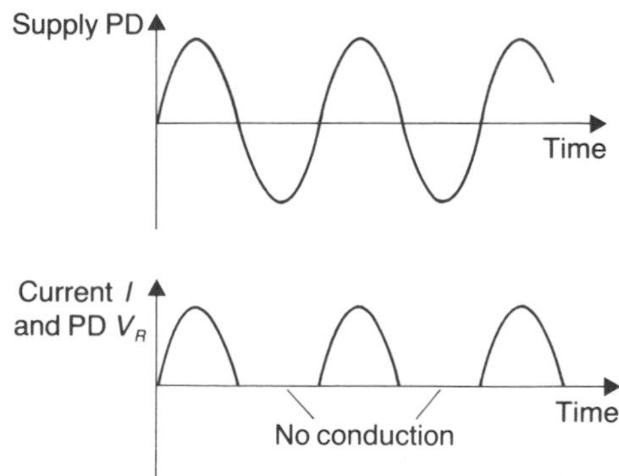

(c) With large smoothing capacitor fitted

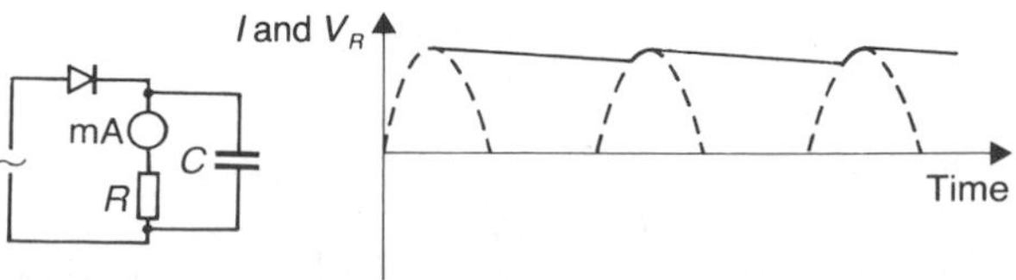

Fig 36.3 Half-wave rectification

The device requiring direct current through it has resistance R.

Unless the voltage supply frequency is exceptionally low, the meter will give a steady reading equal to the average current. Similarly a DC voltmeter fitted across R would read the average value of V_R $(= \bar{I} \times R)$.

With a capacitor C fitted as in Fig 36.3c the capacitor charges to the peak supply (if the diode forward resistance is negligible) and then, when the diode is not conducting, it discharges through R. If the time constant for discharge is large, the PD will be almost constant with time.

Example 2

With reference to Fig 36.3b, if the peak supply PD is 100 V and the forward resistance of the diode is negligible, calculate the RMS and average voltage obtained across the load resistance.

Method

The peak PD for the half-wave is 100 V, so that the RMS value for a half-cycle of conduction is $100/\sqrt{2}$ or 70.7 V.

Therefore the mean square value for this half-cycle is 70.7^2 and for a complete cycle, including the non-conducting part, the mean square value is

$$\frac{70.7^2 + 0}{2} = 2500\ \text{V}^2$$

The RMS value for a whole cycle is $\sqrt{2500}$ which is 50.0 V. Clearly RMS $= \frac{1}{2}$peak for a half-wave rectified waveform.

The average value for the PD is $(2/\pi) \times 100$ (see p.234) for the conducting half-cycle and zero for the other half. Therefore the average for a whole cycle is $100/\pi$ or 31.8 V.

Answer

50.0 V RMS, 31.8 V average.

Example 3

With reference to Fig 36.3a, if $R = 10\ \text{k}\Omega$, and the diode forward resistance is negligible and the sinusoidal voltage supply is 20 V RMS, 50 Hz, calculate (a) the peak current, (b) the DC milliammeter reading. Then if a 16 μF capacitor is fitted in parallel with R, calculate (c) the time constant for the capacitor discharge and (d) the approximate reading of a high-resistance voltmeter fitted across R.

Method

(a) The peak supply PD is $20\sqrt{2} = 28$ V; peak current is $28/10\,000$ A $= 2.8$ mA.

(b) Average current over one or more cycles is

$$2.8 \times \frac{2}{\pi} \times \frac{1}{2} = \frac{2.8}{\pi} = 0.90\ \text{mA}$$

(c) The time constant $= RC = 10\,000 \times 16 \times 10^{-6} =$ 0.16 s. (It is therefore large compared with the 20 ms period of the supply, so that the output looks like Fig 36.3c.)

(d) The voltmeter reading approximates to the peak supply PD, i.e. 28 V, as in Fig 36.3c.

Answer

(a) 2.8 mA, (b) 0.90 mA, (c) 0.16 s, (d) 28 V.

Exercise 123

1 A particular semiconductor sample was rectangular in shape with a thickness of 1.0 mm. When held in a uniform magnetic field parallel to the sample's thickness and of flux density 0.80 T, a current of 1.0 A was passed through it. The Hall voltage obtained was 0.48 mV. Obtain a value for the carrier density in the sample. (Electron charge $= 1.6 \times 10^{-19}$ C.)

2 A simple half-wave recitifier has a load of 500 Ω resistance, a diode with low forward resistance and a sinusoidal voltage supply of 12 V RMS. Calculate the reading expected on a low resistance DC milliammeter connected in series with the load.

Bipolar transistors

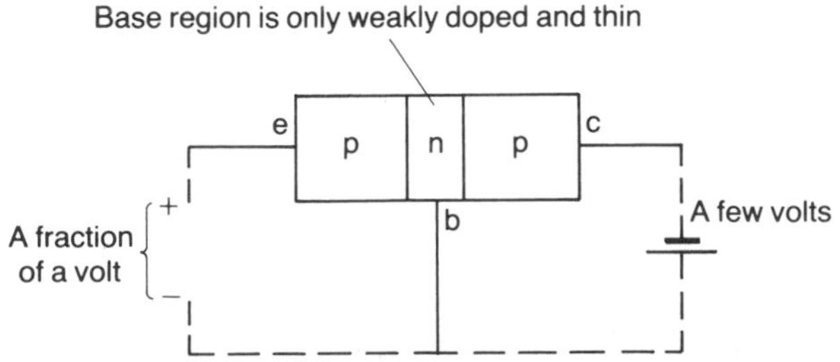

Fig 36.4 A pnp transistor (bipolar type)

These are of the pnp or the npn construction.

Consider the pnp type (Fig 36.4): e, b and c are terminals called the emitter, base and collector respectively and the p, n and p regions are the emitter region, base region and collector region.

The small PD applied between e and b is correct for conduction (i.e. we have 'forward bias'), and holes diffuse from p- to n-regions. (Negligible free electrons diffuse across because the base is only weakly doped n-type material.) This hole current is called I_e.

The bc junction is reverse-biased, with a substantial PD and electric field intensity there. Most of the holes that have entered the base from the emitter quickly reach the bc junction, where they are swept across the junction by the electric field there. These holes amount to a fraction α of I_e. Thus the collector current is

$$I_c = \alpha I_e \qquad (36.2)$$

The base current I_b consists of free electrons flowing into b to replace the small number of electrons that combined with holes in the base, these holes amounting to $I_e - \alpha I_c$.

$$I_b = I_e - \alpha I_e$$

but also, equating currents in and out of the transistor gives

$$I_e = I_b + I_c \qquad (36.3)$$

Using this, we can write 36.2 as

$$I_c = \alpha(I_c + I_b)$$

or $$I_c = \frac{\alpha}{1-\alpha} I_b \text{ or } \alpha' I_b \text{ or } \beta I_b \qquad (36.4)$$

For small changes in currents, if we denote the changes by small letters (i), we have*

$$i_c = \alpha i_e \qquad (36.5a)$$

and $$i_c = \alpha' i_b = \beta i_b \qquad (36.5b)$$

from Equations 36.2 and 36.4.

Also $$\beta = \frac{\alpha}{1-\alpha} \qquad (36.5c)$$

A common-base amplifier

The circuit configuration is shown in Fig 36.5. (This configuration has b common to input and output loops.)

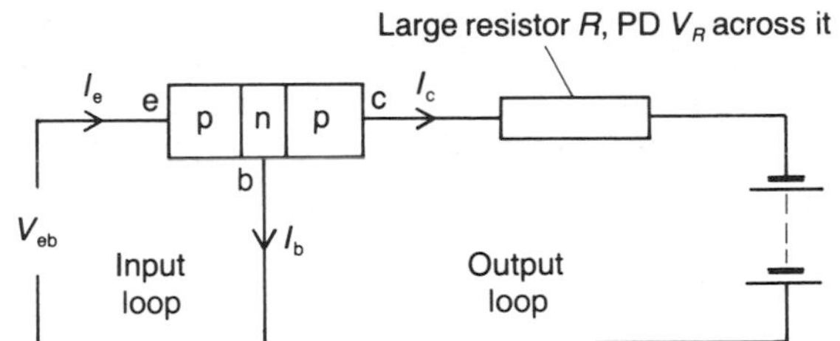

Fig 36.5 Common-base amplifier

The PD across R is $V_R = I_c \times R$.

The purpose of the voltage amplifier is to produce a large change (v_{out}) in V_R when a small change (v_{in}) occurs in V_{eb}. The voltage gain is

$$\frac{v_{out}}{v_{in}} = \frac{i_c R}{i_e r_{in}} = \frac{\alpha R}{r_{in}} \qquad (36.6)$$

where r_{in} is the resistance of the input loop and usually is the resistance of the transistor between e and b (r_{eb}) at the particular eb bias PD used.

The current gain is

$$\frac{i_c}{i_e} = \alpha \text{ (and so is less than 1)} \qquad (36.7)$$

Common-emitter configuration (CE)

The circuit in its simplest form is shown in Fig 36.6.

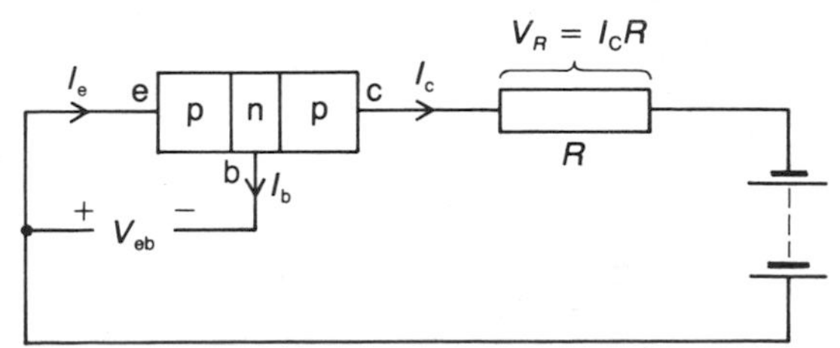

Fig 36.6 Common-emitter amplifier

**The symbols used here do not entirely conform to British Standard 3363 which uses lower case letters for ALL time-varying quantities. It also distinguishes between upper and lower case subscripts.*

The voltage gain is $\frac{v_{out}}{v_{in}} = \frac{i_c R}{i_b r_{in}}$

and using Equation 36.5b we get

$$\text{Voltage gain} = \frac{\beta R}{r_{in}} \qquad (36.8)$$

The current gain is β (and might be about 50 if α is, say, 0.98.)

An extra resistor placed in the base lead would increase r_{in} and so reduce the voltage gain (Equation 36.6) but could make the input resistance r_{eb} negligible. This ensures a constant and easily calculated voltage gain regardless of r_{eb}.

Some practical circuits

Examples of simple voltage-amplifying circuits (pnp, common-emitter types) are shown in Fig 36.7.

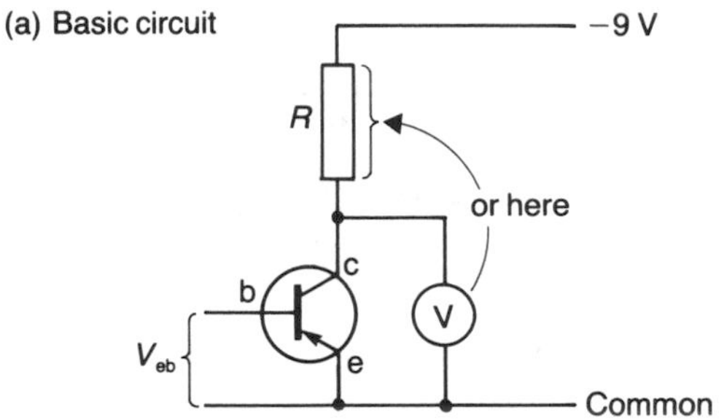

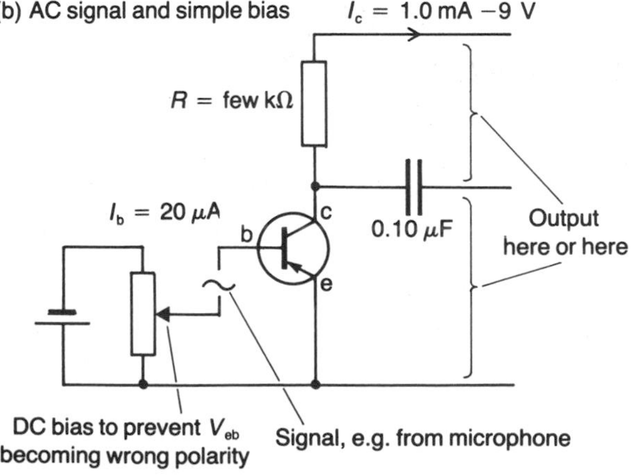

(Fig 36.7 *cont.*)

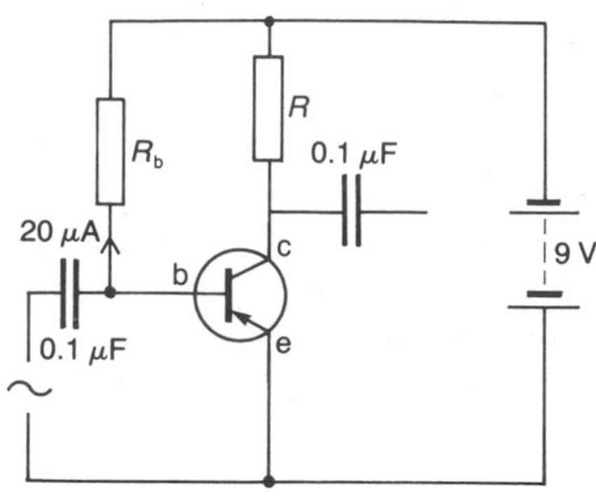

Fig 36.7 Some CE voltage amplifier circuits

Polarities of voltage supplies need to be reversed if the transistor is of npn construction.

The capacitors are used to provide a path for AC signals but not for direct currents.

Hybrid (or *h*-) parameters

These are numbers and quantities which describe the properties of a transistor. For example h_{fb} is the same as the α value of the transistor, $h_{fe} = \beta$, r_{in} for the common-base configuration is h_{ib} and for CE configuration is h_{ie}. These values of h_{fb}, h_{fe}, h_{ib} and h_{ie} (also called h_{21}, h'_{21}, h_{11} and h'_{11} respectively) are related to the characteristic curves of the transistor; see Fig 36.8 for CE characteristics.

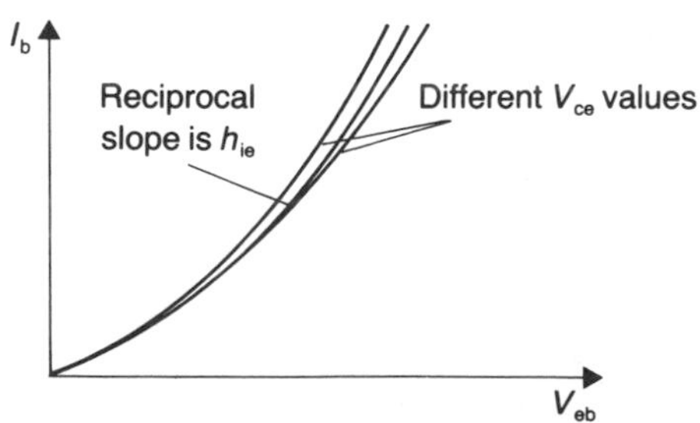

(Fig 36.8 *cont.*)

(b) Collector or output characteristic

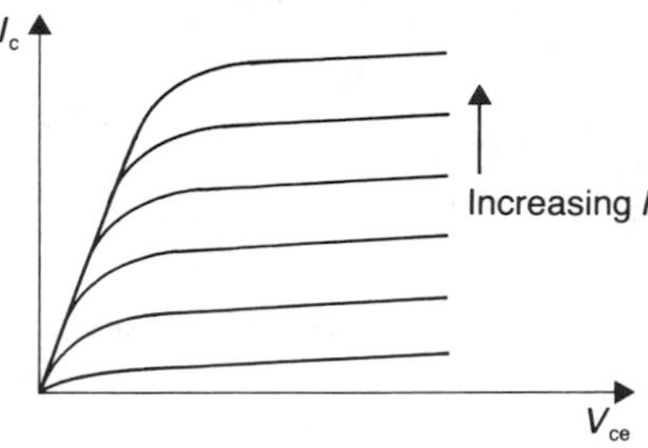

(c) Transfer characteristic

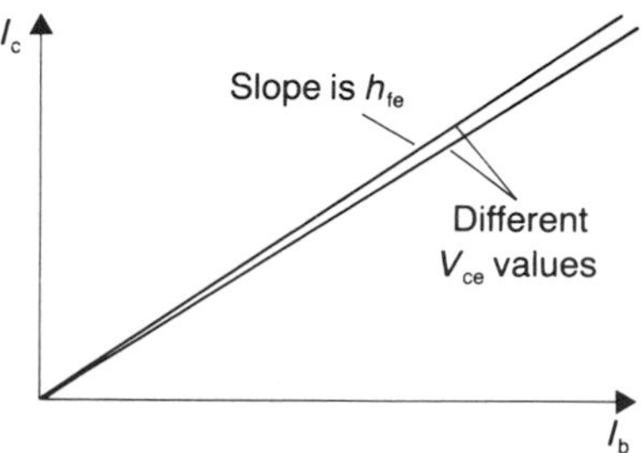

Fig 36.8 Common-emitter, pnp-transistor characteristic curves

The field-effect transistor (FET)

The essential construction features of the ordinary type of FET are illustrated in Fig 36.9a for an n-channel FET.

(a) Construction (simplified)

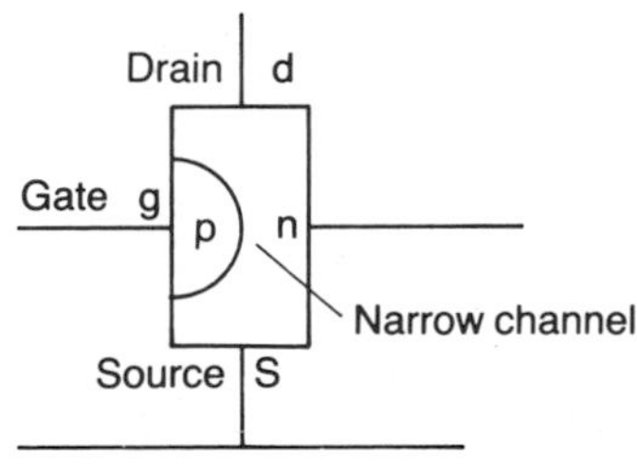

(b) Symbol

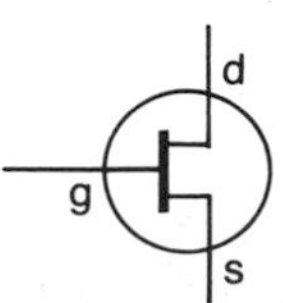

(c) Symbol for IGFET

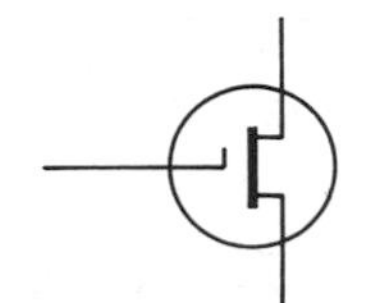

Fig 36.9 Field-effect transistor, n-channel type

When the gate is made negative compared with the source, the gs diode is reversed-biased, so that the pn junction area is deprived of current carriers and so becomes high-resistance. The width of the channel through which conduction can occur is consequently less and the current flow between s and d is reduced.

Small changes in the gate potential can cause large changes in drain current and this can cause large voltage changes across a load resistor R in the drain lead (see Fig 36.10).

(a) Circuit

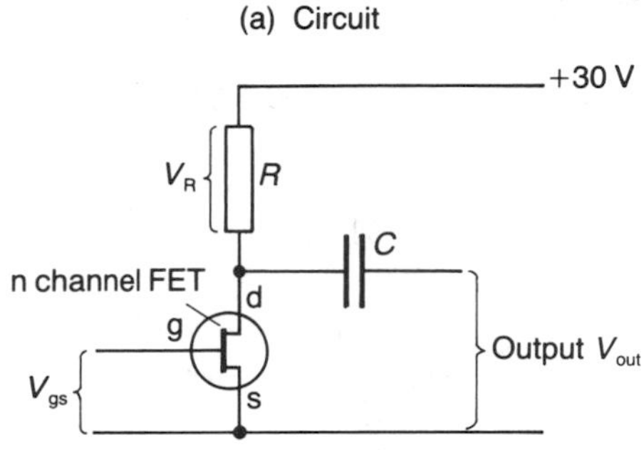

(b) Characteristic curves of the FET

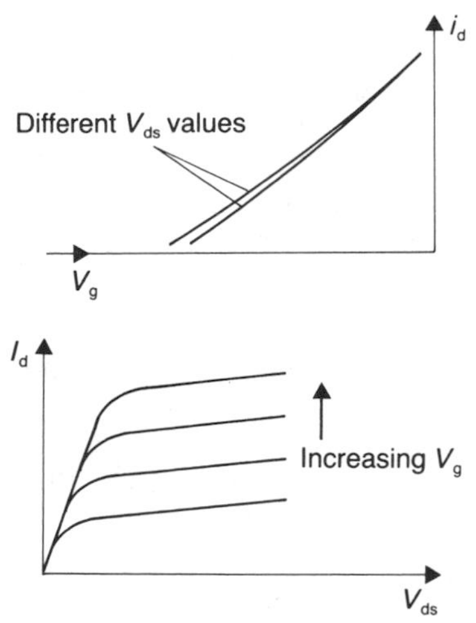

Fig 36.10 FET amplifier

A useful parameter for describing a FET's performance is its mutual conductance g_m which is the slope dI_d/dV_g of the I_d/V_g characteristic (see Fig 36.10b).

The voltage gain is

$$\frac{v_{out}}{v_{in}} = \frac{i_d R}{v_{in}} = \left(\frac{dI_d}{dV_g}\right)R = g_m R \qquad (36.9)$$

where v_{out} is the change in V_d, v_{in} is the change in V_{gs} (or V_g) and i_d is the change in I_d.

The resistance between g and s is very high, so that I_g is zero and the FET does not function as a current amplifier, only as a voltage amplifier.

An insulated-gate FET (IGFET) has special properties and the gate may be made + (increasing I_d) as well as − (to decrease I_d, as in the ordinary-junction FET). The symbol for the IGFET is shown in Fig. 36.9c.

Example 4

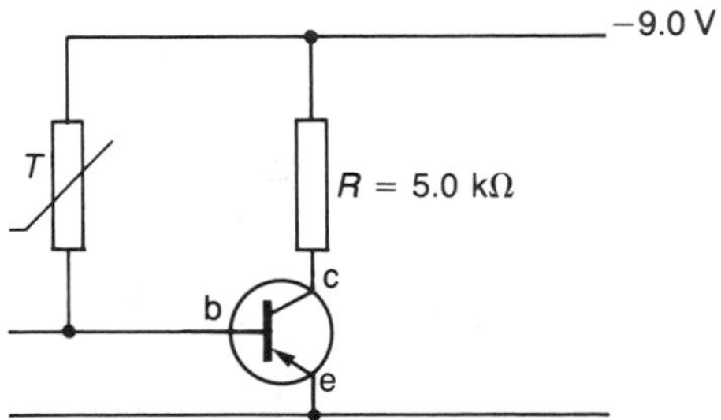

Fig 36.11 Circuit for Example 4

In Fig 36.11 a pnp junction transistor amplifier is being used with a temperature-dependent resistor or thermistor T. If T changes from 0.50 MΩ to 0.33 MΩ, what change occurs in the PD across resistor R, and is it an increase or decrease? What is the change in PD between c and e? The forward current gain h_{fe} = 50. (V_{eb} may be assumed small.)

Method

The base current is initially I_b = 9.0 V/0.50 MΩ (since V_{eb} is small.)

$$\therefore \quad I_b = \frac{9.0}{0.50 \times 10^6} = 18 \times 10^{-6}\,\text{A}$$

The new I_b is

$$I_b = \frac{9.0}{0.33 \times 10^6} = 27 \times 10^{-6}\,\text{A}$$

∴ change in I_c is $i_c = \beta$ (or h_{fe}) times the change in I_b.

$$= 50 \times 9 \times 10^{-6}$$

$$= 450 \times 10^{-6} \quad \text{or} \quad 0.45\,\text{mA}$$

∴ change in PD across R equals

$$i_c \times R = 0.45 \times 10^{-3} \times 5 \times 10^3 = 2.2\,\text{V}$$

I_b has increased, therefore I_c must be greater. Therefore the 2.2 V is an increase across R and there is a decrease of this amount across the transistor between e and c, because the total PD across R and the transistor equals 9.0 V.

Answer

2.2 V increase, 2.2 V decrease.

Example 5

Referring to Fig 36.7b, what is the RMS voltage output if the input signal is 1.0 mV RMS from a low-resistance microphone, resistance r_{be} (or r_{in}) of the transistor is 2.0 kΩ, current gain is 50 and the load resistor is 4.0 kΩ? (The resistance of the voltage divider used to obtain the base bias may be assumed negligible.)

Method

The RMS base current is

$$i_b = \frac{v_{in}}{2.00\,\text{k}\Omega} = \frac{1.0 \times 10^{-3}}{2.0 \times 10^3}$$

$$= 0.50 \times 10^{-6}\,\text{A RMS}$$

∴ the collector current fluctuation, obtained by multiplying the base current by 50, is

$$i_c = 0.50 \times 10^{-6} \times 50 = 25 \times 10^{-6} = 25\,\mu\text{A}$$

and the voltage fluctuation across resistor R (which is 4.0 kΩ) is

$$v = 25 \times 10^{-6} \times 4.0 \times 10^3 = 0.10\,\text{V RMS}$$

Answer

0.10 V RMS.

Example 6

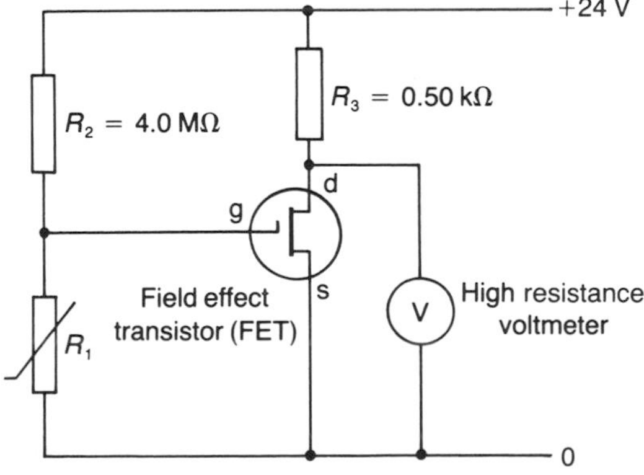

Fig 36.12 Circuit for Example 6

In the circuit shown in Fig 36.12, R_1 is a thermistor (a device whose resistance decreases significantly when its temperature increases). If the temperature changes such that the thermistor's resistance falls from 1.0 MΩ to

0.8 MΩ, calculate the change in voltmeter reading expected, given that the mutual conductance of the FET is 2.5 mA V^{-1}.

Method

R_1 and R_2 form a voltage divider (see Chapter 27) across the 24 V supply. The source-to-gate PD obtained from this voltage divider is

$$24 \times \frac{R_1}{4.0\,\text{M}\Omega + R_1}$$

When R_1 = 1.0 MΩ, this PD is 24 × 1/5 = 4.8 V.

When R_1 = 0.80 MΩ, the PD is 24 × 1/6 = 4.0 V.

(The transistor being used here is an IGFET with its gate +.)

For the 0.8 V difference from 4.8 to 4.0 V the drain current increase is 2.5 mA V^{-1} × 0.8 V or 2.0 mA.

The change in PD across R_3 is R_3 × 2.0 mA, i.e. 0.50 × 10^3 × 2.0 × 10^{-3}, which is 1.0 V. The voltmeter reading, which is the voltage across the FET, falls by 1.0 V.

Answer

1.0 V decrease.

Exercise 124

1 A photoconductive cell (for which the resistance decreases with the amount of light falling on the cell) is used in place of the thermistor in Fig 36.11. If the illumination causes the photocell current to increase by 10 μA, by how much will the PD across R change, given that h_{fe} for the transistor is 60 and the resistor in the drain lead is 500 Ω?

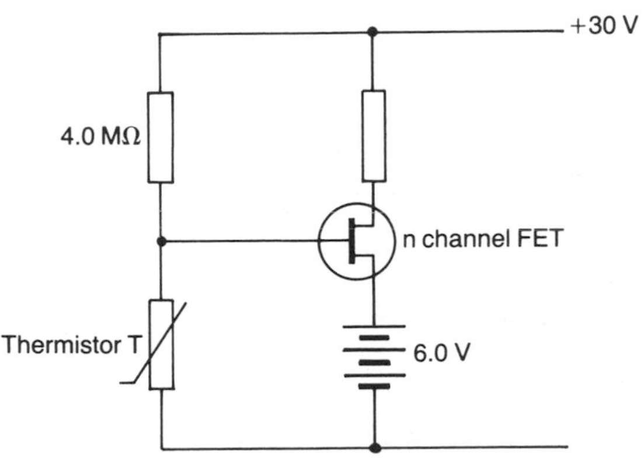

Fig 36.13 Circuit for Question 2

2 In Fig 36.13 the drain current is 4.0 mA when the thermistor resistance is 1.0 MΩ. Calculate the new drain current when the thermistor resistance falls to 0.80 MΩ, given that the mutual conductance is 3.0 mA V^{-1}.

3 In Fig 36.7c (a) calculate a suitable value for resistor R_b if close to 20 μA is required for the base current, and the supply PD is 10 V.
(b) What is the PD across the transistor (from e to c) if R is 5.0 kΩ and I_c is 1.2 mA?
(c) If a sinusoidal signal causes base current fluctuations of 2.0 μA RMS, what is the RMS fluctuation in PD across R (or across the transistor) given that β (or h_{fe}) is 60?

Conduction in liquids

Many compounds that dissolve in water do so with their molecules split up into positive and negative ions. Examples include common salt (sodium chloride) $NaCl$, copper sulphate $CuSO_4$, silver nitrate $AgNO_3$ and sulphuric acid H_2SO_4. Even pure water itself is ionised to a very small extent. The ions formed are respectively* Na^+ and Cl^-, Cu^{2+} and SO_4^{2-}, Ag^+ and NO_3^-, H^+ and SO_4^{2-}, H^+ and OH^-. When two electrodes are immersed in the solution and a PD is applied, the ions move. The positive ones move towards the negative electrode (the cathode) and the negative ones move towards the positive electrode (called the anode). Thus a current flows. The arrival of ions at the electrodes results in chemical changes, so that the conduction process has been called electrolysis. A container for the solution (electrolyte) fitted with the necessary electrodes is called a voltameter.

Charge *q* and mass *m*

Suppose that in t seconds enough ions arrive at one electrode for a mass m of element to be deposited on it or to be liberated as a gas there. Each ion delivers charge ve to the electrode where e is the

Other ions formed are not important here.

electronic charge and v is the number of electronic charges carried by the ion (i.e. its valency). v is 1 for Na^+ and Cl^-, and $v = 2$ for Cu^{2+} and SO_4^{2-}.

Thus mass m is proportional to charge q of the ions that arrive. (This is Faraday's first law of electrolysis). We say that

$$m = zq \quad \text{or} \quad m = zIt \tag{36.10}$$

where m is the mass of whatever number of ions arrive, q their total charge, I the current, t the time of flow and z a characteristic of the type of ion concerned. z is called the electrochemical equivalent and can be measured in kg per coulomb. Its reciprocal is called the charge–mass ratio for the ion.

We should note further that, for the same charge q but different ions, the mass m deposited or liberated is proportional to the ion mass and inversely proportional to the valency of the ion. This is really a statement of Faraday's second law of electrolysis.

The faraday is the name given to the amount of charge carried by Avogadro's number L of monovalent ($v = 1$) ions. The charge per mole is called the Faraday constant ($9.649 \times 10^4\,C\,mol^{-1}$).

If divalent ions are considered, then of course the Avogadro number of ions (i.e. 1 mole of these ions) will carry a charge of 2 faraday.

$$1 \text{ faraday} = Le \tag{36.11}$$

where L is the Avogadro number and e is the electronic charge.

Example 7

Calculate the mass of copper deposited in 30 minutes in a copper sulphate voltameter using a constant current of 2.0 A. (Mass/charge ratio for Cu^{2+} copper ions is $3.3 \times 10^{-8}\,kg\,C^{-1}$.)

Method

$z = (m/q)$ is given as $3.3 \times 10^{-8}\,kg\,C^{-1}$.

Also, using $q = It$, we get $q = 2.0 \times 30 \times 60$, remembering that t must be in seconds.

$$m = 3.3 \times 10^{-8} \times 2.0 \times 30 \times 60 = 12 \times 10^{-5}\,kg$$

Answer

12×10^{-5} kg.

Example 8

The same current flows for the same time through a copper sulphate voltameter and a silver nitrate voltameter. If 0.10 g of copper is deposited on one cathode how much silver is deposited on the other cathode? (Atomic masses of copper and silver relative to the hydrogen atom* are 63.5 and 108 respectively. Valency of copper ions in copper sulphate is 2, of silver ions in the nitrate solution is 1.)

Method

Since the same current flows through both voltameters

$$\text{Mass } m \propto \frac{\text{Atomic mass}}{\text{Valency}}$$

$$\frac{\text{Mass of silver deposited}}{\text{Mass of copper deposited}} = \frac{108/1}{63.5/2}$$

$$\frac{\text{Mass of silver}}{0.1\,g} = \frac{108 \times 2}{63.5}$$

$$\therefore \quad \text{Mass of silver} = \frac{108 \times 2}{63.5} \times 0.10\,g$$

$$= 0.34\,g$$

Answer

0.34 g.

Example 9

Calculate the mass of copper deposited on the cathode by a flow of 0.040 faraday of charge, given that 1 mole of copper atoms has a mass of 63.5 g and that the copper ions in the electrolyte are divalent.

Method

For monovalent ions 1 faraday deposits 1 mole.

For divalent ions 1 faraday deposits $\frac{1}{2}$ mole.

0.040 faraday deposits $0.040 \times \frac{1}{2}$ mole or $0.040 \times \frac{1}{2} \times 63.5\,g$, i.e. 1.27 g.

Answer

1.3 g.

Example 10

Calculate (a) the number of H^+ ions that arrive at the cathode, (b) the mass of hydrogen (H_2) gas produced and (c) the volume of this gas at STP when 2.0 A flows through acidified water for 0.1 hour. (The Faraday constant is $96\,500\,C\,mol^{-1}$, the Avogadro constant is $6.0 \times 10^{23}\,mol^{-1}$, the molar mass of hydrogen gas (H_2) is 2.16 g and 1 mole of any gas occupies $22.4 \times 10^{-3}\,m^3$ at STP.)

*See also p.287

Method

(a) 2.0 A for 0.1 hour (360 seconds) means a charge of 720 C. The number of faradays is 720/96 500, and this is the number of moles of monovalent (H^+) ions that arrive at the cathode:

$$\frac{720}{96\,500} \times 6.0 \times 10^{23} = 4.5 \times 10^{21} \text{ ions}$$

(b) The number of moles of H_2 molecules formed is half of the above number, namely $\frac{1}{2} \times 720/96\,500$ and each of these moles means a mass of 2.16 g. Thus the mass of gas is

$$\frac{1}{2} \times \frac{720}{96\,500} \times 2.16 = 8.1 \times 10^{-3}\text{ g}$$

(c) Each mole of gas occupies $22.4 \times 10^{-3}\text{ m}^3$ at STP and we have $\frac{1}{2} \times 720/96\,500$ moles. Therefore the volume of gas is

$$\frac{1}{2} \times \frac{720}{96\,500} \times 22.4 \times 10^{-3} = 8.4 \times 10^{-5}\text{ m}^3$$

Answer

(a) 4.5×10^{21}, (b) 8.1×10^{-3} g, (c) $8.4 \times 10^{-5}\text{ m}^3$.

Exercise 125

1 Calculate the mass of silver deposited per minute in a silver nitrate voltameter using 0.5 A. The electrochemical equivalent (mass/charge ratio) for silver is $1.1 \times 10^{-6}\text{ kg C}^{-1}$.

2 (a) What volume of hydrogen gas at STP is evolved in 5 minutes at the cathode when acidified water is electrolysed with a current of 0.5 A? The density of hydrogen gas at STP is $9 \times 10^{-2}\text{ kg m}^{-3}$.
(b) What will the volume become at 17 °C and a pressure of 750 mm of mercury?
(Mass/charge ratio for hydrogen ions is $1.05 \times 10^{-8}\text{ kg C}^{-1}$.)

3 The same current flows for the same time through an acidified water voltameter and a copper sulphate voltameter. 60 mg of hydrogen gas is evolved in the first voltameter. How much copper is deposited in the other? (Relative atomic masses are 1 for hydrogen and 63 for the copper ions.)

4 Given that 1 mole of electrons contains 6.02×10^{23} electrons, obtain a value for the Faraday constant.
(Electronic charge $e = 1.6 \times 10^{-19}$ C.)

5 If 1 faraday is 96 500 C, calculate the charge–mass ratio for divalent copper ions whose relative atomic mass is 63. (1 mole of carbon-12 atoms has a mass of 12 g.)

Exercise 126: Questions of A-level standard

1 In a long bar of semiconductor, the concentration of charge carriers is $5.0 \times 10^{20}\text{ m}^{-3}$. If the area of cross-section of the bar is $2.0 \times 10^{-5}\text{ m}^2$ and the velocity of the carriers is 0.5 m s^{-1}, what is the current flowing along the bar? (Electronic charge $e = 1.6 \times 10^{-19}$ C.) [SUJB 82]

2 A sinusoidal alternating voltage of peak value 100 V is connected to a diode in series with a 100 Ω resistor. The forward resistance of the diode may be taken as zero and the backward resistance as infinite. The mean current flowing through the resistor is

A 1 A **B** between 0.5 A and 1 A
C 0.5 A **D** between zero and 0.5 A
E zero [OC 81]

3

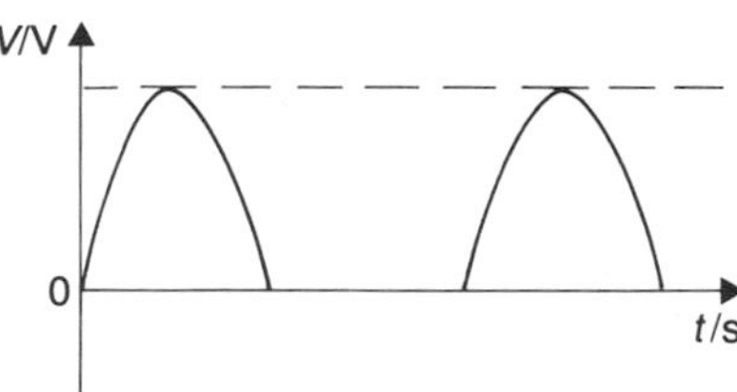

Fig 36.14 Graph for Question 3

Half-wave rectification of a sinusoidal voltage having a peak value of 20 V gives a waveform as shown in Fig 36.14. The RMS value of the rectified voltage is, in volts

A 10 **B** $\frac{10}{\sqrt{2}}$ **C** $\frac{5}{\sqrt{2}}$
D $\frac{20}{\sqrt{2}}$ **E** $\frac{20}{\pi}$

4

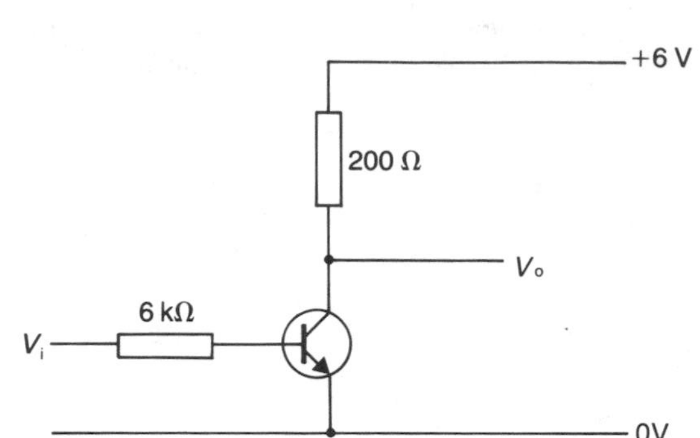

Fig 36.15 Circuit diagram for Question 4

Fig 36.15 shows a circuit incorporating an npn transistor whose current amplification factor h_{fe} (β) is 50. Ignoring the base–emitter voltage, calculate the base current when the input voltage V_i is 1.5 V. What is the output voltage V_o? [O 82, p]

5 In the transistor circuit in Fig 36.16 the potential difference between c and e is 3 V, and that between b and e is negligibly small. The current gain I_c/I_b is 200. Find the base current I_b and the value of the resistance R. [WJEC 80, p]

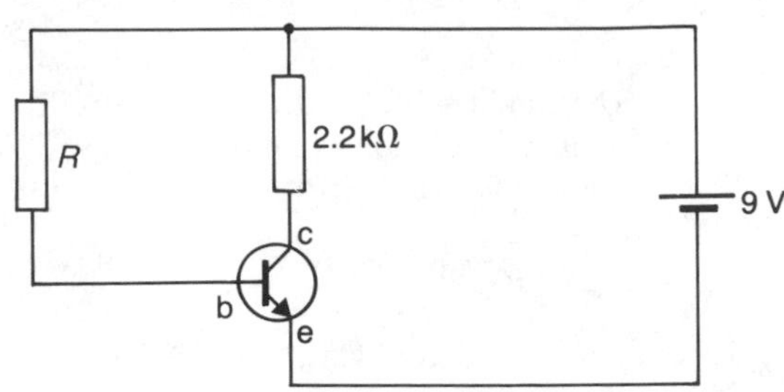

Fig 36.16 Circuit diagram for Question 5

6 (a) 1 mole contains 6.02×10^{23} particles. Calculate the electric charge carried by 1 mole of monovalent ions (i.e. the faraday), given that the electronic charge e is 1.6×10^{-19} C.

(b) When a steady current of 2.0 A flows through a copper sulphate solution (Cu^{2+} and SO_4^{2-} ions), how many Cu^{2+} ions arrive per second at the negative electrode?

37 THE CATHODE RAY OSCILLOSCOPE

Basic principles

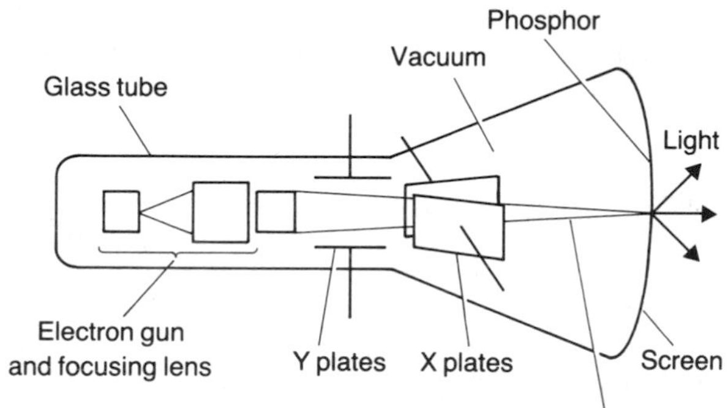

Fig 37.1 A cathode ray oscilloscope tube

The cathode ray oscilloscope (CRO) can be regarded as a very special voltmeter. A PD V to be measured is amplified and then applied to metal plates above and below the electron beam (Y-plates, Fig 37.1).

The beam is deflected up or down depending upon the polarity of V. The size of the deflection is proportional to the size of V.

The time base

When a voltage produced by the 'time base' section of the CRO is correctly applied to the X-plates, the beam moves steadily and repeatedly across the screen. If now an alternating voltage is applied to the Y-plates, the trace seen on the screen is a wave. It is in fact a voltage–time graph.

Measurement of current using a CRO

The technique for this is to let the current flow through a small resistance R and measure the PD V across it with the CRO. Then $I = V/R$.

Frequency measurement

If the number of millimetres in the X-direction occupied by one cycle of a trace is recorded from the screen, then the period can usually be deduced from the time base velocity (marked in seconds per centimetre on the time base selector or control).

With a double-beam CRO the trace of the voltage being examined can be matched in frequency to a known, variable-frequency voltage fed to the second-beam Y-plates and displayed above or below the first trace for easy comparison.

An alternative method makes use of Lissajous' figures. This requires that the time base is not used but the matching frequency voltage is connected to X-input terminals on the CRO. When the two frequencies f_X and f_Y are equal the trace is a single closed loop or a straight line as shown in Fig 37.2.

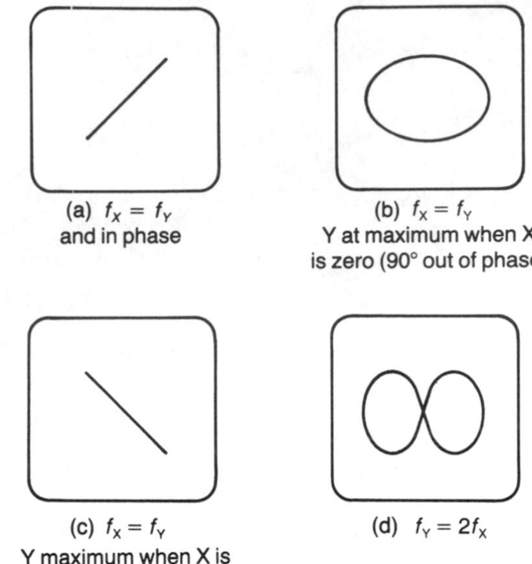

Fig 37.2 Lissajous' figures

Example 1

A 5.0 V RMS, 50 Hz voltage is obtained from a transformer connected to the mains supply and is fed to the Y-plates of a CRO. If the Y sensitivity is set at $10\ \mathrm{V\ cm^{-1}}$ and the time base at $10\ \mathrm{ms\ cm^{-1}}$, what will be the total peak-to-peak height of the trace and how many complete cycles of the voltage will be displayed if the trace is 4.0 cm wide?

Method

If the RMS voltage is 5.0 V, then we find the peak value by multiplying this by $\sqrt{2}$ (see Chapter 35), and to obtain the peak-to-peak value we multiply the peak value by 2.

Peak-to-peak voltage is $5.0 \times \sqrt{2} \times 2 = 14.14\ \mathrm{V}$

Since 10 V is represented by 1.0 cm, then 14.14 V will be represented by $1.0 \times 14.14/10$, i.e. 1.414 cm. (1.4 cm to two significant figures.)

The time occupied by one cycle is one-fiftieth of a second or 20 ms. At 10 ms per cm time base velocity one cycle occupies 2.0 cm. Thus 4.0 cm of screen width will accommodate 2 cycles exactly.

Answer

1.4 cm, 2.0 cycles.

Example 2

An audio-frequency signal generator G operates a loudspeaker S so that sound waves are directed towards the two small microphones M_1 and M_2 as shown in Fig 37.3.

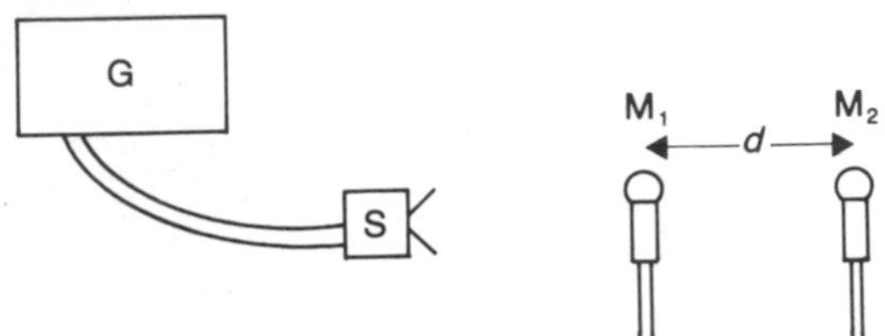

Fig 37.3 Diagram for Example 2

The signals from the two microphones are amplified to identical peak values and fed to the X- and Y-input-terminals of a CRO with the time base switched off. Describe the appearance of the CRO trace as M_2 is slowly moved away from M_1 in the direction of travel of the sound.

If the sound frequency is set to 850 Hz and the smallest increase of d that changes the trace appearance through 1 cycle is 40 cm, calculate the speed of travel of the sound.

Method

When $d = 0$, a straight-line trace is obtained (Fig 37.2a), and as d increases, the phase difference between the signals soon gives a circular trace (Similar to Fig 37.2b), then a straight line (when M_1 and M_2 are exactly out of phase as in Fig 37.2c), then a circle again (M_2 three-quarter period behind M_1), and, when d = wavelength, the original trace is obtained again.

40 cm equals the wavelength λ

Using the equation, speed = frequency × λ (see Chapter 15), we get

$$\begin{aligned}\text{Speed} &= 40\ \mathrm{cm} \times 850\ \mathrm{Hz}\\ &= 34\,000\ \mathrm{cm\ s^{-1}} \quad \text{or} \quad 340\ \mathrm{m\ s^{-1}}.\end{aligned}$$

Answer

$0.34\ \mathrm{km\ s^{-1}}$.

Exercise 127

1 The screen of a cathode ray oscilloscope displays the trace shown in Fig 37.4. The Y sensitivity is set at 10 mV/cm, and the time base is set at 0.20 ms/cm. Obtain values for (a) the peak voltage and (b) the frequency of the alternating signal.

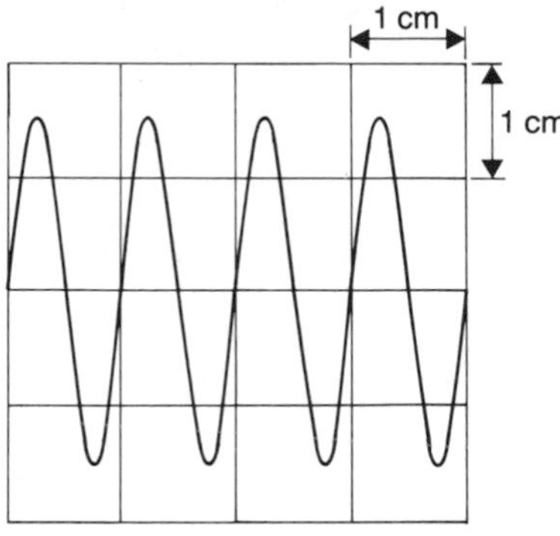

Fig 37.4 Graph for Question 1

2 An oscilloscope is used to measure the time it takes to send a pulse of sound along a 70 cm length of metal rod and back again. Fig 37.5 shows the appearance of the oscilloscope screen. A indicates the original pulse and B the reflected pulse. If the time base speed is 0.10 mm μs^{-1}, what is the speed of travel of the pulse through the rod?

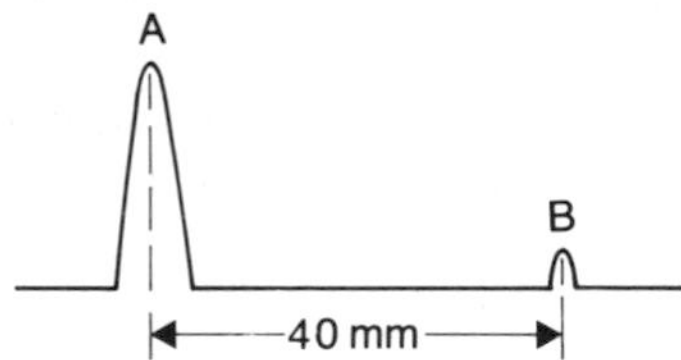

Fig 37.5 Oscilloscope trace for Question 2

3 A Lissajous' figure in the form of an upright figure-of-eight is obtained when a sinusoidal voltage of peak value V and frequency f is applied to the Y-plates of a CRO and a sinusoidal voltage of 10 mV peak and 280 Hz is applied to the X-plates. The figure measures 4.0 cm high by 1.0 cm wide. Without changing the settings of the CRO controls, V is now connected to the X-input-terminals, and the other voltage to the Y-plates. A figure-of-eight lying on its side is obtained which measures 2.0 cm by 2.0 cm. Obtain values for the peak value and frequency of V.

Exercise 128: Questions of A-level standard

1 A cathode ray oscilloscope has its Y sensitivity set to 10 V cm^{-1}. A sinusoidal input is suitably applied to give a steady trace with the time base set so that the electron beam takes 0.01 s to traverse the screen. If the trace seen has a total peak-to-peak height of 4.0 cm and contains 2 complete cycles, what is the RMS voltage and frequency of the input signal? [L 83]

2

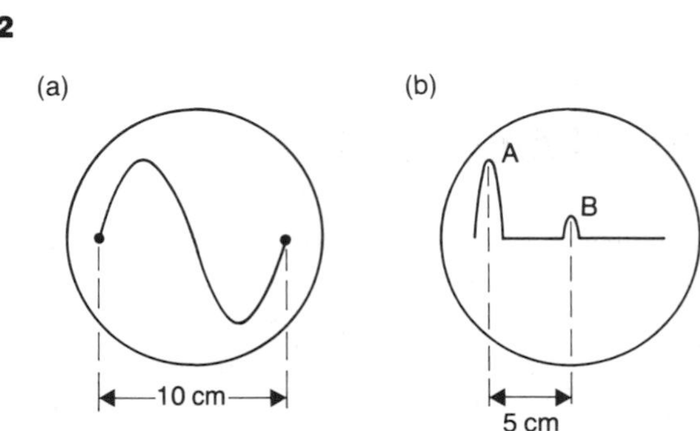

Fig 37.6 Diagrams for Question 2

When a sine-form voltage of frequency 1250 Hz is applied to the Y-plates of a cathode ray oscilloscope, the trace on the tube is as shown in Fig 37.6a. If a radar transmitter sends out short pulses, and at the same time gives a voltage to the Y-plates of the oscilloscope, with the time base setting unchanged, the deflection A is produced as shown in Fig 37.6b. An object reflects the radar pulse which, when received at the transmitter and amplified, gives the deflection B. What is the distance of the object from the transmitter?
(Speed of radar waves = 3×10^8 m s^{-1}.) [L 77]

38 DIGITAL ELECTRONICS AND OPERATIONAL AMPLIFIERS

Digital electronics

The numbers ONE and ZERO can be represented in an electrical circuit by the presence or absence of a potential difference. This idea is used in digital electronic circuits. The voltage chosen is often a nominal +5 V to represent logic 1 while little or no potential difference indicates a 0.

In Fig 38.1 resistor R_A can be chosen so that with $V_A = +5$ V (or even as low as 2.5 V) the transistor conducts well and the current flowing produces a large PD across R with only a small output PD across the transistor.

If instead the input voltage V_A is small or zero the transistor current is small, the PD across R is negligible and the output across the transistor equals the power supply voltage.

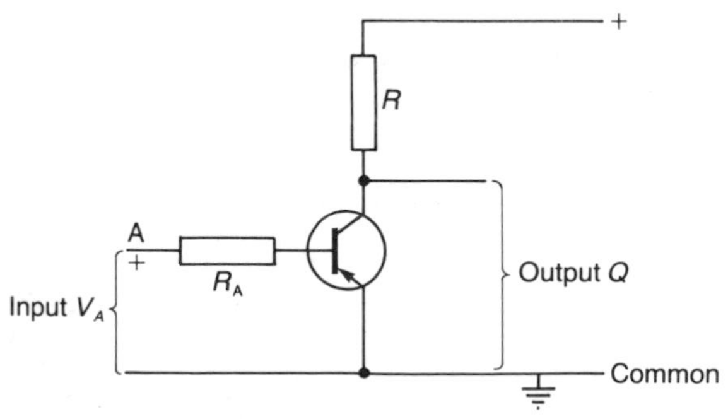

Fig 38.1 A simple amplifier circuit can act as a switch

So the transistor behaves as a switch controlled by V_A.

Circuits where outputs are 1 or 0 according to whether their inputs are 1 or 0 are called logic gates. The switch described above is a logic gate called an inverter and gives an output of 1 when its input is 0 and vice versa. It is also known as a NOT gate, the output being 1 when the input is NOT 1 and zero when the input is NOT zero.

A symbol for the NOT gate is shown in Fig 38.2a.

When gates have two inputs we can denote them by A and B as in Figs 38.2b to f.

Truth tables are tables that can be used to summarise the behaviour of gates as shown in Fig 38.2 where V_A, V_B and Q in the tables represent the input and output states.

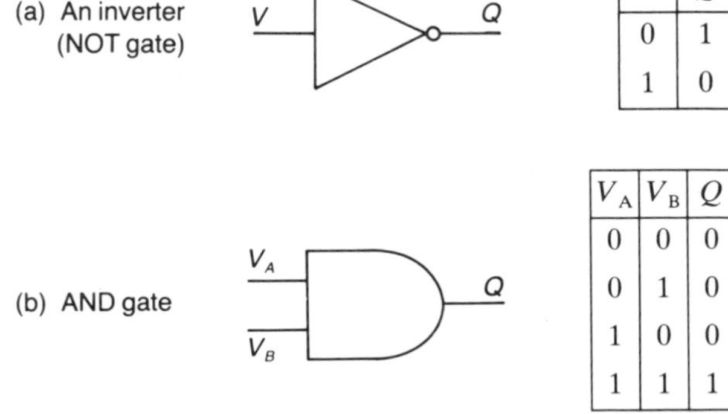

V	Q
0	1
1	0

V_A	V_B	Q
0	0	0
0	1	0
1	0	0
1	1	1

Fig 38.2 *cont.*

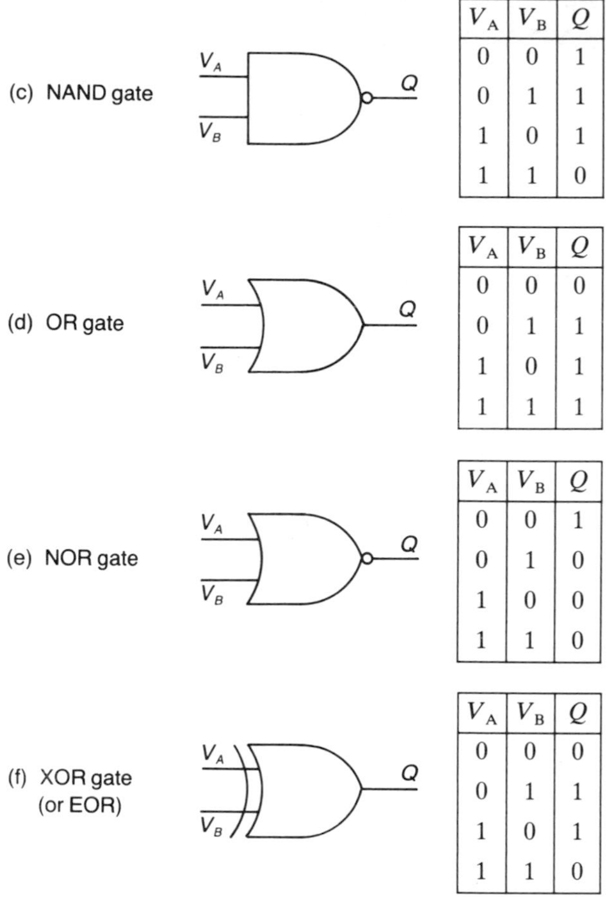

(c) NAND gate

V_A	V_B	Q
0	0	1
0	1	1
1	0	1
1	1	0

(d) OR gate

V_A	V_B	Q
0	0	0
0	1	1
1	0	1
1	1	1

(e) NOR gate

V_A	V_B	Q
0	0	1
0	1	0
1	0	0
1	1	0

(f) XOR gate (or EOR)

V_A	V_B	Q
0	0	0
0	1	1
1	0	1
1	1	0

Fig 38.2 Some logic gates and their truth tables*

Some examples of gates

In Fig 38.1 we could add a second input B connected via a resistor R_B to the transistor's base. Either A or B equal to 1 can make the transistor conduct so that a low output ($Q = 0$) results. In this way we have a NOR gate which is characterised by the truth table shown in Fig 38.2e. An AND gate gives $Q = 1$ only if A and B are both 1. An OR gate gives an output 1 if either A or B is 1. An EXCLUSIVE OR gate gives $Q = 0$ if A and B are both 0, $Q = 1$ if A or B is 1 but, unlike the OR gate, does not give $Q = 1$ if A and B are both 1. A NAND gate (negated AND gate) behaves like an AND gate but with the output reversed (i.e negated), giving zero for the output whenever the AND would give 1. A NOR gate is like an OR gate but with the output reversed.

NAND = AND followed by NOT

NOR = OR followed by NOT

Combinations of gates

The above-mentioned gates can each be made from combinations of the other gates. Also combinations of these basic gates enable devices to be made with all sorts of useful properties.

Example 1

Produce a truth table for the gate combination shown in Fig 38.3 and hence show that it behaves as a NAND gate.

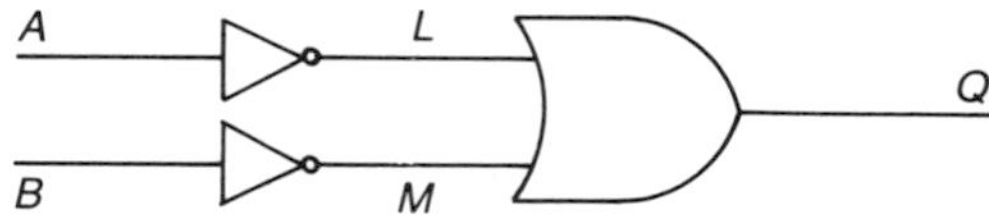

Fig 38.3 Diagram for Example 1

Method

The truth table can be drawn up with columns for A, B, L, M and Q as shown in Fig 38.4. We consider in turn the possibilities for A, B of 0,0, then 0,1, then 1,0 and finally 1,1. Because of the NOT gates the L column will contain ones wherever A = 0 and zeroes where $A = 1$. Similarly M is opposite to B. L and M are the inputs to the OR gate and Q is 1 if L or M is 1 or both are 1.

We can now see from the completed table the relationship between Q and the inputs A and B. The result is the same as for a single NAND gate (see Fig 38.2c).

A	B	L	M	Q
0	0	1	1	1
0	1	1	0	1
1	0	0	1	1
1	1	0	0	0

Fig 38.4 Truth table obtained in Example 1

**A rectangle containing name of gate is often used as the gate symbol for A-level work.*

Example 2

Fig 38.5 shows two NAND gates 'cross coupled' so that the outputs Q and $\overline{Q}$ are used as two of the inputs. Inverters are also included in the input lines.

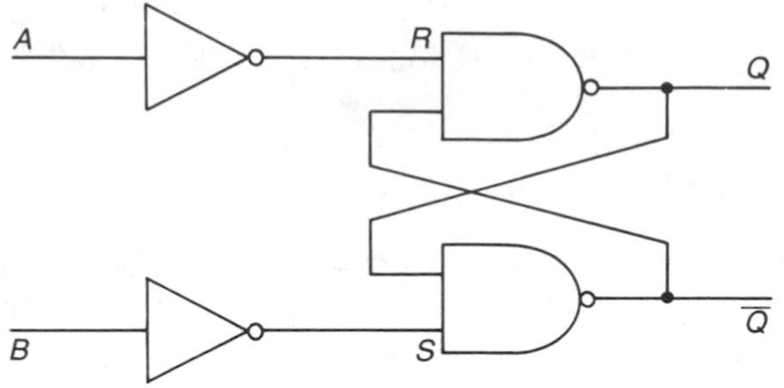

Fig 38.5 Diagram for Example 2

When the inputs are made to follow the sequence $A = 1$, $B = 0$ then $A = 0$, $B = 0$ and then $A = 0$, $B = 1$ what are the corresponding values of Q?

Method

A truth table can be drawn up as shown in Fig 38.6. Initially $A = 1$, $B = 0$ so that $R = 0$, $S = 1$. $R = 0$ makes $Q = 1$. Then $Q = 1$, $S = 1$ makes $\overline{Q} = 0$.

A	B	R	S	Q	$\overline{Q}$
1	0	0	1	1	0
0	0	1	1	1	0
0	1	1	0	0	1

Fig 38.6 Truth table obtained in Example 2

When $A = 0$, $B = 0$ we have $R = 1$, $S = 1$ and it is not obvious what the logic values Q and $\overline{Q}$ will have. So we try say $Q = 1$. Then with $S = 1$, $Q = 1$ we get $\overline{Q} = 0$. Because $\overline{Q} = 0$ and $R = 1$ we get $Q = 1$. If instead we choose $Q = 0$ we find that $Q = 0$, $\overline{Q} = 1$ also agrees with the action of the gates.

However, the state $Q = 1$, $\overline{Q} = 0$ existed already when the inputs became 0,0 and there is no reason for the outputs to change. So $Q = 1$, $\overline{Q} = 0$. (This $Q = 1$ state, set by $A = 1$, remained even when A changed to 0 so that the circuit acts as a latch.)

For $A = 0$, $B = 1$ we have $R = 1$, $S = 0$ so that $\overline{Q} = 1$ and $Q = 0$.

Exercise 129

1 Obtain a truth table for the circuit shown in Fig 38.7. To what basic gate is it equivalent?

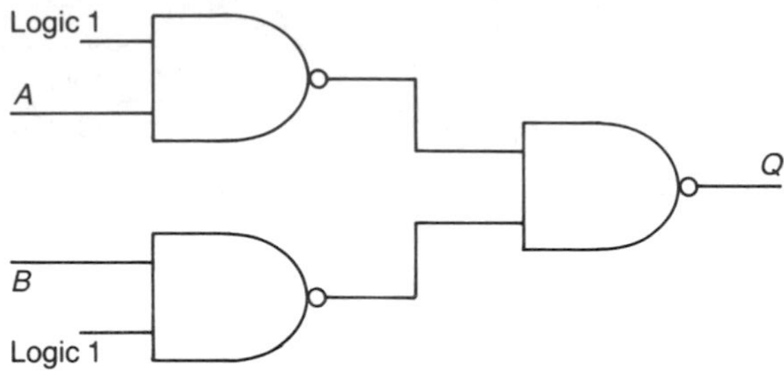

Fig 38.7 Diagram for Question 1

2 The diagram in Fig 38.8 shows two NOR gates and two inverters in combination. The zero volt input is a logic 0 input to the second NOR gate. To what basic gate is it equivalent?

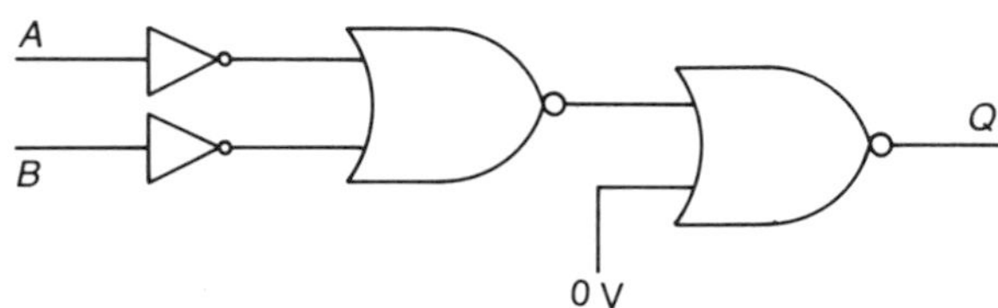

Fig 38.8 Diagram for Question 2

3 For the circuit of Fig 38.9 what are the logic values of Q when the inputs are made to follow the sequence $A = 0$, $B = 1$ then $A = 0$, $B = 0$ and then $A = 1$, $B = 0$? Explain your method.

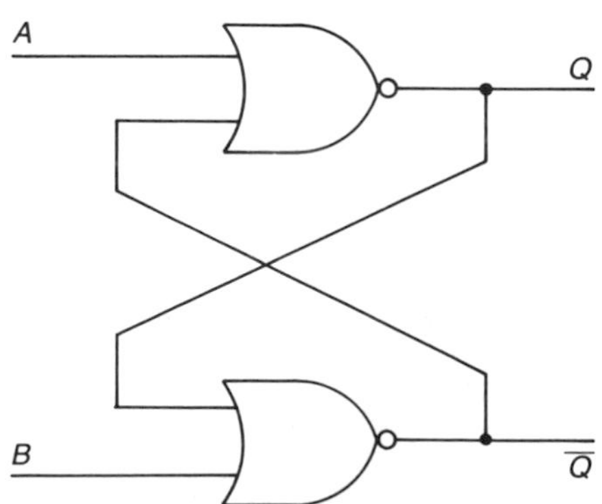

Fig 38.9 Diagram for Question 3

Using flip-flops for counting

Example 2 concerned a circuit having two possible output states ($Q = 1$, $\overline{Q} = 0$ and $Q = 0$, $\overline{Q} = 1$) where input pulses (to A and B) switched the output states and no change in the output state occurred between the pulses (where $A = 0$, $B = 0$). The circuit is a flip-flop. Similar flip-flops are designed so that their two output states change as each pulse arrives at one input terminal. Then Q = logic 1 at every alternate input pulse. Four flip-flops are used in Fig 38.10 below.

Flip-flops are very useful, e.g. in calculators and computers, for handling numbers. These numbers must be binary numbers.

Just as a decimal number consists of a series of digits from right to left telling us the number of ones, tens, hundreds, etc. making up the number, so a binary number's digits tells us how many ones, twos, fours, eights, etc. make up the number. Each digit of a binary number is a one or zero. As an example the binary number 1 0 1 1 is the same as decimal eleven. The eleven contains one 8, no fours, one 2 and a one (which totals to eleven). The binary number 1 0 1 1 can be held by four flip-flops whose Q outputs are at logic 1, 0, 1 and 1.

Adding of binary numbers can be achieved by the use of logic gates. Example 3 is an illustration of this.

Counting can be done with suitable flip-flops arranged in series. When a succession of voltages (pulses) arrive at the input of the first flip-flop it outputs a logic 1 for every second logic 1 input it receives. This output is used as the input for the next flip-flop (Fig 38.10) which sends a logic 1 into the third flip-flop for every two pulses it receives, and so on. At any moment the Q output of each flip-flop is one digit of the binary number indicating the number of pulses received. This kind of counter is called a 'ripple-through' counter.

Example 3

Obtain a truth table for the half-adder circuit shown in Fig 38.11

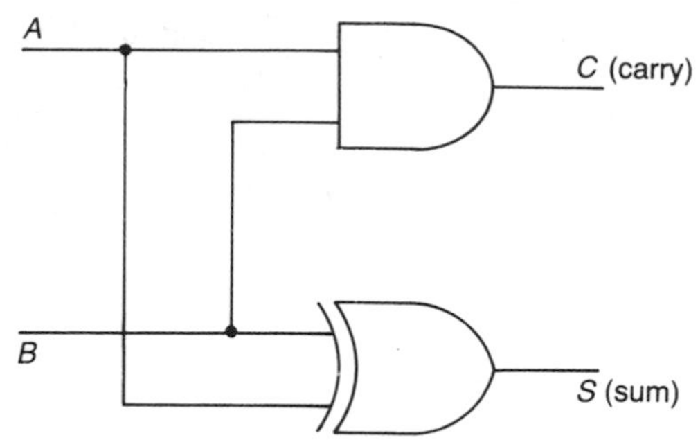

Fig 38.11 The half-adder circuit

Method

A	B	C	S
0	0	0	0
0	1	0	1
1	0	0	1
1	1	1	0

Fig 38.12 Truth table obtained in Example 3

The truth table shown in Fig 38.12 is deduced as follows. When A and B are both logic 0 the AND gate gives $C = 0$. The EXCLUSIVE OR gate gives $S = 0$ because neither of its inputs is 1.

Similarly $A = 0$, $B = 1$ makes $C = 0$ and $S = 1$ while $A = 1$, $B = 0$ gives $C = 0$ and $S = 1$.

$A = 1$, $B = 1$ gives $C = 1$ and $S = 0$ because *both* inputs to the EXCLUSIVE OR gate are logic 1.

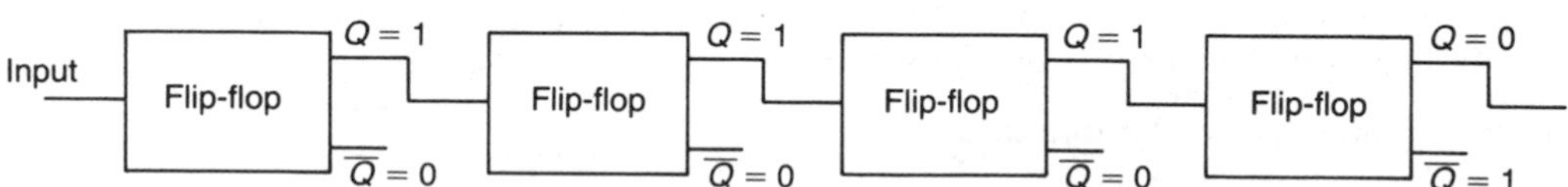

Fig 38.10 A four stage ripple-through counter showing situation after 7 pulses have entered the counter

Exercise 130

1 (a) What decimal number is represented by
(i) 1110 (ii) 10010001?
(b) What is the binary equivalent of thirty-four decimal?

2 If a ripple-through counter (see Fig 38.10) receives 7 input pulses how many pulses are received by (a) the second stage (b) the third stage (c) the fourth stage?

Operational amplifiers (opamps)

A typical opamp is a very high gain amplifier suitable for AC and DC signals and provided with input terminals for two input signals which we can denote by voltages v_1 and v_2. However, the amplifier produces only one output, which we may denote by v_{out}, and this is proportional to the difference of the input voltages. Thus

$$v_{out} = G(v_1 - v_2)$$

where G, the voltage gain, is a constant. The amplifier is thus a difference amplifier.

The two input terminals are labelled − and +. If a zero voltage signal is applied between the + input terminal and the common (usually earthed) terminal while a signal v is applied between the − and common terminals, the output will be of reversed polarity compared with the input signal v. The − terminal is therefore called the inverting terminal. The + terminal is non-inverting.

Note that − indicates inverting, not 'negative'
+ indicates non-inverting, not 'positive'

The symbol recommended for an opamp is shown in Fig 38.13a.

The input resistance of the opamp is very high, its output resistance is low.

(a) Symbol

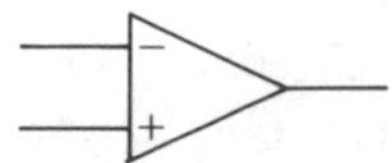

(b) Its use as an inverting amplifier

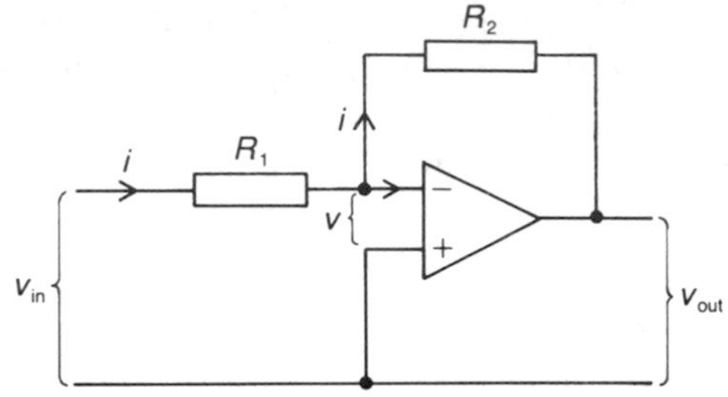

(c) Use as a non-inverting amplifier

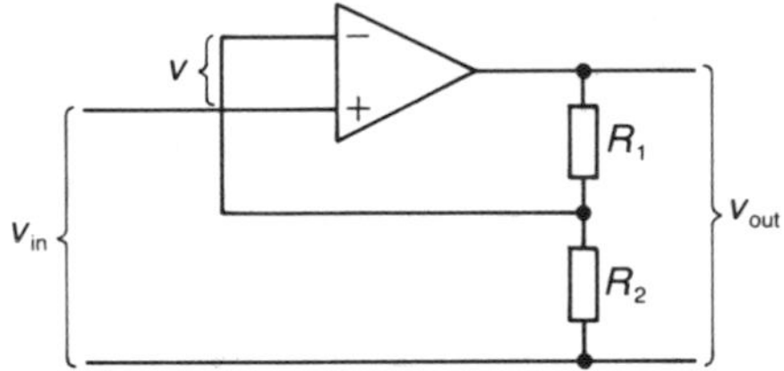

Fig 38.13 An operational amplifier

Two amplifier circuits

An inverting and a non-inverting amplifier are shown in Figs 38.13b and c. For the inverting amplifier, v_{out} will be much greater than v and assuming v is very small

$$\frac{v_{in}}{R_1} = i = -\frac{v_{out}}{R_2}$$

Hence the gain

$$\left(\frac{v_{out}}{v_{in}}\right) = -\frac{R_2}{R_1}$$

For the non-inverting amplifier, v_{in} equals the PD across R_2, i.e.

$$v_{in} = v_{out} \times \frac{R_2}{R_1 + R_2}$$

so that the amplification, or gain, is

$$A = \left(\frac{v_{out}}{v_{in}}\right) = \frac{R_2}{R_1 + R_2}$$

The input resistance (v_{in}/i_{in}) for the inverting amplifier equals R_1. That of the non-inverting amplifier is very high.

Exercise 131

1 What current will be drawn from a DC source of 50 mV open circuit output voltage (i.e. 50 mV EMF) when it is the input to the circuit of Fig 38.13b given that $R_1 = 5.0\,\text{k}\Omega$ and the output resistance (internal resistance) of the source is $5.0\,\text{k}\Omega$?

2 In Fig 38.13b, calculate the output voltage when a sinusoidal input of 10 mV RMS is applied to the input terminals with $R_1 = 5.0\,\text{k}\Omega$ and $R_2 = 50\,\text{k}\Omega$.

Exercise 132: Questions of A-level standard

1 By means of a truth table show that the circuit of Fig 38.14 gives an output only when the two inputs are equal.

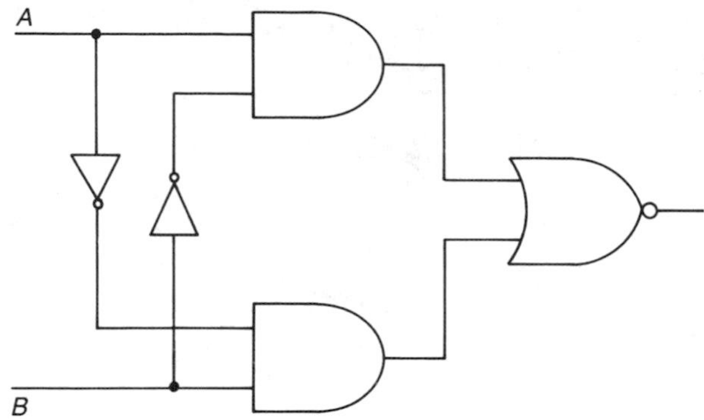

Fig 38.14 Diagram for Question 1

2 If 5 voltage pulses are fed into an 8-stage ripple-through counter which initially reads zero, what binary number should the counter display?

A 00010000 **B** 00011111
C 00000110 **D** 00000101
E 00000111

Section H
Atomic and Nuclear Physics

39

CHARGED PARTICLES — CURRENT IN A VACUUM OR A GAS

Electrons accelerated in a vacuum

If electrons escape from a heated surface (thermionic emission), enter a vacuum with negligible speeds and are then accelerated through a PD V they acquire a kinetic energy given by

$$\frac{1}{2}mv^2 = eV \qquad (39.1)$$

where m is the electron's mass, e is its charge and v its final velocity.

If the anode is a cylinder, the electrons can shoot straight through it as in the electron gun of a cathode ray oscilloscope (Fig 39.1)

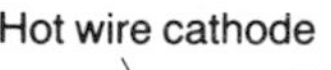

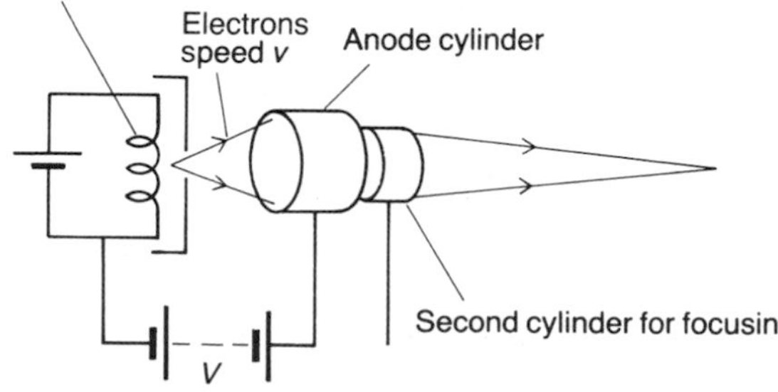

Fig 39.1 Electron gun of a cathode ray oscilloscope

When electrons are subsequently travelling with a steady velocity v, the beam current is given by Equation 27.2 (p.175) as

$$I = nAqv = Nqv \quad \text{or} \quad Nev \qquad (39.2)$$

where N is the number of electrons per metre of beam and q or e the charge on each electron or other charged particle in the beam.

For electrons travelling in a uniform field, e.g. between parallel electrodes, the acceleration is given by a (= force/mass) = Ee/m where E is the electric intensity between the plates.

Time of travel

If the cathode and anode approximate to parallel conductors (as in some radio valves), then the electric intensity E between them is $E = V/d$, where V is the PD between the two electrodes and d is their separation. The force on each electron is Ee or $(V/d) \times e$ and the electron's acceleration is Ee/m or Ve/dm.

If we regard the electron as starting off from the cathode with negligible velocity ($u = 0$) and use the equation $s = ut + \frac{1}{2}at^2$ (see Equation 7.4), we can quite easily deduce t. ($d = s$ here.) Also we can deduce v, if we wish, from $v = u + at$ (Equation 7.1) or $v^2 = u^2 + 2as$ (Equation 7.3).

Example 1

An electron is accelerated in a uniform electric field between two electrodes 2.0 mm apart in a vacuum. It starts off from rest at the cathode. Calculate (a) the time of flight and (b) the final velocity if the PD between the electrodes is 20 V. (Electron charge = -1.6×10^{-19} C, mass = 9.0×10^{-31} kg.)

Method

We have a PD of 20 V over a distance of 2.0 mm or 2.0×10^{-3} m.

Therefore the electric intensity is

$$E \left(= \frac{V}{d}\right) = \frac{20}{2.0 \times 10^{-3}} = 1.0 \times 10^4 \text{ V m}^{-1}$$

therefore the force on the electron is

$$F\ (= Eq) = 10^4 \times 1.6 \times 10^{-19} \text{ N}$$

$$= 1.6 \times 10^{-15} \text{ N}$$

therefore the acceleration

$$a \left(= \frac{F}{m}\right) = \frac{1.6 \times 10^{-15}}{9.0 \times 10^{-31}} = \frac{16}{9} \times 10^{15} \text{ m s}^{-2}$$

But $s = ut + \frac{1}{2}at^2$ (Equation 7.4) and

$$s = d = 2 \times 10^{-3} \text{ m}$$

and the initial velocity u is zero here. $a = (16/9) \times 10^{15}$.

$$\therefore \quad 2.0 \times 10^{-3} = \frac{1}{2} \times \frac{16}{9} \times 10^{15} \times t^2$$

$$\therefore \quad t^2 = \frac{2 \times 2 \times 9}{16} \times 10^{-18}$$

$$\therefore \quad t = 1.5 \times 10^{-9} \text{ s}$$

We can find the final velocity v from the equation

$$v = u + at$$

As before $a = (16/9) \times 10^{15}$ and $u = 0$. Also we have $t = 1.5 \times 10^{-9}$ s.

$$v = \frac{16}{9} \times 10^{15} \times 1.5 \times 10^{-9} = 2.7 \times 10^6 \text{ m s}^{-1}$$

Alternative method

$$\tfrac{1}{2}mv^2 = eV$$

$m = 9.0 \times 10^{-31}$, $e = 1.6 \times 10^{-19}$, $V = 20$.

$$\therefore \quad \tfrac{1}{2} \times 9.0 \times 10^{-31} \times v^2 = 1.6 \times 10^{-19} \times 20$$

$$\therefore \quad v^2 = \frac{1.6 \times 20 \times 2}{9.0} \times 10^{12}$$

$$= \frac{64}{9.0} \times 10^{12}$$

$$\therefore \quad v = 2.7 \times 10^6 \text{ m s}^{-1}$$

Using this value for v, the average velocity is $v/2$ or $2.7 \times 10^6/2$ and equals d/t

$$t = \frac{2 \times 10^{-3} \times 2}{2.7 \times 10^6}$$

$$\therefore \quad t = 1.5 \times 10^{-9} \text{ s}$$

Answer

1.5 ns, 2.7 Mm s^{-1}. Note that the second method is quicker if v only has to be found.

Electrons in gases

When electrons move through a gas they collide with gas atoms and lose kinetic energy to these atoms. Consequently the atoms may be ionised or, with less energy, excited. Excitation means raising an electron within an atom to a higher energy, i.e. to a different, higher energy orbit (see Chapter 40). Ionisation means giving an electron in an atom so much energy that it leaves the atom completely. The energy can also be provided by incident electromagnetic radiation. The energy required to cause an ionisation is called ionisation energy, and the PD through which an electron would have to be accelerated so as to ionise an atom by colliding with it is called the ionisation potential. Similarly we have excitation energies and excitation potentials. Energy given to an atom above the ionisation energy becomes kinetic energy of escape.

The electron-volt (eV)

This is a unit of energy which is particularly useful in particle physics (e.g. atomic and nuclear calculations). It is the energy acquired by an electron freely accelerated (i.e. in vacuum) through a PD of 1 volt. Therefore, since work $W = qV$,

$$1 \text{ eV} = e \text{ joule} \tag{39.3}$$

where e is the electronic charge (1.6×10^{-19} when working in SI units).

$$\therefore \quad 1 \text{ eV} = 1.6 \times 10^{-19} \text{ J}.$$

Example 2

The energy required to ionise an argon atom is 15.8 eV. Express this in joules.
(Electronic charge = 1.60×10^{-19} C.)

Method
Using Equation 39.3,

$$1\,\text{eV} = 1.60 \times 10^{-19}\,\text{J}$$

$$\therefore \quad 15.8\,\text{eV} = 15.8 \times 1.60 \times 10^{-19}\,\text{J}$$

$$= 25.3 \times 10^{-19}\,\text{J}$$

Answer
25.3×10^{-19} J.

Current due to gas ions

Ionisation of atoms produces free electrons (negative ions*); the ionised atoms are positive ions. If positive and negative electrodes are present, the negative ions move to the anode and give up their negative charges to the anode. Similarly positive ions take electrons from the cathode. If n ion pairs per second take part in this process, the current flow (equal to the electron current in the connecting wires) is ne where e is the electronic charge. To this current must be added any current due to particles *emitted* from the cathode (e.g. thermionic electrons) and collected by the anode. Note too that at any point between the electrodes the current equals the sum of electron current in one direction and positive ion current in the other direction.

Example 3
A high potential difference is applied between the electrodes in a hydrogen discharge tube so that the gas ionises; electrons then move towards the positive electrode and protons towards the negative electrode. In each second 5×10^{18} electrons and 2×10^{18} protons pass a cross-section of the tube. The current flowing in the discharge tube is

A 0.10 A **B** 0.48 A **C** 0.80 A
D 1.12 A **E** 1.44 A

(Take the electronic charge e to be 1.60×10^{-19} C.) [O 80]

Method
The current due to electron movement at this cross-section is

$$ne = 5 \times 10^{18} \times 1.60 \times 10^{-19}$$

$$= 0.8\,\text{A}$$

These may become attached to molecules to form larger ions

The proton (+ion) current at this cross-section is

$$ne = 2 \times 10^{18} \times 1.60 \times 10^{-19}$$

$$= 0.32\,\text{A}$$

The total current therefore is

$$0.8 + 0.32 = 1.12\,\text{A}$$

(In this example each ion carries one electron charge. This is the usual case met.)

Answer
D.

Electric and magnetic deflection of charged particles

As shown in Chapter 32, an electron or other charged particle travelling with velocity v perpendicular to a magnetic flux density B experiences a force of

$$F = Bqv \qquad (32.3)$$

where q is the particle's charge. The direction of F is given by the left-hand rule. The path followed in the magnetic field is circular, F being directed towards the centre of the circle whose radius (see Chapter 10) will be given by

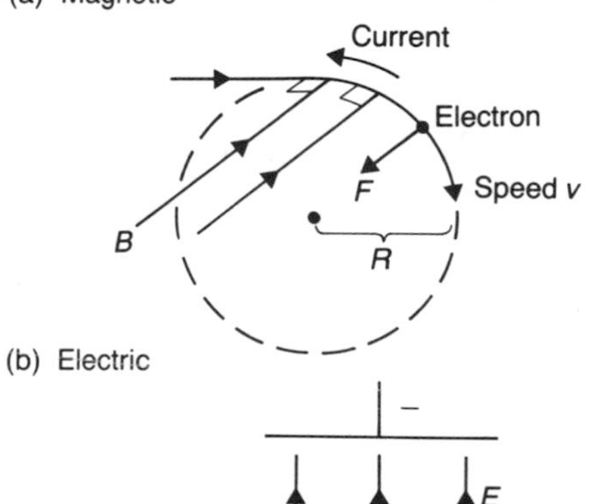

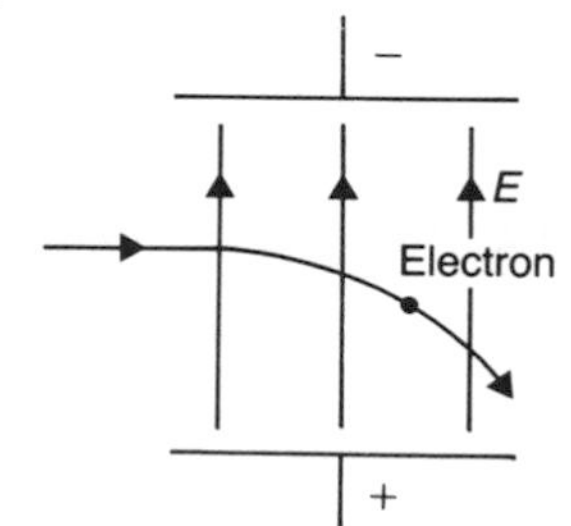

Fig 39.2 Electron deflection

$$F = \frac{mv^2}{R}$$

where m is the particle's mass. Thus

$$\frac{mv^2}{R} = Bqv \qquad (39.4)$$

In a uniform electric field of intensity E the force is Eq. If this force is perpendicular to the original direction of the particle's velocity we get a parabolic path.

Example 4

An electron travelling at $1.0 \times 10^7\,\text{m s}^{-1}$ to the right enters a uniform magnetic field of flux density $5.0 \times 10^{-4}\,\text{T}$ directed into the paper. Calculate the radius of the path followed.

In what direction is the electron first deflected?

(Electronic charge $e = 1.6 \times 10^{-19}\,\text{C}$,
mass $= 9.1 \times 10^{-31}\,\text{kg}$.)

Method

From Equation 39.4

$$Bqv = \frac{mv^2}{R}$$

$B = 5.0 \times 10^{-4}$, $q = 1.6 \times 10^{-19}$, $v = 10^7$,
$m = 9.1 \times 10^{-31}$.

$$\therefore \quad 5.0 \times 10^{-4} \times 1.6 \times 10^{-19} \times 10^7 = \frac{9.1 \times 10^{-31} \times 10^{14}}{R}$$

$$\therefore \quad 8.0 \times 10^{-16} \times R = 9.1 \times 10^{-17}$$

$$\therefore \quad R = \frac{9.1}{8} \times 10^{-1}$$

or 0.11 m or 11 cm

The electron is first deflected downwards as shown in Fig 39.3 (opposite).

Answer

11 cm; downwards.

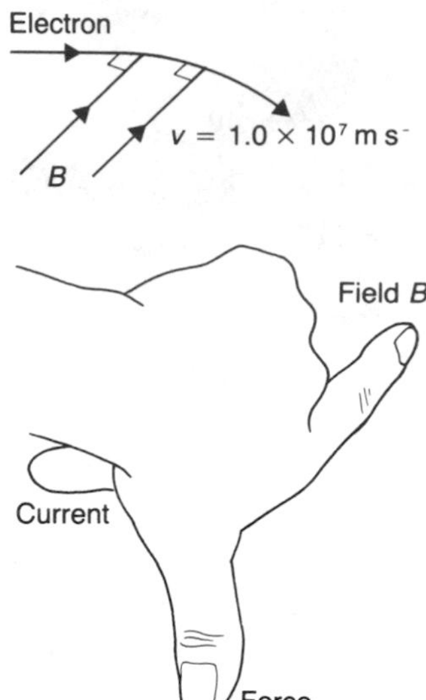

Fig 39.3 Diagram for Example 4

Example 5

A beam of electrons all travelling horizontally at $1.0 \times 10^7\,\text{m s}^{-1}$ pass between two horizontal plates 1.0 cm apart with a PD V between them.

(a) For $V = 100\,\text{V}$ calculate the deflection of the beam when it has travelled a horizontal distance of 2.0 cm between the plates.

(b) Calculate the value of V needed to prevent deflection by a magnetic flux density of $5.0 \times 10^{-4}\,\text{T}$ which is horizontal and at right angles to the beam.

(Electronic charge $= 1.6 \times 10^{-19}\,\text{C}$,
mass $= 9.1 \times 10^{-31}\,\text{kg}$.)

Method

(a) Fig 39.2b is a suitable diagram.

The horizontal velocity is not changed by the vertical electric field. The vertical electric force is $F = qE$.

The time taken to move 2.0 cm horizontally is

$$t = \frac{\text{Distance}}{\text{Speed}} = \frac{2.0 \times 10^{-2}}{10^7} = 2.0 \times 10^{-9}\,\text{s}$$

The electric intensity is

$$E = \frac{V}{d} = \frac{100}{10^{-2}} = 10^4\,\text{V m}^{-1}$$

because the PD is 100 volt and the plate separation is 10^{-2} m.

$$F = eE = 1.6 \times 10^{-19} \times 10^4 = 1.6 \times 10^{-15}\,\text{N}$$

∴ the vertical acceleration

$$\frac{F}{m} = \frac{1.6 \times 10^{-15}}{9.1 \times 10^{-31}}$$

$$= 0.176 \times 10^{16}\ \text{m s}^{-2}$$

We now know the acceleration a and the time t for the vertical deflection. The vertical displacement is given by

$$s = ut + \tfrac{1}{2}at^2 \quad \text{with} \quad u = 0$$

$$\therefore \quad s = 0.5 \times 0.176 \times 10^{16} \times (2 \times 10^{-9})^2$$

$$= 0.352 \times 10^{-2}\ \text{m} \quad \text{or} \quad 0.35\ \text{cm}$$

(b) The magnetic force Bqv must be equalled by the electric force Ee $(= (V/d) \times e)$. $(q = e$ here.)

$$\therefore \quad Bev = Ee = \frac{V}{d} \times e$$

and we require the value of V.

$$\therefore \quad V = Bvd$$

and $B = 5 \times 10^{-4}$, $v = 10^7$, $d = 10^{-2}$

$$\therefore \quad V = 5 \times 10^{-4} \times 10^7 \times 10^{-2}$$

$$= 50\ \text{V}$$

Answer

(a) 0.35 cm, (b) 50 V.

Example 6

The diagram is shown in Fig 39.4

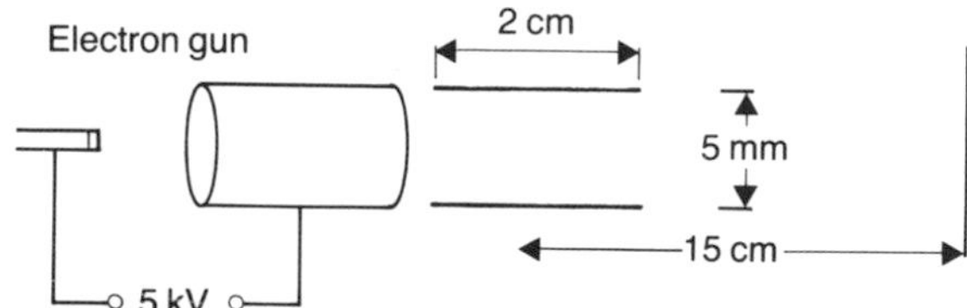

Fig 39.4 Diagram for Example 6

Calculate the deflection sensitivity (deflection of spot in mm per volt potential difference) of a cathode ray tube from the following data:

Electrons are accelerated by a potential difference of 5.0 kV between cathode and anode;

Length of deflector plates = 2.0 cm;

Separation of deflector plates = 5.0 mm;

Distance of midpoint of deflector plates from screen = 15.0 cm. [SUJB 79]

Method

We need the deflection on the screen for a PD of 1.0 V between the plates. If we can determine the vertical velocity of the electrons as they leave the deflector plates, we should be able to calculate the deflection on the screen. We will need to know the time spent in the electric field, and so we require first the horizontal velocity that results from the acceleration.

To find the horizontal velocity we use Equation 39.1:

$$\tfrac{1}{2}mv^2 = Ve$$

$V = 5.0$ kV, i.e. 5000 V, but e and m for the electron are not given in this question. We can only hope that e and m cancel out in the calculation.

$$\tfrac{1}{2}mv^2 = 5000e$$

$$\therefore \quad v = \sqrt{\left(10\,000\frac{e}{m}\right)} \quad \text{or} \quad 100\sqrt{\frac{e}{m}}$$

The time of travel through the deflector plates, a distance s, say, is

$$t = \frac{s}{v} = \frac{2.0 \times 10^{-2}}{100\sqrt{(e/m)}} = \frac{2.0 \times 10^{-4}}{\sqrt{(e/m)}}$$

The deflecting force is $F = Ee = Ve/d$, where V is 1.0 V between the plates, and the distance between the plates d is 5.0 mm or 5×10^{-3} m.

$$\therefore \quad F = \frac{1.0 \times e}{5 \times 10^{-3}} = 0.2 \times 10^3 e$$

$$\text{or} \quad 200e\ \text{newton}$$

This gives an acceleration of

$$a = \frac{F}{m} = \frac{200e}{m}$$

The vertical velocity of the electrons as they leave the plates is obtained by use of the equation $v = u + at$ with $u = 0$, $a = 200e/m$ and $t = 2.0 \times 10^{-4}/\sqrt{(e/m)}$

$$\therefore \quad v = 200\frac{e}{m} \times \frac{2.0 \times 10^{-4}}{\sqrt{(e/m)}}$$

$$= 4 \times 10^{-2}\sqrt{\frac{e}{m}}$$

The time taken to cover the last 14 cm of flight is given by the horizontal velocity divided into the horizontal distance $(14 \times 10^{-2}\ \text{m})$ and equals

$$\frac{14 \times 10^{-2}}{100\sqrt{(e/m)}}$$

The vertical deflection for this time is vertical velocity × time which is

$$4 \times 10^{-2}\sqrt{\frac{e}{m}} \times \frac{14 \times 10^{-2}}{100\sqrt{(e/m)}} \quad \text{or} \quad 56 \times 10^{-6}\ \text{m}$$

But was the beam deflected significantly while between the plates? It left with a vertical velocity of $4 \times 10^{-2}\sqrt{(e/m)}$ but entered with this velocity zero. So the mean value is $2 \times 10^{-2}\sqrt{(e/m)}$ over the time of flight through the plates ($t = 2.0 \times 10^{-4}/\sqrt{(e/m)}$, see above). The resulting deflection is (using distance = velocity × time)

$$2 \times 10^{-2}\sqrt{\frac{e}{m}} \times \frac{2.0 \times 10^{-4}}{\sqrt{(e/m)}} \quad \text{or} \quad 4 \times 10^{-6}\,\text{m}$$

Thus the total deflection at the screen is $(56 + 4)\,\mu$m or 0.06 mm.

Note that, had we assumed the distance of travel with steady vertical velocity to be 15 cm instead of 14 cm, then the deflection within the plates would have been included, i.e. calculation and addition of the 4 μm would not have been needed.

Answer
0.06 mm per volt.

Mass spectrometry

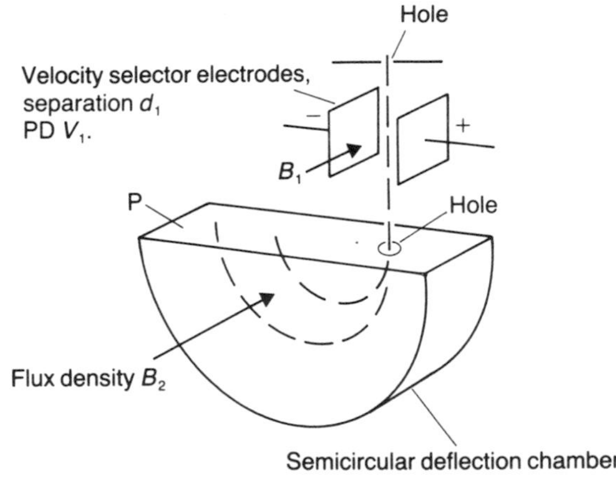

Fig 39.5 Parts of a mass spectrometer

Fig 39.5 illustrates part of one type of mass spectrometer. Ions are accelerated by a suitable PD so that they pass through a hole into the velocity selector and then enter the deflection chamber. The velocity selector subjects the ions of charge q to a magnetic force B_1qv and an electric force Eq or Vq/d.

For ions to suffer no deflection and therefore enter the deflection chamber we have

$$B_1qv = Eq \tag{39.5}$$

or

$$v = \frac{E}{B_1} \quad \text{or} \quad \frac{V}{dB_1}$$

v can be selected by varying V or B_1.

In the deflection chamber the ions travel in circular paths, the radius R of each path being given by

$$\frac{mv^2}{R} = B_2qv \tag{39.6}$$

Thus $R \propto$ mass of ion m for given q, e.g. for singly charged ions. At P a photographic plate records where the ions arrive (so that we have a spectrograph), but a movable electrode could be used to collect the ions and the various currents from it measured (so that we have a spectrometer).

Example 7

What velocity will singly charged ions of mass 2.6×10^{-26} kg have if they travel straight through a velocity selector moving perpendicularly to an electric field of intensity 4.0 kV m^{-1} and a magnetic field of flux density 0.10 T (the fields themselves being at right angles)?

If these ions then enter a region where there is a magnetic flux density of 0.050 T, so that the ions follow a semicircular path perpendicular to the magnetic field direction, what is the radius of the path? (Electronic charge $e = 1.6 \times 10^{-19}$ C.)

Method
In the velocity selector, use Equation 39.5:

$$B_1qv = Eq$$

$B_1 = 0.10$ T, $q = 1.6 \times 10^{-19}$ C, $E = 4.0$ kV m^{-1} or 4.0×10^3 V m^{-1}.

$$\therefore \quad 0.10 \times q \times v = 4.0 \times 10^3 \times q$$

$$\therefore \quad v = \frac{4.0 \times 10^3}{0.10} = 4.0 \times 10^4\,\text{m s}^{-1}$$

For the semicircular path, use Equation 39.6:

$$\frac{mv^2}{R} = B_2qv$$

$m = 2.6 \times 10^{-26}$ kg, $B_2 = 0.05$ T, $v = 4.0 \times 10^4$ m s^{-1}

$$\therefore \quad \frac{2.6 \times 10^{-26} \times (4.0 \times 10^4)^2}{R} = 0.05 \times 1.6 \times 10^{-19} \times (4.0 \times 10^4)$$

$$\therefore \quad R = \frac{2.6 \times 10^{-26} \times 4.0 \times 10^{4}}{0.05 \times 1.6 \times 10^{-19}}$$

$$= 0.13\ \text{m} \quad \text{or} \quad 13\ \text{cm}$$

Answer

4.0×10^4 m s^{-1}, 13 cm.

Exercise 133

1 How many electrons are there in each metre of an electron beam, travelling at 5.0×10^5 m s^{-1}, that delivers 1.0×10^{-3} C of charge per second? (Electronic charge $e = 1.6 \times 10^{-19}$ C.)

2 6.0×10^5 electrons per second are moving from an electrode in a gas and, at the same distance from the electrode, 6.0×10^5 singly charged positive ions are approaching the electrode. How many electrons per second are moving along the wire leading to the electrode, and how many amperes of current are flowing? (Electronic charge $e = 1.6 \times 10^{-19}$ C.)

3 Calculate the magnetic flux density directed at right angles to a beam of electrons travelling at 5.0×10^5 m s^{-1}, that will give the beam a radius of 12 cm. (e/m for electron $= 1.8 \times 10^{11}$ C kg^{-1}.)

4 Electrons are accelerated in a uniform electric field of 9.0 kV m^{-1}. They are initially at rest and move through a distance of 9.0 mm. How long do the electrons take to cover this distance? (Take e/m for the electron to be 1.8×10^{11} C kg^{-1}.)

5 Electrons travelling at 1.0×10^7 m s^{-1} pass between two parallel plate electrodes parallel to the direction of the arriving beam. If the plates are 40 mm long with a uniform electric field of 0.50 kV m^{-1} between them, (a) what is the velocity of the electrons normal to the plates as the electrons emerge and (b) by how much has the beam been deflected normal to the plates at this same stage? (Take e/m for the electron to be 1.8×10^{11} C kg^{-1}.)

6 Calculate the accelerating potential (relative to the cathode filament) that is required in a cathode ray oscilloscope to give electrons a velocity of 2.0×10^7 m s^{-1}. (e/m for electron $= 1.8 \times 10^{11}$ C kg^{-1}.)

7 A uniform electric field is superimposed at right angles to a uniform magnetic field of 4.0×10^{-2} T to form a velocity selector which allows an electron to travel through with velocity 4.0×10^6 m s^{-1}. What is the intensity of the electric field?

Experimental determination of *e/m* of electrons

Well-known methods of determining e/m are illustrated by Examples 8 and 9.

Example 8

A narrow beam of electrons is emitted from a small electron gun within a glass bulb that contains enough gas to glow where the beam travels. Using 300 V to accelerate the electrons, a uniform magnetic field of 7.3×10^{-4} T produces a circle of glow with a radius of 8.0 cm. Obtain a value for e/m of the electrons.

Method

$$\frac{mv^2}{R} = Bev \qquad (39.4)$$

$B = 7.3 \times 10^{-4}$ and $R = 8.0 \times 10^{-2}$.

$$\therefore \quad \frac{mv^2}{8 \times 10^{-2}} = 7.3 \times 10^{-4}\, ev$$

$$v = \frac{e}{m} \times 7.3 \times 10^{-4} \times 8.0 \times 10^{-2}$$

$$= \frac{e}{m} \times 58.4 \times 10^{-6} \qquad \text{(i)}$$

Also $\quad \frac{1}{2}mv^2 = eV \qquad (39.1)$

and $V = 300$

$$\therefore \quad \tfrac{1}{2}mv^2 = 300e$$

$$\therefore \quad v^2 = 600\frac{e}{m} \quad \text{and} \quad v = 24.5\sqrt{\frac{e}{m}} \qquad \text{(ii)}$$

Equating (i) and (ii) we get

$$24.5\sqrt{\frac{e}{m}} = 58.4 \times 10^{-6}\frac{e}{m}$$

$$\therefore \quad \sqrt{\frac{e}{m}} = 0.42 \times 10^{6}$$

and $\quad \dfrac{e}{m} = 1.76 \times 10^{11}\ \text{C kg}^{-1}$

Answer

1.8×10^{11} C kg^{-1}.

Example 9

In an experiment for determining the charge/mass ratio of electrons, the electron beam moved through an arc of radius 25 mm while it moved through the uniform magnetic field of flux density 2.0×10^{-3} T. With this field still present the beam was then made straight by applying an electric field using 200 V potential difference between two parallel plates 12 mm apart. Calculate the beam velocity and the charge/mass ratio.

Method

With the magnetic field alone we have, from Equation 39.4

$$\frac{mv^2}{R} = Bev \qquad \therefore \quad v = \frac{e}{m}BR$$

With zero deflection

$$Bev = \frac{V}{d}e \qquad \therefore \quad v = \frac{V}{Bd}$$

The second equation allows v to be determined, since $V = 200$, $B = 2.0 \times 10^{-3}$, $d = 12 \times 10^{-3}$.

$$v = \frac{200}{2.0 \times 10^{-3} \times 12 \times 10^{-3}} = 8.3 \times 10^6\ \text{m s}^{-1}$$

Using this value in the first equation gives, with $R = 25 \times 10^{-3}$ m,

$$v = 8.3 \times 10^6 = \frac{e}{m} \times 2.0 \times 10^{-3} \times 25 \times 10^{-3}$$

$$\therefore \quad \frac{e}{m} = 1.7 \times 10^{11}\ \text{C kg}^{-1}$$

Answer

8.3×10^6 m s^{-1}, 1.7×10^{11} C kg^{-1}.

The Millikan oil drop experiment

In this famous experiment charged oil droplets of microscopic sizes are sprayed into the space between two horizontal metal plates between which an electric field (intensity E) can be applied. With suitable illumination, a chosen droplet is watched through a microscope and its velocity of fall can be measured.

The forces on the droplet are its weight, a small (usually negligible) upthrust from the surrounding air and a viscous drag force caused by its movement through the air. The latter force is proportional to the velocity of the drop and results in the drop quickly acquiring a steady velocity v_1 with the weight equalled by the drag force.

$$\text{Weight} = \text{Drag force}$$

$$\tfrac{4}{3}\pi R^3 \rho g = kv_1 \qquad (39.7)$$

where R is the radius of the drop and ρ is its density. g is the acceleration of free fall and k is a constant.

When the electric field is applied, an extra force Eq acts on the drop (with charge q) and this may speed it up, slow it down, make it stationary or even make it rise. The new velocity is v_2. For example, for a drop made to rise

$$Eq - \tfrac{4}{3}\pi R^3 \rho g = kv_2 \qquad (39.8)$$

The charge q can be calculated from these equations. It is found that q is always close to a multiple of 1.6×10^{-19} C, showing that the electron charge has this value.

Should the drop be held stationary, then, from Equation 39.8,

$$Eq = \text{Weight of drop} = \tfrac{4}{3}\pi R^3 \rho g$$

i.e. the electrostatic force equals the weight.

Example 10

In a measurement of the electron charge by Millikan's method, a potential difference of 1.5 kV can be applied between horizontal parallel metal plates 12 mm apart. With the field switched off, a drop of oil of mass 1.0×10^{-14} kg is observed to fall with constant velocity 400 μm s^{-1}. When the field is switched on, the drop rises with constant velocity 80 μm s^{-1}. How many electron charges are there on the drop? (You may assume that the air resistance is proportional to the velocity of the drop and that the air buoyancy may be neglected.)

Acceleration of free fall $g = 10$ m s^{-2}.

Electron charge $e = -1.6 \times 10^{-19}$ C. [SUJB 80]

Method

This question like most similar A-level questions has been designed to simplify the calculation needed.

The drop weighs 10^{-14} kg, so that the weight is 10^{-13} N (since $g = 10$ m s^{-2}) and we use this in place of $\frac{4}{3}\pi R^3 \rho g$ in Equation 39.7.

The upthrust (buoyancy) we are told can be neglected.

Equation 39.7 gives

$$\text{Weight} = kv_1$$

$$v_1 = 400 \times 10^{-6}, \text{ weight} = 10^{-13}.$$

$$10^{-13} = k \times 400 \times 10^{-6} \qquad \text{(i)}$$

Equation 39.8 is

$$Eq - \text{Weight} = kv_2, \quad \text{and} \quad E = \frac{V}{d}$$

$E = 1.5 \times 10^3$, $v_2 = 80 \times 10^{-6}$,
$V = 1.5\,\text{kV} = 1.5 \times 10^3$, $d = 12 \times 10^{-3}$.

$$\therefore \quad \frac{1.5 \times 10^3}{12 \times 10^{-3}}q - 1.0 \times 10^{-13} = k \times 80 \times 10^{-6} \qquad \text{(ii)}$$

We see that k is common to (i) and (ii). From (i) it equals 0.25×10^{-9} and using this value in (ii) we get

$$\frac{1.5 \times 10^3}{12 \times 10^{-3}}q - 10^{-13} = 0.25 \times 10^{-9} \times 80 \times 10^{-6}$$

From this q works out to be 9.6×10^{-19} C.

The number of electronic charges on the drop is

$$\frac{9.6 \times 10^{-19}}{1.6 \times 10^{-19}} = 6$$

(The drop has a surplus of 6 electrons or is short of 6 electrons depending on whether it is charged negatively or positively.)

Answer

6.

Exercise 134

1 An electron beam passes between two parallel plates which are 4.0 cm apart in vacuum, and across which there is a potential difference of 600 V. The beam travels straight through without deflection if a magnetic field of flux density 6.0×10^{-4} T is applied in a direction normal to the electric field. Calculate the speed of the electrons.

2 A beam of electrons that have been accelerated from rest through a PD of 225 V passes into a region of uniform magnetic field whose flux density of 5.0×10^{-4} T is perpendicular to the direction of arrival of the beam. Calculate the radius of the circular path taken by the beam in this field. (Take e/m for the electrons to be $1.8 \times 10^{11}\,\text{C kg}^{-1}$.)

3 In a Millikan oil drop experiment the plates were 1.2 cm apart. With no electric field present a drop of mass 6.0×10^{-16} kg fell with a steady velocity of $3.3 \times 10^{-5}\,\text{m s}^{-1}$. With a potential difference of 450 V applied across the plates the drop rose with the same velocity. How many electron charges did the drop carry? (The upthrust due to the surrounding air may be neglected.)
Acceleration of free fall $g = 10\,\text{m s}^{-2}$.
Electron charge $e = -1.6 \times 10^{-19}$ C.

Exercise 135: Questions of A-level standard

1 Electrons are released with negligible speeds from a heated cathode and are accelerated to an anode in vacuum. The accelerating potential difference between the electrodes is 200 V. Calculate the speed of the electrons arriving at the anode. (Electron charge/mass ratio $e/m = 1.8 \times 10^{11}\,\text{C kg}^{-1}$.)

2 An electron is accelerated from rest through a potential difference V, and then enters a region where there is a uniform magnetic field B perpendicular to the direction of motion of the electron. Derive an expression for the radius of the circular orbit of the electron.

What is the period of an electron in its circular orbit in a uniform magnetic field of strength 9.1×10^{-6} T?
(Electron charge $e = -1.6 \times 10^{-19}$ C,
electron mass $m = 9.1 \times 10^{-31}$ kg.) [SUJB 81]

3 In a cathode ray tube, the electrons are accelerated through a potential difference of 500 V and then pass between deflecting plates which are 5×10^{-2} m long. Calculate the time spent in transit between the plates by an individual electron. Explain concisely why the maximum deflection produced by an alternating voltage, of constant amplitude, applied to the deflection plates, will decrease with increasing frequency of that voltage.
The ratio of the charge to the mass of the electron is $1.76 \times 10^{11}\,\text{C kg}^{-1}$. [WJEC 78]

4 In a mass spectrometer, a beam of particles each carrying the same charge and all travelling at the same velocity, $2 \times 10^4\,\mathrm{m\,s^{-1}}$, enters a uniform magnetic field. In this all the particles experience forces of equal magnitude always directed at right angles to their velocities at each instant. They describe a complete semicircle in this field, and finally impinge on a suitably placed detector. Of two kinds of particle, A of mass 6.7×10^{-26} kg and B of mass 8.0×10^{-26} kg, it is found that B describes a semicircle of radius 0.52 m.

(a) What is the radius of the path of A?

(b) How long does each kind of particle spend in the magnetic field? [O 80, p]

5 The plates of a horizontal parallel-plate capacitor are 1 cm apart, and the potential difference is 10 kV. It is found that an oil drop of mass 1.96×10^{-13} kg remains stationary between the plates. How many electronic charges does the oil drop carry. Indicate briefly how the mass of the drop could be determined experimentally assuming its charge is not known. (Electronic charge = 1.6×10^{-19} C; assume $g = 9.8\,\mathrm{m\,s^{-2}}$ at sea level.) [WJEC 80]

6 An electron beam passes between two parallel plates which are separated by 2.0 cm, and across which there is a potential difference of 300 V. The deflection of this beam is just cancelled by a magnetic field of flux density 6.0×10^{-4} T applied in a direction normal to the electric field. Calculate the speed of the electrons. If the radius of curvature of the path of the electrons due to the magnetic field alone is 25 cm, what is the value of the specific charge e/m? [AEB 81]

40 PHOTOELECTRIC EMISSION AND ATOMIC STRUCTURE

Photoelectric emission from the surface of a solid

Electromagnetic radiation is made up of separate ('discrete') quantities of energy which we may describe as light particles (photons). Each photon consists of energy hf joules, where f is the frequency of the light and h is the Planck constant. For an electron to escape from a solid by photoelectric emission it must acquire the energy of an incident photon and use this energy to (1) 'get to the solid's surface' and (2) get through the surface (the energy needed is called the work function energy, WFE), leaving it with (3) some kinetic energy $\frac{1}{2}mv^2$. Thus

$$\underset{(\text{or } hc/\lambda)}{hf} = \begin{pmatrix}\text{Energy}\\ \text{to get to}\\ \text{surface}\end{pmatrix} + \text{WFE} + \tfrac{1}{2}mv^2 \qquad (40.1)$$

where c is velocity of the light, λ the wavelength, m mass of the electron, v velocity of the escaped electron (photoelectron).

If $hf <$ WFE then no electron emission occurs.

Of all the electrons escaping the *fastest* will be those which did not have to use energy to reach the surface so that, for them,

$$hf\left(\text{or } \frac{hc}{\lambda}\right) = \text{WFE} + \frac{1}{2}mv^2 \qquad (40.2)$$

If an electrode is placed near the emitting surface and is made negative by V volts, then the photoelectrons can be repelled back to the surface. Even the fastest electrons that aim directly at the negative electrode will be prevented from reaching it if the retarding PD V equals or exceeds the value given by

$$eV = \frac{1}{2}mv^2 \quad \text{which equals} \quad \frac{hc}{\lambda} - \text{WFE} \qquad (40.3)$$

where eV (the work to be done in reaching the electrode) is the electron charge × PD.

Work function energy can be quoted in joules or electron-volts. The work function *voltage* is the PD needed to accelerate electrons to such an energy (see definition of electron-volt, Chapter 39).

Example 1

Electromagnetic radiation of frequency 0.88×10^{15} Hz falls upon a surface whose work function is 2.5 V.

(a) Calculate the maximum kinetic energy of photoelectrons released from the surface.

(b) If a nearby electrode is made negative with respect to the first surface using a PD V, what value is required for V if it is to be just sufficient to stop any of the photoelectrons from reaching the negative electrode?

(Planck constant $h = 6.6 \times 10^{-34}$ J s, electron charge $e = -1.6 \times 10^{-19}$ C.)

Method

(a) Using Equation 40.1 or 40.2,

$$hf = \text{WFE} + \begin{pmatrix}\text{Kinetic energy of} \\ \text{fastest photoelectrons}\end{pmatrix}$$

we have

$$6.6 \times 10^{-34} \times 0.88 \times 10^{15} = 2.5 \times 1.6 \times 10^{-19} + E_{max}$$

(2.5 is multiplied by 1.6×10^{-19} here in order to convert the 2.5 eV energy to joules.)

So

$$E_{max} = 5.8 \times 10^{-19} - 4.0 \times 10^{-19} = 1.8 \times 10^{-19}\,\text{J}$$

In electron-volts,

$$E_{max} = \frac{1.8 \times 10^{-19}}{1.6 \times 10^{-19}} = 1.125\,\text{eV} \quad \text{or} \quad 1.1\,\text{eV}$$

(b) Working in joules again (our equations are all written for SI units) we have, from Equation 40.3

$$eV = \tfrac{1}{2}mv^2$$

or $\quad eV = E_{max} = 1.8 \times 10^{-19}\,\text{J}$

$$\therefore \quad V = \frac{1.8 \times 10^{-19}}{1.6 \times 10^{-19}} = \frac{1.8}{1.6} = 1.125 \quad \text{or} \quad 1.1\,\text{V}$$

More simply, $E_{max} = 1.1$ eV and retarding PD = 1.1 V.

Answer

1.1 eV, 1.1 V.

De Broglie wavelength for a particle of matter

Light and other electromagnetic radiations must be regarded as waves but also as particles (quanta of energy), i.e. photons. Photons, like all other things, have mass $m = E/c^2$ where E is the energy of the photon and c is the velocity of light (see also Chapter 43). Using $E = hc/\lambda$ for the photon:

$$m = \frac{hc/\lambda}{c^2} \quad \text{or} \quad mc = \frac{h}{\lambda}$$

Thus

$$\text{Momentum } mc = \frac{h}{\lambda} \qquad (40.4)$$

De Broglie proposed that any particle of *matter*, e.g. an electron or proton, has, like a photon, both wave and particle properties, so that it has a wavelength given by Equation 40.4.

Example 2

Calculate the wavelength of electrons that have been accelerated from rest through a PD of 100 V. What kind of electromagnetic radiation has wavelengths similar to this value?

(Electron mass $m = 9.1 \times 10^{-31}$ kg,
electron charge $e = -1.6 \times 10^{-19}$ C,
Planck constant $h = 6.6 \times 10^{-34}$ J s.)

Method

From Equation 40.4, wavelength = h/momentum.

To find the electron's momentum we use Equation 39.1:

$$\tfrac{1}{2}mv^2 = eV$$

$$\therefore \quad (mv)^2 = 2meV$$

$$\therefore \quad \text{Momentum} = mv = \sqrt{(2meV)}$$

so

$$\lambda = \frac{h}{\text{Momentum}} = \frac{h}{\sqrt{(2meV)}}$$

$h = 6.6 \times 10^{-34}$, $m = 9.1 \times 10^{-31}$, $e = 1.6 \times 10^{-19}$, $V = 100$.

$$\therefore \quad \lambda = \frac{6.6 \times 10^{-34}}{\sqrt{(2 \times 9.1 \times 10^{-31} \times 1.6 \times 10^{-19} \times 100)}} = 1.22 \times 10^{-10}\,\text{m}$$

In the electromagnetic spectrum this would be X-radiation.

Answer

1.2×10^{-10} m, X-radiation.

Example 3

A monochromatic source emits a narrow, parallel beam of light of wavelength 546 nm, the power in the beam being 0.080 W. How many photons leave the source per second? If this beam falls on the cathode of a photocell, what is the photocell current, assuming that 15% of the photons incident on the cathode liberate electrons?

(Planck constant = 6.6×10^{-34} J s,
velocity of light in vacuum = 3.0×10^{8} m s^{-1},
electronic charge = 1.6×10^{-19} C.)

Method

At 546 nm, i.e. 546×10^{-9} m wavelength the photon energy is hc/λ and equals

$$\frac{6.6 \times 10^{-34} \times 3 \times 10^{8}}{546 \times 10^{-9}} \quad \text{or} \quad 3.626 \times 10^{-19}\,\text{J}$$

The number of photons per second

$$= \frac{\text{Joules per second}}{\text{Photon energy}} = \frac{0.08}{3.626 \times 10^{-19}}$$

$$= 2.2 \times 10^{17}$$

The number of electrons liberated per second

$$= \frac{15}{100} \times 2.2 \times 10^{17} = 33 \times 10^{15}$$

The current

$$= \text{Electrons per second} \times \text{Electronic charge}$$

$$= 33 \times 10^{15} \times 1.6 \times 10^{-19}$$

$$= 52.8 \times 10^{-4}\,\text{A} \quad \text{or} \quad 5.3\,\text{mA}$$

Answer

2.2×10^{17}, 5.3 mA.

Circular orbits

For an electron (mass m, charge e) in circular orbit, radius R, about a nucleus containing Z protons we have

Electrostatic force of attraction (equation 30.1) is

$$F = \frac{q_1 q_2}{4\pi\varepsilon_0 R^2} = \frac{Ze \times e}{4\pi\varepsilon_0 R^2}$$

and this must equal mv^2/R (see Chapter 10) where v is the electron's speed.

$$\frac{mv^2}{R} = \frac{Ze^2}{4\pi\varepsilon_0 R^2} \qquad (40.5)$$

In addition it is found that the distance round the orbit, $2\pi R$, must equal a whole number n of electron wavelengths, so that

$$2\pi R = n\lambda = \frac{nh}{mv}$$

(using the de Broglie relation, 40.4)

$$\therefore \quad 2\pi mvR = nh \qquad (40.6)$$

where n may be 1, 2, 3, . . . The above two equations can be solved to find R for each n value.

For each of these allowed orbits we can calculate the electron's energy E. These allowed energy values are called energy levels.

The circular orbit closest to the nucleus has $n = 1$, and the energy of an electron here is lower than for $n = 2, 3, \ldots$ The hydrogen atom has one electron only and this will normally reside in this innermost orbit, i.e. this is its ground state. The hydrogen atom is simple also because the energies for its elliptical orbits are near enough the same as for the circular ones.

Atomic electrons may be classified in groups called shells. In the hydrogen atom all electrons of the same shell have the same energy and same n value.

Excitation

An atomic electron if it acquires sufficient energy, e.g. from a colliding particle (see p.268) or from incident electromagnetic radiation (a photon), can move within the atom to a higher energy level. Typically it will stay at this 'excited' level for only a minute fraction of a second and then 'fall back' to its ground state or other lower energy level, giving out energy as a photon of electromagnetic radiation (often visible light).

$$E - E' = hf \qquad (40.7)$$

where E is electron energy in the higher state and E' electron energy in the lower state; h is the Planck constant, f the frequency of radiation emitted as a photon (energy hf). This process is called *excitation*.

Example 4

The three lowest energy levels of the electron in the hydrogen atom have energies

$$E_1 = -21.8 \times 10^{-19}\,\text{J}$$

$$E_2 = -5.45 \times 10^{-19}\,\text{J}$$

$$E_3 = -2.43 \times 10^{-19}\,\text{J}$$

The zero of energy is taken to be when the electron is at rest at a great distance from the nucleus. What is the wavelength of the H_α line in the hydrogen spectrum which arises from transitions between the levels E_3 and E_2?

Through what potential difference must an electron be accelerated if it is to be capable of (a) ionising a hydrogen atom, (b) causing emission of H_α radiation from a normal hydrogen atom?

(Planck constant = 6.6×10^{-34} J s,
speed of light in vacuum = 3×10^8 m s^{-1},
electron charge = 1.6×10^{-19} C.) [SUJB 82]

Method

(Note that the energy values given are negative because of the zero chosen.)

For the E_3 to E_2 transition, using Equation 40.7:

$$(-2.43 \times 10^{-19}) - (-5.45 \times 10^{-19}) = hf = \frac{hc}{\lambda}$$

$$3.02 \times 10^{-19} = \frac{6.6 \times 10^{-34} \times 3 \times 10^8}{\lambda}$$

$$\therefore \quad \lambda = \frac{6.6 \times 3}{3.02} \times 10^{-7}\,\text{m}$$

$$= 6.556 \times 10^{-7}\,\text{m} \quad \text{or} \quad 656\,\text{nm}$$

(but note that c was given to one-figure accuracy only).

Accelerating PD needed:

(a) For ionisation of normal atom, transition is from ground state (E_1) to outside of the atom (our zero of energy)

$$eV \; (= \tfrac{1}{2}mv^2) = (0) - (-21.8 \times 10^{-19})$$

$$V = \frac{21.8 \times 10^{-19}}{1.6 \times 10^{-19}} = 13.6\,\text{V}$$

(b) For emission of H_α radiation an electron must be moved from ground state (E_1) to E_3 so that it can fall to E_2.

$$\therefore \quad eV = (-2.43 \times 10^{-19}) - (-21.8 \times 10^{-19})$$

$$= 19.4 \times 10^{-19}$$

$$\therefore \quad V = \frac{19.4 \times 10^{-19}}{1.6 \times 10^{-19}} = 12.1\,\text{V}$$

Answer

6.56×10^{-7} m, 13.6 V, 12.1 V.

Exercise 136

(Where necessary take
velocity of light in vacuum $c = 3.0 \times 10^8$ m s^{-1},
Planck constant $h = 6.6 \times 10^{-34}$ J s.)

1 Calculate the frequency and the photon energy for blue light of wavelength 4.0×10^{-7} m.

2 Through what potential difference must electrons be accelerated to be able to produce visible light of wavelength 6.0×10^{-7} m?
(Electron charge $e = -1.6 \times 10^{-19}$ C.)

3 A metal surface is illuminated with monochromatic light and it becomes charged to a steady positive potential of 1.0 V relative to its surroundings. The work function energy of the metal surface is 3.0 eV, and the electron charge e is 1.6×10^{-19} C. Calculate the frequency of the light.

4 A clean surface is irradiated with light of wavelength 5.5×10^{-7} m and electrons are just able to escape from the surface. When light of wavelength 5.0×10^{-7} m is used, electrons emerge with energies of up to 3.6×10^{-20} J. Obtain a value for the Planck constant h.

5 The beam of light from a certain laser has a power of 1.0 mW and a wavelength of 633 nm. How many photons are emitted per second by this laser?

6 Calculate the de Broglie wavelength of 300 V electrons.
(Mass of electron = 9.1×10^{-31} kg,
electron charge = -1.6×10^{-19} C)

7 The first excitation energy of the hydrogen atom is 10.2 eV. Calculate the speed of the slowest electron that can excite a hydrogen atom.
(Electron charge/mass ratio e/m
= 1.7×10^{11} C kg^{-1}.)

Exercise 137: Questions of A-level standard

1 Light of wavelength 0.50 μm incident on a metal surface ejects electrons with kinetic energies up to a maximum value of 2.0×10^{-19} J. What is the energy required to remove an electron from the metal? If a beam of light causes *no* electrons to be emitted however great its intensity, what condition must be satisfied by its wavelength?
(Planck constant $h = 6.6 \times 10^{-34}$ J s,
speed of light $c = 3 \times 10^8$ m s^{-1}.) [SUJB 80]

2 The cut-off voltage (the reverse voltage that will just prevent electron flow) for a photoelectric cell is measured for light of two frequencies. The following results are obtained:

Cut-off voltage/V	0.93	1.41
Frequency/10^{14} Hz	6.10	7.30

Use this data to determine a value for the Planck constant.
(Electron charge $e = -1.6 \times 10^{-19}$ C.) [L 82]

3 A clean magnesium surface is supported in a vacuum as the cathode of a photocell. A wire-mesh electrode surrounds the cathode and is used as the anode.
When the cathode is illuminated with ultraviolet radiation of wavelength 254 nm, the anode current can be reduced to zero by making it negative with respect to the cathode using a PD of 1.2 V or greater. Obtain a value for the work function of magnesium in electron-volts.
(Planck constant $h = 6.6 \times 10^{-34}$ J s,
velocity of light $c = 3.0 \times 10^{8}$ m s^{-1},
electron charge $e = -1.6 \times 10^{-19}$ C.)

4 A 50 W discharge lamp radiates in the visible region (400 to 700 nm wavelength). Estimate the number of photons produced each second by the lamp.
(Speed of light $c = 3 \times 10^{8}$ m s^{-1}; Planck constant, $h = 6.6 \times 10^{-34}$ J s.)

5 The de Broglie wavelength of a hydrogen atom of mass 1.7×10^{-27} kg, if it is moving at 300 m s^{-1}, is

A $\frac{51}{66} \times 10^{5}$ m **B** $\frac{66}{51} \times 10^{-9}$ m **C** $\frac{66}{51} \times 10^{9}$ m
D $\frac{51}{66} \times 10^{-9}$ m **E** $\frac{66}{51} \times 10^{-5}$ m
(Planck constant $h = 6.6 \times 10^{-34}$ J s.)

6 Fig 40.1 shows three energy levels of the thallium atom.

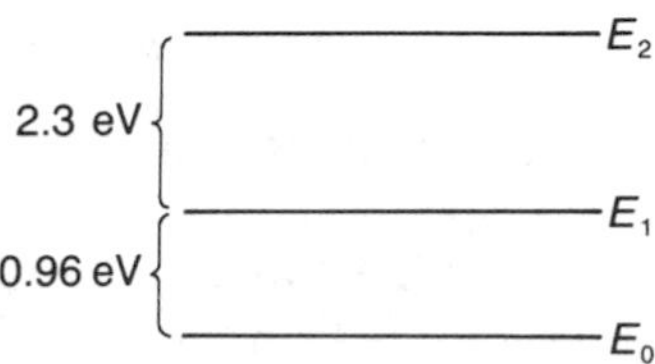

Fig 40.1 Diagram for Question 6

To excite electrons from energy level E_0 to E_2 requires light of wavelength 378 nm. What wavelength of light is emitted when electrons fall from level E_2 to E_1?

41
X-RAYS

X-radiation

This radiation is usually produced by accelerating electrons to high speeds in an evacuated glass tube and letting them strike a target or anode (Fig 41.1). The X-radiation results from two processes. One is the rapid slowing down of the electrons, as they enter the target's surface. The other is excitation of the target atoms.

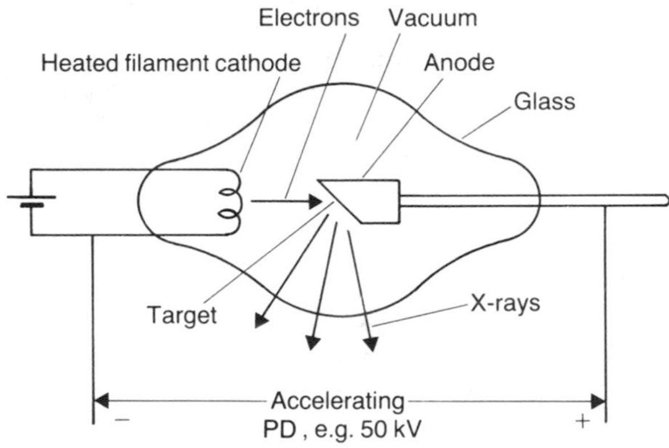

Fig 41.1 An X-ray tube

The slowing process gives X-rays over a range of wavelengths, i.e. a continuous spectrum. This radiation is called 'white'. The shortest wavelength (λ_{min}) obtained (highest photon energy) is due to incident electrons stopping abruptly, so that all the electron's kinetic energy becomes one photon of X-radiation. Thus

$$\frac{hc}{\lambda_{min}} = hf_{max} = \tfrac{1}{2}mv^2 = eV \qquad (41.1)$$

where λ_{min} is the shortest wavelength of X-radiation produced, f_{max} is its frequency, v is the velocity of the accelerated electrons.

V is the potential difference used for the acceleration. The excitation process gives X-rays at a few definite wavelengths called 'characteristic X-rays' and decided by the allowed transitions in the target atoms, according to the equation below:

$$E - E' = hf = \frac{hc}{\lambda} \qquad (40.7)$$

In this process an accelerated electron removes an inner electron of the atom at energy level E', Then an electron at higher energy level E can fall to level E' and so emit X-radiation.

Fig 41.2 shows a typical X-radiation spectrum from such an X-ray tube.

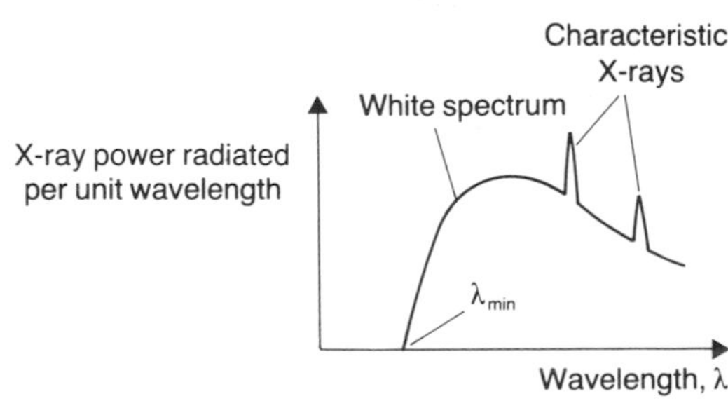

Fig 41.2 Typical X-ray spectrum

Example 1

The accelerating voltage across an X-ray tube is 33.0 kV. Explain why the frequency of the X-radiation cannot exceed a certain value, and calculate this maximum frequency. (Planck constant = 6.6×10^{-34} J s, charge e on an electron = 1.6×10^{-19} C.) [AEB 81]

Method

We use Equation 41.1:

$$\frac{hc}{\lambda_{min}} = eV \quad \text{or} \quad hf_{max} = eV$$

$h = 6.6 \times 10^{-34}$, $e = 1.6 \times 10^{-19}$, $V = 33 \times 10^3$.

$$\therefore \quad f_{max} = \frac{eV}{h} = \frac{1.6 \times 10^{-19} \times 33 \times 10^3}{6.6 \times 10^{-34}}$$

$$= 8.0 \times 10^{18}\,\text{Hz}$$

The whole of the accelerated electron's energy has been used to produce a photon of this frequency. A higher frequency is not possible.

Answer

8.0×10^{18} Hz.

Example 2

Characteristic X-radiation is described as K_α if it is due to an electron transition to the $n = 1$ shell (K shell) from the $n = 2$ (L shell). K_β radiation is due to $n = 3$ (M shell) to K shell transitions. For molybdenum the K_α wavelength is 0.071 nm and K_β is 0.063 nm. Calculate the difference between (a) the K and L energy levels and (b) the L and M energy levels.

(Planck constant $h = 6.6 \times 10^{-34}$ J s, velocity of electromagnetic radiation in vacuum $c = 3.0 \times 10^8\,\text{m s}^{-1}$.)

Method

$$E - E' = \frac{hc}{\lambda} \tag{40.7}$$

(a) For K_α, $\lambda = 0.071 \times 10^{-9}$ m; $h = 6.6 \times 10^{-34}$ J s, $c = 3.0 \times 10^8\,\text{m s}^{-1}$.

$$\therefore \quad E_L - E_K = \frac{6.6 \times 10^{-34} \times 3.0 \times 10^8}{0.071 \times 10^{-9}}$$

$$= 278.9 \times 10^{-17}\,\text{J} \quad \text{or} \quad 2.8 \times 10^{-15}\,\text{J}$$

(b) For K_β, $\lambda = 0.063 \times 10^{-9}$ m;

$$\therefore \quad E_M - E_K = \frac{6.6 \times 10^{-34} \times 3.0 \times 10^8}{0.063 \times 10^{-9}}$$

$$= 314.3 \times 10^{-17}\,\text{J}$$

and

$$E_M - E_L = (E_M - E_K) - (E_L - E_K)$$

$$= 314.3 \times 10^{-17} - 278.9 \times 10^{-17}$$

$$= 35.4 \times 10^{-17}\,\text{J} \quad \text{or} \quad 3.5 \times 10^{-16}\,\text{J}$$

Answer

(a) 2.8×10^{-15} J, (b) 3.5×10^{-16} J.

Exercise 138

1 Calculate the highest frequency of X-radiation that can be obtained from an X-ray tube operated with a PD of 17 kV. If the target of such a tube is made of molybdenum, will the K_α (0.071 nm) and K_β (0.063 nm) characteristic radiations be obtained? (Electron charge $e = -1.6 \times 10^{-19}$ C; Planck constant $h = 6.6 \times 10^{-34}$ J s; velocity of electromagnetic radiation in vacuum $c = 3.0 \times 10^8\,\text{m s}^{-1}$.)

2 An X-ray tube has an electron beam current of 10 mA, and the accelerating voltage is 50 kV. The efficiency (i.e. percentage of input power converted into X-ray power) is 0.5%. Calculate (a) the input power, (b) the power lost in the tube as heat, (c) the minimum wavelength of X-rays produced.
(Electron charge $e = -1.6 \times 10^{-19}$ C; Planck constant $h = 6.6 \times 10^{-34}$ J s; velocity of electromagnetic radiation in vacuum $c = 3.0 \times 10^8\,\text{m s}^{-1}$.)

3 The K_α and K_β characteristic X-radiations from a copper target have wavelengths 0.154 nm and 0.139 nm, and are due to electron transitions to the $n = 1$ shell from the $n = 2$ and $n = 3$ shells respectively. Calculate a value for the energy difference between (a) the $n = 1$ and $n = 3$ shells, (b) the $n = 2$ and $n = 3$ shells.
($h = 6.6 \times 10^{-34}$ J s; $c = 3.0 \times 10^8\,\text{m s}^{-1}$.)

X-ray crystallography

The atoms of a crystal may be regarded as forming layers (i.e. planes) in the manner indicated in Fig 41.3.

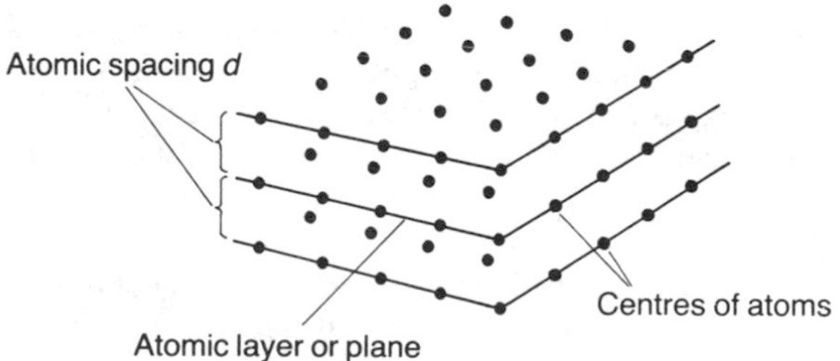

Fig 41.3 Atomic planes

When an X-ray beam falls upon the crystal the possible directions of emergence of the X-rays must satisfy two conditions:

(1) the emergent beam is inclined to the planes at an angle θ equal to the glancing angle of incidence (see Fig 41.4) i.e. they are 're-flected';

(2) $$2d \sin \theta = n\lambda \quad (41.2)$$

where d is the distance between the planes (lattice spacing), n is any whole number, 0, 1, 2 . . ., and λ is the X-ray wavelength.

Note the $2d$, not d as in a diffraction grating where light arrives perpendicular to the grating.

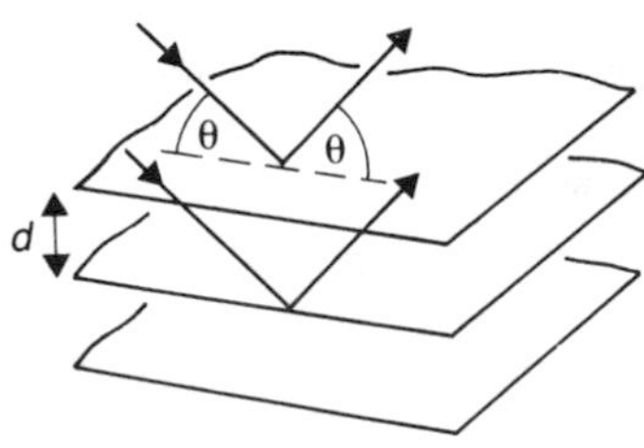

Fig 41.4 'Reflection' of X-rays

Equation 41.2 is known as the Bragg equation and it shows that X-ray measurements can determine the atomic spacing d.

Example 3

A beam of X-radiation largely composed of K_α and K_β radiations from a copper target is incident upon a crystal in a Bragg spectrometer. The beam is first inclined at a small angle θ to atomic planes for which the separation is 2.40×10^{-10} m; then the angle θ is gradually increased by rotating the crystal. A Geiger–Müller counter moves so as to make an angle 2θ with the X-ray beam direction of incidence on the crystal (i.e. at equal angle θ to the crystal surface). At what angles to the incident beam direction are peaks expected in the resulting spectrum?

(Wavelengths are 0.154 nm for K_α, 0.139 nm for K_β.)

Method

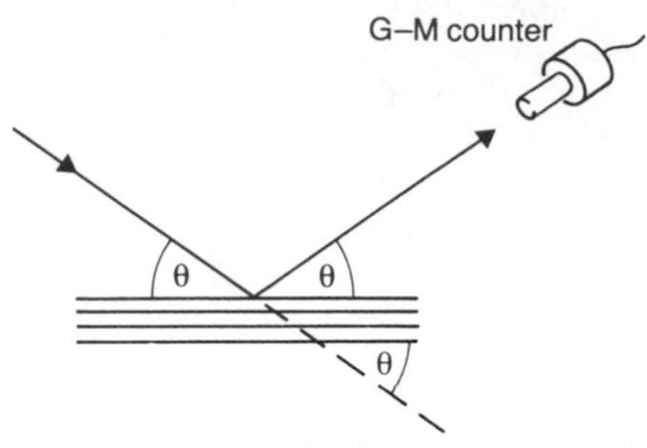

Fig 41.5 Diagram for Example 3

We use Equation 41.2:

$$2d \sin \theta = n\lambda$$

$$d = 2.4 \times 10^{-10}\,\text{m}$$

For $\lambda = 0.154 \times 10^{-9}$ m, with $n = 1$, we have

$$2 \times 2.4 \times 10^{-10} \times \sin \theta = 1 \times \lambda$$

$$\therefore \quad \sin \theta = \frac{0.154 \times 10}{4.8} = 0.321$$

$$\therefore \quad \theta = 18.7° \quad \text{and} \quad 2\theta = 37.4°$$

Similarly with $n = 2$, $\sin \theta = 0.642$ and $\theta = 39.9°$, $2\theta = 79.9°$

with $n = 3$, $\sin \theta = 0.963$ and $\theta = 74.4°$, $2\theta = 148.8°$

with $n = 4$, $\sin \theta = 1.283$ (impossible)

For $\lambda = 0.139 \times 10^{-9}$, with $n = 1$, $\sin \theta = 0.290$ and $\theta = 16.9°$, $2\theta = 33.7°$

with $n = 2$, $\sin \theta = 0.580$ and $\theta = 35.5°$, $2\theta = 70.9°$

with $n = 3$, $\sin \theta = 0.869$ and $\theta = 60.3°$, $2\theta = 121°$

with $n = 4$, $\sin \theta = 1.16$ (impossible)

For $n = 0$, $\sin \theta = 0$ and $\theta = 0$ of course. So the spectrum consists of the zero-order (straight-through) beam plus three K_α and three K_β beams on each side of the zero-order. Note that the maximum value for $\sin \theta$ is 1.00, giving $\theta = 90°$ or $2\theta = 180°$, and the emergent beam would be travelling back in the direction of incidence, i.e. reversed.

Answer

37.4°, 79.9°, 148.8°, 33.7°, 70.9°, 121°.

Exercise 139

1 If the first-order ($n = 1$) X-ray reflection from crystal planes 0.24 nm apart occurs for a glancing angle of incidence of 5.0°, what is the wavelength of the X-radiation?

At what angle would the second-order reflection be obtained? Note that *glancing* angle means the angle made with a surface or plane, not the angle made to the normal of the plane.

2 A collimated X-ray beam consists of mainly copper K_α (0.154 nm wavelength) and K_β (0.139 nm) X-radiations, and is incident upon a crystal starting at zero glancing angle of incidence with a chosen set of parallel crystal planes. The crystal is then gradually rotated to increase the angle of incidence. At what angles to the incident beam will the first-order reflections be obtained if the distance between the planes is 0.28 nm? How many beams of the K_α-radiation emerge from each side of the zero-order ('straight-through') beam?

Exercise 140: Questions of A-level standard

1 Electrons are accelerated from rest through a potential difference of 10 000 V in an X-ray tube. Calculate
(a) the resultant energy of the electrons in eV;
(b) the wavelength of the associated electron waves;
(c) the maximum energy and the minimum wavelength of the X-radiation generated.

Charge of electron = 1.6×10^{-19} C.
Mass of electron = 9.11×10^{-31} kg
Planck's constant = 6.62×10^{-34} J s.
Speed of electromagnetic radiation *in vacuo* = 3.00×10^{8} m s^{-1}.

[JMB 80]

2 The energy required to remove an electron from the various shells of the nickel atom is:

K shell	1.36×10^{-15} J
L shell	0.16×10^{-15} J
M shell	0.08×10^{-15} J

An X-ray tube with a nickel target emits the X-ray K radiation of nickel. What is (a) the minimum potential difference across the tube, (b) the energy of the X-ray quantum of longest wavelength in the K spectrum of nickel?
(Electron charge $e = 1.6 \times 10^{-19}$ C.) [SUJB 79]

3 The spacing between a certain set of planes of atoms in a crystal is 0.300 nm. The crystal is used in an X-ray spectrometer. How many orders of diffraction could be observed from these planes if the incident X-rays were of wavelength 0.154 nm? [C 77]

4 A beam of X-rays of wavelength 0.154 nm is diffracted by a crystal. For first-order diffraction by a certain set of planes it is found that the X-ray beam is deviated by 32.0°. What angle does the incident X-ray beam make with these planes? Find the spacing of the planes. [C 79]

42 RADIOACTIVITY

The nucleus

The nucleus of an atom consists of a number Z of protons and a number N of neutrons. Z is called the atomic number or proton number. The sum of the proton number and the neutron number is the total number of nuclear particles (or nucleons). It is usually denoted by A and is often called the mass number. For example $^{238}_{92}U$ denotes a uranium atom for which $Z = 92$ and $Z + N = 238$.

Some nuclei, because of the particular numbers of protons and neutrons they contain are unstable and may decay (or 'disintegrate') at any time. These changes usually amount to emission of α, β^- or β^+ particles, often followed immediately by emission of a photon of γ-radiation. For a large number n of nuclei all of the same kind (i.e. the same nuclide) the number of decays per second, or (what is the same thing) the reduction in the value of n occurring in a second, is given by $-dn/dt$, and is proportional to the number of nuclei waiting to decay i.e. $-dn/dt \propto n$, or

$$-\frac{dn}{dt} = \lambda n \qquad (42.1)$$

where λ is a constant called the decay constant whose value is characteristic of the particular nuclide concerned. This equation can be integrated to give

$$n = n_0 e^{-\lambda t} \qquad (42.2)$$

where n_0 is the initial number of nuclei at the time $t = 0$.

The decay graph relating n and t is shown in Fig 42.1.

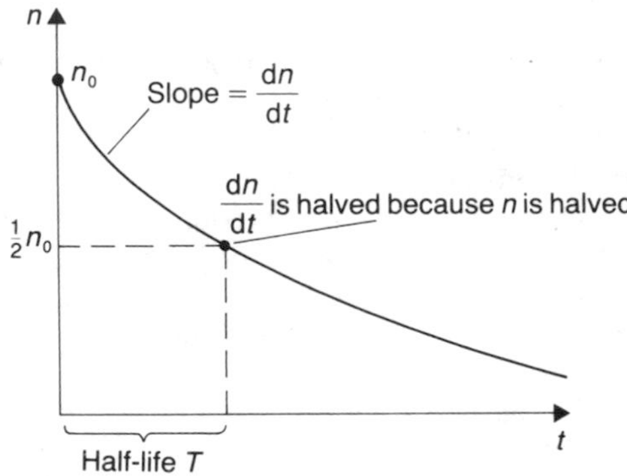

Fig 42.1 Graph for radioactive decay

The time T required for n to fall from n_0 to $\frac{1}{2}n_0$ is called the half-life of the nuclide, and from Equation 42.2 we can get

$$T = \frac{\ln 2}{\lambda} \quad \text{or} \quad \frac{0.693}{\lambda} \qquad (42.3)$$

An alternative and more obvious formula for n is

$$n = \frac{n_0}{2^Y} \qquad (42.4)$$

where $Y = t/T$ = the number of half-lives. Thus if the time elapsed is T, then $Y = 1$ and $n = \frac{1}{2}n_0$; if $2T$, then the answer is halved again to give $\frac{1}{4}n_0$, because $Y = 2$.

Activity of a nuclide

This is the number of decays per unit time occurring in a sample. An activity of one decay per second is a becquerel (Bq). An earlier unit for activity was the curie (Ci) and 1 Ci = 3.7×10^{10} Bq.

Since each decay means the reduction by one in the value of n

$$\text{Activity } A = -\frac{dn}{dt} = \lambda n \qquad (42.5)$$

Also we can write Equations 42.2 and 42.4 in terms of activities:

$$\lambda \times n = \lambda \times n_0 \, e^{-\lambda t}$$

$$\therefore \quad A = A_0 \, e^{-\lambda t} \qquad (42.6)$$

and similarly

$$A = A_0/2^Y \quad \text{where} \quad Y = t/T \quad (42.7)$$

The activity is of course decided by the mass of radioactive material and its type (since these decide n and λ).

Example 1

A radioactive sample, whose disintegration product is non-radioactive, was known to have an activity of 5×10^{11} Bq at a certain time. At this time an α-particle detector showed a count rate of $32\,s^{-1}$. Exactly 10 days later the count rate had dropped to $8\,s^{-1}$. Find the half-life, the decay constant and the number of radioactive nuclei present in the sample at the start. How many radioactive nuclei remain 100 days from the start? [WJEC 81]

Method

Note, first of all, that the count rate is smaller than the activity of the sample since only a small fraction of the emitted particles are detected by the counter.

To find the half-life T we may use $A = A_0/2^Y$ where $Y = t/T$ (Equation 42.7):

$$\frac{A}{A_0} = \frac{8}{32}$$

$$\therefore \quad 8 = \frac{32}{2^Y} \quad \therefore \quad 2^Y = \frac{32}{8} = 4 \quad \therefore \quad Y = 2$$

and, using $Y = t/T$,

$$t = 10 \text{ days}, \quad Y = 2;$$

$$\therefore \quad 2 = \frac{10}{T} \quad \therefore \quad T = 5 \text{ days}$$

To find the decay constant λ we use

$$\lambda = \frac{\ln 2}{T}$$

$$\therefore \quad \lambda = \frac{0.693}{5} = 0.1386 \text{ day}^{-1}$$

(The equation $A = A_0 e^{-\lambda t}$ could be used to find λ but the calculation is more tedious.)

To find n_0 we have $A = \lambda n$ (Equation 42.5) from which $A_0 = \lambda n_0$.

$A_0 = 5 \times 10^{11}$ Bq (or s^{-1}) and $\lambda = 0.1386 \text{ day}^{-1}$ (i.e. 'per day') or $0.1386/(24 \times 60 \times 60)$ per second, which is $1.6 \times 10^{-6}\,s^{-1}$.

$$\therefore \quad n_0 = \frac{A_0}{\lambda} = \frac{5 \times 10^{11}}{1.6 \times 10^{-6}} = 3.125 \times 10^{17}$$

After 100 days (i.e. 20 half-lives):

$$n = \frac{n_0}{2^{20}} = \frac{3.125 \times 10^{17}}{1.05 \times 10^6} = 3.0 \times 10^{11}$$

Answer

5 days, 0.14 day^{-1} (or $1.6 \times 10^{-6}\,s^{-1}$), 3.1×10^{17}, 3.0×10^{11}.

Exercise 141

1 The half-life of radium is 1620 years. How long will it take for the initial activity of a radium compound to fall to a fifth of its original value? (Remember $\log a^b = b \log a$.)

2 The activity of a radioactive source decreases by seven-eighths of its initial value in 30 hours. Calculate the half-life and the decay constant for this source. ($\log_e 2 = 0.693$.)

Atomic mass unit

The unified atomic mass unit (u) is defined as equal to one-twelfth of the mass of the carbon-12 atom.

Since the Avogadro number is defined as the number of atoms in one gram of carbon-12 atoms, we see that 12 g of these atoms is one mole. For *any* particle the relative atomic mass is its mass relative to this atomic mass unit, and so is numerically equal to its mass in u. 1 mole of the particles equals the relative atomic mass expressed in grams.

Mass of nuclide related to activity

The mass of a nuclide needed to provide a certain activity A can be calculated as follows:

$$A = \lambda n = \lambda \times \frac{\text{Mass of sample}}{\text{Mass of atom}} \qquad (42.8)$$

The mass of the atom may be given in kg or in unified atomic mass units (u). Since the mass of a nucleon is approximately 1 u, the mass of an atom in u is approximately equal to its nucleon number (mass number).

Example 2

Calculate the activity of 1 g of pure $^{238}_{92}U$ given that its half-life is 4.5×10^9 year. ($1\,u = 1.66 \times 10^{-27}$ kg.)

Method

$$A = \lambda n = \lambda \times \frac{\text{Mass of sample}}{\text{Mass of atom}} \qquad (42.8)$$

$\lambda = \ln 2/T$, $T = 4.5 \times 10^9$ year, mass of atom = 238 u, mass of sample = 10^{-3} kg

$$\therefore \quad A = \frac{0.693}{4.5 \times 10^9} \times \frac{10^{-3}}{238 \times 1.66 \times 10^{-27}}$$

$$= 3.9 \times 10^{11}\ \text{year}^{-1}$$

or $\dfrac{3.9 \times 10^{11}}{365 \times 24 \times 60 \times 60}\ s^{-1}$,

which is 12 kBq.

Answer

12 kBq.

Isotopes

These are different nuclides of the same element. For example $^{238}_{92}U$ and $^{235}_{92}U$ are isotopes. They differ only in the number of neutrons they contain.

Example 3

Calculate the atomic mass for natural uranium which consists of 99.3% ^{238}U and 0.7% ^{235}U. The atomic masses of ^{238}U and ^{235}U are 238.051 u and 235.044 u.

Method

The atomic mass required is the mean for all its atoms.

Out of every 100 atoms we would have 99.3 weighing 238.051 u each and 0.7 weighing 235.044 u each. Therefore the average mass is

$$\frac{(99.3 \times 238.051) + (0.7 \times 235.044)}{100}$$

and equals 238.03 u.

Note that we can say the *relative* atomic mass is 238.03 or 238.03 g per mole.

Answer

238.03 u.

Exercise 142

1 The atomic mass of ordinary chlorine gas is 35.5 u, while it is known that this gas is made up of two isotopes whose atomic masses are 35 u and 37 u. Calculate the ratio of the number of the heavier atoms to the number of lighter ones.

Nuclear changes

An alpha particle consists of 2 protons and 2 neutrons holding together, like a helium nucleus, and travelling very fast. When an α-emission occurs, the proton number falls by 2 and the nucleon number falls by 4. For β^--emission the nucleus changes one of its neutrons into a proton plus a negative electron. This electron is emitted from the nucleus at extremely high speed as the β^--particle. Thus the nucleon number does not change, but the proton number increases by one. β^+-emission causes decrease of the proton number by one, a β^+ particle being a positive electron.

Example 4

A radioactive isotope of cobalt $^{60}_{27}Co$ decays by emission of a beta particle followed by emission of one or more gamma photons. Thus we may write

$$^{60}_{27}Co \rightarrow P + \beta + \gamma$$

where P is the product nuclide, β is the beta particle and γ represents one or more gamma photons. The atomic number of P is

A 29 **B** 28 **C** 27 **D** 26 **E** 25

Method

Beta emission increases the atomic number (i.e. proton number) by 1, so that the atomic number of P is 27+1 or 28. Note that β means β^- unless otherwise stated.

Answer

B.

Radioactive dating

A good example of this concerns the decay of the carbon $^{14}_{6}C$ isotope which is a β^--emitter and is found in a small, known concentration in atmospheric carbon dioxide. Plant or other material contains carbon taken from the atmosphere at the time of the material's formation. When the living material dies it will subsequently contain a decreasing proportion of $^{14}_{6}C$ to normal $^{12}_{6}C$ atoms as the carbon-14 slowly decays.

Example 5

The half-life of $^{14}_{6}C$ is 5570 years. (a) What is its decay constant? (b) How many disintegrations per second are obtained from 1 g of carbon if 1 carbon atom in 10^{12} is of the radioactive $^{14}_{6}C$ type? (c) after what time will the activity per gram have fallen to 3 disintegrations per minute? ($1\ u = 1.66 \times 10^{-27}$ kg.)

Method

(a) $\lambda = \dfrac{\ln 2}{T}$ (Equation 42.3) $= \dfrac{0.693}{5570}\ \text{year}^{-1}$

$$\therefore \quad \lambda = \frac{0.693}{5570 \times 365 \times 24 \times 60 \times 60}\ \text{s}^{-1}$$

$$\therefore \quad \lambda = 3.94 \times 10^{-12}\ \text{s}^{-1}$$

(b) For 1 g carbon:

$$\text{Number of carbon atoms} = \frac{\text{Mass}}{\text{Mass of atom}} = \frac{10^{-3}\ \text{kg}}{12 \times 1.66 \times 10^{-27}}$$

(assuming all atoms are carbon $^{12}_{6}C$ — almost true)

$$= 5.02 \times 10^{22}$$

∴ number of carbon-14 atoms

$$= 5.02 \times 10^{22} \times \frac{1}{10^{12}}$$

$$= 5.02 \times 10^{10}$$

The (carbon-14) activity is

$$A = \lambda n = 3.94 \times 10^{-12} \times 5.02 \times 10^{10}$$

$$= 0.198\ \text{s}^{-1}$$

Note: This means 12 disintegrations per minute.

(c) Using $A = A_0/2^Y$ (Equation 42.7) we get

$$2^Y = \frac{12}{3} = 4 \qquad \therefore \quad Y = 2$$

∴ time of decay $t = 2 \times$ half-life, i.e. 2×5570 years or 11 140 years.

Answer

(a) $3.94 \times 10^{-12}\ \text{s}^{-1}$, (b) $0.198\ \text{s}^{-1}$, (c) 11.1×10^3 years.

Inverse square law

(a)

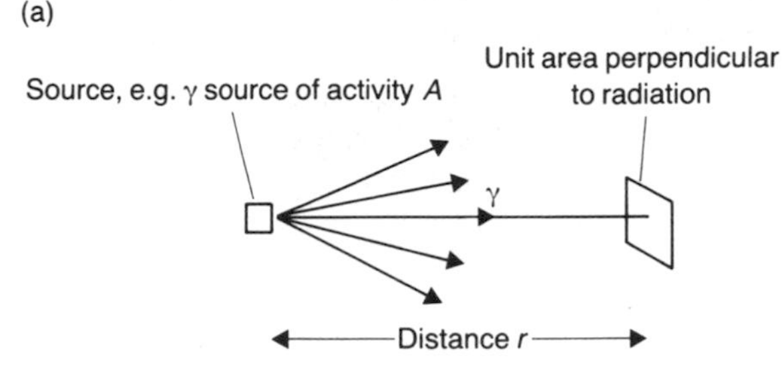

(b)

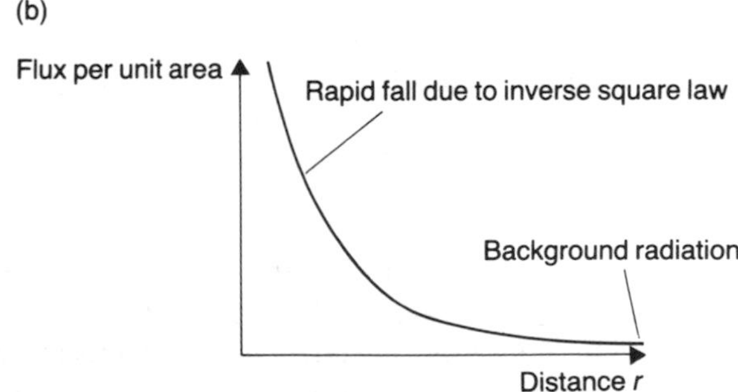

Fig 42.2 The inverse square law

For a small γ-source (dimensions $\ll r$) in a low density medium, e.g. air, there will be little loss of the emitted photons, but they spread as they travel so that the number of particles per second (the particle flux) received per unit area at distance r is proportional to $1/r^2$. Hence the term 'inverse square law'. This follows the fact that, at distance r, the total flux F is spread over a spherical area $4\pi r^2$ so that the flux per unit area is

$$\frac{F}{\text{area}} = \frac{\text{Activity}}{4\pi r^2} \qquad (42.9)$$

assuming that each disintegration of the source produces one particle (i.e. one emergent γ-photon from the γ-source). For α- and β-particles the flux per unit area is also reduced by intervening air.

Absorption

When γ-photons pass into any medium their number will begin to decrease because of the ionising of atoms and other processes. (A photon is used up when an ionisation is produced.)

The half-thickness of a medium is that thickness which halves the flux. We can denote it by T. The emergent flux F is given by

$$F = \frac{F_0}{2^Y} \qquad (42.10)$$

where $Y = t/T$, t being the thickness of the absorbing medium, T the half-thickness; F_0 is the initial flux entering. We can also write

$$F = F_0 \, e^{-\mu t} \qquad (42.11)$$

where μ is the absorption coefficient for the medium. Also

$$\mu = \frac{\ln 2}{T} \qquad (42.12)$$

Note that Equations 42.10, 11 and 12 are analogous to Equations 42.4, 2 and 3 for decay; this makes learning the equations easier. The same formulae happen to apply to absorption of β-radiation from radioactive sources.

Example 6

A point source of γ-radiation has a half-life of 30 minutes. The initial count rate, recorded by a Geiger counter placed 2.0 m from the source, is $360\,s^{-1}$. The distance between the counter and the source is altered. After 1.5 h the count rate recorded is $5\,s^{-1}$. What is the new distance between the counter and the source?

[L 77]

Method

Here the count rate falls due to both decay and the change of distance.

As regards decay the expected count rate R, say, after 1.5 h can be deduced from Equation 42.7:

$$A = \frac{A_0}{2^Y}$$

Since the count rate is proportional to the activity we can now write

$$R = \frac{360}{2^Y}$$

and $$Y = \frac{t}{T} = \frac{1.5\text{ h}}{0.5\text{ h}} = 3$$

$$\therefore \quad R = \frac{360}{2^3} = \frac{360}{8} = 45\,s^{-1}$$

The reduction from $45\,s^{-1}$ to $5\,s^{-1}$ must be due to increased distance, and since flux and therefore count rate are proportional to $1/r^2$ (the inverse square law),

$$\frac{\text{First count rate}}{\text{Second count rate}} = \frac{r_2^2}{r_1^2}$$

$$\therefore \quad \frac{45}{5} = \frac{r_2^2}{2^2}$$

$$\therefore \quad r_2^2 = 4 \times 9 = 36 \quad \text{and} \quad r_2 = 6\text{ m}$$

Answer

6 m.

Example 7

The half-thickness of a certain material for β-radiation, from a particular source is 3.0 mm. What thickness is needed to reduce the β-radiation flux by 90%?

A 15 mm **B** 5.4 mm **C** 10 mm
D 0.6 mm **E** 0.15 mm

Method

An approximate value for the thickness can be obtained as follows:

One half-thickness gives a reduction to 50% of the initial flux; two half-thicknesses gives a further reduction to half of this, i.e. 25% of the initial flux would emerge; three gives $12\frac{1}{2}$% and four gives 6.25%. Therefore 90% reduction, i.e. 10% emerging, requires a thickness between three and four times the half-thickness 3 mm, i.e. between 9 mm and 12 mm. In fact this argument does tell us that the only answer which can be correct is **C**.

The alternative approach is to use Equation 42.10, as follows:

$$F = \frac{F_0}{2^Y} \quad \text{and} \quad Y = \frac{t}{T}$$

F is the flux surviving and is to be 10%, F_0 is the original flux, i.e. 100%, half-thickness $T = 3.0$ mm and t is the thickness to be found.

$$\therefore \quad 10 = \frac{100}{2^Y} \quad \therefore \quad 2^Y = 10$$

$$\therefore \quad \log_{10} 2^Y = \log_{10} 10 = 1$$

But $\log 2^Y = Y \log 2$ so that

$$Y \log 2 = 1 \quad \therefore \quad Y = \frac{1}{\log 2} = \frac{1}{0.3010} = 3.22$$

$$\therefore \quad \frac{t}{T} (= Y) = 3.22$$

$$\therefore \quad t = 3.22 \times 3.0\ \text{mm}$$

$$= 9.66\ \text{mm} \quad \text{or} \quad 10\ \text{mm}$$

Answer

C.

Exercise 143

1 A certain radioactive nucleus $^{222}_{86}X$ decays by alpha emission. What are the nucleon number and the proton number of the daughter nucleus?

2 If the γ-radiation flux per unit area is acceptably low at a distance of 1.0 m from a certain small source, at what distance from the source will the same radiation level be obtained when the source is enclosed in a container whose walls have a thickness equal to 2 half-thicknesses?

3 A very old wooden tool is found to have a carbon-14 activity of 0.07 Bq per gram of total carbon content. The half-life for carbon-14 is 5570 years and the carbon dioxide of the atmosphere, from which the wood's carbon was obtained, gives 19 disintegrations per minute per gram of carbon. What is the age of the tool?

Exercise 144: Questions of A-level standard

1 The count rate recorded by a Geiger counter 10 cm from a small source of radiation drops from 4000 to 1000 counts per minute in two hours. What will be the count rate per minute recorded one hour later by the counter at 20 cm from the source?

A 62.5 **B** 125 **C** 250 **D** 375
E 500 [L 83]

2 Radon-222 is a radioactive gas for which the decay constant is $2.1 \times 10^{-6}\ \text{s}^{-1}$. If the initial decay rate is $5.6 \times 10^{10}\ \text{s}^{-1}$, calculate (a) the initial number of radioactive atoms present and (b) the time which will elapse before the activity is reduced to one-quarter of its initial value. ($\ln 2 = 0.693$.) [AEB 81]

3 In order to determine the volume of a patient's blood, 2 ml of a solution containing a radioactive sodium isotope of half-life 15 hours is injected into his bloodstream. The activity of the solution injected is 3000 counts per minute, and the activity of a blood sample of volume 12 ml taken 3 hours later is found to be 6 counts per minute.

(a) Calculate the volume of the patient's blood given by these results.

(b) Discuss any assumptions made in the determination and suggest how they might be checked. [O 81]

4 The potassium isotope $^{42}_{19}K$ has a half-life of 12 h, and disintegrates with the emission of a γ-ray to form the calcium isotope $^{42}_{20}Ca$. What other radiation besides the γ-rays must be emitted? How many electrons, protons and neutrons are there in an atom of the calcium isotope?

The amount of radiation received in unit time by a person working near a radioactive source, commonly called the dose rate, is measured in rem h^{-1}. The safety regulations forbid dose rates in excess of 7.5×10^{-4} rem h^{-1}. The γ dose rate from the $^{42}_{19}K$ source is found to be 3×10^{-3} rem h^{-1} at a distance of 1 m. What is the minimum distance from this source at which it is safe to work?

After how long will it be safe to work at a distance of 1 m from this source? [SUJB 79]

5 The relative atomic mass of rubidium (atomic number 37) is found by chemical methods to be 85.5. Mass spectroscopy shows that rubidium has isotopes of mass 85 and 87. How many atoms of $^{87}_{37}Rb$ are present in 5.0×10^{-6} kg of the natural element? (Mass of H atom $= 1.66 \times 10^{-27}$ kg.)

What elementary particles, and how many of each, comprise the $^{87}_{37}Rb$ atom?

$^{87}_{37}Rb$ is β-radioactive, with a half-life of 1.6×10^{18} s. What is the rate of emission of β-particles from 5.0×10^{-6} kg of the natural element? Note: $dN/dt = 0.693\, N/T_{1/2}$, where $T_{½}$ is the half-life.

[SUJB 82]

6 In a certain meteorite rock the ratio of the number of uranium $^{238}_{92}U$ atoms to the number of lead $^{206}_{82}Pb$ atoms attributable to radioactive decay of the $^{238}_{92}U$ is 1.17. Given that the half-life of the uranium atoms is 4.5×10^{9} year, estimate the age of the rock.

43 NUCLEAR REACTIONS

The Einstein mass–energy relationship

The mass m of any body, defined by the equation $F = ma$ (see Chapter 7), and the total energy E of the body are related by the equation

$$E = mc^2 \qquad (43.1)$$

where c = velocity of light in vacuum.

For a body at rest m is the rest mass and the corresponding energy E (= mc^2) is the rest mass energy. If the body were then to move, E would increase on account of the body's acquiring kinetic energy and so m increases also.

In nuclear reactions the energy changes are sufficient for the mass changes to be significant.

If a nuclear change, i.e. reaction, occurs with no supply of energy from outside, then we have a spontaneous reaction such as a radioactive decay. The potential energy of the nucleus must fall, and so the rest mass must decrease. The energy lost escapes from the nucleus usually as the kinetic energy of an emitted particle or as a γ-photon, or both.

The loss of rest mass in a reaction is denoted by Q (the 'Q value' of the reaction).

Example 1

Alpha-particle emission from $^{238}_{92}U$ can be described by the equation

$$^{238}_{92}U = {}^{234}_{90}Th + {}^{4}_{2}He + Q$$

where $^{234}_{90}Th$ is the product or daughter nucleus, $^{4}_{2}He$ represents the α-particle and Q is the energy that becomes the kinetic energy of the emitted α-particle. The masses of the *atoms* are 238.0509 u, 234.0436 u and 4.0026 u for ^{238}U, ^{234}Th and ^{4}He respectively. Calculate the value of Q (a) in joules, (b) in electron-volts. Take $1\,u = 1.7 \times 10^{-27}$ kg, speed of light $c = 3.0 \times 10^8\,m\,s^{-1}$ and electron charge $e = 1.6 \times 10^{-19}$ C.

Method

(By using the atomic masses rather than nuclear masses we make the same error — 92 electrons included in this example — on both sides of the equation and the errors cancel.*)

The total rest mass on the right-hand side of the equation is

$$234.0436 + 4.0026 \text{ or } 238.0462\,u$$

The rest mass for the left-hand side is 238.0509 u.

The difference, which is the mass of the energy Q, is

$$238.0509 - 238.0462 \quad \text{or} \quad 0.0047\,u$$

In kilograms this is $0.0047 \times 1.7 \times 10^{-27}$ or 0.80×10^{-29} kg. In joules it is, using $E = mc^2$,

$$E = 0.80 \times 10^{-29} \times (3.0 \times 10^8)^2$$
$$= 7.2 \times 10^{-13}\,J$$

In electron-volts, using 1 eV = e joules, we get

$$E = \frac{7.2 \times 10^{-13}}{1.6 \times 10^{-19}}$$
$$= 4.5 \times 10^6\,eV$$

(*Check*: The number of eV is large because the eV is such a small energy.)

Answer

(a) 7.2×10^{-13} J, (b) 4.5 MeV.

**Exceptions need not concern us.*

Equations for nuclear reactions

As in the above example, the right-hand side of the equation contains the reaction products. On the left is the particle or particles at the start. The equation can also include Q, the net loss of rest mass energy or gain in other energy.

Note too that the sum of the mass numbers on left and right must be equal (there is no way in which nucleons can be created or destroyed) and so the proton number must add up to the same total on each side.

To conform with these rules we write the negative electron, the positive electron and the neutron as ${}^{0}_{-1}e$, ${}^{0}_{1}e$ and ${}^{1}_{0}n$ respectively.

Method

The total rest mass of the reacting particles is

$$2.0141 + 6.0155 \quad \text{or} \quad 8.0296\,\text{u}$$

The total rest mass of the products is

$$2 \times 4.0026 \quad \text{or} \quad 8.0052\,\text{u}$$

The kinetic energy of the products is Q, given by

$$Q = 8.0296 - 8.0052$$

$$= 0.0244\,\text{u}$$

$$= 0.0244 \times 1.66 \times 10^{-27}\,\text{kg} = 4.0 \times 10^{-29}\,\text{kg}$$

and, using $E = mc^2$,

$$E \text{ or } Q = 4.0 \times 10^{-29} \times (3.00 \times 10^8)^2$$

$$= 3.6 \times 10^{-12}\,\text{J}$$

The energy *per α-particle* is half of 3.6×10^{-12} J, i.e. 1.8×10^{-12} J.

Answer

1.8×10^{-12} J.

Induced nuclear reactions

Some nuclear reactions are brought about by the nucleus interacting with a bombarding particle. These reactions are not ruled out by a negative value of Q in the reaction equation because the bombarding particle can arrive at the nucleus with a useful amount of kinetic energy.

It seems that these reactions consist of two stages, the first where the bombarding particle joins into the nucleus to form a compound nucleus and the second where the new nucleus reorganises itself, usually emitting a particle of some kind.

Note that all nuclear reactions must of course conserve momentum as well as mass/energy.

Example 2

When a deuteron of mass 2.0141 u and negligible kinetic energy is absorbed by a lithium nucleus of mass 6.0155 u, the compound nucleus disintegrates spontaneously into two α-particles, each of mass 4.0026 u. Calculate the energy, in joules, given to each α-particle.
($1\,\text{u} = 1.66 \times 10^{-27}$ kg. Speed of light in vacuum $= 3.00 \times 10^8\,\text{m s}^{-1}$.) [L 81]

Example 3

A uranium atom (U), travelling with a velocity of $5.00 \times 10^5\,\text{m s}^{-1}$ relative to the containing tube, breaks up into krypton (Kr) and barium (Ba). The krypton atom is ejected directly backwards at a velocity of $2.35 \times 10^6\,\text{m s}^{-1}$ relative to the barium after separation. With what velocity does the barium atom move forward relative to the tube? You may assume that no other particles are produced and that relativistic corrections are small. What is the velocity of the krypton atom relative to the containing tube? (Kr, 95.0 u; Ba, 140 u; U, 235 u.)

Method

Let the velocities of Kr and Ba atoms after separation be v_K *backwards* and v_B *forwards* relative to the tube. Then

$$v_K + v_B = 2.35 \times 10^6 \qquad \text{(i)}$$

Also, conservation of momentum requires

$$m_U v_U = m_B v_B - m_K v_K$$

$$\therefore \quad 235 \times 5.00 \times 10^5 = 140 v_B - 95 v_K \qquad \text{(ii)}$$

We now have two simultaneous equations (see Chapter 2) and the two unknowns can therefore be worked out. We shall do this by writing $v_K = 2.35 \times 10^6 - v_B$ from (i) and substitute this expression for v_K in (ii) so as to eliminate v_K.

$$235 \times 5.00 \times 10^5 = 140v_B - 95(2.35 \times 10^6 - v_B)$$

$$\therefore \quad 1175 \times 10^5 = 140v_B - 2232.5 \times 10^5 + 95v_B$$

$$\therefore \quad 235v_B = 3407 \times 10^5$$

$$\therefore \quad v_B = 14.5 \times 10^5 \text{ m s}^{-1}$$

and v_K $(= 2.35 \times 10^6 - v_B$ from (i))

$$= 23.5 \times 10^5 - 14.5 \times 10^5$$

$$= 9.00 \times 10^5 \text{ m s}^{-1}$$

Answer

14.5×10^5 m s^{-1}, 9.00×10^5 m s^{-1}.

Nuclear binding energy

Nucleons hold together in a nucleus only because of attractive *nuclear forces*. The potential energy of the nucleons is lower when they are in a nucleus close together than when they are separated from each other. When this potential energy decreases, the rest mass energy decreases and energy Q is released in the form of γ-radiation or kinetic energy. Conversely, to break up a nucleus into separate nucleons requires an equal energy to be provided, and this is called the binding energy of the nucleus. The binding energy of a nucleus divided by its nucleon number gives the 'binding energy per nucleon'.

Example 4

The mass of the isotope $^7_3$Li is 7.018 u. Find its binding energy, given that the mass of $^1_1$H is 1.008 u, the mass of the neutron = 1.009 u and 1 u = 931 MeV.

[WJEC 79, p]

Method

The $^7_3$Li nucleus contains three protons ($^1_1$H nuclei) and four neutrons. The total mass of these particles is

$$(3 \times 1.008) + (4 \times 1.009) \quad \text{or} \quad 7.060 \text{ u}$$

The mass of $^7_3$Li is 7.018 u.

The Q value (or energy released during formation) is the difference between the above masses.

$$\therefore \quad Q = 7.060 - 7.018 \quad \text{or} \quad 0.042 \text{ u}$$

Converted to MeV we have 0.042×931 or 39 MeV.

Answer

39 MeV.

Nuclear fission

The binding energy per nucleon is smaller (potential energy is bigger) for nucleons in large nuclei such as uranium nuclei and is greatest for nucleons in medium sized nuclei. Consequently splitting, or fission, can occur where a large nucleus divides into two approximately equal-size nuclei, and energy is released.

Example 5

Calculate the energy released in the following fission reaction

$$^{235}_{92}\text{U} + ^1_0\text{n} = ^{92}_{36}\text{Kr} + ^{141}_{56}\text{Ba} + 3^1_0\text{n} + Q$$

The atomic masses are

U 235	235.0439 u
neutron	1.0087 u
Kr 92	91.8976 u
Ba 141	140.9136 u

(1 u = 931 MeV.)

Method

The sum of the rest masses on the left-hand side of the equation is

$$235.0439 + 1.0087 \quad \text{or} \quad 236.0526 \text{ u}$$

and on the right-hand side is

$$91.8976 + 140.9136 + (3 \times 1.0087)$$

or 235.8373 u

Q is the difference between the above two masses.

$$\therefore \quad Q = 236.0526 - 235.8373 = 0.2153 \text{ u}$$

and, converted to MeV, we have 0.2153×931 or 200 MeV.

Answer

200 MeV.

Nuclear fusion

This is the joining together of two small nuclei to produce a nucleus for which the binding energy per nucleon is larger. Energy is released as kinetic energy of a product particle (if any), or otherwise as gamma radiation.

Example 6

Calculate the energy in MeV released by fusing two protons and two neutrons to form a helium nucleus. (This reaction is difficult to achieve.)

The atomic masses of hydrogen (^1_1H) and helium (^4_2He) are 1.007 825 u and 4.002 604 u respectively. The mass of the neutron is 1.008 665 u. 1 u = 931 MeV.

Method

The reaction is

$$2\,^1_1H + 2\,^1_0n = \,^4_2He + Q$$

The sum of the masses for the left-hand side is

$$(2 \times 1.007\,825) + (2 \times 1.008\,665)$$

or $\quad 4.032\,98$ u

(In addition to the protons, neutrons and helium nucleus, two electrons are included in each side of the equation, one in each hydrogen atom and two in the helium atom. These electron masses cancel.)

Subtracting the mass on the right from that on the left gives Q.

$$\therefore \quad Q = 4.032\,98 - 4.002\,604 = 0.030\,376 \text{ u}$$

and converted to MeV,

$$Q = 0.030\,376 \times 931 = 28.3 \text{ MeV}$$

per fusion.

Answer

28.3 MeV.

α-particle scattering and the structure of the nucleus

If an α-particle approaches the (positive) nucleus of an atom, it may be scattered, i.e. deflected (by the repulsion of the positive charges) and the nucleus is unchanged. It is an elastic collision or interaction.

The scattering is found to be consistent with the inverse square law of repulsion between the two positive charges. This shows that the particle does not penetrate the nucleus. The distance r to which an α-particle of charge $2e$ and speed v can approach the centre of the nucleus can be calculated. If Z is the atomic number of the nucleus, its charge is Ze. Equating the kinetic energy of the particle to the work done in approaching the positive nucleus:

$$\tfrac{1}{2}mv^2 = \begin{pmatrix}\text{Potential at}\\ \text{distance } r\\ \text{from nucleus}\end{pmatrix} \times \begin{pmatrix}\text{Charge on}\\ \alpha\text{-particle}\end{pmatrix}$$

$$= \frac{Ze}{4\pi\varepsilon_0 r} \times 2e \qquad (43.2)$$

(We use here the equation $V = Q/4\pi\varepsilon_0 r$, see Chapter 30.)

In experiments, r is found to be very small, showing that the nucleus is very small and agreeing with the planetary model of the atom.

Exercise 145

1 An α-particle having a speed of 1.2×10^6 m s^{-1} collides with a stationary proton. The proton acquires a speed of 1.9×10^6 m s^{-1} in the direction in which the α-particle was travelling.

What is the speed of the α-particle immediately after the collision?

How much kinetic energy is gained by the proton in the collision? How much kinetic energy is lost by the α-particle? (Mass of α-particle = 6.64×10^{-27} kg, mass of proton = 1.66×10^{-27} kg.)

2 Calculate the energy released per fission in the following fission reaction:

$$^{235}_{92}U + \,^1_0n \rightarrow \,^{95}_{42}Mo + \,^{139}_{57}La + 2\,^1_0n + 7\,^{0}_{-1}e$$

(^{235}U, 235.0439 u; neutron, 1.0087 u; ^{95}Mo, 94.9060 u; ^{139}La, 138.9060 u; 1 u = 931 MeV. The masses of the seven electrons may be neglected here.)

3 Calculate the energy in MeV released in the fusion reaction

$$^2_1H + \,^2_1H \rightarrow \,^3_1H + \,^1_1H + Q$$

The atomic masses are

deuterium	^2_1H,	2.014 102 u
tritium	^3_1H,	3.016 049 u
hydrogen	^1_1H,	1.007 825 u

(1 u = 931 MeV.)

4 Use equation 43.2 to calculate the closest approach possible for an 8 MeV α-particle to a gold nucleus, given that the atomic number of gold is 79, the electronic charge is 1.60×10^{-19} C, the permittivity of free space is 8.855×10^{-12} F m^{-1}.

Exercise 146: Questions of A-level standard

1 The nucleus of a radioactive isotope of thorium decays by the emission of an α-particle ($A = 4$) to an isotope of radium ($A = 226$). Calculate the ratio of the speeds of the α-particle and radium nucleus, and hence find the recoil kinetic energy in MeV of the radium nucleus if the ejected α-particle has a kinetic energy of 4.61 MeV. Assume the thorium nucleus was at rest. [L 80]

2 What is meant by the *binding energy* of the nucleus? Using the values given below calculate its value for $^{238}_{92}\text{U}$.
The following atomic masses may be used:

$^{238}_{92}\text{U} = 238.0508\ \text{u}$
$^{1}_{0}\text{n} = 1.008\,665\ \text{u}$ $^{1}_{1}\text{p} = 1.007\,825\ \text{u}$

(1 u = 931 MeV.) [L 77, p]

3 When a nucleus of deuterium (hydrogen-2) fuses with a nucleus of tritium (hydrogen-3) to give a helium nucleus and a neutron, 2.88×10^{-12} J of energy are released.

The equation of the reaction is

$$^{2}_{1}\text{H} + {}^{3}_{1}\text{H} \rightarrow {}^{4}_{2}\text{He} + {}^{1}_{0}\text{n}$$

Calculate the mass of the helium nucleus produced.

Mass of deuterium nucleus $= 3.345 \times 10^{-27}$ kg.
Mass of tritium nucleus $= 5.008 \times 10^{-27}$ kg.
Mass of a neutron $= 1.675 \times 10^{-27}$ kg.
Speed of light in vacuum $= 3.00 \times 10^{8}\ \text{m s}^{-1}$.
[L 83]

4 Calculate the closest possible approach of 4.0 MeV α-particles to carbon nuclei.

Proton number for carbon = 6.
Proton number for α-particles = 2.
Electronic charge e $= 1.6 \times 10^{-19}$ C.
Permittivity of free space $\varepsilon_0 = 8.9 \times 10^{-12}\ \text{F m}^{-1}$.

Section I Calculations Involving Graphs

44 QUESTIONS INVOLVING GRAPHS

Introduction

The way in which the variation of one quantity affects another can be expressed as a graph as an alternative to an equation. One advantage of a graph is the quickness with which its information can be grasped.

Unless other symbols are preferred, we use x to denote the quantity plotted on the horizontal axis (abscissa) and y for the other quantity (ordinate).

The units employed for x and y and/or the scale markings on the axes must be chosen to approximately fill the space available for the graph. The origin is the place where the axes meet, and often we can make both x and y start from zero at the origin, i.e. the origin is (0, 0). However there may be good reason to do otherwise.

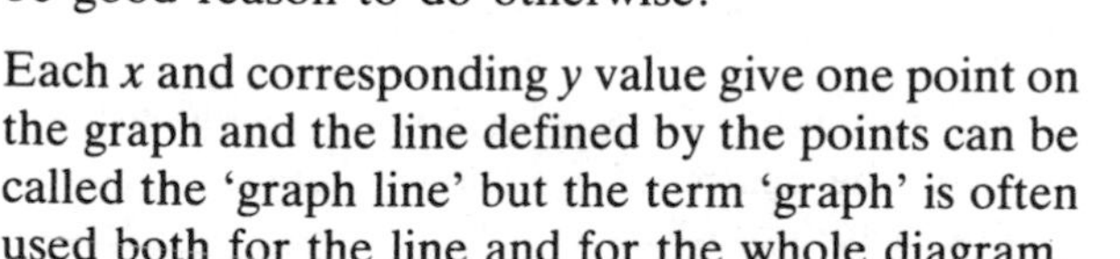

Each x and corresponding y value give one point on the graph and the line defined by the points can be called the 'graph line' but the term 'graph' is often used both for the line and for the whole diagram.

Some common graphs

If $y \propto x$, i.e. $y = mx$ where m is a constant (not affected by variation of x and y), then we get a straight line which passes through the point $x = 0$, $y = 0$ (Fig 44.1a). Other common examples are also shown in Fig 44.1.

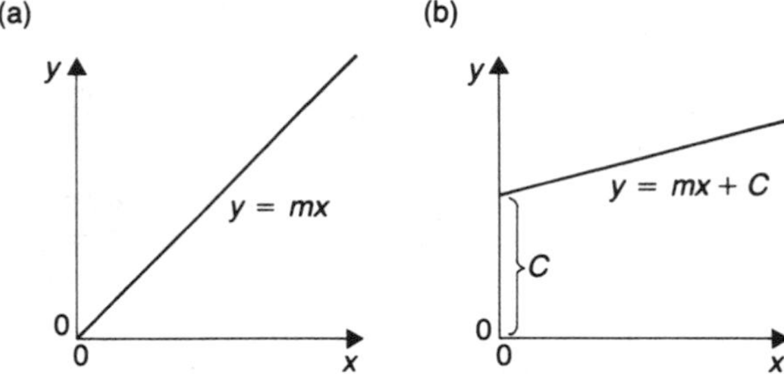

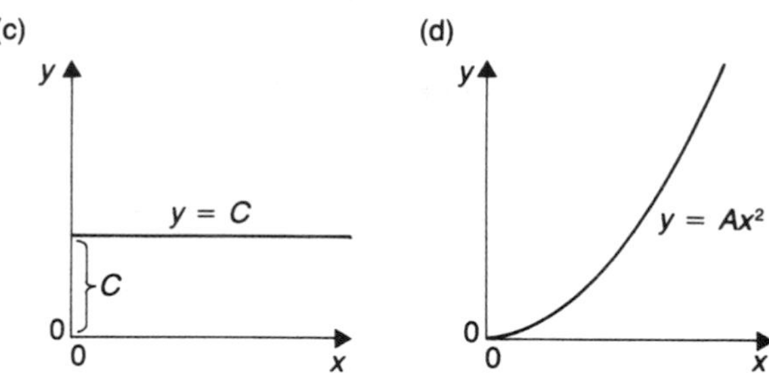

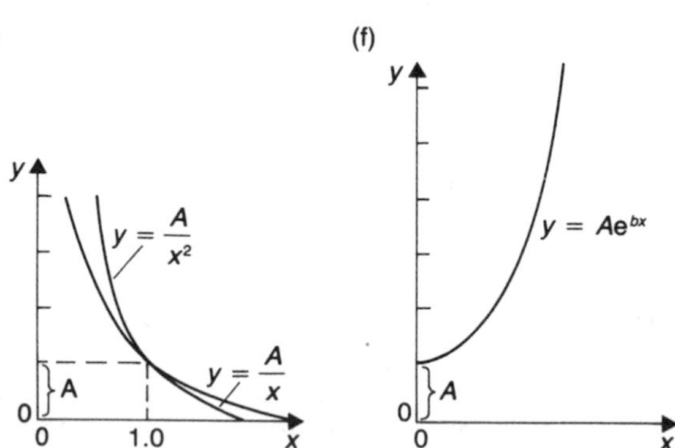

Fig 44.1 ***cont.***

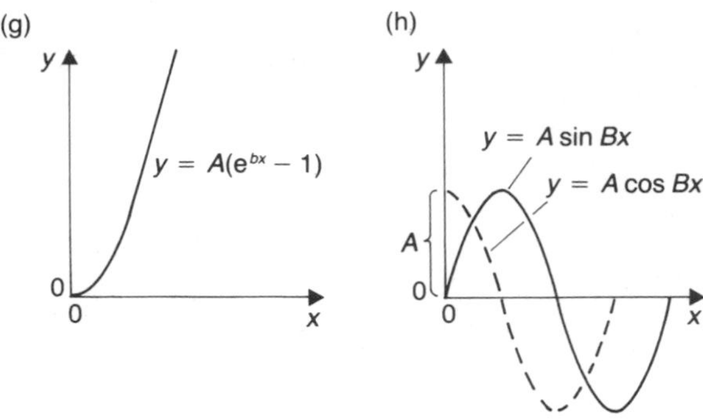

Fig 44.1 Some common graphs

The most important of these examples is $y = mx + C$.

To decide the general shape of the graph for a given equation it may be necessary to consider a few simple values of x, like $x = 0, 1, 2, \infty, -1$. The corresponding y values are worked out and are quickly plotted in a rough graph. For example $y = Ax^2 + B$ gives $y = B$ for $x = 0$, $A + B$ for $x = 1$, $4A + B$ for $x = 2$ and ∞ for $x = \infty$, $A + B$ again for $x = -1$. The resulting rough sketch is shown in Fig 44.2.

It should be noted that if only even powers of x are concerned, then y values for negative and positive x values are the same, as we see in the example below.

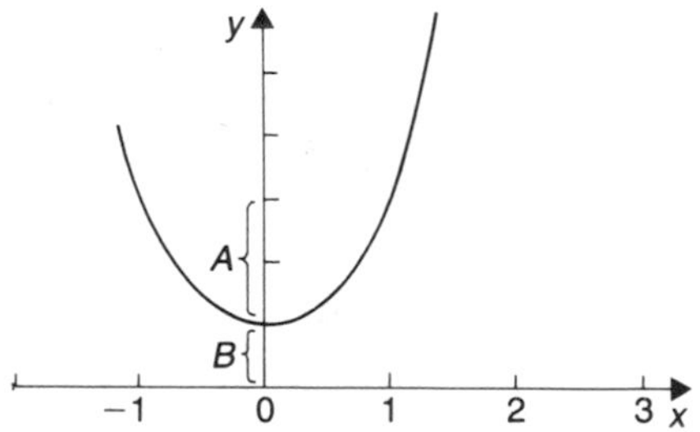

Fig 44.2 Graph of $y = Ax^2 + B$

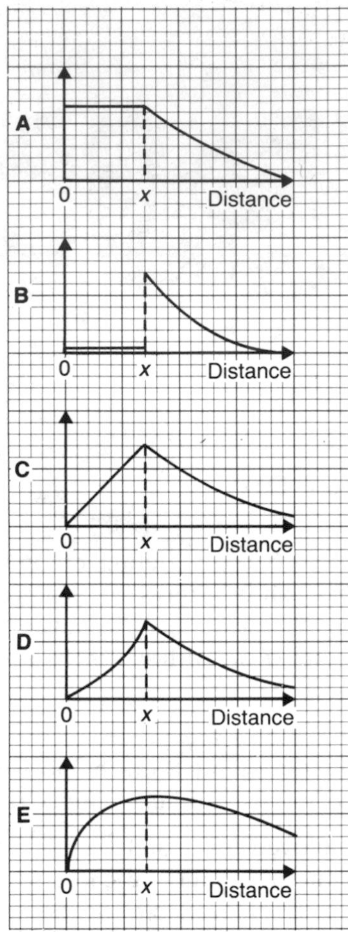

Fig 44.3 Graphs for Examples 1–3

Example 1

The five graphs (Fig 44.3) labelled **A** to **E** show the variation of some physical quantity with distance from a point O.

Which of them best illustrates:

(a) the variation of electrical intensity with distance from the centre of a hollow, charged, spherical conductor of radius x;

(b) the variation of electrical potential with distance from the centre of a hollow, charged, spherical conductor of radius x;

(c) the variation with distance of the velocity of a train accelerating uniformly from rest for a distance x and then being acted upon by a uniform braking force? [L 83]

Method

(a) We know that the intensity E is zero inside the conductor, is large just outside and then decreases with $E \propto 1/\text{distance}^2$. The decrease must resemble Fig 44.1e. The answer is **B**.

(b) We know $V = Q/4\pi\varepsilon r$ (see Equation 30.5) outside the sphere and V must fall like Fig 44.1e. V is highest at the surface of the sphere and is equally high inside the sphere. The answer is **A**.

(c) For the uniform acceleration from rest ($u = 0$) we have $v^2 = 2as$ (see Equation 7.3).

Suppose, for simplicity, that $2a = 1$ and consider simple values 0, 1, 2, 3 for s:

s	0	1	2	3
v^2	0	1	2	3
v	0	1	1.4	1.7

The rate of increase of v with s becomes less as s increases. Only answer **E** agrees with this.

Answer

(a) **B** (b) **A** (c) **E**.

Small changes

If symbols x and y are used to denote the values of two related quantities, then it is usual to denote small corresponding increases in x and y by δx and δy (where δ is pronounced 'delta'). Small decreases would be $-\delta x$ and $-\delta y$.

If now x were made extremely small (too small to measure it), i.e. $\delta x \to 0$, we denote the value of $\delta y/\delta x$ when $\delta x \to 0$ by dy/dx (pronounced 'dee y by dee x').

Slope of a graph

The gradient or slope s of a graph at a chosen point is

$$s = \frac{\text{Small change in } y}{\text{Corresponding small change in } x} = \frac{\delta y}{\delta x}$$

For a straight line graph s is the same for all parts of the graph, and so

$$s = \frac{\text{Any change in } y}{\text{Corresponding change in } x} = \frac{a}{b} \quad \text{in Fig 44.4a.}$$

Note that for $y \propto x$, we can write

$$s = \frac{y}{x}$$

To measure s at a point on a curve it is not too difficult to draw a tangent to the curve and measure its slope, which equals the required slope (Fig 44.4b).

Note that for measuring the slope of a graph the choice of origin is not important.

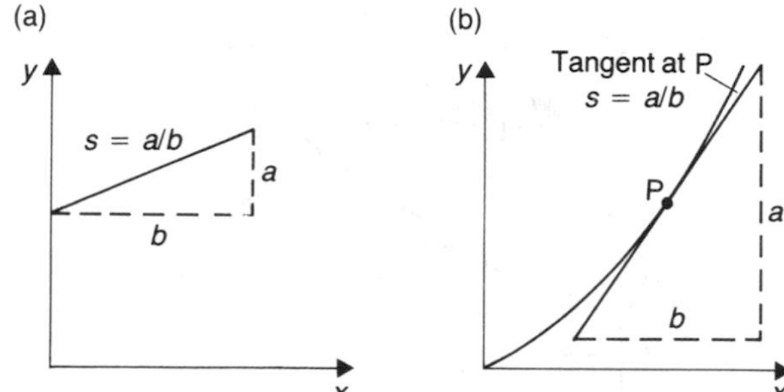

Fig 44.4 Slope of a graph

The intercepts of a graph

The y intercept is the value of y when $x = 0$, and the x intercept is x when $y = 0$. Often we need to make use of only one of the two intercepts and we normally use the y intercept. The graph $y = mx + C$ has a y intercept C (Fig 44.1b). For a straight-line (i.e. linear) graph the x intercept and y intercept are related by

$$\text{Slope } s = \frac{\text{Magnitude of } y \text{ intercept}}{\text{Magnitude of } x \text{ intercept}}$$

(see Fig 44.5)

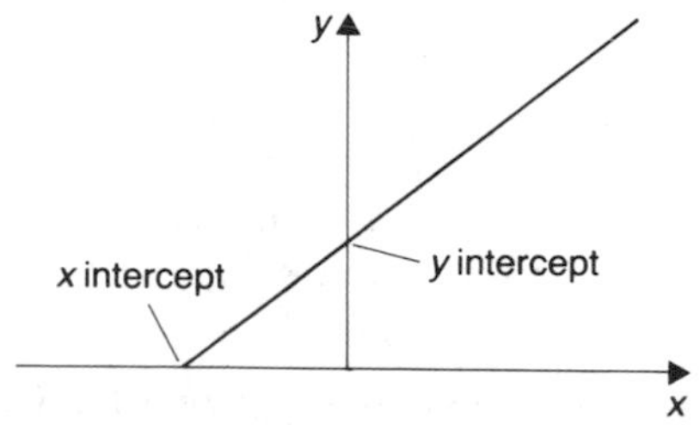

Fig 44.5 Intercepts

Example 2

The following values of resistance R and corresponding Celsius temperature θ conform to the equation $R = R_0(1 + \alpha\theta)$ where R_0 and α are constants. Plot a straight line graph from these results and hence determine R_0 and α.

θ/°C	10	30	60	90
R/Ω	10.3	11.0	12.0	13.0

Method

The graph is plotted as shown in Fig 44.6

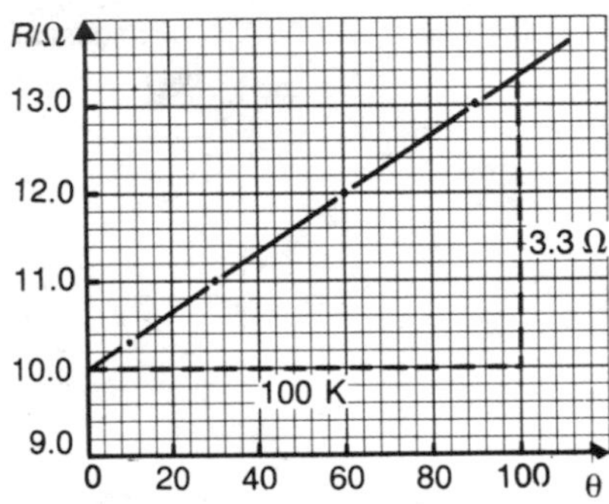

Fig 44.6 Graph for Example 2

To find R_0 we first note the resemblance between $R = R_0(1 + \alpha\theta)$ and $y = C + mx$. If we rewrite the R equation as $R = R_0 + R_0\alpha\theta$, it is seen that R_0 is the intercept on the R axis and $R_0\alpha$ is the slope.

From the graph the slope is 3.3/100 or $0.033\,\Omega\,K^{-1}$ and the intercept is $10.0\,\Omega$. So $R_0 = 10.0\,\Omega$ and $R_0\alpha = 0.033\,\Omega\,K^{-1}$, whence α $(= 0.033/R_0)$ $= 0.0033\,K^{-1}$.

Answer

$R_0 = 10.0\ \Omega$, $\alpha = 0.0033\ K^{-1}$

Plotting log *y* versus log *x*

If $y = Ax^p$ then

$$\log_{10}y = \log_{10}A + \log_{10}x^p$$
$$= \log_{10}A + p\log_{10}x.$$

The same is true if we use $\log_e$ (i.e. $\ln y$, etc.) instead.

We can plot $\log_{10}y$ versus $\log_{10}x$ to obtain a straight line graph whose slope equals the power p. The intercept is $\log_{10}A$ from which A can be calculated.

Example 3

Corresponding values of volume (V) and pressure (P) are given below for a fixed mass of air at constant temperature. Given that $P = AV^p$, where A and p are constants, deduce the values of p and A.

P/kPa	100	200	300	400	500	600
V/cm^3	100	50	32	25	20	16

Method

Plotting V versus P gives a graph like Fig 44.1e and the power p would not be determined. Now $\log P = \log A + p \log V$ and p is the slope of the $\log P$ versus $\log V$ graph. So we need log values:

$\log_{10}P$	2.0	2.3	2.5	2.6	2.7	2.8
$\log_{10}V$	2.0	1.7	1.5	1.4	1.3	1.2

The graph is plotted in Fig 44.7.

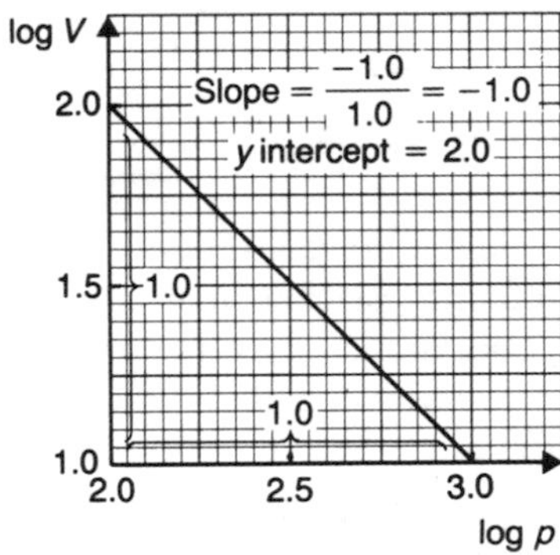

Fig 44.7 Graph of slope for Example 3

The slope is found to be −1.0, i.e. $p = -1$ (Boyle's law of course).

The y intercept is $\log_{10}A$ and equals 2.0, i.e. $A = 100$.

The unit for A is the same as for P/V^p, i.e. pascal/cm^{3p} or Pa cm^{-3p}. Here $p = -1$, so that $A = 100$ Pa cm^3.

Answer

$p = -1$, $A = 100$ Pa cm^3.

Plotting log *y* or ln *y* versus *x*

If $dy/dx = my$, then $\ln y = mx + C$ or

$$y = e^{mx+C} = e^C \times e^{mx}$$

Also when $x = 0$, $y = e^C$ denoted by y_0. Thus $y = y_0 e^{mx}$.

(The radioactive decay formulae $dN/dt = -\lambda N$ and $N = N_0 e^{-\lambda t}$ fit this description.)

So a graph of $\ln y$ versus x can be plotted and the slope m can be determined and hence, for example, $-\lambda$ can be deduced. Alternatively $\log_{10} y$ may be plotted, which equals $2.303 \ln y$, so that $\log_{10} y = 2.303mx + 2.303C$.

Example 4

For a hot object cooling in a draught the excess temperatures θ recorded at times t were as follows:

t min^{-1}	0	5	10	15	20	25	30
θ/°C	60	50	40.5	31.5	25.8	20.1	16.5

Show that the results agree with

$$\frac{d\theta}{dt} = -A\theta \quad \text{or} \quad \theta = \theta_0 e^{-At}$$

and deduce the values of the constants A and θ_0.

Method

A graph of $\ln \theta$ versus t should give a straight line because $\ln \theta = \ln \theta_0 - At$.

t min^{-1}	0	5	10	15	20	25	30
$\ln \theta$	4.09	3.91	3.70	3.45	3.25	3.00	2.80

The graph obtained is shown in Fig 44.8.

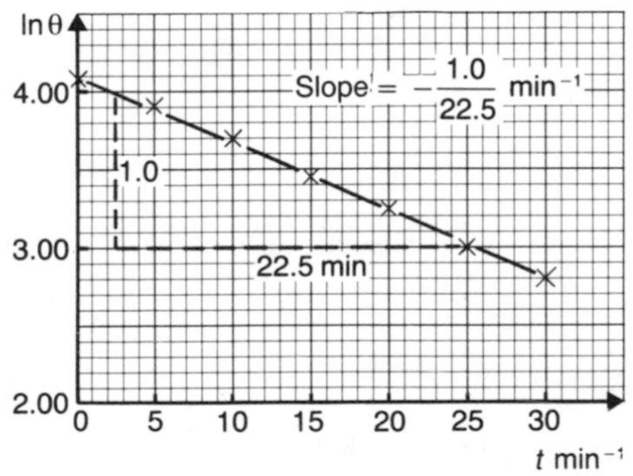

Fig 44.8 Graph for Example 4

The slope is found to be -0.044 min^{-1} so $A = 0.044$ min^{-1}.

When $t = 0$, $\theta = \theta_0$ and is equal to 60 °C.

Answer

$A = 0.044$ min^{-1}, $\theta_0 = 60$ °C.

Plotting graphs from experimental results

Once the units and scales have been chosen, the axes should be labelled, e.g. t/min (or time/min) for the abscissa in Fig 44.8. As pointed out in Chapter 3, t/min is a number. This agrees with our marking the axis with numbers (0, 5, 10, etc). t/min could be replaced by t/60 s or $t \times 60$/h if s or h are preferred as units and this does not cause ambiguity. Similarly if a graph requires, say, distance d to be plotted between 0 and 5×10^{-9} m, we might like to mark off the axis with 50 steps each of 10^{-10} m but labelled 0 to 50 and the axis would be labelled with '$d/10^{-10}$ m' or '$d \times 10^{10}$/m'.

When the points are plotted they will, we hope, indicate the graph line, but will not necessarily lie exactly on a perfectly straight line or smooth curve because of small experimental errors. The best straight line or smooth curve is drawn as close as possible to all the points.

Slope of a graph as a method of averaging

In Fig 44.9 the resistance of the conductor is $R = V/I$.

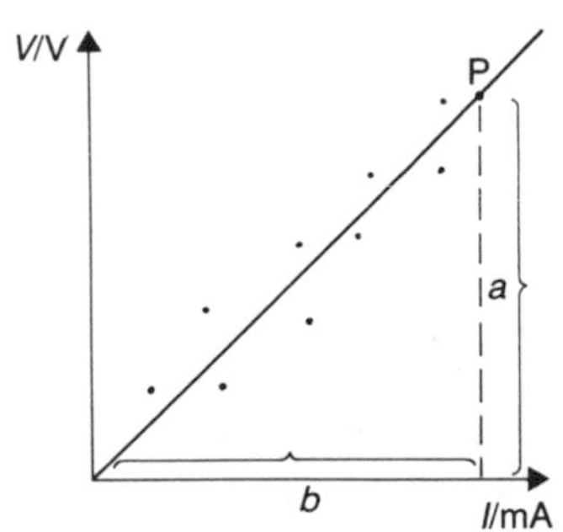

Fig 44.9 Graph of *V* versus *I* for an ohmic conductor

No particular pair of results can be regarded as the correct one for calculating R but the line drawn is an average for the results. For any point on the line such as P the x and y values (i.e. the coordinates of P) can be regarded as average values and R is calculated as $R = a/b$, which is the slope of the graph.

Area under a graph

Measuring the area under a graph can be useful. For example the area under a velocity—time graph equals the distance covered. The area under a graph of electric intensity versus distance equals the potential difference across the distance covered.

Example 5

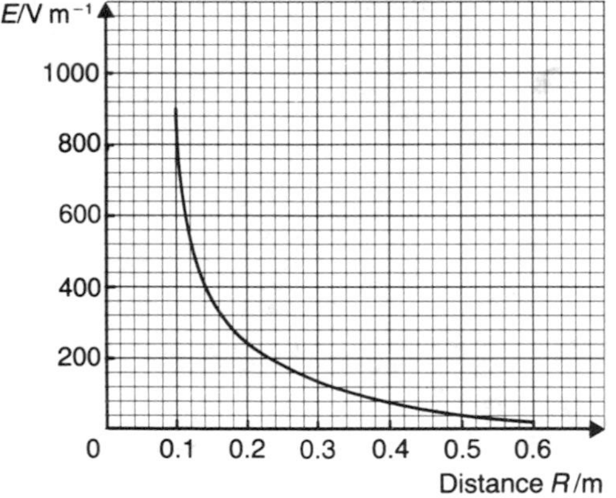

Fig 44.10 Graph for Example 5

Fig 44.10 shows a graph of intensity versus distance from a point charge. Obtain a value for the potential 0.15 m from the charge.

Method

The number of small squares of the graph paper is counted for the area under the graph between $R = 0.15$ m and $R = 0.6$ m or more. The answer is about 80. The area of each of these squares is 40 V m^{-1} by 0.02 m, i.e. 0.8 V.

$$\therefore \quad \text{Total area is } 80 \times 0.8 = 64\text{ V}$$

Answer

64 V approximately.

Exercise 147

1

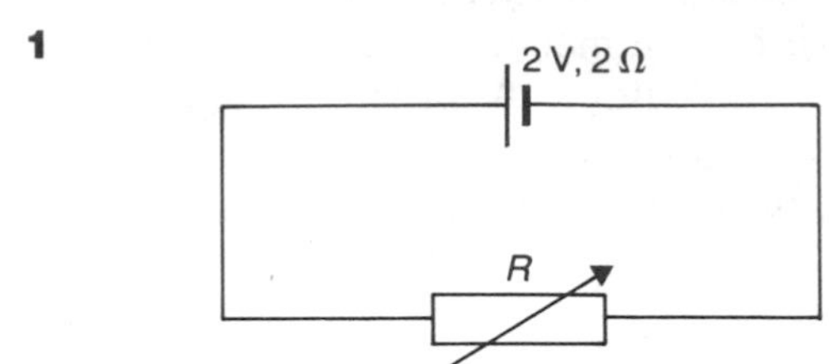

Fig 44.11 Diagram for Question 1

The power P dissipated in the variable resistance R (Fig 44.11) is given by $V \times I$ where V is the PD across R and I is the current through it. Which of the graphs **A** to **E** (below) describes correctly the variation of P with R?

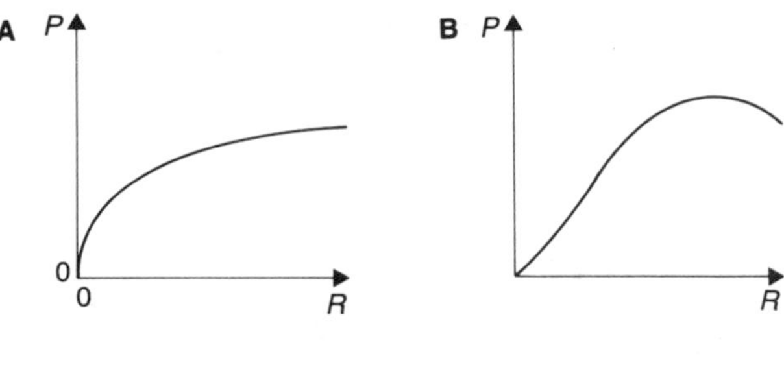

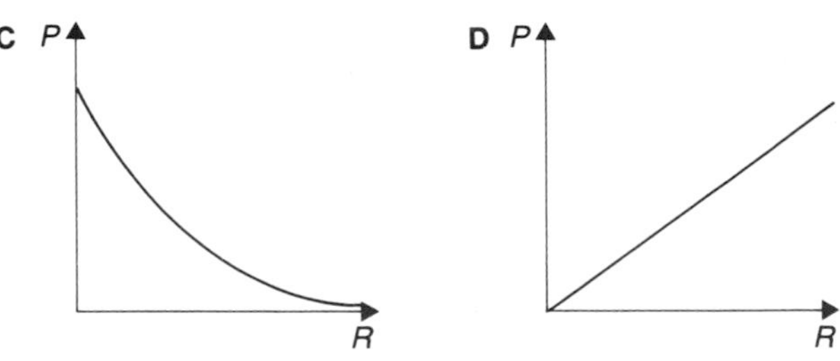

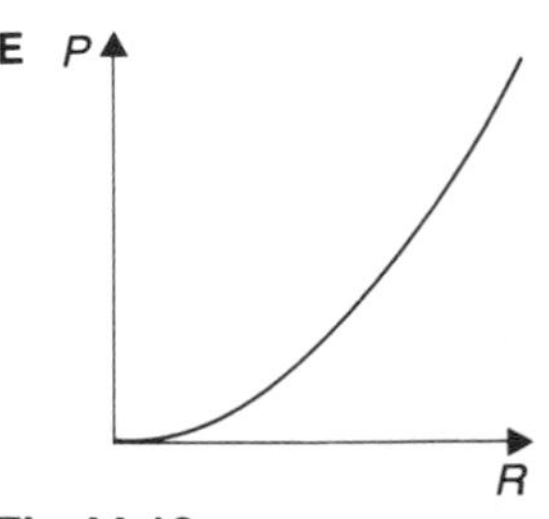

Fig 44.12

2 Plot the graph of $y = 2x^2 + 4$, where y is in kg and x is in cm and measure the slope at $y = 12.0$ kg.

3 The following readings were taken of power P supplied to evaporate a liquid and the resulting mass per second of liquid evaporated m.

P/W	2.0	4.0	6.0	8.0	10
$m/10^{-5}\,\mathrm{kg\,s^{-1}}$	0.3	1.0	1.65	2.3	3.0

Plot a graph to check agreement of these results with the equation $P = mL + h$, where h is a constant rate of heat loss and L is constant. Hence evaluate L and h.

4 The following values were obtained for the electric current I and potential difference V for a certain electronic device.

V/V	1.0	2.0	3.0	4.0	5.0
I/A	3.5	20	55	112	196

It is suggested that $I = AV^p$ where A and p are constants. Plot a graph to show that this suggestion is correct and evaluate A and p.

5 The activity A of a radioactive sample decreases with time t giving the following results:

t/day	0	20	40	60	80	100
A/Bq	1000	830	665	545	446	368

If $A = A_0 e^{-\lambda t}$ where A_0 and λ are constants, plot a suitable straight line graph and from it determine λ.

Exercise 148: Questions of A-level standard

In Questions 1–4, the graphs **A** to **E** represent relationships in mechanics.

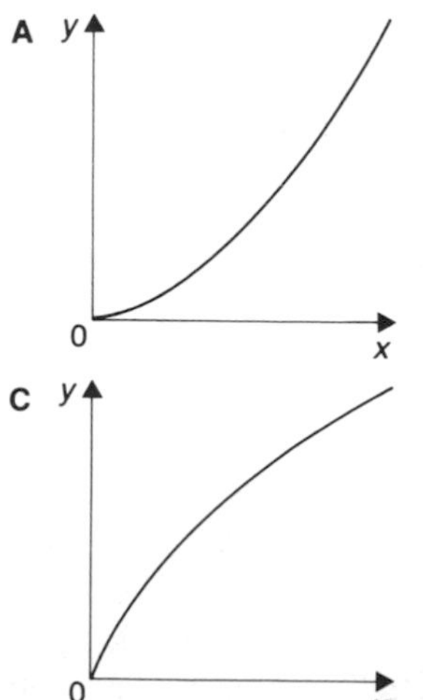

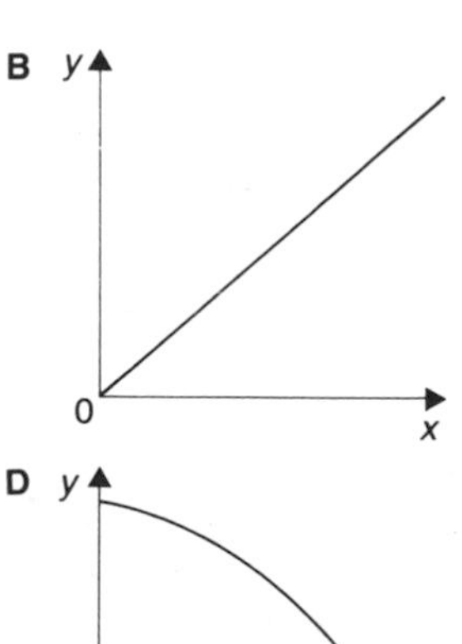

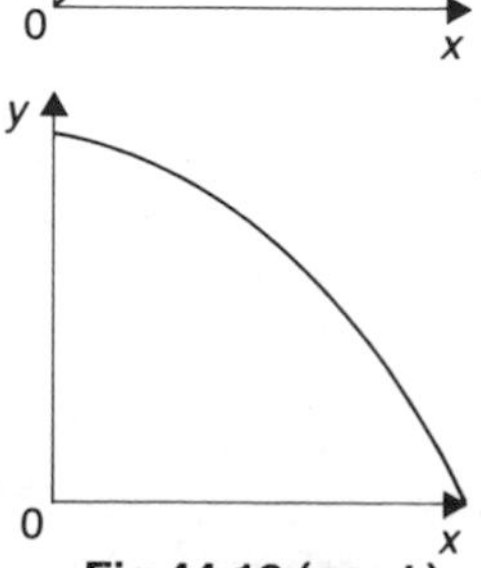

Fig 44.13 (cont.)

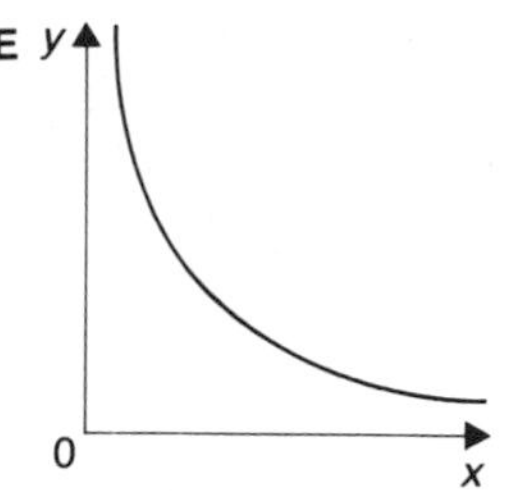

Fig 44.13

Which graph represents most closely the relation between the following pairs of variables?

	y	x
1	The frictional torque at the bearings of a spinning flywheel.	The number of revolutions the flywheel makes in coming to rest from a given angular velocity.
2	The angular acceleration of a pendulum bob executing simple harmonic motion.	The angle the string makes with the vertical during small-amplitude oscillations.
3	The tension in the cord supporting a picture as shown in Fig 44.14.	The angle θ.

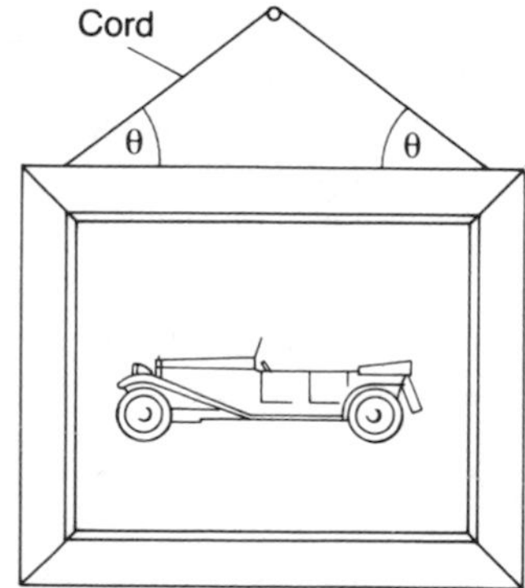

Fig 44.14

4	The angular velocity of a coin rolling from rest down an inclined plane.	The distance the coin has rolled.

[L 80]

5

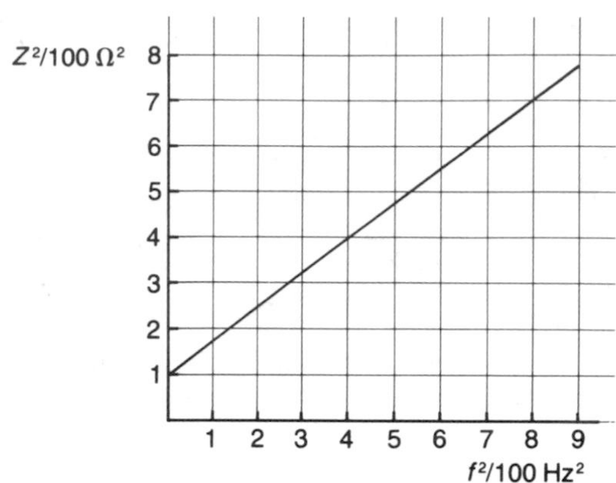

Fig 44.15 Graph for Question 5

A coil of wire which has resistance R and inductance L has an impedance Z given by

$$Z^2 = R^2 + 4\pi^2L^2f^2$$

where f is the frequency of the alternating current flowing in the coil.

Fig 44.15 shows the results of measurements of the impedance over a range of frequencies from 0 to 30 Hz. Use the graph to determine the resistance and inductance of the coil. [SUJB 81]

6 Define *electric potential at a point* and *electric field strength.*

A strip of stiff, conducting paper, pivoted at its centre of gravity, is placed between two large parallel fixed metal plates. The plates are connected to a high-voltage supply. The paper is observed first to oscillate, then come to rest pointing towards the plates. Account for this.

Values of the high voltage V and period of oscillation T are as follows:

V/kV	52	67	98	198
T/s	1.92	1.49	1.02	0.51

By means of an appropriate logarithmic graph, or otherwise, determine the law relating V with T.

Assuming that induced charge is proportional to inducing field strength, show that the torque for a small angular displacement of the strip is proportional to V^2 and hence that the form of the law you have found is to be expected theoretically. [C 79]

7 Show that N kg^{-1} is a valid unit for g, the acceleration due to gravity.

Draw a graph showing how g varies with distance from the earth's centre. Start your graph from the earth's surface and assume that $g = 10\ \text{m s}^{-2}$ at the surface. Take as your unit along your distance axis the earth's radius (6.4×10^6 m) and extend the axis to six radii.

Estimate from your graph the loss in potential energy as a body of mass 1 kg falls from 2.56×10^7 m to 1.92×10^7 m from the earth's centre. [WJEC 79, p]

8 A ball is thrown upwards and is caught as it descends. Which of the graphs **A** to **E** (below) best describes the variation of upward velocity with time, taking upwards velocity as positive?

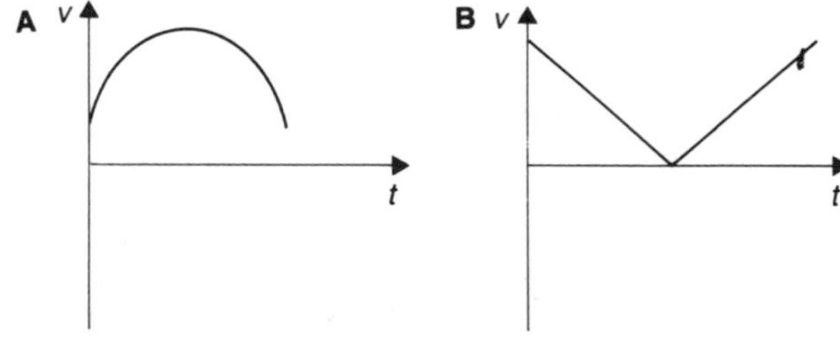

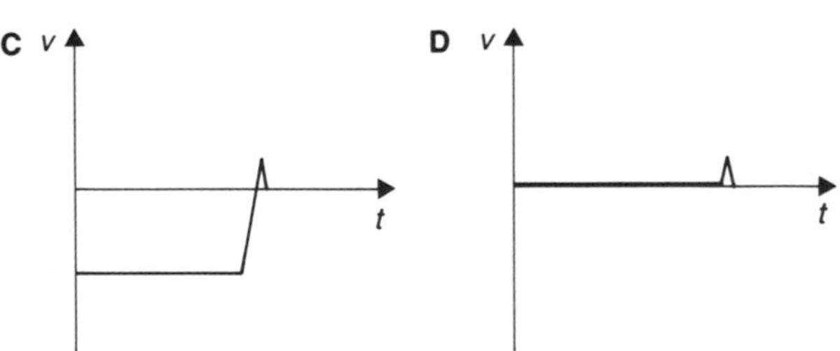

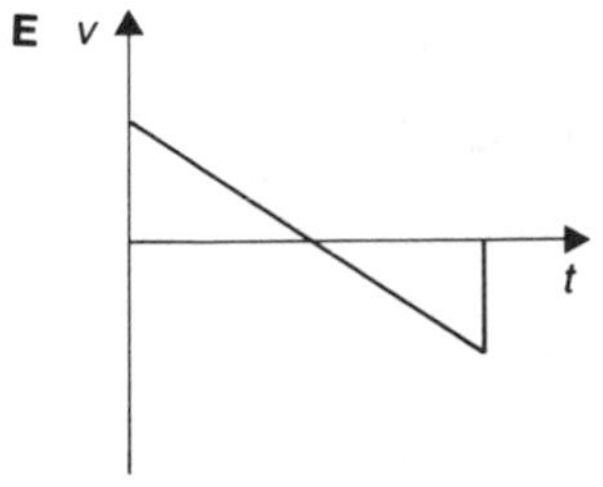

Fig 44.16

Appendices

HINTS FOR QUESTIONS OF A-LEVEL STANDARD

Exercise 7 (Chapter 3)

1 $F = ma$ (Probably worth memorising MLT^{-2} for force)
2 $W = Fd$
3 $P = F/A$
4 $C = Fd$
5 Stress = F/A (see Chapter 13)
6 $F = GM_1M_2/R^2$ (see Chapter 12)
7 See Example 3

Exercise 9 (Chapter 4)

1 Part (c): Evaluate percentage errors in D^2 and d^2, then actual errors. Hence actual error for bracket, then percentage. Add percentage errors for product
2 $\frac{1}{2} \times 0.6$, 1×0.3, $(\frac{3}{2}) \times 0.4$. Add

Exercise 12 (Chapter 6)

1 (a) 20 N in direction OA. See Example 2
(b) Acceleration = (resultant force) ÷ M. See Example 1

2 KE $= \frac{1}{2}mv^2$

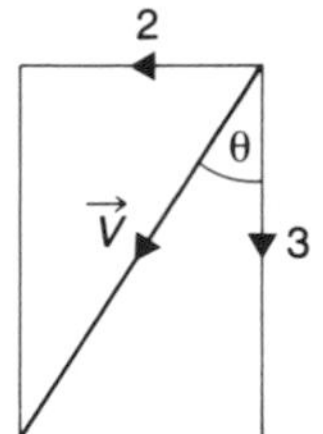

3 $v^2 = x^2 + y^2 + z^2$
4 R_{max} = sum, R_{min} = difference between vectors
5 See Example 5
6 Resolve vertically for T, then horizontally for T'

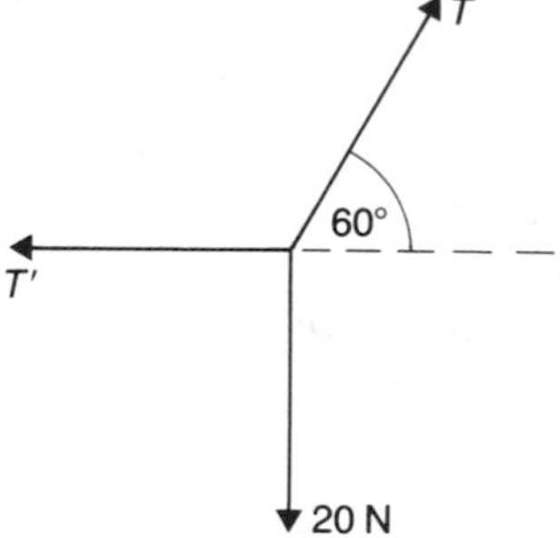

Exercise 19 (Chapter 7)

1 Calculate a, then $v = 0$, $u = 15$, hence s
2 Reaction distance $s_1 = 0.7 \times 13$. If 'retarding' distance = s_2, total = $s_1 + s_2$
3 (a) Upwards motion, $t = 1.5$ s, find u
(b) See Example 7
4 (a) See Example 6
(b) Find speed on hitting sand, then see Example 11
5 $u = 10$, $a = +1.6$, $s = 120$
6 Net acceleration = 4 m s^{-2}, $u = 0$, $s = 2$

7 Object projected horizontally at 20 m s^{-1} at height 180 m, see Example 8
8 Net force is 'weight' of 0.2 kg mass
9 Consider *trailer*, resistive force is 1000 N
(a) $F - 1000 = ma$
(b) $a = 0$
10 (a) Force = pressure (Pa) × area of cross-section (m^2)
(b) $u = 0$, $s = 0.5$, find v

Exercise 24 (Chapter 8)

1 $\frac{1}{2}m(2v)^2$
2 See Equation 8.1
3 Change in KE
4 A = amplitude
Find change in h, hence change in mgh

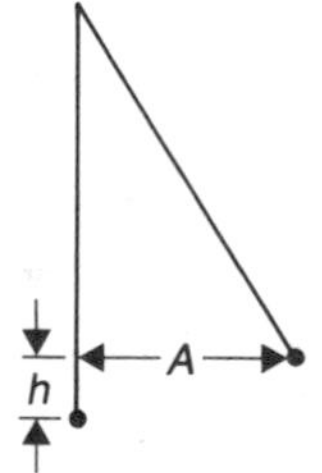

5 Vertical height ⩽ 50 m
See Example 4c for velocity
6 Mass per second = $\pi \times (\text{radius})^2 \times$ speed × density.
See Example 8
7 $M = \pi r^2 v \times \rho$, KE $= \frac{1}{2}mv^2$
8 (a) KE $= \frac{1}{2}mv^2$ (b) $p = F \times v$
9 72 km h^{-1} = 20 m s^{-1}. See Example 11

Exercise 30 (Chapter 9)

1 See Example 2a
2 Final mass = 6 m
3 Spheres do not stick together. Use Equation 9.2
See Example 3
4 See Example 4
5 Show $v = \frac{1}{4}$ of original velocity
6 See Example 5. PE is total KE of riders
7 $234v_1 = 4v_2$. Require ratio $4v_2{}^2 : 234v_1{}^2$
8 Speed of astronaut = 0.5 m s^{-1}. See 'Explosions'
Energy = total KE of astronaut + pack

9 Momentum gain = impulse = area under curve
10 Use Equation 9.6 with F = 8100 N
For t = 1 s, m = mass of air = area × velocity v × Density
Power = KE given to air per second

Exercise 34 (Chapter 10)

1 (a) $\omega = 2\pi/T$ (b) $v = r\omega$ (c) $a = v^2/r$
2 $v = 2\pi r/t$
3 Find velocity due to rotation. Add vectorially to 70 m s^{-1} to find total velocity. $F = mr\omega^2$.
4 $F = mr\omega^2$
5 Let mass = m, extension required = e'. When hanging, $mg = ke$ where $e = 0.02$. When rotating, $mr\omega^2 = ke'$ where $r = 0.20 + e'$. Divide to remove m and k.
6 See Example 4, $R = mg - mv^2/r$
7 (a) Equation 10.4
(b) $T \sin \theta$
(c) See Example 5
8 (a) $T \cos 30 = mv^2/r = mr\omega^2$
(b) $T \sin 30 = L - mg$
$r = 8 \cos 30$, $\omega = \pi/2$

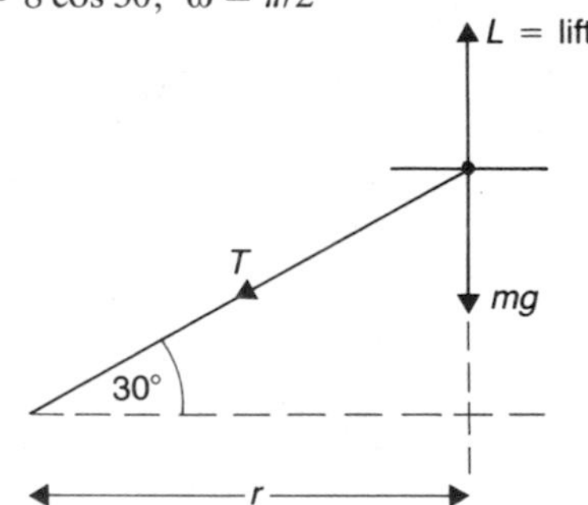

Exercise 40 (Chapter 11)

1 (a) Equation 11.5
(b) $\Gamma = I\alpha$.
Increase in moment of inertia $= mr^2 = 50 \times 3^2$.
Use Equation 11.10
2 (a) Equation 11.5
(b) Equation 11.6 or 11.7; 1 rev = 2π rad
(c) $\Gamma = I\alpha$
3 Note: 1 rev = 2π rad. Equation 11.6, $\Gamma = I\alpha$
4 (a) $\Gamma = F \times d$, $d = 0.015$. Use equations of motion. In (c) $F' = \Gamma'/d'$, $d' = 0.25$
5 Work done = initial KE, 1 rev = 2π rad
6 Angular KE $= \frac{1}{2}I\omega^2$, I given, $\omega = v/r$

7 (a) Work done = force × distance and equals KE gained
(b) KE = $\frac{1}{2}I\omega^2$
(c) $I\omega$
(d) Equation 11.5, $\Gamma = I\alpha$
(e) Equation 11.6. Note: 1 rev = 2π rad

8 (a) 1 rev = 2π rad
(b) In 1 second, work done (50 J) = Γ × Angular displacement (20 rev)

9
$$\begin{pmatrix}\text{Gravitational}\\ \text{PE}\end{pmatrix} \rightarrow \begin{pmatrix}\text{Translational}\\ \text{KE}\end{pmatrix} + \begin{pmatrix}\text{Rotational}\\ \text{KE}\end{pmatrix}$$
$$\underset{\text{(mass)}}{mgh} = \underset{\text{(flywheel)}}{\tfrac{1}{2}I\omega^2} + \underset{\text{(mass)}}{\tfrac{1}{2}mv^2}$$

10 1 revolution in 24 h

11 I of hoop = mass × (radius)^2. Combined moment of inertia = *sum* of separate moments of inertia

Exercise 44 (Chapter 12)

1 Equate mass ÷ (distance)^2
2 See Equation 12.2 and rearrange
3 $g \propto 1/r^2$
4 (i) $M = \frac{4}{3}\pi r^3\rho$
(ii) See Equation 12.3
5 See Example 4
6 See Example 4
7 $U \propto 1/r$. $U_Y > U_X$

Exercise 49 (Chapter 13)

1 (1) $l = 1.00$ m, $e = 0.02$ m
(2) $A = 4 \times 10^{-6}$ m^2

2 (a)

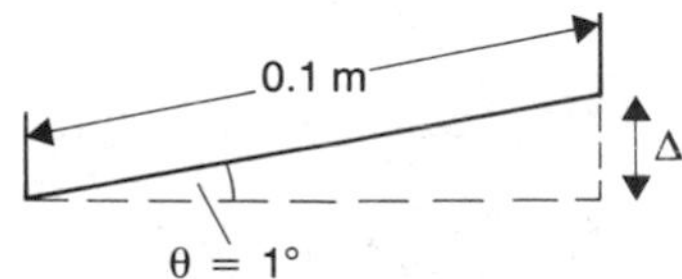

Difference in extensions = Δ

(b) Tensions are equal
(c) See Equation 13.4
(d) $e = 1.0$ mm + Δ
(e) $\frac{1}{2}Fe$

3 F and l the same: $e \propto 1/E$ and $e \propto 1/A$
Note: $A \propto$ (diameter)^2

4 $\epsilon = 0.1$; $e = \epsilon l$, energy = $\frac{1}{2}Fe$
5 See Example 6
6 (b) See Equations 13.1 to 13.5

7 (a) Stiffest has highest E value
(b) Largest $\sigma \times \epsilon$

8 Let length be l, this cancels out. See Example 7

9 (a), (b) See Example 6
(c) Heat energy loss = mass × specific heat capacity × temperature change
Note: mass = volume × density
(d) Calculate new e and F value. Hence find elastic energy stored before and after cooling; subtract

Exercise 54 (Chapter 14)

1 $r\omega^2 = g$
2 $k = F/e$, $\omega^2 = k/m$
3 $T \propto \sqrt{m}$
4 $T = 2\pi\sqrt{(l/g)}$,
(a), (b), $mgh \rightarrow \frac{1}{2}mv^2$

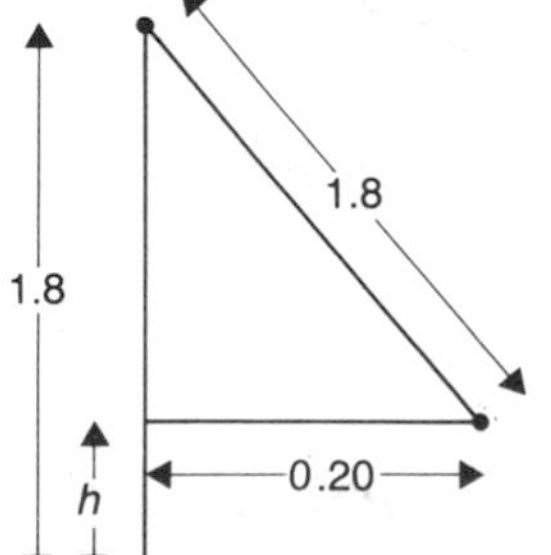

Use four significant figures to calculate h

5 See Example 5
6 $r\omega$, $\omega = 2\pi f$
7 See Example 3, $y = r\cos\omega t$ since $y = r$ when $t = 0$
8 (a) See Example 7
(b) Equation 14.7 with $y = \pm 0.01$
(c) $\frac{1}{2}mr^2\omega^2$
9 $\frac{1}{2}mr^2\omega^2$, $\omega = 2\pi/T$
10 $v \propto r$, $E \propto r^2$, $a \propto r$ for fixed f.

Exercise 57 (Chapter 15)

1 (Length of wave train) ÷ λ
2 Find λ and c, hence f
3 Ship receives $(c + 5) \div \lambda$ waves per second (=f), $f = 1/T$ Find λ. See also Chapter 19
4 Calculate λ. See Example 4
5 $\omega = 2\pi f$, $k = 2\pi/\lambda$. Double f, halve λ

Exercise 63 (Chapter 16)

1 YP = 2.50 m
2 For constructive interference, path difference = $n\lambda$
3 $y = \lambda D/a$
4 $y \propto 1/n$
5 160 fringes, $y = \lambda/2\alpha$, $T = \alpha L$
6 See Example 10

Exercise 65 (Chapter 17)

1 Use four significant figures
2 (a) $\sin\theta = (1/d)\lambda$; plot $\sin\theta$ versus λ, $(1/d)$ is slope. Note: right minus left = 2θ for given λ
(b) (i) $n = 2$ (ii) See Example 1
3 See Example 1c
4 See Example 2, use four significant figures
5 $d\sin 46° = 3\lambda_1 = 4\lambda_2$
6 $(d\sin\theta =)\ n_1\lambda_1 = n_2\lambda_2$
7 (a) See Example 3; $\lambda_B < \lambda_Y < \lambda_R$
(b) $\theta_B = 18.97°$, $n = 3$

Exercise 70 (Chapter 18)

1 $f = c/\lambda$
2 $\lambda = 2L$
3 $(1/2L\sqrt{(T/m)}) = (1/L')\sqrt{(T'/m)}$
4 Equation 14.5 to find tension. Then length = 3 m and see Example 3
5 $f \propto \sqrt{T}$ and $T = W$, $W_1 \propto$ density of iron and $W_2 \propto$ (density iron − density water)
6 See Example 6
7 $f_2 = 3f_1$
8 $\lambda = 2(L_2 - L_1)$
9 $L_1 = \lambda_1/2$; $\lambda_2 = 2L_2$
10 $\lambda_{pipe} = 2L$; $f_{fork} = f_{pipe\ at\ 0\ °C} + f_B$; then $f_{pipe} \propto \sqrt{T}$ as Example 9

Exercise 73 (Chapter 19)

1 Equation 19.1
2 See Example 3
3 $600 = f_s\left(\dfrac{340}{340 - u_s}\right)$, $500 = f_s\left(\dfrac{340}{340 + u_s}\right)$. Solve for u_s.

Solve for f_s then use Equation 19.4 with negative signs first, then positive signs second

Exercise 77 (Chapter 20)

1 $m = 1$ so $u = v = 0.30$. Use Equation 20.1
2 See Example 3. Use mm
3 See Example 5, $v_1 = 30.0$ cm, so $u_2 = -10$. Find v_2
4 See Example 6. Sign not required in final answer

Exercise 80 (Chapter 21)

1 $v = -15$, $D = 15$, $M = D/u$
2

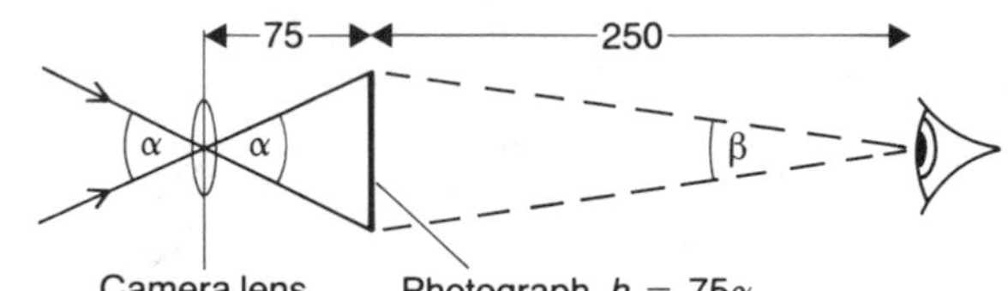

$$M = \frac{\beta}{\alpha} \qquad \beta = \frac{h}{250}$$

(a), (b) For total magnification of 1 we require $\beta = \alpha$, so magnifying lens must be placed 75 mm from photograph, so $u = 75$ mm. From given v, find f
3 See Example 4
4 $\beta = 0.051$, $M = f_o/f_e$; hence α.
Diameter = $\alpha \times 3.8 \times 10^5$ km
5 (a) $S = f_o + f_e$, $M = f_o/f_e$
(b) For eyepiece $v = 150$, $f = 12$, find u. Original $u = f_e$
(c) See Example 5
6 Using notation of Example 5, find m; hence h since H known. Then $h = \alpha f_o$, hence α

Exercise 83 (Chapter 22)

1 Equation 22.1
2 See Example 3. Should get $\dfrac{50a + 2.5a}{100a + 10a} \times 100$

3 Straightforward use of equation given. Then Boyle's law
4 I = EMF/Resistance

8 See Example 7
9 See Example 8; 50 °C not needed
10 $\sqrt{(\overline{c^2})} \propto \sqrt{T}$
11 Temperature is doubled; $\sqrt{(\overline{c^2})} \propto \sqrt{T}$

Exercise 86 (Chapter 23)

1 $s = H/m\theta$ (Equation 23.1) therefore J kg^{-1} K^{-1}; work energy or heat in J or N m; MLT^{-2} for force
2 Water equivalent = mass of water having same thermal capacity. Similar to Example 5
3 When steady, power supply = power loss. When supply removed, power loss = $mc\,(\mathrm{d}\theta/\mathrm{d}t)$
4 See Example 4
5 Power minus loss per second = ml
6 Pressure due to air = total minus SVP. Also is $\propto T$
7 When stabilised, supply power = heat loss per second. Loss per second $\propto$ excess temperature. 25 °C is mean temperature for the 10 K rise

Exercise 88 (Chapter 24)

1 $Q = kA(\theta_1 - \theta_2)/l$. Q in watt, i.e. J s^{-1}
2 $n = \dfrac{Q_g}{Q_b} = \dfrac{k_g A(\theta_1 - \theta_2)l_b}{k_b A(\theta_1 - \theta_2)l_g}$
3 Usual equation (24.1) for first part. In second part layers are in series. New Q is same for both. Use 24.1 for each. Could calculate thermal resistance
4 $Q = \dfrac{k_1 A(100 - x)}{5l_2} = \dfrac{1.5k_1 A(x - 15)}{l_2}$
5 See Example 2

Exercise 92 (Chapter 25)

1 $p_1 = 10$; $p_2 = 10 +$ depth
2 $p_1 = h + 50$; $p_2 = h + 150$
3 Find volume at 1.0×10^5 Pa
4 Equation 25.6; use kg
5 Equation 25.5; $n = 1$. Find volume occupied by molecules, hence empty space. (Use four significant figures)
6 See Example 5. Calculate M_g if $p_1 = 9.5$ and $p_2 = 10$. Mass escaping = $19 - M_g$
7 Apply Equation 25.7 to each cylinder; $p_1 = p_2$

Exercise 96 (Chapter 26)

1 $\Delta W = p(V_2 - V_1)$
2 $\Delta Q = mL$; see Example 2
3 $\gamma = C_p/C_v$, $C_p - C_v = R$
4 $\Delta U = C_v \Delta T$ (1 mol), so $\Delta U = c_v \Delta T$ for 1 kg. $\Delta T = 300$ K
5 See Examples 6 and 7
6 See Equation 26.10
7 $C_p - C_v = R$, hence C_p, $\gamma = C_p/C_v$, then Equation 26.9
8 (a) $pV = nRT$ and number = nL
(b) Equation 25.11
(c) Equation 26.10
9 (a) Find V_1/V_2 from Equation 26.10
(b) Equation 26.9 to find p_2

Exercise 99 (Chapter 27)

1 $R = \rho l/A$, $I = V/R$, $I = nAvq$
2 $R = \rho l/A$. Need R/l. 0.163 Ω m^{-1} for 1 strand. To reduce to 0.01 need to multiply number of strands by 16.3
3 $I = P/V$. Heating power is I^2R
4 (a) $R = \rho l/A$, (b) $R = R_0(1 + 20\alpha)$, $P = V^2/R$
5 Treat ABC as voltage divider to obtain V_{AB} and ADC as voltage divider to find V_{AD}. For $V_{BD} = 0$ need $R_{AD} = 3\Omega$; use resistor in parallel with 12 Ω
6 (a) 9 V across 18 Ω, (b) 9 V is shared by 10 Ω and 8.0 Ω. 9 V is shared between 4.0 Ω and 2.0 Ω
7 In parallel with R between X and Y there is resistance R in series with 'R in parallel with $2R$'.
8 (a) EMF = number of cells in row × EMF per cell
Number of cells in each row is N/p
Internal resistance of row = number of cells × r
Divide by p to obtain effective internal resistance of battery
(b) $I = E/(R + r)$
(c) Try $\rho = 1, 2, 3, 4$
9 $E = V + Vr/R$ (from $E = V + Ir$ and $I = V/R$). Two simultaneous equations

10 EMFs add. Power and PD give current I. Lost volts $= Ir$
(a) Each cell $r/2$, (b) $P = I^2 \times$ internal resistance

11 (a) See Example 8
(b) Power $= I^2R$
(c) $V = E - Ir$

Exercise 102 (Chapter 28)

1 Calculate FSD PD. (a) Series resistor R. PD = meter current $\times$ (R + meter resistance). (b) Shunt. Shunt PD = meter PD

2 For $7I$ total, have I through meter, $6I$ through shunt. Use 'same PD across shunt and meter' or '6 times smaller resistance carries 6 times greater current, for same PD'

3 Can deduce FSD PD. $R = R_0(1 + \alpha\theta)$ FSD PD unchanged

4 (a) $V = E - Ir$ and $I = V/R$
(b) 10 Ω in parallel with 10 Ω and repeat (a)

Exercise 105 (Chapter 29)

1 Top voltage divider decides potential at P. Lower voltage divider decides potential at Q

2 (a) 95 Ω in parallel with R_1. Calculate PD across this combination using voltage divider principle or by deducing circuit current. This is PD across meter
(b) Simple balanced potentiometer
(c) At balance no current flows through test cell, therefore resistance not important
(d) Direct reading means calibrated. Note *small* EMF (could be thermocouple EMF)

3 $E = V - Ir = V - Vr/R$ and can deduce E. Also 5.6 mV for 100 °C. Assume $E \propto$ temperature difference

4 Calculate EMF per metre and hence current through wire. Also, current equals 2 V/total circuit resistance

5 (a) Use equations 29.3, 27.7
(b) $S = \rho l/A$

6 (a) Terminal PD = EMF = 2V. $I_A = 2/(7 + 6)$, $I_B = 2/(5 + 5)$
(b) Can use $V = IR$ to deduce PDs across R_3 and R_4

Exercise 107 (Chapter 30)

1 $F = \text{Constant} \times Q_1Q_2/R^2$

2 Work done = change in potential energy $= Q \times \text{PD} = Q \times Ed$
Work done = 0 for movement perpendicular to field lines

3 For intensity, B and D cancel but we have $q/4\pi\varepsilon(x/\sqrt{2})^2$ due to q at C and same due to q at A

4 PE is value of charge $\times$ potential of the field in which it is placed. (a) Equate sum of KEs with PE. (b) Must be negative and half way between identical charges. (iii) Force between A and C now greatest

5 $E = Q/4\pi\varepsilon R^2$, $V = Q/4\pi\varepsilon R$ $\therefore$ $V/E = R$

6 I constant. $Q/4\pi\varepsilon R$ = potential relative to earth $= IR$. $t = Q/I$

7 I = charge per second = belt width $\times$ charge per m^2 $\times$ velocity. Power = PD $\times$ current

Exercise 110 (Chapter 31)

1 $C = Q/V$, $E = \frac{1}{2}QV$

2 $\frac{1}{2}CV^2 = mgh$

3 $I = fQ = fCV$

4 $Q = C_1V$, $8 = Q(C_1 + C_2)$

5 $Q = 2 \times 120\,\mu\text{C}$, $C = C_1 + C_2$, $V = Q/C$

6 $C_1 = \varepsilon Ad = K/2$, $C_2 = K/5$, $C = C_1$ in parallel with C_2, $Q = CV$

7 $E = V/d$, $C = \varepsilon A/d$, $W = \frac{1}{2}CV^2$, $C \propto \varepsilon$, Q fixed (capacitor isolated), $V(= Q/C) \propto 1/\varepsilon$

8 $Q = CV$, $Q = It$.
(a) $C = \varepsilon A/d$, $C = C_1 \times C_2/(C_1 + C_2)$,
(b) $Q = CV$, $V = Q/C_A$

9 (a) $Q = CV$, $C = 4\pi\varepsilon R$,
(b) Energy $E = \frac{1}{2}CV^2$,
(c) Intensity $E = Q/4\pi\varepsilon R^2$

Exercise 112 (Chapter 32)

1 See Equation 32.1

2 Resultant of $\mu I_1I_2l/2\pi d$ and B_0I_1l or B_0I_2l

3 Left-hand rule. Equate components of BIl and mg up and down rails respectively

4 $BAIn = k\theta$, $I = E/(R + r)$

5 $BAIn = k\theta$. (a) See Example 1, Chapter 28,
(b) See Example 2, Chapter 28

Exercise 116 (Chapter 33)

1 $E = Blv$, $B = B_0 \tan 70$
2 100/60 rev per s. See Example 2
3 (a) $E = d\phi/dt$ and $\phi = BAn$. $I = E/R$
(b) $P = EI$, energy $= Pt$
4 (a) $I = V/R$, (b) PD $= IR = 0$, PD $V = LdI/dt$,
(c) $LdI/dt = 1.0$ V
5 $E = MdI/dt$
6 (a) $\Phi = BAn$, (b) $Q = BAn/R$

Exercise 118 (Chapter 34)

1 (b) Equation 34.1, (c) $F = Bqv$
2 (a) Fields $\mu In/2R$ and B_{HOR} ($= 2 \times 10^{-5}$ T) subtract. $F = BIl$
(b) Fields add
3 Equation 34.2 and 32.4
4 Equations 34.3 and 33.10

Exercise 122 (Chapter 35)

1 $I_{RMS}{}^2 \times R/2 = I_{DC}{}^2 R$
2 RMS = peak/$\sqrt{2}$ for a sine wave
3 $Z^2 = R^2 + (\omega L)^2$. Simultaneous equations. True power $= I_{RMS}{}^2 R$
4 $V_R = IR$, $V_C = I/\omega C$
5 Try looking at ratio of V_L to V_C (RMS values), remembering that $V_L = IX_L$, etc. and $f = \omega/2\pi$
6 (a) Normal current is $I \times P/V$. Resistance $\times V/I$
(b) $Z = V/I = \sqrt{[R^2 + (\omega L)^2]}$
7 An inductor, not a capacitor because direct current was obtainable.
Deduce resistance R of circuit. Deduce impedance Z of circuit. $Z^2 = R^2 + (\omega L)^2$ or $1/\omega C = 2\omega L$
With capacitor use equation 35.8
8 Equations 35.13 and 35.14 or $V_1 I_1 = V_2 I_2$
9 (a) $I = P/V$
(b) Heating power $= I^2R$
(c) V = total power/current
(d) Efficiency = useful power/total power
(e) Transformer output power = 96% of input power

Exercise 126 (Chapter 36)

1 Equation 27.2
2 See Example 2
3 See Example 2
4 $I_c = 50I_b$. PD across load will be 2.5 V
5 Deduce PD across load and hence current through load. $I_b = I_c/200$
6 (a) $F = ne$, (b) $Q = It$ and Q = number × twice electronic charge

Exercise 128 (Chapter 37)

1 See Example 1
2 Time elapsed is for twice the required distance

Exercise 132 (Chapter 38)

1 Use truth table with columns for A, B, two other inputs to AND gates, two inputs to NOR gate and output.
2 5 equals a 4 (digit 3 from right) plus a 1 (first digit from right).

Exercise 135 (Chapter 39)

1 Work done on electron is V_e. Kinetic energy acquired is $\frac{1}{2}mv^2$. See equation 39.1
2 $\frac{1}{2}mv^2$ = work done Ve. $Bev = mv^2/R$. (Equation 39.4.) Period $= 2\pi R/v$
3 Calculate electron velocity first.
Time = distance/velocity
4 Equation 39.6 shows radius ∝ mass for given velocity. Time = distance/velocity and distance $= \pi \times$ radius
5 Weight equals electric force
6 Electric force Ee or Ve/d and magnetic force Bev (see Example 9)

Exercise 137 (Chapter 40)

1 Photon energy hc/λ. See Equation 40.2
2 Equation 40.3. Simultaneous equations
3 Equation 40.3
4 Use mean wavelength, e.g. 550 nm. Each photon has energy hc/λ
5 $\lambda = h/\text{momentum}$
6 Energy change = photon energy = hc/λ

Exercise 140 (Chapter 41)

1 (a) See definition of electron-volt (in this chapter)
(b) Energy = $\frac{1}{2}mv^2$. Deduce momentum, hence wavelength (see Chapter 40, Example 2)
(c) Equation 41.1
2 (a) A K shell electron must be ejected
(b) L to K transition
3 $2d \sin \theta = n\lambda$. Consider n needed for θ = maximum value of 90°
4 Deviation is twice angle between beam and planes. Equation 41.2

Exercise 144 (Chapter 42)

1 Find half-life which is 1 hour. Use inverse square law
2 (a) Equation 42.5, (b) Equation 42.3 may be useful
3 For blood volume V ml (or $V\,\text{cm}^3$), 12 ml has a fraction $12/V$ of the total activity. Can use Equation 42.7. (2 ml is unnecessary information)
4 Need reduction by factor of 4. Dose $\propto$ 1/distance^2. Time is 2 half-lives.
5 Read 'Atomic mass unit' (in this chapter). Mass of H atom is approximately 1 u. Can show first that ratio of numbers is 3 to 1 (see Example 3) using, say, x for percentage of ^{87}Rb. Can equate average masses to find total number (of ^{87}Rb plus ^{85}Rb). Hence number of ^{87}Rb atoms.
To find activity use formula given in Equation 42.8
6 Ratio of initial number of U atoms to final number is 2.17/1.17

Exercise 146 (Chapter 43)

1 Conservation of momentum.
KE = $\frac{1}{2}mv^2$. (Consider ratio of kinetic energies)
2 See Example 4
3 See Example 6
4 Use Equation 43.2. Kinetic energy here is 5 MeV. (See Chapter 42 for electron-volt and conversion to joules)

Exercise 148 (Chapter 44)

1 Torque × perimeter × revolutions = work, therefore torque $\propto$ 1/rev
2 For SHM acceleration $\propto$ displacement
3 $2T \sin \theta$ = weight therefore $T \propto 1/\sin \theta$
4 $\frac{1}{2}I\omega^2 + \frac{1}{2}mv^2 = mgh = mgd \sin \theta$, $v = R\omega$ therefore $\omega^2 \propto d$
5 R^2 = intercept on Z^2 axis. Slope = $4\pi^2L^2$. Measured slope is $0.75\,\Omega^2\,\text{Hz}^{-2}$
6 $T = AV^p$ gives $\log T = \log A + P \log V$
$\log T$ (as Y) versus $\log V$ (as X) gives slope of -1.0, therefore $T \propto 1/V$. Use a pair of T and V values to find A
Torque $\tau \propto$ charge Q and $\propto$ intensity E. $Q = k_1V$, $E = k_2V$, $\tau = k_1k_2V^2$
7 Area between R = 3 earth radii to R = 4 earth radii equals average g × distance fallen h. Therefore PE loss mgh is 1 kg times area
8 Velocity reverses sign and is zero at half-way stage

ANSWERS

The answers given are the responsibility of the authors and where questions have been taken from past GCE examinations the examination boards concerned are in no way responsible for these.

Chapter 2

Exercise 1

1 (a) 3 (b) 33 (c) 6
(d) $6\frac{1}{2}$ (e) $-1\frac{1}{2}$

2 (a) 10 (b) 10^{-1} or 0.1 (c) 10^{16}
(d) 10^{-15} (e) 10^{-20} (f) 10^{10}
(g) 10^{6} (h) 100^{6} or 10^{12} (i) 10
(j) 10^{2} or 100 (k) 10^{2} or 100 (l) 3×10^{2} or 300

3 (a) 2 (b) 9 (c) -2

4 (a) (i) 5 (ii) 1 or log 10 (iii) 1 or log 10
(b) (i) 1.301 (ii) 3.301

5 4.6

6 (a) 1.097 (b) 5.697

Exercise 2

1 (a) (i) $r_1 + r_2 = \alpha$ (ii) $r_1 + r_2 = \gamma$
(b) (i) $\angle BCD + \gamma = 180°$
(ii) $\angle BCD + \alpha = 180°$

2 (a) $\gamma = i$ (b) $\alpha = 2i$

3 $\beta = \alpha + 2i$, $\gamma = \alpha + i$

4 21 cm (or 20.94 cm)

5 (a) $\pi/3$ or 1.047 (b) $\pi/4$ or 0.7855
(c) 8π or 25.14 (d) $\pi/2$ or 1.571
(e) π or 3.142

Exercise 3

1 (a) $x^2 + 6x + 8$ (b) $x^2 - 2x - 8$
(c) $x^2 - 6x + 8$ (d) $x^2 - 8x + 16$
(e) $3x^2 + x - 10$

2 (a) 4, -2 (b) $-2/3$, $2\frac{1}{2}$ (c) $+3$, -3

3 $x = 3$, $y = 7$

Chapter 3

Exercise 4

1 (a) ML^{-3} (b) L^2
(c) L^3T^{-1} (d) ML^2T^{-3}

2 (a) T (b) MLT^{-1}
(c) T^{-1}

3 $MQ^{-1}T^{-1}$

4 $ML^2T^{-2}\theta^{-1}$ (θ is temperature)

5 MT^{-2}

Exercise 5

1 $\alpha = 1$, $\beta = 2$

Exercise 6

1 (a) $8\frac{1}{3}$ or 8.3 m s^{-1} (b) 10^4 mm^2
(c) 0.4 μm (d) 2000 s^{-1}

2 0.000 01 mΩ^{-1}

Exercise 7

1 B

2 A

3 C

4 A

5 C

6 C

7 LT^{-1}, LT^{-2}, $M^{-1}L^3T^{-2}$
$a = 0$, $b = \frac{1}{2}$, $c = -\frac{1}{2}$
$T \propto (r)^{3/2}$

Chapter 4

Exercise 8

1 B

2 4%

3 4%

4 Between 0.73 and 0.75%

Exercise 9

1 (a) 2.5%, 1.7%, 2.0% (b) 1.4×10^3 mm^3
(c) 23%

2 D

Chapter 6

Exercise 10

1 (a) 24 N at 12° to 15 N (b) 36 N at 52° to 50 N
2 67.1 N at 63.4° to 30 N
3 79.4 N at 40.9° to 30 N

Exercise 11

1 (a) $300\ \mathrm{m\,s^{-1}}$ (b) 53°
2 (a) 5.3 N down plane (b) 9.7 N up plane
3 (a) 30 N (b) 4.6 kg

Exercise 12

1 (a) 25 N at 37° to OA
(b) $\frac{2}{3}\ \mathrm{m\,s^{-2}}$ at 60° to each 2 N force
2 33.7°, 32.5×10^{-7} J
3 C
4 (a) Yes (b) Yes (c) No
5 20 N
6 23 N in CD, 11.5 N in AB

Chapter 7

Exercise 13

1 (a) π s
(b) $8/\pi\ \mathrm{m\,s^{-1}}$ to left
(c) $8.0\ \mathrm{m\,s^{-1}}$ upwards
2 $\sqrt{4.5}\ \mathrm{m\,s^{-1}}$ at 135° to original direction

Exercise 14

1 (a) $12\ \mathrm{m\,s^{-1}}$ (b) 24 m
2 (a) $2.5\ \mathrm{m\,s^{-2}}$ (b) 8.0 s, 128 m (c) 10 s, 125 m

Exercise 15

1 $360\ \mathrm{m\,s^{-2}}$
2 (a) 3.0 s (b) 63 m
(c) t = 6.0 s (for s increasing) and t = 14 s (for s decreasing)
3 $8.0 \times 10^{15}\ \mathrm{m\,s^{-2}}$

Exercise 16

1 (a) 45 m (b) $30\ \mathrm{m\,s^{-1}}$
2 $17\ \mathrm{m\,s^{-1}}$
3 4.0 s

Exercise 17

1 (a) 4.5 s (b) 18 m
(c) $45\ \mathrm{m\,s^{-1}}$, $4.0\ \mathrm{m\,s^{-1}}$
2 1.25m
3 (a) 40.0 s (b) 13.9 km (c) 2.00 km

Exercise 18

1 (a) 7.5 N (b) $0.50\ \mathrm{m\,s^{-2}}$ (c) 3.0 kg
2 (a) $4.0\ \mathrm{m\,s^{-2}}$ (b) 50 m
3 (a) 5.6 kN (b) 3.2 kN (c) 1.6 kN
4 (a) $42\ \mathrm{m\,s^{-1}}$ (b) 7.5 kN
5 3.6 kN

Exercise 19

1 A
2 2.9 m beyond stop line
3 (a) $15\ \mathrm{m\,s^{-1}}$ (b) 11 m
4 (a) 45 m (b) 400 N
5 B
6 A
7 (a) 6.0 s (b) 0.12 km
8 $4.0\ \mathrm{m\,s^{-2}}$, upwards
9 (a) 1750 N (b) 1000 N
10 (b) $1.8 \times 10^{6}\ \mathrm{m\,s^{-2}}$ (c) $1.3\ \mathrm{km\,s^{-1}}$

Chapter 8

Exercise 20

1 (a) 1.8×10^{5} J (b) 4.0×10^{8} J
(c) 1.8×10^{-16} J
2 (a) $15\ \mathrm{m\,s^{-1}}$ (b) 22.5 m
(c) 337.5 J (d) 337.5 J
3 72 N
4 $4.6\ \mathrm{m\,s^{-1}}$

Exercise 21

1 (a) 24 J (b) $13\ \mathrm{m\,s^{-1}}$
2 (a) 5.0 J (b) 0.20 m
3 6.0 J

4 $1.6\ m\ s^{-1}$
5 (a) 0 (b) $1.5\ m\ s^{-1}$ (c) $1.8\ m\ s^{-1}$

Exercise 22

1 (a) 0.42 kJ (b) 28 m
2 18 J
3 (a) 18 J (b) 4.4 N

Exercise 23

1 (a) 150 W (b) 300 W (c) 214 W
2 22.5 kW
3 (a) $10 \times 10^3\ kg\ s^{-1}$ (b) $20 \times 10^3\ kg\ s^{-1}$
4 (a) 18 kW (b) 36 kW (c) 25 kW
5 (a) 23 kW (b) 46 kW (c) 33 kW
6 (a) 0.64 kN (b) 1.0 kN
7 (a) $13.3\ m\ s^{-1}$ (b) $8.0\ m\ s^{-1}$

Exercise 24

1 E
2 D
3 0.36 kJ
4 7.5 mJ
5 87 m, $24\ m\ s^{-1}$
6 1.74 MW
7 D
8 (a) D (b) B
9 (b) (i) $0.40\ MJ\ s^{-1}$
(ii) 0.66 MW

Chapter 9

Exercise 25

1 (a) $7.5\ kg\ m\ s^{-1}$ (b) $4.0\ m\ s^{-1}$
2 (a) $-10\ kg\ m\ s^{-1}$ (b) $10\ kg\ m\ s^{-1}$ (c) 0

Exercise 26

1 $3.3\ m\ s^{-1}$
2 $-0.67\ m\ s^{-1}$
3 $19\ m\ s^{-1}$

Exercise 27

1 0.33 kJ
2 16.3 kJ
3 2.4 J

Exercise 28

1 $0.40\ m\ s^{-1}$
2 $0.42\ m\ s^{-1}$
3 $807\ km\ s^{-1}$ forwards

Exercise 29

1 6 N
2 60 N

Exercise 30

1 E
2 D
3 $1.0\ m\ s^{-1}$, 2.3 J
4 1 kg ball is $-4/3\ m\ s^{-1}$; 2 kg ball is $+8/3\ m\ s^{-1}$
5 $\frac{3}{4}$ of original KE is lost
6 $0.09\ m\ s^{-1}$, 1.1 mJ
7 58.5
8 $1.5\ m\ s^{-1}$, 37.5 J
9 (a) 50 N s (b) 87.5 N s (c) 100 N s
10 $15\ m\ s^{-1}$, 61 kW

Chapter 10

Exercise 31

1 (a) $4.7\ rad\ s^{-1}$ (b) $0.66\ m\ s^{-1}$
2 $0.033\ rad\ s^{-1}$

Exercise 32

1 1.5 kN, road friction
2 252 N, 132 N, $10\ m\ s^{-1}$ at bottom
3 90 m, leaves road

Exercise 33

1 (a) 1.8 s (b) 5.8 N

Exercise 34

1 (a) 7.3×10^{-5} rad s^{-1} (b) 0.47 km s^{-1}
(c) 0.034 m s^{-2}
2 D
3 94 m s^{-1}, 1.6 N
4 3.2 rad s^{-1}
5 22 cm
6 5.5×10^3 N, downwards
7 (a) 25 N (b) 15 N (c) 1.3 s
8 (a) 8.0 N (b) 8.0 N

Exercise 40

1 (a) 0.40 rad s^{-2} (b) 3.6 kN m, 0.19 rad s^{-1}
2 (a) −200 rad s^{-2} (b) 14×10^3
(c) −16 kN m
3 0.044 N m
4 (a) 4.0 rad s^{-2} (b) 16 rad s^{-1}, 29 J
(c) 1.8 N
5 E
6 D
7 (a) 0.24 J (b) 31 rad s^{-1}
(c) 1.55×10^{-2} kg m^2 rad s^{-1} (d) 1.55×10^{-4} Nm
(e) 0.25×10^3
8 (a) $4.0\pi^2 \times 10^3$ J (b) 0.40 N m
9 13 kg m^2
10 (a) 5.8×10^{33} kg m^2 rad s^{-1} (b) 2.1×10^{29} J
11 1.5×10^{-3} kg m^2, 9.2×10^{-3} J

Chapter 11

Exercise 35

1 (a) 5.0 (b) 14 (c) 6.0
2 (a) 2.5 rad s^{-2} (b) 50 rad s^{-1} (c) 500 rad
3 (a) 9.4 rad s^{-1} (b) −0.21 rad s^{-2} (c) −0.084 N m
(d) 118 rad

Exercise 36

1 (a) +3.0 (b) −2.5 (c) −2.0
(d) +2.0
2 (a) 3.0π N m (b) 7.5π N m
3 (a) 30 s (b) 180 rad (c) 6 rad s^{-1}
4 (a) 5.0π rad s^{-2} (b) 0.95 kg m^2
5 (a) 40 N m (b) (i) 33 s (ii) 1.7×10^3 rad

Exercise 37

1 (a) 22 rad s^{-1} (b) 214 rev min^{-1}
2 (a) 4.8π m s^{-1} (b) 8.2 kJ

Exercise 38

1 (a) 60 J (b) 8.7 rad s^{-1}
2 6.1 m, 25 rad s^{-1}

Exercise 39

1 (a) 9/8 kg m^2 (b) 27 J
2 (a) 0.50 kg m^2 (b) 27 J (c) 20 J
(d) 7 J

Chapter 12

Exercise 41

1 27×10^{-9} N
2 49 N
3 36 kg
4 3.4×10^8 m from earth

Exercise 42

1 (a) 7.3×10^{22} kg (b) 0.67 N kg^{-1}
2 (a) 2.45 m s^{-2} (b) 19.6 m s^{-2} (c) 9.8 m s^{-2}
3 0.54 m s^{-2}

Exercise 43

1 12×10^3 J
2 (a) 3.0×10^6 J kg^{-1} (b) 4.5×10^9 J
3 0.65 km s^{-1}
4 5.8×10^7 m s^{-1}

Exercise 44

1 C
2 E
3 D
4 (a) 4.8×10^{17} kg m^{-3} (b) 1.3×10^{12} N kg^{-1}
5 D
6 E
7 D

Chapter 13

Exercise 45

1 25 kN
2 0.50 mm
3 (a) 7.5×10^7 N (b) 3.0×10^{-4}

Exercise 46

1 (a) 57 MN m^{-2} (b) 0.45×10^{-3}
(c) 13×10^{10} N m^{-2}
2 7.5×10^5 N
3 (a) 1.4 mm (b) 42 N

Exercise 47

1 (a) 4.0×10^{-2} J (b) 14×10^{-2} J
2 (a) 15 J (b) 30 J
3 (a) 0.19 mm (b) 11 mJ
4 3.6 kJ

Exercise 48

1 0.58 kN
2 (a) 0.38 MN (b) 0.69 kJ

Exercise 49

1 1, 2 and 3 correct
2 (a) 1.7×10^{-3} m (b) 50 N
(c) – (d) 1.0×10^{11} N m^{-2}
(e) 25 mJ
3 2 mm copper, 4 mm steel
4 7.0 MN m^{-2}, 0.14 J
5 (a) 50 N (b) 0.50 J
6 (b) 15×10^7 N, 2.3 mm, 17×10^{-2} J
7 (a) steel (b) rubber
8 0.28 MN
9 (a) 0.41 m (b) 0.74 kJ (c) 47 kJ
(d) 0.12 kJ

Chapter 14

Exercise 50

1 (a) 4.0 Hz (b) $\pm 1.3\pi^2$ m s^{-2}, 0
(c) $-0.32\pi^2$ m s^{-2}
2 (a) 32×10^3 m s^{-2} (b) 25 kN
3 (a) 1.6 Hz (b) 0.63 s
4 2.9 Hz

Exercise 51

1 (a) 0.20 m (b) 0.89 s (c) 68
2 (a) 1.5 s, 0.65 Hz (b) 0, ± 0.33 m s^{-2}
3 (a) 0.25 m (b) 0.063 m (c) 60
4 1.0 m, 0.84 m

Exercise 52

1

t/s	0	0.25	0.50
$y/10^{-3}$ m	0	28	40
v/m s^{-1}	0.13	0.089	0
a/m s^{-2}	0	−0.28	−0.39

The remaining values follow by 'symmetry'
2 (a) 0.30 m s^{-2} (b) 0.094 m s^{-1}
(c) 0.099 m s^{-2}, 0.089 m s^{-1}, $r = 0.030$, $\omega = \pi$

Exercise 53

1 0.87 Hz
2 18 mJ
3 (a) 8.0 mJ (b) 0.5 mJ (c) 32 mJ

Exercise 54

1 2.2 Hz, at top
2 D
3 B
4 2.7 s
(a) 0.24 J (b) 0.46 m s^{-1}
5 50.0
6 E
7 (a) 0.89 s (b) $y = 0.04 \cos 7.1t$
8 (a) 0.021 m s^{-1}, 0.037 m s^{-2}
(b) 0.67 s
(c) 74 μJ, No
9 3.9 mJ
10 1 and 3

Chapter 15

Exercise 55

1 (a) 5.0×10^{14} Hz (b) 1.5 km
2 17 °C

3 $316\ m\ s^{-1}$
4 $71\ m\ s^{-1}$
5 (a) $0.10\ km\ s^{-1}$ (b) $2.5 \times 10^{-4}\ kg\ m^{-1}$

Exercise 56

1 (a) +4.3 cm, −4.3 cm
(b) $2\pi/3$ rad, $10\pi/3$ rad
2 $3\frac{1}{3}$ cm
3 (a) 0.50 Hz (b) 12 m, 2.4 m, for example

Exercise 57

1 12×10^6
2 50 Hz
3 **B**
4 **B**
5 $a = 2 \times 10^{-3}$ m, $\omega = 100\pi\ rad\ s^{-1}$, $k = 2\pi/3\ m^{-1}$, +sign. $c = 150\ m\ s^{-1}$, ω and k both doubled

Chapter 16

Exercise 58

1 (a) nothing (b) a double amplitude sound
2 3.16 cm
3 (a) 1.0 m, 340 Hz; 0.50 m, 680 Hz
(b) 2.0 m, 170 Hz; $\frac{2}{3}$ m, 510 Hz
4 (a)567 Hz (b) 283 Hz

Exercise 59

1 0.49 mm
2 0.42 mm
3 (a) 0.51 mm (b) 0.80 mm (c) 1.6 mm

Exercise 60

1 (a) $0.70\ \mu m$ (b) $3.5\ \mu m$
2 (a) 2.0×10^{-3} rad (b) 0.24 mm
3 0.30 mm
4 0.75 mm
5 (a) $0.90\ \mu m$ (b) 1.0×10^{-4} rad

Exercise 61

1 1.33
2 0.54 mm
3 1.5(3)

Exercise 62

1 (a) 0.10 s (b) 430 Hz, 450 Hz
2 262 Hz
3 (a) 1.59 Hz, 1.42 Hz (b) 5.95 s
4 (a) 0.042 Hz (b) 0.500 Hz (c) 0.542 Hz

Exercise 63

1 (a) 120 mm (b) zero
2 1
3 1.18 mm
4 **D**
5 1.79×10^{-3} rad, $63.8\ \mu m$
6 **B**

Chapter 17

Exercise 64

1 638 nm
2 17.5°, 36.9°, 64.2°
3 1.0°, 2
4 7.4°

Exercise 65

1 $2.000\ \mu m$
2 (a) 5.51×10^5 (b) (i) 621 nm (ii) 4
3 **C**
4 0.19°
5 450 nm, 4.00×10^5
6 **B**
7 433 nm

Chapter 18

Exercise 66

1 (a) 0.213 m (b) $1.1\ m\ s^{-1}$
2 2.8 cm, 11 GHz

Exercise 67

1 (a) 1.8 m, 59 Hz; 0.90 m, 117 Hz; 0.60 m, 176 Hz
(b) 0.90 m, 117 Hz; 0.45 m, 234 Hz; 0.30 m, 351 Hz

2 (a) 1.6×10^{-3} kg m^{-1} (b) 0.19 km s^{-1} (c) 1.6 m (d) 0.81 m
3 (a) 260 Hz (b) 300 Hz (c) 520 Hz
4 241 Hz

Exercise 68

1 1.328 m, 256 Hz; 0.664 m, 512 Hz; 0.443 m, 768 Hz
2 (a) 0.304 m
(b) 0.607 m, 560 Hz; 0.304 m, 1.12 kHz
3 0.166 m
4 (a) 1.0 cm (b) 50.8 cm
5 545 Hz
6 16 Hz

Exercise 69

1 (a) 276 Hz (b) 25 °C
2 (a) 30 Hz (b) 66 °C

Exercise 70

1 10 GHz
2 448 m s^{-1}
3 $\sqrt{2}$ times as long
4 22.4 m s^{-1}
6.0 m, 3.7 Hz; 3.0 m, 7.5 Hz; 2.0 m, 11 Hz; 1.5 m, 15 Hz; 1.2 m, 19 Hz
5 7.95×10^3 kg m^{-3}
6 165 Hz and 495 Hz
7 E
8 D
9 1.75 m, 16 Hz
10 8 °C

Chapter 19

Exercise 71

1 (a) 710 Hz (b) 700 Hz, 690 Hz, 680 Hz
2 (a) 600 Hz (b) 567 Hz
3 449 Hz
4 (a) 2.09 m s^{-1} (b) 252 Hz, 248 Hz
5 (a) 30 m s^{-1} (b) 600 Hz

Exercise 72

1 (a) 239 Hz (b) 219 Hz (c) 232 Hz
2 400 Hz

Exercise 73

1 D
2 1.02 kHz
3 30.9 m s^{-1}; 582 Hz, 515 Hz

Chapter 20

Exercise 74

1 +12 cm
2 (a) −30 cm (b) +15 cm
3 (a) +5 cm (b) −10 cm
4 −24 cm
5 (a) 11.1 cm (b) 10.2 cm (c) 0.9 cm

Exercise 75

1 (a) v = 120 mm, real, h = 9.00 mm
(b) 300 mm from lens 2, real, h = 45.0 mm
(c) 15 times
2 −30 cm
3 (a) $v = f$ = 20 cm, real
(b) 60 cm from diverging lens, real

Exercise 76

1 −80/3 cm

Exercise 77

1 C
2 16 mm
3 C
4 E

Chapter 21

Exercise 78

1 (a) 27 mm (b) (−) 9.3 times
2 (a) 30 mm (b) (−) 8.3 times
3 (a) 26 mm (b) (−) 7.0 times

Exercise 79

1 (a) 20.0 (b) 63.0 cm (c) 40.0×10^{-3} rad
2 (a) 22.4 (b) 62.7 cm (c) 44.8×10^{-3} rad
3 (a) 18 (b) 5.6×10^{-3} rad (c) 2.1×10^3 km

Exercise 80

1 $x = 15/4$ cm, $M = 4.0$, $u_{max} = 5.0$ cm
2 0.30 times; (a) 75 mm (b) 107 mm
3 50 mm
4 (b) 2.1×10^3 km
5 (a) 912 mm, 75 times
(b) Move 1.0 mm away from objective lens
(c) 93 mm
6 9.1×10^{-3} rad

Chapter 22

Exercise 81

1 (a) 6.0 cm from bulb (b) 1.2 cm from bulb
2 (a) 33 °C (b) −14 °C
3 67 °C

Exercise 82

1 (a) 5.8 Ω (b) 25 °C
2 45 °C

Exercise 83

1 E
2 47.7 °C
3 364 K, 368 K
4 0.27 mA

Chapter 23

Exercise 84

1 5.0 K
2 7.1 V
3 1800 J kg^{-1} K^{-1}
4 39 °C
5 2.0×10^3 J kg^{-1} K^{-1}
6 (a) 0.38 kJ kg^{-1} K^{-1} (b)(i) 32 W (ii) 8.0 W

Exercise 85

1 377 W
2 (a) 438 kJ kg^{-1} (b) 1.61 kJ (c) 59.4 mg s^{-1}
3 0.84×10^5 N m^{-2}

Exercise 86

1 C
2 42 °C
3 2.7 kJ kg^{-1}K^{-1}
4 2.6×10^3 J kg^{-1} K^{-1}
5 B
6 3.77×10^5 Pa, 118 N
7 (b) (i) 1.3 W (ii) 0.96 kJ kg^{-1} K^{-1}

Chapter 24

Exercise 87

1 3.6 W
2 38 W, 36 °C
3 (a) 0.33 kW (b) 0.038 kW
4 8.6 K
5 16 cm
6 0.21 W m^{-1} K^{-1}

Exercise 88

1 C
2 E
3 533 W, 252 W, 9.5 °C
4 25 °C
5 400 W m^{-1} K^{-1}

Chapter 25

Exercise 89

1 (a) 280 (b) 983 (c) 193 (d) 74
2 327 °C
3 63×10^{-3} m^3

4 210 °C
5 170 cm^3
6 98×10^{-3} kg

Exercise 90

1 22.5×10^{-3} m^3
2 (a) 22.7 (b) 1.37×10^{25}
3 (a) 0.100 (b) 28.0×10^{-4} kg
(c) 2.19×10^{-3} m^3
4 2.30 kg, 73×10^{-3} m^3
5 1.26×10^5 N m^{-2}

Exercise 91

1 579 m s^{-1}
2 (a) 30 m s^{-1} (b) 34 m s^{-1} (c) 35 m s^{-1}
3 0.179 kg m^{-3}
4 1.12 km s^{-1}
5 666 m s^{-1}

Exercise 92

1 A
2 A
3 40
4 C
5 24×10^{-3} m^3; 0.9995
6 0.33 kg
7 $\frac{1}{6}$ m
8 B
9 B
10 A
11 D

Chapter 26

Exercise 93

1 -1.5×10^2 J
2 (a) −50 J (b) −70 J
3 (a) 50×10^{-3} kg (b) 113 kJ
(c) 8.4 kJ (d) 105 kJ

Exercise 94

1 1.4
2 (a) 33.3 J mol^{-1} K^{-1}, 25.0 J mol^{-1} K^{-1}
(b) 1.33
(c) 9.98 kJ
3 (a) 1.66 (b) 0.31 kJ kg^{-1} K^{-1}
(c) 25 kJ
4 (a) 2.32 kJ (b) 6.65×10^{-3} m^3 (c) 665 J
(d) 1.66 kJ (e) 1.66 kJ

Exercise 95

1 (a) 6.73×10^5 N m^{-2}, 290 K
(b) 14.4×10^5 N m^{-2}, 619 K
2 (a) 15.0×10^{-4} m^3, 280 K
(b) 9.47×10^{-4} m^3. 177 K
3 232 K, 1.29 m Hg

Exercise 96

1 C
2 (a) 2.26×10^6 J (b) 1.69×10^5 J
(c) 2.09×10^6 J
3 E
4 9.0×10^2 J
5 4.40×10^5 Pa, 288 K
7.66×10^5 Pa, 501 K
6 B
7 B
8 (a) 2.5×10^{22} (b) 6.0×10^{-21} J (c) 1.40
9 (a) Compression ratio = 15.6, (b) 250 J

Chapter 27

Exercise 97

1 12.5×10^{12}
2 1.25×10^{22}
3 0.4 A, 80 V; 0.08 A, 160 V; 0.32 A, 160 V
4 1.1 Ω
5 18×10^{-8} Ω m
6 240 J
7 (a) 0.02 Ω (b) 300 A
(c) 1.8 kJ s^{-1} (1.8 kW)

Exercise 98

1 (a) $3 = 2I_3 + 1.0I_1$
(b) $1.5 = 5.0I_2 + 1.0I_1$
(c) $1.5 = 2.0I_3 - 5.0I_2$
2 See Example 3, p.178

Exercise 99

1 3.9 mm s^{-1}
2 (a) 0.016 Ω m^{-1} (b) 17

3 C
4 (a) 1.125 Ω (b) 3.3 W
5 2 V, 4 Ω in parallel
6 0.5 A, 1.0 V
7 C
8 (a) NE/p, Nr/p^2 (b) $\dfrac{NEp}{Nr + Rp^2}$ (c) $p = 3$
(d) (i) 2.3 A (ii) 6.7 W (iii) 6.9 W
9 3.0 V, 2.0 Ω
10 (a) 1.0 Ω (b) 3.8 J
11 (b) 4.2 W, 2.1×10^{-3} kW h
(c) (i) 8.2 V (ii) 3.6 V

Chapter 28

Exercise 100

1 (a) 10.0 mV (b) 10.0 mA
2 (a) 0.040 Ω shunt (b) 2.0 kΩ in series
(c) 10 Ω in series
3 (a) 0.010 shunt (b) 20 kΩ in series
4 C

Exercise 101

1 4.0 V, 5.0 V
2 (a) 5.0 A (b) 3.65 A
3 3.96 and 3.92 divisions

Exercise 102

1 (a) 10 kΩ in series (b) 2.5×10^{-3} Ω in parallel
2 C
3 0.076 Ω shunt, 0.91 A
4 (a) 10 V (b) 8.6 V

Chapter 29

Exercise 103

1 (a) 4.0 V (b) 3.4 V
2 (a) 1.1 V (b) 0.46 A, 4.00 Ω
3 20 Ω, 0.08 W, 48 mW
4 (a) 0.50 mA (b) 3.0 mV

Exercise 104

1 (a) 10 Ω (b) 0.2 W, (5 Ω), 0.1 W, (10 Ω)
2 3.3 mA
3 (a) 7.5 Ω (b) 0.80 V (c) 0.40 mA
4 D

Exercise 105

1 C
2 (a) 1 mA (b) 0.1 V
(c) Same, i.e. 0.1 V
(d) $R_1 = 393\ \Omega$, $R_2 = 405\ \Omega$
3 55 °C
4 D
5 (a) 175 Ω (b) 75×10^{-8} Ω m
6 (a) 0.15 A, 0.15 A, 0.20 A, 0.20 A
(b) 0.08 V
(c) 0.71 W
1.03 mA, 4.3 Ω, 17.5 Ω

Chapter 30

Exercise 106

1 (a) 0.12 μN (b) 0.40 nC
2 (a) (i) 1.8 kV m^{-1} (ii) zero (b) 36 V
3 2.0 V
4 8.9 kV, 8.6 mJ
5 zero, 2.8 kV m^{-1} to the right
6 (a) 8.0 W (b) 2.0×10^{10} Ω

Exercise 107

1 A
2 C
3 B
4 4.4×10^{-18} J
(a) 63 km s^{-1}
(b) (i) -0.56×10^{-19} C midway between (ii) zero
(iii) A and C come together, B moves away
5 300 V
6 5.6 s
7 D

Chapter 31

Exercise 108

1 (a) 0.34×10^{-10} F (b) 0.17×10^{-10} F
2 8.0×10^{7} mm C^{-1}, 3.0 μF, 0.010 μF
3 2.4×10^{-10} F, 9.0×10^{-12} F m^{-1}
4 8.0 mA

Exercise 109

1 **B**
2 (a) and (b) 3.0 μC, 2.3 μJ, 1.5 V
(c) 2.0 μC, 1.0 μJ, 1.0 V
3 (a) 0.80 mJ (b) 13 V (c) 0.53 mJ
4 6.7×10^{-10} F
5 **A**
6 (a) 20 V s^{-1} (b) 7.3 V s^{-1}
7 (a) 1.1 MV (b) 2.2 MV m^{-1}

Exercise 110

1 0.2 μF, 0.25 mJ
2 C
3 2.0 μF
4 2.0 μF
5 40 V
6 0.57 V, 1.4 V; 0.29 kV m^{-1}, 0.29 kV m^{-1}
7 0.35 kV, 0.17 kV, 44×10^{-13} F, 0.27 μJ
(a) 23 V (b) 66×10^{-12} F (c) 0.018 μJ
8 100*s* (a) 4.5 pF (b) 22 kV m^{-1}
9 (a) 1.1 mC (b) 2.8 kJ (c) 2.2 MV m^{-1}

Chapter 32

Exercise 111

1 12 μN
2 4.0 mN
3 (a) 0.05 N
(b) (i) 3.0 cm (ii) 10 cm from pivot towards wire
4 10 μA
5 (a) 2.4 mN m (b) 30 rad s^{-2}

Exercise 112

1 9×10^{-6} N
2 Zero (20 A wire) 1.0×10^{-3} N, West
3 21.8°
4 **B**
5 9.5×10^{-6} N rad^{-1}, (a) 0.06 Ω shunt
(b) 15 kΩ in series

Chapter 33

Exercise 113

1 (a) 5.0 mV (b) zero (c) 4.3 mV
2 (a) 25 mV (b) 2.5 mA
3 6.0×10^{-7} Wb

Exercise 114

1 (a) 1.6×10^{-4} Wb (b) 80×10^{-4} Wb
(c) 16 mV
2 1.0×10^{-2} H
3 10 mV

Exercise 115

1 90°: (a) 1.1×10^{-4} C (b) 1.1×10^{-4} C;
180°: (a) 2.2×10^{-4} C (b) zero
2 (a) 1.8 mC (b) 0.9 mC
3 (a) 13 mV (b) 12 mV

Exercise 116

1 0.26 V
2 2.6 mV
3 0.7 A, 7.8×10^{-4} J
4 (a) 4.0 A (b) 1.0 A s^{-1} (c) 0.50 A s^{-1}
5 2.0 A s^{-1}
6 25×10^{-4} Wb, 25 μC

Chapter 34

Exercise 117

1 2.4×10^{-5} T
2 (a) 2.4 μV (b) 2.4 μH
3 1.7×10^{-5} T,
4 30×10^{-5} H, 6.0 μJ

Exercise 118

1 (b) 4.0×10^{-4} T (c) 3.2×10^{-19} N
2 12×10^{-6} N, 28×10^{-6} N
3 5.3 A, 3.7×10^{-6} N m^{-1}
4 27 divisions

Chapter 35

Exercise 119

1 (a) 2.8 V (b) 0.14 A
2 (a) 1/400 s or 2.5 ms (b) 0.83 ms
3 (a) 0.10 kΩ (b) 1.0 A (c) 87 V
(d) 0.5 A
4 (a) 25 Ω (b) 22 Ω
5 (a) 74 Ω (b) 2.8 V
6 2.9 H

Exercise 120

1 0.25 kHz
2 5.1 μF, 2.4 A
3 (a) 0.36 kHz (b) 0.20 A (c) 89 V

Exercise 121

1 10 V RMS, 0.020 A RMS
2 50 kW, 0.56 W

Exercise 122

1 E
2 C
3 (b) 11 mH (c) 1/2.6
4 2.0 V
5 40 Hz, 1.0 A
6 (a) 40 Ω (b) 69 mH
7 Resistance of 10 Ω, inductance of 55 mH; 92 μF
8 C
9 (b) 0.39 MW (c) 1.5 kV (d) 27%
(f) 0.67 kW (g) 95.6%

Chapter 36

Exercise 123

1 1.0×10^{25} m^{-3}
2 11 mA

Exercise 124

1 0.30 V
2 1.0 mA
3 (a) 0.50 MΩ (b) 4.0 V (c) 0.60 V

Exercise 125

1 33 mg
2 (a) 17 cm^3 (b) 19 cm^3
3 1.9 g
4 96×10^3 C
5 3.1 MC kg^{-1}

Exercise 126

1 0.80 mA
2 D
3 A
4 0.25 mA, 3.5 V
5 14 μA, 0.66 MΩ
6 (a) 96×10^3 C (b) 6.2×10^{18}

Chapter 37

Exercise 127

1 15 mV, 5.0 kHz
2 3.5 km s^{-1}
3 20 mV, 140 Hz

Exercise 128

1 14 V, 0.2 kHz
2 60 km

Chapter 38

Exercise 129

1 OR gate
2 NAND
3 1 then 1 then 0

Exercise 130

1 (a) (i) 14 (ii) 145 (b) 0010 0010
2 (a) 3 (b) 1 (c) zero

Exercise 131

1 5 μA
2 100 mV RMS with reversed phase

Exercise 132

2 **D**

Chapter 39

Exercise 133

1 1.2×10^{10}
2 12×10^{5}, 1.9×10^{-13} A
3 2.3×10^{-5} T
4 3.3 ns
5 (a) 36×10^{4} m s^{-1} (b) 0.72 mm
6 1.1 kV
7 1.6×10^{5} V m^{-1}

Exercise 134

1 2.5×10^{7} m s^{-1}
2 0.10 m
3 2

Exercise 135

1 2.5×10^{7} m s^{-1}
2 $R = \sqrt{(2mV/B^2e)}$, 3.9 μs
3 3.8×10^{-9} s
4 (a) 0.44 m (b) 82 μs, 68 μs
5 12
6 25×10^{6} m s^{-1}, 1.7×10^{11} C kg^{-1}

Chapter 40

Exercise 136

1 0.75×10^{15} Hz, 5.0×10^{-19} J
2 2.1 V
3 0.97×10^{15} Hz
4 6.6×10^{-34} J s
5 32×10^{14}
6 0.071 nm
7 1.9×10^{6} m s^{-1}

Exercise 137

1 2.0×10^{-19} J, $hc/\lambda <$ WFE ($\lambda > 1.0$ μm)
2 6.4×10^{-34} J s
3 3.7 eV
4 1.4×10^{20}
5 **B**
6 535 nm

Chapter 41

Exercise 138

1 4.1×10^{18} Hz, No
2 (a) 500 W (b) 497.5 W (0.50 kW)
(c) 0.025 nm
3 (a) 1.4×10^{-15} J (b) 1.4×10^{-16} J

Exercise 139

1 0.042 nm, 10°
2 32°, 29°; 3

Exercise 140

1 (a) 10 000 eV (b) 1.2×10^{-11} m
(c) 1.6×10^{-15} J, 0.12 nm
2 (a) 8.5 kV (b) 1.2×10^{-15} J
3 3 orders (excluding zero order)
4 16°, 0.28 nm

Chapter 42

Exercise 141

1 3762 years
2 10 h, 0.069 h^{-1}

Exercise 142

1 1/3

Exercise 143

1 218, 84
2 0.5 m
3 12000 years

Exercise 144

1 **B**
2 (a) 2.7×10^{16} (b) 0.66×10^{6} s
3 (a) 5.2 litre
4 β^-; 20, 20, 22; 2.0 m; 24 h
5 88×10^{17}; 37 electrons, 37 protons, 50 neutrons; 3.8 Bq
6 4.0×10^{9} year

Chapter 43

Exercise 145

1 0.73×10^{6} m s^{-1}, 0.44×10^{-15} J
2 0.95 GeV
3 4.03 MeV
4 2.8×10^{-14} m

Exercise 146

1 56.5, 1.44×10^{3} MeV
2 1800 MeV (1.8 GeV)
3 6.646×10^{-27} kg
4 4.3×10^{-15} m

Chapter 44

Exercise 147

1 **B**
2 8 kg cm^{-1}
3 $L = 3.0 \times 10^{5}$ J kg^{-1}, $h = 1.0$ W
4 3.5 AV^{-25}, 2.5
5 0.010 day^{-1}

Exercise 148

1 **E**
2 **B**
3 **E**
4 **C**
5 10 Ω, 0.14 H
6 $T = 100/V$
7 5 MJ approximately
8 **E**

Table I

ABBREVIATIONS AND SYMBOLS

Abbreviations for SI units	
kilogram (mass)	kg
metre (length)	m
second (time)	s
newton (force)	N
joule (work or energy)	J
watt (power)	W
kelvin (temperature)	K
volt (potential difference)	V
ampere (current)	A
ohm (resistance)	Ω
farad (capacitance)	F
henry (inductance)	H
tesla (magnetic flux density)	T
weber (magnetic flux)	Wb
coulomb (charge)	C
becquerel (activity)	Bq

Table II

MULTIPLES AND SUBMULTIPLES FOR UNITS
The following prefixes are commonly used and their values are given below

Symbol	*Prefix*	*Value*
p	pico	10^{-12}
n	nono	10^{-9}
μ	micro	10^{-6}
m	milli	10^{-3}
c	centi	10^{-2}
k	kilo	10^{3}
M	mega	10^{6}
G	giga	10^{9}

NOTE: A selection of electrical circuit symbols to BS 3939 are shown on the next two pages.

ELECTRIC CIRCUIT SYMBOLS conforming to BS 3939

Component	Symbols	Notes
Conductor, wire or lead		Implies zero resistance
Crossing wires	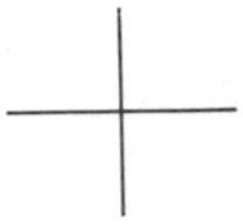	No electrical connection between wires
Junction of two or more conductors		
Earth connection		
Switch (single way and two way)	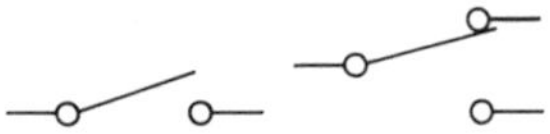	Called 'single pole' since it 'breaks' only one conductor
Battery (one cell and multi cell)		Positive terminal is longer line. Used for any DC source
Resistor	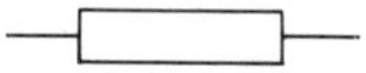	
Variable resistor	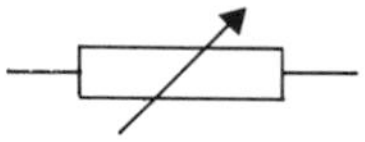	
Potential or voltage divider	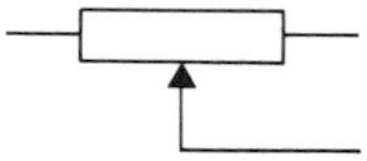	
Alternative method of showing voltage divider	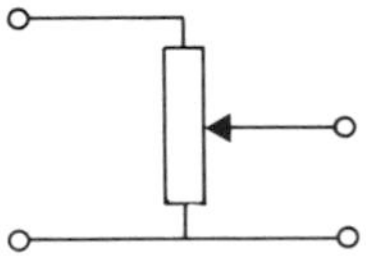	Often referred to as a potentiometer or a 'pot'
Capacitor		
Inductor (coil)		
Diode or Rectifier	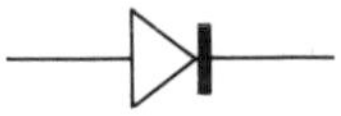	Arrow indicates 'easy' direction of current flow
Bulb (lamp)		Second symbol is used for indicator lamps

Component	Symbols	Notes
Fuse link	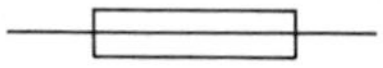	Rating is usually shown by a number
Voltmeter and Ammeter	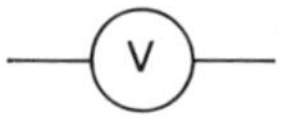 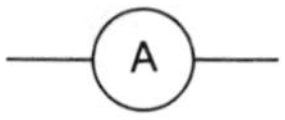 V A	For millivoltmeter and milliammeter letters mV and mA are used
Galvanometer		
Thermistor	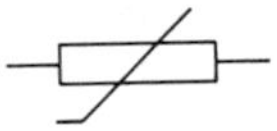	
Bipolar transistor	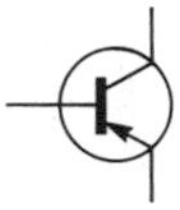	*pnp*
	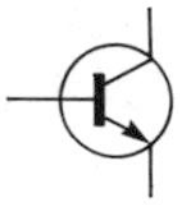	*npn*
Field effect transistor	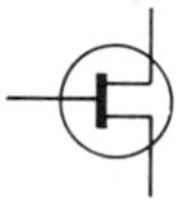	FET
Insulated gate	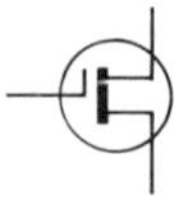	FET
Transformer	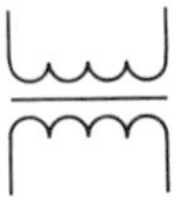	
Photoconductive cell	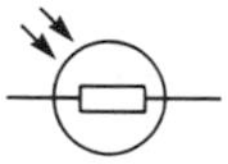	
Amplifier	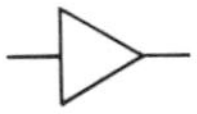	

INDEX